AF896333

Biosystems & Biorobotics

Volume 9

Series Editor

Eugenio Guglielmelli, Campus Bio-Medico University of Rome, Rome, Italy
e-mail: e.guglielmelli@unicampus.it

Editorial Board

Dino Accoto, Campus Bio-Medico University of Rome, Rome, Italy
Sunil Agrawal, Columbia University, New York, USA
Fabio Babiloni, Sapienza University of Rome, Rome, Italy
Jose M. Carmena, University of California, Berkeley, CA, USA
Maria Chiara Carrozza, Scuola Superiore Sant'Anna, Pisa, Italy
Paolo Dario, Scuola Superiore Sant'Anna, Pisa, Italy
Arturo Forner-Cordero, University of Sao Paolo, São Paulo, Brazil
Masakatsu G. Fujie, Waseda University, Tokyo, Japan
Nicolas Garcia, Miguel Hernández University of Elche, Elche, Spain
Neville Hogan, Massachusetts Institute of Technology, Cambridge, MA, USA
Hermano Igo Krebs, Massachusetts Institute of Technology, Cambridge, MA, USA
Dirk Lefeber, Vrije Universiteit Brussel, Brussels, Belgium
Rui Loureiro, University College London, UK
Marko Munih, University of Ljubljana, Ljubljana, Slovenia
Paolo M. Rossini, University Cattolica del Sacro Cuore, Rome, Italy
Atsuo Takanishi, Waseda University, Tokyo, Japan
Russell H. Taylor, The Johns Hopkins University, Baltimore, MD, USA
David A. Weitz, Harvard University, Cambridge, MA, USA
Loredana Zollo, Campus Bio-Medico University of Rome, Rome, Italy

Aims & Scope

Biosystems & Biorobotics publishes the latest research developments in three main areas: 1) understanding biological systems from a bioengineering point of view, i.e. the study of biosystems by exploiting engineering methods and tools to unveil their functioning principles and unrivalled performance; 2) design and development of biologically inspired machines and systems to be used for different purposes and in a variety of application contexts. The series welcomes contributions on novel design approaches, methods and tools as well as case studies on specific bioinspired systems; 3) design and developments of nano-, micro-, macro- devices and systems for biomedical applications, i.e. technologies that can improve modern healthcare and welfare by enabling novel solutions for prevention, diagnosis, surgery, prosthetics, rehabilitation and independent living.

On one side, the series focuses on recent methods and technologies which allow multiscale, multi-physics, high-resolution analysis and modeling of biological systems. A special emphasis on this side is given to the use of mechatronic and robotic systems as a tool for basic research in biology. On the other side, the series authoritatively reports on current theoretical and experimental challenges and developments related to the "biomechatronic" design of novel biorobotic machines. A special emphasis on this side is given to human-machine interaction and interfacing, and also to the ethical and social implications of this emerging research area, as key challenges for the acceptability and sustainability of biorobotics technology.

The main target of the series are engineers interested in biology and medicine, and specifically bioengineers and bioroboticists. Volume published in the series comprise monographs, edited volumes, lecture notes, as well as selected conference proceedings and PhD theses. The series also publishes books purposely devoted to support education in bioengineering, biomedical engineering, biomechatronics and biorobotics at graduate and post-graduate levels.

About the Cover

The cover of the book series Biosystems & Biorobotics features a robotic hand prosthesis. This looks like a natural hand and is ready to be implanted on a human amputee to help them recover their physical capabilities. This picture was chosen to represent a variety of concepts and disciplines: from the understanding of biological systems to biomechatronics, bioinspiration and biomimetics; and from the concept of human-robot and human-machine interaction to the use of robots and, more generally, of engineering techniques for biological research and in healthcare. The picture also points to the social impact of bioengineering research and to its potential for improving human health and the quality of life of all individuals, including those with special needs. The picture was taken during the LIFEHAND experimental trials run at Università Campus Bio-Medico of Rome (Italy) in 2008. The LIFEHAND project tested the ability of an amputee patient to control the Cyberhand, a robotic prosthesis developed at Scuola Superiore Sant'Anna in Pisa (Italy), using the tf-LIFE electrodes developed at the Fraunhofer Institute for Biomedical Engineering (IBMT, Germany), which were implanted in the patient's arm. The implanted tf-LIFE electrodes were shown to enable bidirectional communication (from brain to hand and vice versa) between the brain and the Cyberhand. As a result, the patient was able to control complex movements of the prosthesis, while receiving sensory feedback in the form of direct neurostimulation. For more information please visit http://www.biorobotics.it or contact the Series Editor.

More information about this series at http://www.springer.com/series/10421

Hanjoong Jo · Ho-Wook Jun
Jennifer Shin · SangHoon Lee
Editors

Biomedical Engineering: Frontier Research and Converging Technologies

Springer

Editors
Hanjoong Jo
Coulter Department of Biomedical Engineering
Georgia Tech and Emory University
Atlanta, GA
USA

Ho-Wook Jun
Department of Biomedical Engineering
The University of Alabama at Birmingham
Birmingham, AL
USA

Jennifer Shin
Department of Mechanical Engineering Soft Biomechanics & Biomaterials Lab
Korea Advanced Institute of Science and Technology
Daejeon (KAIST)
Korea, Republic of (South Korea)

SangHoon Lee
Korea University
Department of Biomedical Engineering, College of Health Science
Seoul
Korea, Republic of (South Korea)

ISSN 2195-3562 ISSN 2195-3570 (electronic)
Biosystems & Biorobotics
ISBN 978-3-319-21812-0 ISBN 978-3-319-21813-7 (eBook)
DOI 10.1007/978-3-319-21813-7

Library of Congress Control Number: 2015944754

Springer Cham Heidelberg New York Dordrecht London
© Springer International Publishing Switzerland 2016
This work is subject to copyright. All rights are reserved by the Publisher, whether the whole or part of the material is concerned, specifically the rights of translation, reprinting, reuse of illustrations, recitation, broadcasting, reproduction on microfilms or in any other physical way, and transmission or information storage and retrieval, electronic adaptation, computer software, or by similar or dissimilar methodology now known or hereafter developed.
The use of general descriptive names, registered names, trademarks, service marks, etc. in this publication does not imply, even in the absence of a specific statement, that such names are exempt from the relevant protective laws and regulations and therefore free for general use.
The publisher, the authors and the editors are safe to assume that the advice and information in this book are believed to be true and accurate at the date of publication. Neither the publisher nor the authors or the editors give a warranty, express or implied, with respect to the material contained herein or for any errors or omissions that may have been made.

Printed on acid-free paper

Springer International Publishing AG Switzerland is part of Springer Science+Business Media (www.springer.com)

Preface

Solving problems in human health, well-being, and disease increasingly requires convergence of scientific and engineering fields with medicine. While some of the last frontiers of biomedicine, such as neuroscience and regenerative medicine, critically demands new ideas and tools from other disciplines, paradigm-shifting technological innovations in information science, nanotechnology, and robotics could open new opportunities in healthcare. At the same time, a new generation of engineers, "fluent" in many different languages of science, are creating entirely new fields to view the old questions with a fresh look. This book aims to provide timely state-of-the-art reviews written by leading biomedical engineers for students, fellows, faculties, and industry investigators in broad biomedical engineering fields.

Acknowledgement

The editors are grateful for overwhelimg support by all authors who have contributed their valuable time and effort to share their insights into the biomedical engineering, an exciting field of convergent technology. We also thank Drs. Kookjoon Ahn and Jae H. Kim for their support and encouragement to publish this book. In addition, Ms. Leontina Cecco Di in Springer for her support to get this book published in time.

Contents

About the Editors

Hanjoong Jo

Dr. Hanjoong Jo is John and Jan Portman Professor in the Coulter Department of Biomedical Engineering (BME) at Georgia Tech and Emory University, and Professor of Medicine at Emory University. He is also the Associate Chair of Emory in BME Department. Upon graduation from Korea University, Dr. Jo received PhD under the co-mentorship of Professors John Tarbell (Chemical Engineering) and Ted Hollis (Physiology) at Pennsylvania State University in 1989. Following postdoctoral training in Jay McDonald Lab at Washington University in St. Louis and University of Alabama at Birmingham, he became Assistant Professor in Pathology and BME. Dr. Jo joined the BME Department at Georgia Tech and Emory University in 2000. He directs the Cardiovascular Mechanobiology and Nanomedicine lab. His lab studies how mechanical force associated with blood flow regulates vascular biology and cardiovascular disease, especially atherosclerosis, aortic valve (AV) calcification, and abdominal aortic aneurysms. He has published more than 140 peer-reviewed papers and edited a book. He developed the mouse model of atherosclerosis, known as partial carotid ligation model, induced by disturbed flow. His work led to the discovery of several genes (mechanosensitive genes and microRNAs) and epigenetic controlling mechanisms that are regulated by bad blood flow and play key roles in atherosclerosis and AAA. By targeting some of these mechanosensitive genes, his lab has been able to treat atherosclerosis and AAA in mice. His lab is now working on nanotechnologies to developing targeted gene and drug therapies in an effort to translate mouse studies toward clinical application. He is an elected fellow of American Institute of Medical and Biological Engineering, Biomedical Engineering Society, American Heart Association and American Physiological Society. He serves as associate editors and editorial board members of several cardiovascular and biomedical engineering journals including Scientific Reports, Circulation Research, Atherosclerosis

Thrombosis Vascular Biology, Am J Physiology, Cell Molecular Bioengineering and Cardiovascular Engineering and Technology. He also has been serving as reviewers and chairs of study sections of the NIH, NSF, Veterans Administration and Am Heart Association. He also organized several international meetings, and in 2012, he served as the Chair of the Annual BME Society Meeting. He is also the founding President of Korean-American Biomedical Engineering Society (KBMES) and Chairs of US-Korea Annual BMES Workshops since 2013. He has been a Distinguished Visiting Professor at Ewha Womans University and Chunbuk National University.

SangHoon Lee

Dr. SangHoon Lee is Professor in the Department of Biomedical Engineering (BME) at Korea University. Dr. Lee received PhD under the Advisor of Professors B.G. Min (Biomedical Engineering) at Seoul National University in 1992. After PhD, he became Assistant Professor in Department of Biomedical Engineering at the medical school of Dankook university in Korea. He also worked as director of BME Department at Dankook university hospital from 1992, and managed the medical device and information system of hospital. At 2006, he moved to medical school of Korea University, and at 2008, he established Department of BME at college of health science. He is now operating 'Intelligent BioMEMS Lab', and his major research area is tissue engineering and regeneration medicine using microtechnology, organ-on-a-chip and electronic skin. He has published more than 150 peer-reviewed papers including "Nature Materials", "Nature Communications", "Advanced Materials", "Biomaterials", "Stem Cells", "Small", "Lab on a Chip" and so forth. He has a great interest in engineering and regeneration of liver, brain, and pancreas. Using diverse primary cell sources, he successfully fabricated complicate liver-ultra structures. He also demonstrated 3D brain tissues networked using nervous system. Another interesting topic is the brain and liver on a chip to simulate liver and CNS disease on the microfluidic platform. Another topic is the electronic skin which is highly stretchable and flexible like skin, and this e-skin is continuously wearable over one week. He is chief editor of 'Biomedical Engineering Letters', and organized several international conferences. He served as committee members of microTAS, ISMM and EMBC. He was dean of 'College of Health Science' at Korea University, and he received several awards due to his academic and industrial achievements.

Ho-Wook Jun

Dr. Ho-Wook Jun is an Associate Professor in the Department of Biomedical Engineering of School of Engineering and School of Medicine at the University of Alabama at Birmingham. Dr. Jun received Ph.D. in Bioengineering from Rice University in 2004. From 2004-2006, he was the Peter and Ruth Nicholas Postdoctoral Fellow at the Richard Smalley Institute for Nanoscale Science and Technology and in the Department of Chemistry at Rice University. In 2006, he joined the faculty at the University of Alabama at Birmingham. Dr. Jun is also a Co-Founder and Chief Scientific Officer of Endomimetics LLC, a company that develops bionanomatrix coatings for cardiovascular stents and other medical implants.

Dr. Jun group focus on nanostructured biomaterials and stem cell based tissue regeneration to develop an innovative strategy to treat various diseases. His translational research includes 1) a biomimetic hybrid nanosack to increase the efficacy of pancreatic islet transplantation to treat type I diabetes, 2) a native endothelium mimicking nanomatrix coating for stents and cardiovascular devices, 3) a biphasic composite nanomatrix gel for treating severe limb injuries caused by high energy blasts and gunshots in military operations, 4) chondral defect repair using amnionic fluid derived stem cells delivered within a biomimetic nanomatrix gel.

Dr. Jun has published more than 40 peer-reviewed papers and filed 7 patents. Dr. Jun has been recognized by various awards including the Early Career Award from the Wallace H. Coulter Foundation in 2007, the Innovation Award from the American Diabetes Association in 2009, and the Early Career Award from the National Science Foundation in 2010, Dean's Award for Excellence in Mentorship in 2010, Finalist of Alabama Launchpad Governor's Business Plan Competition, and Wallace H. Coulter Fellow in 2011.

Dr. Jun is a Vice President of Korean-American Biomedical Engineering Society (KBMES).

Jennifer H. Shin

Dr. Jennifer H. Shin is an associate professor in Mechanical Engineering Department at Korea Advanced Institute of Science and Technology (KAIST). She obtained B.S., M.S., and Ph.D. in the field of biophysics in Mechanical Engineering Department at MIT in 1998, 2000, and 2004, respectively. After that, she worked as a postdoctoral associate in School of Engineering and Applied Sciences (SEAS) at Harvard University before she started her career at KAIST in 2005. Dr. Shin's research interests center around the emerging interdisciplinary mechanobiology. At the center of mechanobiology, is the process of mechanotransduction, the way cells sense and respond to various physical stimulations from the microenvironments. Her laboratory focuses on the development of physical stimulation systems that are suitable for biological and biophysical assays to investigate cellular level mechanobiology. Her lab focuses on finding links between cellular biomechanics and physiological changes associated with diseases and differentiation states in light of diagnostics and therapeutics of the diseases, as well as tissue engineering applications. Her lab also studies biological motility at various levels from a single cell, collective cellular clusters, to an organism based on force visualization techniques such as traction force microscopy and monolayer stress microscopy.

She served as a member of the Presidential Advisory Council on Science and Technology (2008-2012) and has been an active board member for 7 academic societies across the fields of core mechanical engineering and biomedical engineering. She has been a leading member to organize several international symposiums and workshops, and also served a committee member of international conferences including Biofabrication, microTAS, and IBEC. She is one of the founding members of Korean-American Biomedical Engineering Society (KBMES) and US-Korea Annual BMES Workshops since 2013. In particular, Dr. Shin displays genuine passion and enthusiasm for teaching students and was recognized with several teaching excellence awards both at the departmental and institutional levels. As a woman faculty in mechanical engineering, she has shown commitments to supporting female students in the field of science and technology through her active contributions at Center for Women in Science, Engineering and Technology (WISET) and Federation of Women's Science and Technology Association in Korea.

Part I
Nanotechnology and Nanomedicine

Nanoparticles for Imaging and Non-viral Gene Therapy

Yoonjee Park

Abstract Gene therapy has not been investigated as much as pharmacotherapy because of immunogenic issues when a virus was used as a gene delivery vector. Despite the challenges, gene therapy still has attractive aspects. It has less side effects and is more target-specific compared to pharmacotherapy, and it also has potential for generic disease treatment or personalized medicine. Therefore, it would be truly beneficial if safe and reliable vectors are used and targeted for area of interest. Interest in multifunctional nanomedicine for diagnostics and therapeutics has been increasing. For this reason, non-viral gene delivery has been studied, combined with molecular imaging to visualize targeting. In this review, complex nanoparticle systems designed for molecular imaging and gene delivery are discussed. There are design criteria which need to be considered for the nanoparticle complex systems. The criteria are as follows: i) the nanoparticle complex should be stable; ii) it should have efficient targeting capability; iii) controlled release of genes should be available; iv) molecular imaging should be possible; and lastly, v) there should be noticeable therapeutic efficacy. Examples on nanoparticle complex which meet these criteria are described in the review.

Keywords: Molecular imaging · Gene therapy · Nanomedicine · Contrast agents · Non-viral gene delivery · Theranostics

1 Introduction

Gene therapy often requires vectors to deliver genes into cells and the vectors are mainly categorized into two types: viral vectors and non-viral vectors. Although viral vectors, such as a retrovirus and an adenovirus, have high transfection

Y. Park(✉)
Department of Biomedical, Chemical & Environmental Engineering,
University of Cincinnati, Cincinnati, USA
e-mail: yoonjeep@gmail.com

© Springer International Publishing Switzerland 2016

H. Jo et al. (eds.), *Biomedical Engineering: Frontier Research and Converging Technologies*,
Biosystems & Biorobotics 9,
DOI: 10.1007/978-3-319-21813-7_1

efficiency, many critical safety issues and concerns still remain [1]. Because of the immunogenic issues, non-viral vectors recently have drawn attention. In addition, in order to visualize the expression of gene transduction in target tissue or specific organs of the body, molecular imaging has been combined with gene therapy.

This chapter highlights the current gene delivery and molecular imaging technologies, involved in novel non-viral nanoparticle complexes as multifunctional "nanomedicine". Nanoparticles as a gene carrier form a complex where nucleic acid molecules like exogenous DNA are encapsulated or absorbed. The particle size of the nanoparticles is usually 10-100 nm. A high surface-to-volume ratio makes the nanoparticles have outstanding adsorption, which is advantageous for delivering exogenous genes. Non-viral vectors have greater payload capacity than viral vectors [2]. Various types of nanoparticles as a gene carrier will be discussed in Section 2. Molecular imaging techniques will provide a practical and clinically useful way to identify successful gene transduction and expression in patients undergoing gene therapy in a non-invasive manner [3-5].

To achieve molecular imaging, targeted imaging contrast agents are required and the contrast agents are often nanoparticles. Therefore, it is beneficial to develop a nanoparticle complex which has multiple functions for targeted gene therapy and molecular imaging. The design criteria for the nanoparticle complex are as follows: i) the nanoparticle complex should be stable; ii) it should have efficient targeting capability; iii) controlled release of genes should be available; iv) molecular imaging should be possible; and lastly, v) there should be noticeable therapeutic efficacy. Examples on nanoparticle complexes which meet these criteria are described in the chapter.

2 Nanoparticles as a Gene Carrier

2.1 Liposomes

Liposomes are spherical vesicles composed of a lamellar phase lipid bilayer and these are formed when energy is supplied, such as heat and stirring, or sonication. Because liposomes are incorporated well within cell membrane and relatively easy to be manipulated, liposomes have been used widely in the analytical sciences as well as for drug and gene delivery. Cationic lipids are usually used for delivering genes which are negatively charged. The positively charged head group allows efficient DNA compaction due to attractive electrostatic forces that occur between the phosphate backbone of DNA and the cationic head groups of the lipids [6]. Loading genes on nanoparticles stably is important; however, at the same time, weak electrostatic interaction will be beneficial to deliver genes [7].

2.2 Polymers

With the same reason as described above in Section 2.1., cationic polymer has been shown as a promising carrier among non-viral gene delivery systems.

Because polycation-DNA complexes generally are stable, a number of cationic polymers have been investigated as gene carriers, including polylysine [8], polyethyleneimine [9], polyamidoamine dendrimers [10], poly(a-(4-aminobutyl)-L-glycolic acid) [11] etc.

Block copolymers composed of a cationic segment and a hydrophilic segment spontaneously associate with polyanionic DNA to form block copolymer micelles. The distinct feature of the structure is that the core of the polyion complex between DNA and the polycation is coated by a layer of the hydrophilic polymer. The characteristic core-shell structure endows the complex a high colloidal stability and reduced interaction with blood components [12].

There is no concrete comparison that can easily be made to suggest that one carrier is better than another for all cell types, environments, and applications. While some of the carriers presented above were originally found to yield little to no cytotoxicity for a given cell type, the observation does not necessarily hold true when they are applied to different cell types [13-15]

2.3 *Inorganic Nanoparticles*

Inorganic nanoparticles, such as gold, silica, iron oxide (ex. magnetofection) and calcium phosphates have been shown to be capable of gene delivery [16]. Some of the benefits of inorganic vectors are in their storage stability, low manufacturing cost and often time, low immunogenicity, and resistance to microbial attack. Nanosized materials less than 100 nm have been shown to efficiently trap the DNA or RNA and allow its escape from the endosome without degradation. Inorganics have also been shown to exhibit improved in vitro transfection for attached cell lines due to their increased density and preferential location on the base of the culture dish. Mirkin and co-workers attached siRNA (small interfering RNA) molecules to the surface of AuNPs via a thiol group [17]. The polyvalent siRNA/nanoparticle conjugates showed a 6 times greater half-life and prolonged gene knockdown compared to free RNA duplexes [18]. Quantum dots have also been used successfully and permit the coupling of gene therapy with a stable fluorescence marker.

2.4 *Other Carriers*

Chitosan [19], dendrimers [20, 21], and cell-penetrating peptides [22, 23] as a form of nanoparticles have been investigated for potential gene-delivering carriers. Chitosan is a biodegradable polysaccharide [24] extracted from crustacean shells. It has been shown to be non-toxic in a range of toxicity tests, both in experimental animals [25] and humans. Mao et al. [19] demonstrated that the chitosan-DNA nanoparticle system effectively delivered genes to the small intestine of mice. When used to orally deliver pCMVArh2 plasmid, which encoded a dominant anaphylaxis-inducing antigen identified in mice sensitized with peanut, nanoparticle immunization elicited a relatively high IgG2a immune response and protected the experiment animals from an allergic challenge. A dendrimer is a highly branched macromolecule with a spherical shape. When in the presence of genetic material such as DNA or RNA, a temporary association of the nucleic acid with the cationic dendrimer occurs. Cell-penetrating peptides (CPPs), also known as peptide transduction domains (PTDs), are short peptides (< 40 amino acids) that efficiently pass through cell membranes while being covalently or non-covalently bound to various molecules, thus facilitating these molecules' entry into cells [26]. CPP-decorated LNP (CPP-LNP) for siRNA delivery demonstrated that siRNA in CPP-LNP was efficiently internalized into B16F10 murine melanoma cells in a time-dependent manner, although siRNA in LNP without CPP was hardly internalized into these cells.

2.5 *Hybrids*

Hybrid systems also have been developed to combine advantages from the materials described above. Elbakry et al. first developed the PEI/siRNA/PEIAuNP system to deliver siRNA into cells and knockdown the expression of target gene based on the self-assembly layer-by-layer technology [27]. PEI (poly ethyleneimine), which has strong escape capacity from the endosome and is usually a gold standard of a polymeric transfection agent, was deposited on the gold nanoparticles to bind siRNA. Guo et al. developled charge-reversal functional gold nanoparticles to reduce high binding affinity between gold nanoparticles and siRNA to release siRNA in cellular cytoplasm easily [28]. When the vector (ternary complexes) is entrapped into the acidic intracellular organelles such as endosomes or lysosomes (pH 5-6), charge conversion facilitates the endosomal escape of the polyplexes through membrane disruption.

3 Nanoparticles as a Molecular Contrast Agent

Over the past decade, there has been an increasing interest and effort in developing molecular imaging. Molecular imaging can be broadly defined as the 'non-invasive and repetitive imaging of targeted macromolecules and biological processes in living organisms.' This field of study has advanced rapidly in recent years, in part due to the application of nanotechnology. The versatility of different imaging modalities has been significantly enhanced by innovative nanoparticle development. These nanoparticles can be used to image specific cells and tissues within a whole organism. An ideal contrast agent should selectively accumulate at the site of interest to be able to interact physically, chemically, biochemically, and functionally with the target and enhance contrast of imaging. Some of the nanoparticles under development may be useful to measure biological processes associated with human disease and help monitor how these change with treatment. In vivo molecular imaging using nanoparticles promises to revolutionize diagnosis and treatment of human diseases, but some practical limitations still need to be resolved.

3.1 Imaging Contrast Agents Used for Gene Delivery

Molecular imaging techniques will provide a practical and clinically useful way to identify successful gene transduction and expression in patients undergoing gene therapy in a non-invasive manner. This section will introduce various types of imaging contrast agents which have been combined with gene delivery.

3.1.1 MRI Contrast Agents

MRI (magnetic resonance imaging) is based on the response of proton spin in the presence of an external magnetic field when triggered with a radio frequency (RF) pulse. Under the influence of an external magnetic field, protons align in one direction. On application of the RF pulse, aligned protons are perturbed and subsequently relax to their original state. There are two independent relaxation processes: longitudinal (T_1) and transverse (T_2) relaxation, which are typically used to generate the MR images. For T_1-weighted contrast agents, gadolinium- (Gd^{3+}) or manganese- (Mn^{2+}) chelates are often used. These are paramagnetic, which increases the T_1 relaxation time, resulting in bright contrast T_1-weighted images. For T_2-weighted contrast agents, superparamagnetic Fe_3O_4 nanoparticles are used and these reduce T_2 relaxation times, giving rise to dark contrast T_2-weighted images [29].

Gd^{3+} ion on/in the nanoparticles reduced toxicity of Gd^{3+}, and enhanced T_1-weighted MR signal [30]. It can also increase the cellular uptake of Gd^{3+} ions through size and shape tuning of the vehicle nanoparticles. Furthermore, biological functional groups can be conjugated on the nanoparticles surfaces for studying dynamic biomolecular phenomena through suitable biochemical reactions and signaling. Superparamagnetic iron oxide nanoparticles (SPIONs) provide the most change in signal per unit of metal, in particular on T_2-weighted images. SPIONs are composed of biodegradable iron, which is biocompatible and can thus be reused/recycled by cells using normal biochemical pathways for iron metabolism and can be magnetically manipulated [29].

Dixit et al. reported phospholipid micelle encapsulating gadolinium oxide (Gd_2O_3) nanoparticles for molecular imaging and gene delivery [31]. These gadolinium oxide nanoparticles within phospholipid micelles as a novel low cytotoxic T_1-weighted MRI imaging contrast agent (MGdNPs) can also deliver small molecules such as DNA plasmids. MGdNPs show relatively good MRI relaxivity values, negligible cytotoxicity, excellent cellular uptake and expression of DNA plasmids in vivo. Biodistribution studies in mice show that intranasal and intraperitoneal administration of MGdNPs can effectively target specific organs.

As introduced in Section 2.5, cationic polymers have been combined with SPIONs via electrostatic interaction to deliver genes [32]. The transfection efficiency of SPION-gene complexes can be enhanced by magnetic field (magnetofection). Transfection efficiency of SPIONs with PEI on pulsed magnetic fields in vitro was 40 times higher than in the absence of a magnetic field [33]. Lee et al. conjugated siRNA onto SPION surfaces using disulfide bonds via a thiol-functionalized PEG (Figure 1) [34]. Targeting and stability of the SPIONs *in vivo* were enhanced.

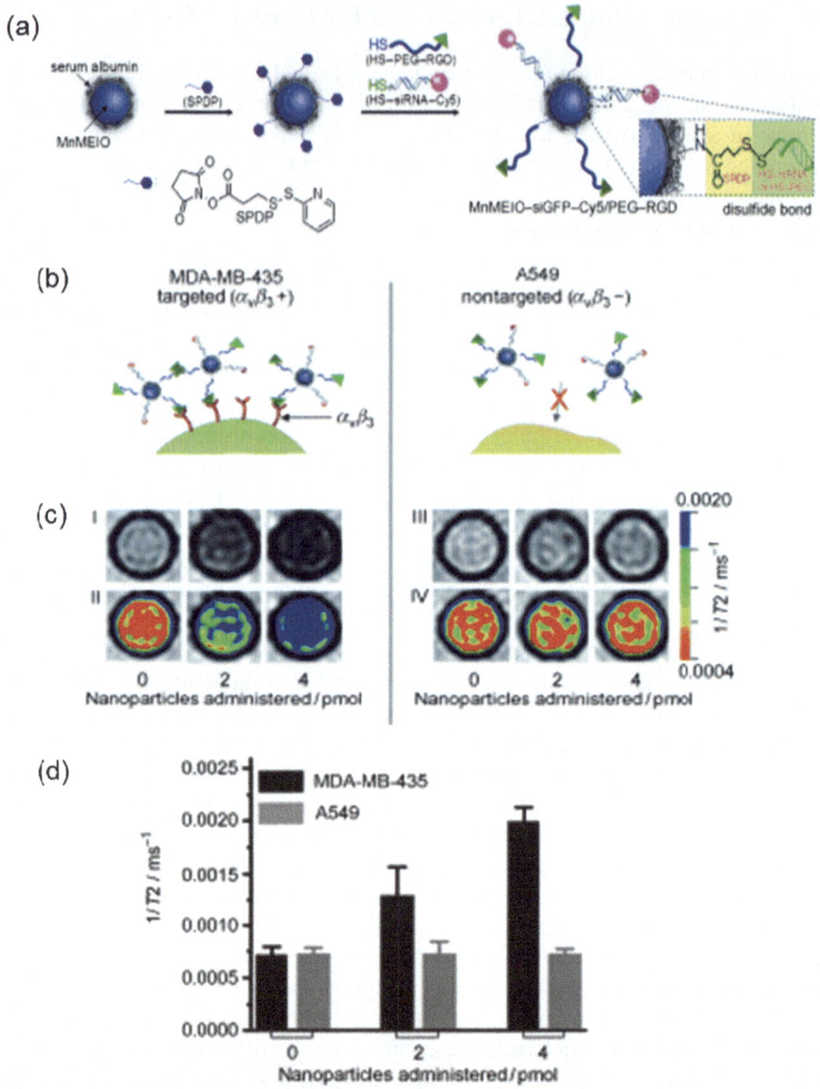

Fig. 1 (a) Fabrication of multimodal MnMEIO–siGFP–Cy5/PEG–RGD. Schematic of MnMEIO–siGFP–Cy5/PEG–RGD. SPDP, which forms disulfide bonds, was used as a linker. (b) Schematic of the target-specific binding of the nanoparticle to αvβ3 integrin positive cells. (c) T2*-weighted MR images and their color maps of I,II) MDA-MB-435 and III,IV) A549 cells treated with MnMEIO–siGFP–Cy5/PEG–RGD. Red corresponds to untreated cells and nanoparticle-treated A549 cells, and green and green-blue correspond to treated MDA-MB-435 cells. c) Values 1/T2 for cells treated with an increasing amount of nanoparticles. d) Confocal microscopy images showing the distribution of MnMEIO–siGFP–Cy5/PEG–RGD nanoparticles in cells. Nanoparticles appear red, endosomes stained with lysotracker appear blue, merged red and blue regions appear pink. The cell morphology is outlined on the images [34].

(e)

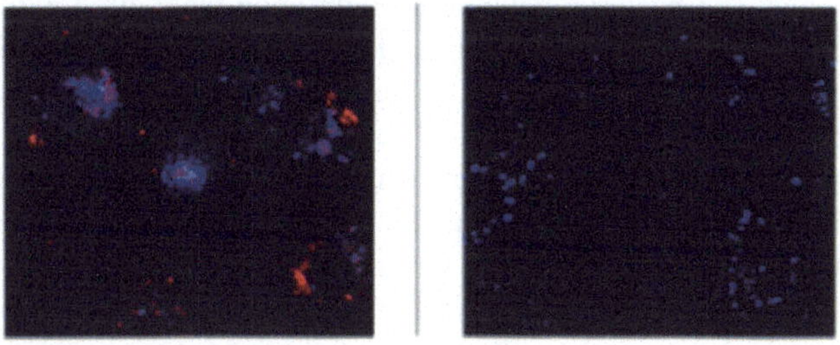

Fig. 1 (*continued*)

3.1.2 CT Contrast Agents

Computed tomography (CT) is an X-ray-based whole-body imaging technique that is widely used in medicine. Clinically approved contrast agents for CT are iodinated small molecules or barium suspensions. The development of nanoparticle-based X-ray contrast agents is providing an increasing contribution to the field of diagnostic and molecular imaging. The utilization of nanoparticles provides several advantages over the widely used iodinated contrast agent solutions, similar to the Gd^{3+} cases described in the previous section 3.1.1. For example, their surface can be modified to enhance their specificity by attaching targeting moieties, increase their circulation half-life by adding appropriate coatings (e.g. polymers, silica), and improve their functionality by adding other components, including fluorescent markers and therapeutic agents. Gold nanoparticles are a good candidate for CT contrast agents because they induce X-ray attenuation [35]. X-ray attenuation coefficient is determined by the atomic number and electron density of the tissue. The atomic number and electron density of gold (79 and 19.32 g/cm3, respectively) are much higher than those of the currently used iodine (53 and 4.9 g/cm^3). In addition, gold nanoparticles provide a high degree of flexibility in terms of ease of producing, modifying the surface for conjugating functional groups, and have also proved to be nontoxic and biocompatible *in vivo*.

3.1.3 Optical Contrast Agents

Fluorescent proteins (FP) have been used as a fluorescent marker for gene activity, protein expression, subcellular components, cell status/fate, among others [36]. Even though there are many advantages of FP-based systems, such as relatively easy and non-cytotoxic expression, this technique does not give sufficient signal for in vivo imaging. Low signal to noise (S/N) ratios due to poor tissue penetration of the excitation light and, more importantly, tissue autofluorescence has restricted the utility of most FP-based systems to imaging of cultured cells or superficial tissues [37]

Bioluminescence imaging of luciferase reporter genes is much more sensitive for in vivo imaging applications due to the lack of background bioluminescence signal [38]. Luciferases are photoenzymes isolated from a wide variety of insects, marine organisms, and bacteria, which emit light upon interaction with a specific substrate. Zhou et al. tracked expression of triple fusion (TF) gene which was loaded in liposome nanoparticles by using Renila luciferase bioluminescence imaging (Figure 2) [39]. Triple fusion (TF) plasmid, pcDNA3.1-RFP-Rluc-HSV-ttk, consisting of red fluorescence protein (RFP), renilla luciferase (Rluc), and herpes simplex virus truncated thymidine kinase (HSV-ttk) reporter genes22 were used for imaging and gene therapy. As is shown in Figures 2 (c), the TF gene entrapped in liposome nanoparticles was found to be expressed at a considerably higher level, in which no significantly prolonged duration was observed.

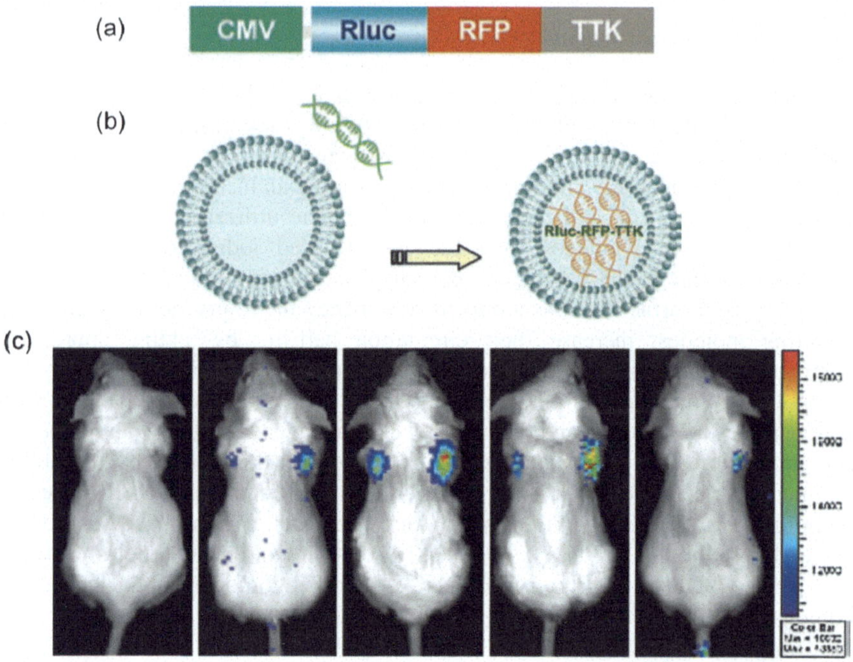

Fig. 2 Schematic of liposome loaded with a tri-fusion reporter gene (TF). (A) Schematic representation of the tri-fusion reporter gene (TF) containing Renila luciferase (Rluc), red fluorescent protein (RFP) and HSV-ttk driven by a CMV promoter. (B) Schematic representation of liposome nanoparticle and liposome-DNA coupling reaction. Therapeutic liposomes were loaded with Rluc-RFP-ttk gene. (C) The expression of TF transgene was tracked and analyzed by imaging and immunofluorescence. Imaging of Renilla luciferase intensity from one representative BALB/c mouse [39].

3.1.4 Plasmonic Nanoparticles

Metallic nanoparticles, usually made of gold or silver, scatter light with remarkable efficiency by surface plamsonic resonance effect. Typical scattering cross sections of metal nanoparticles greatly exceed the absorption cross-section of fluorescent dyes [40] and of fluorescent proteins [41]. The photostability, water solubility and nontoxicity of gold nanoparticles make these probes advantageous for biological imaging. Conjugation strategies to attach targeting or delivery moieties are well developed [42]. Gold nanoparticles can provide optical contrast via either absorption, scattering or luminescence. Surface plasmon resonance peak of gold nanoparticles can be tuned to the NIR spectral region (700–850 nm), where tissue is the most transparent, by modifying particle size, shape, material composition or interparticle spacing. [43]. It has been demonstrated that targeted gold nanoparticles can be used to provide high contrast images of cancer cells using a variety of optical modalities to measure light scattered by the nanoparticles, including reflectance confocal microscopy [44], dark-field microscopy[45], phase-sensitive optical coherence tomography [46] and photoacoustic imaging [47]. Aaron et al. demonstrated that scattering spectrum of EGFR-targeted nanoparticles shifts from green to yellow to red as the receptor tracks [44] (Figure 3).

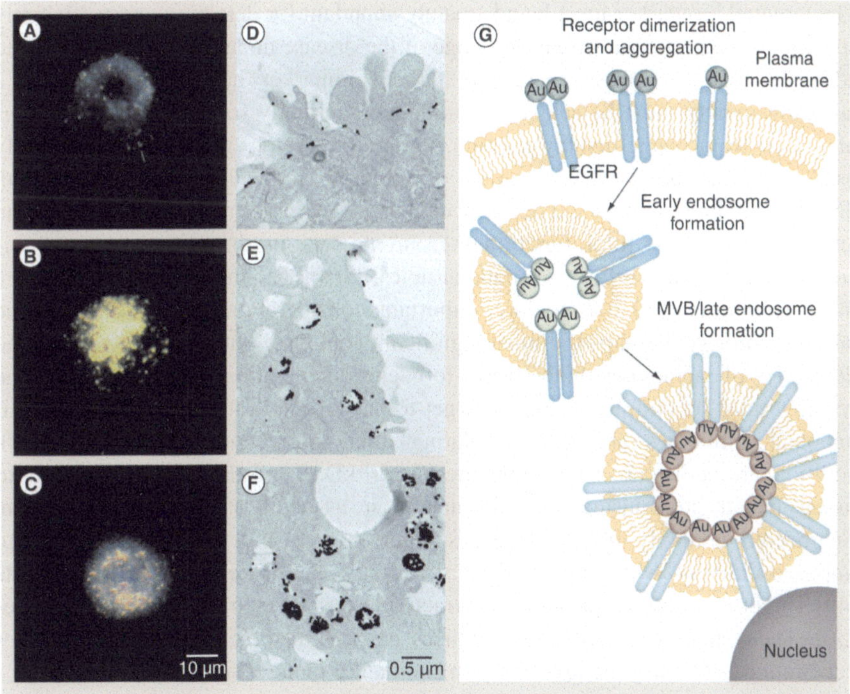

Fig. 3 Gold nanoparticles for dynamic imaging of EGF receptor trafficking in live cells. Dark-field and transmission electron microscopy images of cells labeled with 25-nm anti-EGF receptor (EGFR)-targeted gold nanoparticle conjugates at (A & D) 4°C, (B & E) 25°C and (C & F) 37°C. Labeling at these temperatures arrests the EGFR regulatory process at critical points, with receptors located on the cell membrane at 4°C, endosomal internalization at 25°C and MVB sorting at 37°C. (G) The relationship between EGFR regulation state and the optical signature of the gold nanoparticles in that arrangement. MVB: Multivesicular body [44].

Gold nanorods exhibit intense two-photon luminescence. Two-photon imaging of tissue phantoms treated with 50 × 15 nm gold nanorods functionalized with EGFR antibodies increased intensity by three orders of magnitude compared with cellular autofluorescence with 760 nm excitation [48].

4 Targeting and Controlled Release of Genes

4.1 Targeting

Targeting is achieved by using a targeting moiety on the surface of a delivery vehicle. Targeting is critical not only for payload delivery to the target to minimize side effects due to non-specific delivery but also for molecular imaging or image-guided delivery using imaging contrast agents. The targeting moiety is an important determinant of the specificity and sensitivity of the contrast agent. The targeted biomarker must be adequately abundant for detection and sufficiently specific to the particular disease or stage of the disease under examination to yield adequate image contrast. Once the target biomarker has been identified, a targeting moiety must be selected. Antibodies, antibody fragments, and peptides identified from phage-display peptide libraries are some of the most commonly used targeting moieties; each class of targeting molecule is associated with different binding kinetics and delivery challenges. A targeting moiety, often via an amine, carboxyl or thiol functional group present in the targeting protein is conjugated to the surface of nanoparticles. Stability, the potential to image multiple targets, and toxicity are important considerations when selecting the targeting moiety. Perhaps the most significant challenge with this approach is the difficulty of achieving high target-to-background ratios. The imaging time window must be optimized, as the target-to-background ratio may remain high if unbound antibodies are still circulating after intravenous (iv.) injection. In addition, antibodies are large molecules (~150 kDa), and this presents barriers to tissue penetration during iv. administration as well as topical application. Engineered antibody fragments are smaller in size and have the potential to improve delivery while maintaining specificity and lowering production costs [49].Targeting peptides present an attractive alternative to the use of antibodies [50]. Their small size (<10 kDa) reduces barriers to topical delivery and tumor penetration. Also they are less likely to elicit an immune response [51].

4.2 Controlled Release of Genes

As mentioned throughout Section 2, release of genes at the target of interest is as important as carrying genes safely. Even though a technique of electrostatic binding of genes has been studied extensively, a main issue of releasing genes still remains. A strong binding between the abundant positive charge on the polymer and the negatively charged phosphonic acid of the nucleic acid hinders the release of genes within the cytoplasm thus lowering the efficacy of the gene expression [52]. New materials have been developed to have capability of release via a trigger, including temperature [53], pH [54], redox potential [55], light [56], electric pulse [57], enzymatic degradation [58], and salt levels [59], in order to increase intracellular delivery efficiency. In addition, ultrasound exposure in the presence of microbubbles (e.g. contrast agents used to enhance ultrasound imaging) increases gene transfection efficiency in vitro by several orders of magnitude [60]. Park et al. demonstrated via direct observation of vessels-on-a-chip that cavitating microbubbles enhance targeted drug intracellular delivery efficiency. Ultrasound enhanced gene transfer has also been successfully achieved in vivo, with reports of spatially restricted and therapeutically relevant levels of transgene expression [60].

In addition, in order for gene therapy to be applicable in clinical medicine, it is imperative that a suitable method for stable controlled release of the required amount of the vector delivered over the desired period of time be developed. A magnetic field has been used with magnetic particles to facilitate gene delivery for a desired period of time, as shown in Figure 4 [61]. Advantages of the method include the rapid accumulation of the therapeutic agent at the target site, thereby reducing the transfection time and increasing the efficiency of transfection while decreasing cytotoxicity. For in vivo magnetofection, the magnetic nanoparticles carrying therapeutic genes are generally injected intravenously. As the particles flow through the bloodstream, they remain at the target site because of the application of a very strong high-gradient external magnetic field. Once they are captured at the site, the magnetic particles carrying the therapeutic gene are taken up by the cells in the area, followed by the release of the gene via enzymatic cleavage of cross-linked molecules or degradation of the polymer matrix. If DNA is embedded inside or within the coating material, the magnetic field must be applied to heat the particles and release the gene from the magnetic carrier [62]. In this therapy MNPs were subjected to an oscillating magnetic field, which leads to the generation of heat via two mechanisms depending on the size of the particles: (a) Brownian mode: for MNPs < 100 nm in diameter, heat production is due to the friction between the oscillating particles, and (b) Neel mode: `for larger MNPs the heat is produced via the rotation of the magnetic moment with individual field oscillation [63]. Moreover, it would be desirable to be able to stop delivery and expression of the gene as soon as it is decided to stop the treatment and to minimize the significant side effects of this type of therapy.

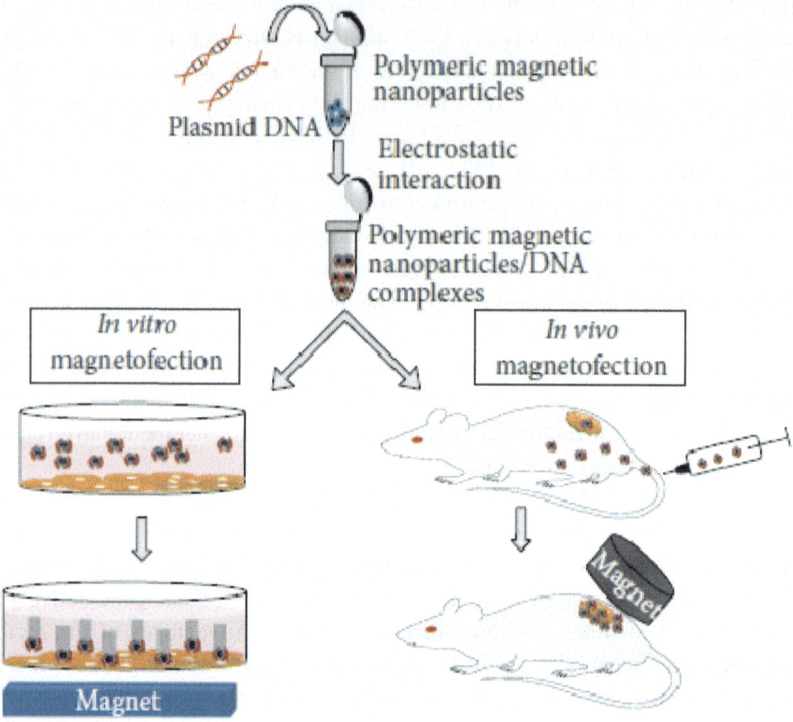

Fig. 4 Schematic representation of in vitro and in vivo gene delivery using magnetofection (grey color pattern on the left corner of the image represents the direction of movement of MNPs under the influence of magnet) [61]

5 Therapeutic Efficacy and Safety

5.1 Current Status of Non-viral Gene Delivery with Molecular Imaging

Up to date, approximately 70% of gene therapy clinical trials have used modified viruses such as retroviruses, lentiviruses, adenoviruses and adeno-associated viruses (AAVs) to deliver genes. However, developments in material sciences, such as new polymers and lipids, as well as in nanotechnology, nanoparticles for gene delivery have shown a rapid progress. siRNA, miRNA, and DNA delivery have been tried clinically using non-viral vectors. Table 1 shows non-viral gene delivery systems which completed clinical trials. Mostly the systems consist of liposomes or polymers. One study which used PEI as a plasmid DNA delivery vehicle for intermediate-risk superficial bladder cancer showed serious adverse events for 3 patients out of 47 (6.38%) (Clinicaltrials.gov identifier: NCT00595088). Therefore, development of new materials which can deliver genes into cytosol without cytotoxicity is critical. Truong et al. developed degradable poly(2-dimethylaminoethyl acrylate) (PDMAEA), which enables the release of siRNA after endocytosis. The polymer changes to a negatively charged and non-toxic polymer after the release of small interfering RNA, presenting potential for multiple repeat doses and long-term treatment of diseases [64].

None of clinical trials which involve non-viral gene therapy with contrast agents has been found. This might be in part due to lack of signal intensity for combined systems. Development of new imaging contrast agents which can guide gene delivery and enable us to track gene expression over time without toxicity is necessary. Overall, ideal materials for nanoparticle complexes should be biocompatible, nonimmunogenic, and multifunctional for carrying and releasing genes and for molecular imaging.

Table 1 Non-viral gene delivery systems which completed clinical trials

Delivery system	Gene type	Condition	phase	Clinicaltrials.gov identifier
Cationic Liposome	cystic fibrosis gene (pGT-1)	Cystic fibrosis	I	NCT00004471
Cholesterol-fus1 Liposome Complex (DOTAP:Chol-fus1)	fus1 plasmid DNA	Non-small cell lung cancer	I	NCT00059605
DOTMA/Cholesterol [Ratio 1:0.5(-/+)] Liposomes	interleukin-2 gene	Head and neck cancer cells	II	NCT00006033
LErafAON (liposomal c-raf antisense oligonucleotide	c-raf antisense oligonucleotide	Neoplasms	I	NCT00024661
PEG-PEI-Cholesterol Lipopolymer	IL-12 Plasmid EGEN-001	Peritoneal Carcinoma/Recurrent Ovarian Carcinoma	II	NCT01118052
ALN-PCS02 Liposomes	siRNA KIF11 and VEGF	Solid tumours	I	NCT01158079
ALN-PCS02 Liposomes	siRNA PCSK9	Hypercholesterolemia	I	NCT01437059

References

1. Yi, Y., Noh, M.J., Lee, K.H.: Current Advances in Retroviral Gene Therapy. Current Gene Therapy **11**(3), 218–228 (2011)
2. Pack, D.W., et al.: Design and development of polymers for gene delivery. Nature Reviews Drug Discovery **4**(7), 581–593 (2005)
3. Blasberg, R.G., Tjuvajev, A.G.: Molecular-genetic imaging: current and future perspectives. Journal of Clinical Investigation **111**(11), 1620–1629 (2003)

4. Jacobs, A., et al.: Positron-emission tomography of vector-mediated gene expression in gene therapy for gliomas. Lancet **358**(9283), 727–729 (2001)
5. Jolly, D.: Viral Vector Systems for Gene-Therapy. Cancer Gene Therapy **1**(1), 51–64 (1994)
6. Godbey, D.A.B.A.: Liposomes for Use in Gene Delivery. Journal of Drug Delivery **2011**, 1–12 (2011)
7. Fattahi, A., et al.: Preparation and characterization of oligochitosan-tragacanth nanoparticles as a novel gene carrier. Carbohydrate Polymers **97**(2), 277–283 (2013)
8. Choi, Y.H., et al.: Lactose-poly(ethylene glycol)-grafted poly-L-lysine as hepatoma cell-targeted gene carrier. Bioconjugate Chemistry **9**(6), 708–718 (1998)
9. Boussif, O., et al.: A Versatile Vector for Gene and Oligonucleotide Transfer into Cells in Culture and in-Vivo - Polyethylenimine. Proceedings of the National Academy of Sciences of the United States of America **92**(16), 7297–7301 (1995)
10. Haensler, J., Szoka, F.C.: Polyamidoamine Cascade Polymers Mediate Efficient Transfection of Cells in Culture. Bioconjugate Chemistry **4**(5), 372–379 (1993)
11. Lim, Y.B., et al.: Biodegradable polyester, poly[alpha-(4 aminobutyl)-L-glycolic acid], as a non-toxic gene carrier. Pharmaceutical Research **17**(7), 811–816 (2000)
12. Kakizawa, Y., Kataoka, K.: Block copolymer self-assembly into monodispersive nanoparticles with hybrid core of antisense DNA and calcium phosphate. Langmuir **18**(12), 4539–4543 (2002)
13. Leventis, R., Silvius, J.R.: Interactions of Mammalian-Cells with Lipid Dispersions Containing Novel Metabolizable Cationic Amphiphiles. Biochimica Et Biophysica Acta **1023**(1), 124–132 (1990)
14. Gao, X., Huang, L.: A Novel Cationic Liposome Reagent for Efficient Transfection of Mammalian-Cells. Biochemical and Biophysical Research Communications **179**(1), 280–285 (1991)
15. Behr, J.P., et al.: Efficient Gene-Transfer into Mammalian Primary Endocrine-Cells with Lipopolyamine-Coated DNA. Proceedings of the National Academy of Sciences of the United States of America **86**(18), 6982–6986 (1989)
16. Wagner, D.E., Bhaduri, S.B.: Progress and Outlook of Inorganic Nanoparticles for Delivery of Nucleic Acid Sequences Related to Orthopedic Pathologies: A Review. Tissue Engineering Part B-Reviews **18**(1), 1–14 (2012)
17. Zheng, D., et al.: Topical delivery of siRNA-based spherical nucleic acid nanoparticle conjugates for gene regulation. Proceedings of the National Academy of Sciences of the United States of America **109**(30), 11975–11980 (2012)
18. Giljohann, D.A., et al.: Gene Regulation with Polyvalent siRNA-Nanoparticle Conjugates. Journal of the American Chemical Society **131**(6), 2072 (2009)
19. Mao, H.Q., et al.: Chitosan-DNA nanoparticles as gene carriers: synthesis, characterization and transfection efficiency. Journal of Controlled Release **70**(3), 399–421 (2001)
20. Lakshminarayanan, A., et al.: Efficient Dendrimer-DNA Complexation and Gene Delivery Vector Properties of Nitrogen-Core Poly(propyl ether imine) Dendrimer in Mammalian Cells. Bioconjugate Chemistry **24**(9), 1612–1623 (2013)
21. Maiti, P.K., Bagchi, B.: Structure and dynamics of DNA-dendrimer complexation: Role of counterions, water, and base pair sequence. Nano Letters **6**(11), 2478–2485 (2006)
22. Copolovici, D.M., et al.: Cell-Penetrating Peptides: Design, Synthesis, and Applications. ACS Nano **8**(3), 1972–1994 (2014)
23. Palm-Apergi, C., Lonn, P., Dowdy, S.F.: Do Cell-Penetrating Peptides Actually "Penetrate" Cellular Membranes? Molecular Therapy **20**(4), 695–697 (2012)

24. Onishi, H., Machida, Y.: Biodegradation and distribution of water-soluble chitosan in mice. Biomaterials **20**(2), 175–182 (1999)
25. Rao, S.B., Sharma, C.P.: Use of chitosan as a biomaterial: Studies on its safety and hemostatic potential. Journal of Biomedical Materials Research **34**(1), 21–28 (1997)
26. Asai, T., et al.: Cell-penetrating peptide-conjugated lipid nanoparticles for siRNA delivery. Biochemical and Biophysical Research Communications **444**(4), 599–604 (2014)
27. Elbakry, A., et al.: Layer-by-Layer Assembled Gold Nanoparticles for siRNA Delivery. Nano Letters **9**(5), 2059–2064 (2009)
28. Guo, S.T., et al.: Enhanced Gene Delivery and siRNA Silencing by Gold Nanoparticles Coated with Charge-Reversal Polyelectrolyte. ACS Nano **4**(9), 5505–5511 (2010)
29. Bulte, J.W.M., Kraitchman, D.L.: Iron oxide MR contrast agents for molecular and cellular imaging. Nmr in Biomedicine **17**(7), 484–499 (2004)
30. Zhu, D.R., et al.: Nanoparticle-Based Systems for T-1-Weighted Magnetic Resonance Imaging Contrast Agents. International Journal of Molecular Sciences **14**(5), 10591–10607 (2013)
31. Dixit, S., et al.: *Phospholipid micelle encapsulated gadolinium oxide nanoparticles for imaging and gene delivery*. Rsc Advances **3**(8), 2727–2735 (2013)
32. Pan, B.F., et al.: Dendrimer-modified magnetic nanoparticles enhance efficiency of gene delivery system. Cancer Research **67**(17), 8156–8163 (2007)
33. Kamau, S.W., et al.: Enhancement of the efficiency of non-viral gene delivery by application of pulsed magnetic field. Nucleic Acids Research **34**(5) (2006)
34. Lee, J.H., et al.: All-in-One Target-Cell-Specific Magnetic Nanoparticles for Simultaneous Molecular Imaging and siRNA Delivery. Angewandte Chemie-International Edition **48**(23), 4174–4179 (2009)
35. Popovtzer, R., et al.: Targeted Gold Nanoparticles Enable Molecular CT Imaging of Cancer. Nano Letters **8**(12), 4593–4596 (2008)
36. Tsien, R.Y.: The green fluorescent protein. Annual Review of Biochemistry **67**, 509–544 (1998)
37. Hernandez, R., Orbay, H., Cai, W.: Molecular Imaging Strategies for In Vivo Tracking of MicroRNAs: A Comprehensive Review. Current Medicinal Chemistry **20**(29), 3594–3603 (2013)
38. Contag, C.H., Bachmann, M.H.: Advances in vivo bioluminescence imaging of gene expression. Annual Review of Biomedical Engineering **4**, 235–260 (2002)
39. Zhou, M.Q., et al.: Assessment of Therapeutic Efficacy of Liposomal Nanoparticles Mediated Gene Delivery by Molecular Imaging for Cancer Therapy. Journal of Biomedical Nanotechnology **8**(5), 742–750 (2012)
40. Yguerabide, J., Yguerabide, E.E.: Light-scattering submicroscopic particles as highly fluorescent analogs and their use as tracer labels in clinical and biological applications - II. Experimental Characterization. Analytical Biochemistry **262**(2), 157–176 (1998)
41. Hellebust, A., Richards-Kortum, R.: Advances in molecular imaging: targeted optical contrast agents for cancer diagnostics. Nanomedicine **7**(3), 429–445 (2012)
42. Horisberger, M., Rosset, J., Bauer, H.: Colloidal Gold Granules as Markers for Cell-Surface Receptors in Scanning Electron-Microscope. Experientia **31**(10), 1147–1149 (1975)
43. Timothy, L., et al.: Nanoparticles for targeted therapeutics and diagnostics. In: Handbook of Biomedical Optics, pp. 697–722. CRC Press. (2011)

44. Aaron, J., et al.: Plasmon resonance coupling of metal nanoparticles for molecular imaging of carcinogenesis in vivo. Journal of Biomedical Optics **12**(3) (2007)
45. El-Sayed, I.H., Huang, X.H., El-Sayed, M.A.: Surface plasmon resonance scattering and absorption of anti-EGFR antibody conjugated gold nanoparticles in cancer diagnostics: Applications in oral cancer. Nano Letters **5**(5), 829–834 (2005)
46. Skala, M.C., et al.: Photothermal Optical Coherence Tomography of Epidermal Growth Factor Receptor in Live Cells Using Immunotargeted Gold Nanospheres. Nano Letters **8**(10), 3461–3467 (2008)
47. Yang, X.M., et al.: Photoacoustic tomography of a rat cerebral cortex in vivo with au nanocages as an optical contrast agent. Nano Letters **7**(12), 3798–3802 (2007)
48. Durr, N.J., et al.: Two-photon luminescence imaging of cancer cells using molecularly targeted gold nanorods. Nano Letters **7**(4), 941–945 (2007)
49. Holliger, P., Hudson, P.J.: Engineered antibody fragments and the rise of single domains. Nature Biotechnology **23**(9), 1126–1136 (2005)
50. Lee, S., Xie, J., Chen, X.Y.: Peptide-Based Probes for Targeted Molecular Imaging. Biochemistry **49**(7), 1364–1376 (2010)
51. Reubi, J.C., Maecke, H.R.: Peptide-Based Probes for Cancer Imaging. Journal of Nuclear Medicine **49**(11), 1735–1738 (2008)
52. Lin, G.M., et al.: Biodegradable Nanocapsules as siRNA Carriers for Mutant K-Ras Gene Silencing of Human Pancreatic Carcinoma Cells. Small **9**(16), 2757–2763 (2013)
53. Lee, S.Y., et al.: Shell cross-linked polyethylenimine-modified micelles for temperature-triggered drug release and gene delivery. Rsc Advances **4**(101), 57702–57708 (2014)
54. Mehrotra, S., Lee, I., Chan, C.: Multilayer mediated forward and patterned siRNA transfection using linear-PEI at extended N/P ratios. Acta Biomaterialia **5**(5), 1474–1488 (2009)
55. McKenzie, D.L., et al.: Low molecular weight disulfide cross-linking peptides as nonviral gene delivery carriers. Bioconjugate Chemistry **11**(6), 901–909 (2000)
56. Handwerger, R.G., Diamond, S.L.: Biotinylated photocleavable polyethylenimine: Capture and triggered release of nucleic acids from solid supports. Bioconjugate Chemistry **18**(3), 717–723 (2007)
57. Dieguez, L., et al.: Electrochemical tuning of the stability of PLL/DNA multilayers. Soft Matter **5**(12), 2415–2421 (2009)
58. Saurer, E.M., et al.: Assembly of erodible, DNA-containing thin films on the surfaces of polymer microparticles: Toward a layer-by-layer approach to the delivery of DNA to antigen-presenting cells. Acta Biomaterialia **5**(3), 913–924 (2009)
59. Schuler, C., Caruso, F.: Decomposable hollow biopolymer-based capsules. Biomacromolecules **2**(3), 921–926 (2001)
60. Newman, C.M.H., Bettinger, T.: Gene therapy progress and prospects: Ultrasound for gene transfer. Gene Therapy **14**(6), 465–475 (2007)
61. Uthaman, S., et al.: Polysaccharide-Coated Magnetic Nanoparticles for Imaging and Gene Therapy. BioMed Research International
62. Dobson, J.: Gene therapy progress and prospects: magnetic nanoparticle-based gene delivery. Gene Therapy **13**(4), 283–287 (2006)
63. Cherukuri, P., Glazer, E.S., Curleya, S.A.: Targeted hyperthermia using metal nanoparticles. Advanced Drug Delivery Reviews **62**(3), 339–345 (2010)
64. Truong, N.P., et al.: An influenza virus-inspired polymer system for the timed release of siRNA. Nature Communications **4** (2013)

Nanoparticles for Pancreatic Islet Imaging

Min Jun Kim, Yong Hwa Hwang and Dong Yun Lee

Abstract While clinical islet transplantation is being investigated as a useful method to cure diabetes mellitus (DM), the outcome of this therapy remains imperfect. This is mainly due to early graft rejection of transplanted islets by the instant blood-mediated inflammatory reaction (IBMIR) or islet ischemia. Therefore, it is important to develop imaging tools for transplanted islets, so that real-time treatment can be used to protect the islets from host immune reactions if required. Recently, synthetic nanoparticles are emerging in the area of biomedical imaging. Innovative nanoparticle developments have significantly enhanced the versatility of different imaging modalities. Nanoparticles represent potential probes to monitor transplanted islets through magnetic resonance imaging (MRI), positron emission tomography (PET), and optical imaging. Nanoparticles can be delivered to transplanted islets by conjugation onto the cell membrane, intracellular delivery, cellular membrane receptor or transporter targeting. Visualizing the survival, function, and biodistribution of transplanted islets may suggest an appropriate treatment, which will enhance islet graft survival. This chapter focuses on recent advances and applications of various nanoparticle-based imaging techniques used in molecular imaging to monitor transplanted islets.

Keywords: Nanoparticle · Islet · Diabetes · Transplantation · MRI · PET

1 Introduction

Pancreatic islet transplantation is a promising therapeutic method for diabetes mellitus (DM), including pancreatomized DM and type 1 DM [1, 2]. Recent

M.J. Kim · Y.H. Hwang · D.Y. Lee(✉)
Departments of Bioengineering, College of Engineering, and Institute for Bioengineering and Biopharmaceutical Research, Hanyang University, Seoul 133-791, Republic of Korea
e-mail: dongyunlee@hanyang.ac.kr

M.J. Kim · Y.H, Hwang · D.Y. Lee
BK21 PLUS Future Biopharmaceutical Human Resources Training and Research Team, Hanyang University, Seoul 133-791, Republic of Korea

© Springer International Publishing Switzerland 2016

H. Jo et al. (eds.), *Biomedical Engineering: Frontier Research and Converging Technologies*, Biosystems & Biorobotics 9,
DOI: 10.1007/978-3-319-21813-7_2

studies in experimental models and clinical trials have demonstrated that islet transplantation represents an alternative approach to insulin therapy since islet transplantation provides a better normalization of metabolic control [3]. For clinical islet transplantation, the procedure involves infusion of isolated islets in the liver via the portal vein. This procedure is relatively fast and safe, and can be performed under anesthesia. Even though significant improvements have been achieved in this area, many challenges are still present, such as preservation of islet function and improvement of islet engraftment [4]. The worst expected outcomes of intrahepatic islet transplantation are islet graft rejection and islet ischemia during the early stage of transplantation [5]. In addition, islet graft loss is accelerated by a thrombotic and nonspecific inflammatory reaction called the instant blood-mediated inflammatory reaction (IBMIR) [6]. IBMIR involves the activation of the coagulation and complement systems, which results in blood clot formation and infiltration of immune cells into the islets, thereby leading to islet destruction [6]. Therefore, treatment with immunosuppressive agents to prevent islet graft rejection is necessary to enhance the survival rate of transplanted islets in the liver. Currently, co-administration of several immunosuppressive drugs, such as tacrolimus, sirolimus and daclizumab has shown to enable maintenance of normoglycemia in more than 85% of type 1 DM patients one year after human islet transplantation [7]. However, less than 10% of patients received islets was able to maintain normal blood glucose levels 5 years after islet transplantation [8]. Long-term administration of immunosuppressive drugs can induce many side effects including neurotoxicity, nephrotoxicity, or hypertension [9-11]. Also, immunosuppressive drugs lead to reduced insulin secretion from the islets by decreasing ATP production and glycolysis [12]. Thus, developing more effective and less detrimental approaches of immunosuppressive drug treatment is required to lower the drug toxicity to the transplanted islets [13, 14]. To do so, real-time assessment of transplanted islets is necessary for optimal treatment of immunosuppressive drugs. Conventional assessment methods of transplanted islets include indicators of metabolic parameters, such as blood glucose levels, glucose tolerance testing, serum C-peptide levels, Hemoglobin A1c (HbA1c) levels, and insulin secretion. However, all of these parameters only provide information about the late stages (end-point assessment) of islet graft rejections. Therefore, it is important to develop non-invasive imaging methods for early detection of islet graft rejection. Successful monitoring of the transplanted islet grafts by non-invasive islet imaging would allow us to evaluate the functionality and viability of the graft via monitoring of islet cell mass. It could also provide a superior understanding of various mechanisms related to islet graft loss and enable us to rescue islets from damage by finding appropriate treatment.

The earliest non-invasive imaging method for islet grafts is the optical imaging technique based on bioluminescence for visualization of transplanted islets [15]. This bioluminescence imaging (BLI) system utilizes the reaction between luciferase enzymes and luciferin substrates. The luciferase enzyme generates visible light by the oxidation of substrate in the presence of oxygen. The advantages of BLI include high sensitivity with low background from mammalian tissues and capable of continuous data evaluation in a natural biological environment. Transfected islets expressing luciferase enzyme were transplanted into mice and successfully imaged by BLI for more than 1 year [15-17]. Although BLI is effectively utilized for the detection of transplanted islets, it is not clinically suitable because the emission wavelength by the enzyme reaction is limited in tissues with a depth of over 5 mm [18].

An alternative non-invasive imaging modality is magnetic resonance imaging (MRI). MRI has various advantages such as tomographic properties, the highest resolution among non-invasive imaging modalities, unlimited depth of penetration, and soft tissue contrast. Thus, MRI has been the most widely used non-invasive technique to evaluate islet graft survival. It has been reported that islet damage from allo-rejection can be monitored using MRI [19]. Therefore, MRI is a powerful tool to monitor transplanted islets.

Another alternative non-invasive imaging method is positron emission tomography (PET). PET is a nuclear medical imaging method used to evaluate the functional processes in the body. It has good sensitivity and high resolution. PET images are obtained by the cellular consumption of molecules that are labeled with positron-emitting isotopes. The first trial to detect islets used 2-[^{18}F] fluoro-2-deoxy-D-glucose (FDG) [20]. FDG was internalized into the pancreatic islet via glucose transporters, thereby enabling labeling and imaging of islets. In one study, Lu *et al* visualized transplanted rat and human islets for 90 days after transplantation [21].

Nanotechnology has rapidly advanced over the past decade, which provides new opportunities to enhance the efficiency of current non-invasive molecular imaging systems. In the case of MRI, magnetic nanoparticles as contrast agents can enhance the sensitivity of a native MR image, which provides more accurate distinctions between the pancreatic islets and the surrounding non-islet tissues. When islets are labeled with magnetic nanoparticles, the labeled islets show a marked decay in transverse magnetization of tissue due to the high magnetic property of the magnetic nanoparticles, thereby darkening signals on MRI. The hypointensity of the magnetic nanoparticles makes the labeled islets distinguishable from the non-islet cells or other tissues. On the other hand, in the case of PET imaging, nanoparticles can provide islet targeting specificity and half-life of tracer. Collectively, the introduction of nanoparticle-based molecular imaging techniques has the great potential to successfully achieve non-invasive monitoring the status of viability and functionality of the islet grafts after transplantation. In this chapter, various nanoparticle-based imaging approaches will be introduced.

2 Intracellular Delivery of Nanoparticles Into Islet Cells

To visualize graft location and survival of transplanted islets, several types of nanoparticles including magnetic nanoparticles, gold nanoparticles, and fluorophores have been delivered into islets by random uptake before transplantation *ex vivo* [22-26]. In this section, several islet-labeling strategies - (1) non-specific nanoparticle uptake, (2) uptake of cationic material-conjugated nanoparticles, (3) uptake of antibody-conjugated nanoparticles, (4) sulfonylurea receptor-mediated uptake of nanoparticles, (5) glucagon like peptide-1 receptor-mediated uptake of nanoparticles, (6) glucose transporter-mediated uptake of nanoparticles, and (7) vesicular monoamine transporter-mediated uptake of nanoparticles - will be introduced.

2.1 Nanoparticle Uptake for Islet Labeling

Magnetic nanoparticles have recently gained significant attention among the broad spectrum of nanoscale materials due to their intrinsic magnetic properties, which enable tracking the nanoparticles through magnetic resonance imaging. Magnetic nanoparticles based on metals such as iron have been developed in various ways in modern technology. In particular, superparamagnetic iron oxide (SPIO) nanoparticles have been studied in many biomedical applications such as biological separation, biosensors, magnetic resonance imaging (MRI) contrast enhancement, targeted drug delivery, catalysis, hyperthermia, and diagnostic medical devices; because it is the only clinically-approved metal oxide nanoparticle available today [27, 28]. In addition, SPIO nanoparticles provide many advantages for biomedical applications such as the generation of magnetic fields and the ability to be guided to a specific organ or tissue by an external magnetic field.

SPIO nanoparticles are generally categorized into three types based on the hydrodynamic diameter: oral SPIO (300 nm-3.5 μm), standard SPIO (50-150 nm), and ultra-small SPIO (<50 nm). Usually SPIO nanoparticles of 10-100 nm are considered optimal for intravenous administration because particles (>200 nm or <10 nm) are removed by the spleen or thorough renal clearance, respectively [29]. These nanoparticles are mostly fabricated by chemical synthesis to obtain uniform composition and size [30]. SPIO nanoparticles tend to aggregate due to surface-to-volume ratio and dipole-dipole interaction. Aggregation reduces their intrinsic superparamagnetic property. Therefore, surface modification of SPIO nanoparticles with polysaccharides or synthetic polymers provides several advantages of their properties including prevention of agglomeration of the iron oxide coreand improvement of stability, solubility, and biocompatibility (Fig. 1.) [31]. Currently, a number of SPIO nanoparticles are applied in early clinical trials or experimental study stages [32-34]. Several formulations for surface modification of SPIO nanoparticles have been approved for clinical use for medical imaging and therapeutic applications such as Lumiren® (Silicon coated SPIO, AMI-121; Gastromark®) for bowel imaging [35], Feridex IV® (Dextran coated SPIO, AMI-25; Feridex) for liver and spleen imaging [36], Combidex® (Dextran coated USPIO, AMI-227; Sinerem®) for lymph node metastasis imaging [37], and most recently, Ferumoxytol® for iron replacement therapy [38].

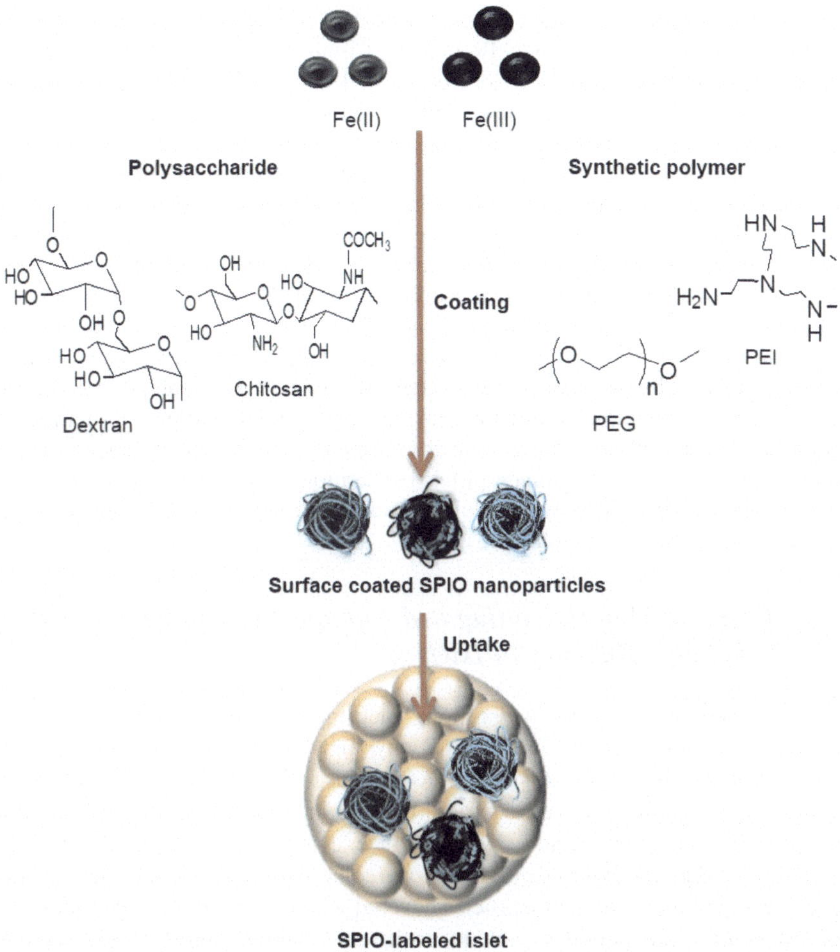

Fig. 1 Surface coating of SPIO nanoparticles using various polysaccharides and synthetic polymers

Non-invasive imaging of transplanted islets using SPIO nanoparticles for MR imaging could help visualize the islets because pancreatic islets have little MRI-sensitivity, resulting in no *in vivo* visualization. Therefore, SPIO nanoparticles have been used for labeling via endocytosis pathways to enhance the visualization of transplanted pancreatic islets in MR imaging. There are several validated strategies for optimal SPIO nanoparticle uptake into pancreatic islets which ensure (1) the lack of toxicity, (2) the appropriate amount of iron uptake, (3) the rapid clearance of the agent after cell destruction, and (4) a correlation between the image and the reality [39].

Ferridex®, a dextran-coated SPIO nanoparticle, simply labels islets by co-incubation for more than 24 hours at a nanoparticle concentration of 250 μg/ml [23]. Since SPIO nanoparticles have highly magnetic properties, labeled islets induce a marked decay in transverse magnetization of tissue (T_2/T_2* effect), which results in darkening signals on MR images. Compared to non-labeled islets, the signal intensity of labeled islets decreased and appeared hypointense (dark) by accumulation of SPIO nanoparticles. When human pancreatic islets labeled with SPIO nanoparticles were transplanted into the left renal subcapsular membrane or liver organ, MR images showed a significant decrease in signal intensity at the transplanted site for a long time. Cellular internalization of SPIO nanoparticles into islets did not affect their insulin secretion during the time that the transplanted islets were monitored. However, this strategy has not yet been applied in clinical islet transplantation because labeling efficacy of SPIO nanoparticles is random (10~80%). Therefore, quantification of the SPIO-labeled islet mass is impossible using an MRI scanner. Moreover, intracellular labeling of SPIO nanoparticles could induce cellular side effects such as membrane leakage of lactate dehydrogenase, impaired mitochondrial function, inflammation, formation of apoptotic bodies, chromosome condensation, generation of reactive oxygen species (ROS), and DNA damage [40-43]. Therefore, it is necessary to develop a new approach that will reduce cellular toxicity and enhance MRI sensitivity in pancreatic islet transplantation.

2.2 Cationic Materialconjugated Nanoparticles to Enhance the Labeling Efficiency of Islets

As mentioned in the previous section, islet intracellular labeling with SPIO nanoparticles was simply carried out by incubating them simultaneously. However, there is no driving force to enhance the internalization of SPIO nanoparticles into the islet cells. In addition, the internalization efficacy of SPIO nanoparticles was low because their surface charges as well as the cellular membrane charge are both negative. To overcome these limitations, some groups have developed magnetic nanoparticles covered with cell penetrating peptides [24-26]. Most of these peptides include positively charged amino acids, such as arginine and lysine.

Cationic liposomes could increase the internalization efficacy of SPIO nanoparticles into islets [24]. Liposomes are widely used to allow DNA incorporation into cells. Also, it has been reported that liposomes can promote SPIO uptake in various types of cells, such as mesenchymal stem cells [44] and myoblasts [45]. To fabricate SPIO nanoparticle-loaded liposomes, ferucarbotran (Resovist) nanoparticles, a commercially used contrast agent, were mixed with liposomes for 20 minutes [24]. Ferucarbotran nanoparticles are negatively charged and thus easily entrapped by cationic liposomes. This combination of ferucarbotran particles with liposomes enables the iron nanoparticles to be more lipophilic. After 24 hours of incubation with islets, ferucarbotran-loaded liposomes improved islet- labeling efficiency at each SPIO concentration when compared to ferucarbotran itself without liposomes. [24]. Thus, cationic liposomes can be used to enhance SPIO uptake into islets.

A synthesized cationic poly(ethylene imine) (PEI) polymer was also introduced to improve islet-labeling efficiency of SPIO nanoparticles [26]. When PEI polymer binded to the negatively charged surface of SPIO nanoparticles, it was able to interact with anionic proteoglycan at the islet surface, thereby facilitating internalization of the SPIO-PEI complex into islets via the endocytosis pathway. To this end, SPIO nanoparticles could be simply coated with PEI polymer through short-term incubation of their mixture for 20 minutes at room temperature. Then, islets were incubated with PEI-coated SPIO nanoparticles for 24 hours at 37°C for cellular labeling. PEI-coated SPIO nanoparticles significantly improved the labeling efficiency of islets by 5.5-fold. The enhancement of intracellular labeling of PEI-coated SPIO nanoparticles was mainly observed in the islet cells located at the periphery of the islet, while the central area of the islet was slightly labeled with PEI-coated SPIO nanoparticles. This is most likely due to the tight junctions between islet cells, which could affect the penetration of the SPIO nanoparticles. In the case of syngeneic islet transplantation, islets labeled with PEI-coated SPIO nanoparticles were continuously detected for 3 months after islet implantation. This result demonstrated that PEI-coated SPIO nanoparticles could be stably located in the islets and visualized for 3 months using an MRI scanner.

Chitosan-coated SPIO nanoparticles were also developed for islet intracellular labeling [25]. Chitosan is an *N*-deacetylated form of chitin which is a polysaccharide that is an abundant in nature. Chitosan is specifically applied in metal nanoparticle synthesis because of its interaction with metal oxide nanoparticles. Chitosan has primary amines that enable it to chelate metal ions and therefore immobilize nanoparticles [46]. Thus, chitosan is a good dispersant for many different nanoparticles such as gold, platinum and silver nanoparticles [47], and single-walled carbon nanotubes [48]. However, the low solubility of chitosan at neutral or high pH can limit its application for nanoparticle synthesis. Depolymerization of relative high-molecular-weight chitosan can improve its water solubility. Thus, depolymerized low-molecular weight chitosan with enhanced water solubility can be better for its application on nanoparticle synthesis. Co-60 gamma-ray irradiation of chitosan can reduce its molecular weight by breaking glycosidic bonds without damaging the polymeric backbone [49]. Thus, to fabricate chitosan-coated SPIO nanoparticles, iron chloride salts (Fe^{2+} and Fe^{3+}) were coprecipitated inside a matrix of chitosan by irradiation with Co-60 gamma-ray [50]. Chitosan-coated SPIO nanoparticles showed a higher content of magnetite and monodispersed clusters [50]. For labeling, islets were incubated with chitosan-coated SPIO nanoparticles overnight [25]. Chitosan effectively increased labeling efficiency of SPIO nanoparticles. Specifically, chitosan-coated SPIO nanoparticles were detected for 18 weeks after islet transplantation [25].

2.3 *Antibody-conjugated Nanoparticles to Enhance the Labeling Efficiency of Islets*

B-cell lymphoma (Bcl)-2 monoclonal antibody-conjugated magnetic nanoparticles were developed to improve the labeling efficiency of SPIO nanoparticles [22]. Bcl-2 is an apoptosis-regulating protein that can promote cell survival (Bcl-xL, Bcl-w and Bcl-2) or facilitate cell death (BAD, BAX, and BAK) [51]. The overexpression of Bcl-2 can protect pancreatic islets against apoptotic cell death [52]. Bcl-2 stabilizes the mitochondrial membrane and blocks the activation of the cytochrome c (Cyt c)/apoptotic protease activating factor (Apaf)-1 pathway, thereby promoting islet survival [22]. Bcl-2 has three advantages in labeling islets: first, the expression of Bcl-2 increases under higher magnetic nanoparticle concentrations, which enables improved labeling efficiency; second, Bcl-2 is a membrane protein that is localized in the endoplasmic reticulum, mitochondria, and nuclear envelope, thereby helping the SPIO nanoparticles to internalize through cell membranes; third, Bcl-2 has a relatively small molecular weight (26 kDa) and shows low cytotoxicity compared with other probes [22]. To conjugate SPIO nanoparticles with the Bcl-2 antibody, COOH-Polyethylene glycol (PEG)-SPIO was produced by evaporating SPIO and COOH-PEG-OE together, then collecting it by centrifugation [22]. The conjugation of COOH-PEG-SPIO with the Bcl-2 antibody was processed by the *N*-hydroxysuccinimide (NHS)/ethyl(diethylaminopropyl) carbodiimide (EDC) method [2]. To label, islets were cultured with Bcl-2 antibody-conjugated SPIO nanoparticles for 24 hours [22]. When islets were treated with Bcl-2 antibody-conjugated SPIO nanoparticles, the expression of Bcl-2 increased [22]. Collectively, Bcl-2 antibody conjugation to SPIO nanoparticles can successfully enhance the labeling efficiency of islets.

2.4 *Sulfonylurea Receptor-Targeted Nanoparticles*

Insulin secretion is controlled by the membrane potential of the beta cell, which relies on the activity of adenosine triphosphate-sensitive potassium ion channels (K_{ATP} channel) in the plasma membrane. The K_{ATP} channel has a hetero-octameric complex of two protein subunits, an inwardly rectifying K^+ channel (Kir6.1 or Kir6.2) and a sulfonylurea receptor (SUR) [53]. The sulfonylurea receptor is the target for sulfonylurea agents that facilitate insulin release from islets, thereby improving glycemic control in type 2 diabetics [54]. Sulfonylurea receptor-1 (SUR1) is highly expressed in the plasma membrane of pancreatic islets but not in the exocrine cells [55]. Therefore, radiolabeled sulfonylurea receptor targeting agents have been used for the visualization of pancreatic islets using single photon emission computed tomography (SPECT) or positron emission tomography (PET) [56, 57].

Glibenclamide is a second-generation sulfonylurea drug that strongly binds to SUR1. Thus, radiolabelled glibenclamide targets and allows for imaging of pancreatic islets. However, labelling with radioactive molecules can affect the biological behavior of glibenclamide (lipophilicity, binding affinity). When fluorine-18 (^{18}F) was introduced into the glibenclamide molecule, binding affinity to SUR1 was well preserved compared with the original glibenclamide. However, lipophilicity was increased because of altered structural elements of the glibenclamide [56]. Lipophilic derivatives lead to high plasma protein binding and nonspecific uptake of the tracer into the liver and kidney in both animals and humans, which results in a poor signal-to-noise ratio [56]. Thus, a reliable visualization and assessment of the pancreatic islet could not be observed with an ^{18}F-labelled glibenclamide derivative [56]. To overcome the lipophilicity-derived problems, a hydrophilic technetium-99m (^{99m}Tc) labelled glibenclamide derivative was developed, due to its relatively lower lipophilicity [56]. ^{99m}Tc has been widely studied for the labeling of radiopharmaceuticals due to its favorable physical properties and low cost. This ^{99m}Tc-labelled glibenclamide derivative showed not only a high binding affinity to SUR1, but also an approximately 20-fold lower lipophilicity when compared with the original glibenclamide or the ^{8}F-labelled glibenclamide [56]. Moreover, after injection into healthy rats, plasma protein binding and liver uptake were decreased by three orders of magnitude compared to that of the 18F-labelled glibenclamide, thereby increasing the signal-to noise ratio [56]. Thus, ^{99m}Tc-labelled glibenclamide derivatives could be used for non-invasive imaging of pancreatic islets.

The clearance kinetics and excretion routes of radiopharmaceuticals are important for high islet/background ratios in diagnostic imaging by PET. The introduction of carbohydrates can increase the hydrophilicity of the tracer and improve the signal-to-noise ratio [58]. When glucose molecules were conjugated to glibenclamide, the hydrophilicity was significantly increased (12-fold) [57]. The glucose-conjugated glibenclamide derivative was more quickly cleared from the blood flow via the kidneys, which indicates reduced plasma protein binding [57]. Therefore, glucose-conjugated glibenclamide allows for better detection of the pancreatic islets *in vivo* [57].

2.5 Glucagon Like Peptide-1 Receptor-Targeted Nanoparticles

The glucagon-like peptide 1 receptor (GLP-1R) is abundantly expressed in mouse, rat, and human pancreatic beta cells [59]. Rare expression of the GLP-1R has been reported in the pancreatic alpha, delta, and pp cells, and exocrine tissues, and other organs (stomach, lung, brain, heart, and duodenum) [59]. Thus, GLP-1R ligand is considered an ideal probe for pancreatic beta cell imaging. Glucagon-like peptide 1 (GLP-1) is an incretin hormone that regulates glucose levels via binding to a G protein-coupled receptor (GPCR) and promoting glucose dependent insulin synthesis or secretion, thereby inducing glucagon secretion suppression [60]. Native GLP-1 is rapidly degraded by the enzyme dipeptidyl peptidase-4 (DPP-IV) in the circulation, which makes the half-life of GLP-1 extremely short and limits its application [61]. Exendin-4, a half-life enhanced GLP-1 analog, shares 53% homology and shows similar biological characteristics as GLP-1, and is more metabolically stable for beta cell targeting. Therefore, GLP-1R-targeted exendin-4 has been widely studied for beta cell imaging [62-69].

Radiolabeled-exendin analogues have been widely studied for beta cell imaging using PET or SPECT [62-67]. Preclinical studies with ^{64}Cu- or ^{18}F-labelled exendin demonstrated its ability to visualize transplanted islets in rodents [63, 64]. However, ^{18}F-labeled exendin-3 (9-39) did not show a correlation between its uptake and pancreatic beta cell mass in a rat model [67]. This is probably because exendin-3 (9-39) does not have a strong enough affinity for GLP-1R *in vivo* [66], which indicates that agonistic activity is needed for successful visualization of beta cells *in vivo*. Importantly, Brom *et al* have recently demonstrated *in vivo* monitoring of beta cells with ^{11}In-labelled exendin-4 using SPECT [65]. Wu et al developed the ^{64}Cu-labeled dimeric exendin-4 [62]. Interestingly, dimeric exendin-4 showed about 3 times higher binding affinity compared to the exendin-4 monomer, thereby providing clearer PET images [62]. Collectively, radiolabeled exendin-4 has great potential for GLP-1R-mediated beta cell imaging using PET or SPECT.

Exendin-4 functionalized SPIO nanoparticles were developed for targeting and imaging of pancreatic beta cells using MRI [68]. To conjugate SPIO nanoparticles with exendin-4, PEG-SPIO nanoparticles were synthesized in advance. A mixture of $FeCl_2(H_2O)_4$, $FeCl_3(H_2O)_6$, and PEG was heated, following by step-by-step addition of 2,2-(ethylenedioxy)bis(ethylamine) (EDEA) and (3-aminopropyl)trimethoxysilane (APS). EDEA was used as a precipitating agent of nanoparticles while APS facilitated silanization via self-assembly. Then, PEG-SPIO was conjugated with exendin-4 using an NHS/EDC method. Results showed that exendin-4-conjugated SPIO nanoparticles were able to specifically bind to and label GLP-1R-expressing INS-1 cells with high labeling efficiency, compared with non-targeted SPIO nanoparticles. Remarkably, exendin-4-conjugated SPIO nanoparticles could differentially label islets in beta cell grafts or pancreatic slices *in vitro*. Also, systemic delivery of exendin-4-conjugated SPIO nanoparticles resulted in the accumulation of the nanoparticles in insulinoma (derived from INS-1 cells), thereby generating a strong magnetic resonance image by a clinical MRI scanner. Even though nanoparticles were unavoidably captured in the liver and spleen, they had a tendency to accumulate in pancreatic islets. Collectively, exendin-4-conjugated SPIO nanoparticles have potential as targeted imaging agents for transplanted islets.

An exendin-4-conjugated bimodal PET/fluorescence imaging agent was introduced in order to overcome the limitations of PET and optical fluorescence imaging technologies [69]. PET imaging has high sensitivity, however, it can provide only spatial resolutions in the range of millimeters [70]. On the other hand, optical fluorescence can provide subcellular resolution for *in vivo* window chamber imaging or *ex vivo* imaging for preclinical studies [71]. However, it has a lower tissue penetration than PET because of the strong absorption of near-infrared (NIR) and visible light in living tissues. Thus, the combination of PET and fluorescence imaging modalities can provide high sensitivity, high resolution and deep tissue penetration. Results showed that the bimodal-imaging agent enabled the visualization of small xenografts (<2 mm) with PET as well as the detection of individual pancreatic islets by the fluorescent properties of the probe [69]. Collectively, this multimodal imaging agent could be a powerful alternative to overcome the current limitations of islet imaging.

2.6 Glucose Transporter-Targeted Nanoparticles

Non-invasive imaging of pancreatic beta cells using glucose transporters, especially glucose transporter 2 (GLUT2), has gained attention since suitable analogs of D-mannoheptulose were proposed about 10 years ago as feasible tools for imaging GLUT2-expressing insulin-producing cells [72]. Glucose transporters are membrane proteins that facilitate the transport of glucose across a plasma membrane. GLUT2, also known as soluble carrier family 2, is a transmembrane carrier protein expressed in pancreatic endocrine beta cells and hepatocytes. GLUT2 facilitates the transfer of glucose and has a high capacity but a low affinity for glucose [73]. For non-invasive imaging, the application of D-mannoheptulose analogs using ^{19}F-heptuloses was recently considered as an alternative for using ^{1}H MRI which is often blurry and lacks quantification. ^{19}F-heptuloses provides highly specific contrast and is directly quantifiable [74]. ^{19}F MRI with ^{19}F compounds as "tracers" has been performed for over 20 years [75]. ^{19}F has several properties as an MRI tracer including high relative sensitivity, 100% natural abundance, and an extremely broad chemical shift. Moreover, ^{19}F has a T_1 sensitivity to oxygen tension that might permit its use as an *in vivo* sensor, and barely presence in natural tissues which does not induce a background signal. In contrast to ^{1}H MRI, its imaging can be carried out in a longitudinal manner for ^{19}F (specifically the labeled cells) and ^{1}H (anatomy) without a "before" image. Furthermore, quantification of the ^{19}F signal can be calculated from *in vivo* data, which provides tracking information for both the localization and number of cells. ^{19}F-heptuloses are efficiently taken up and inhibit glucose-stimulated insulin release from isolated rat islets [76]. This result is derived from an intervention with GLUT2, which transports ^{19}F-heptuloses into cells. Moreover, ^{19}F NMR and ^{19}F MRI experiments indicate that the local concentration of ^{19}F-labeled sugars achieved in cells is sufficient for non-invasive imaging and allow MRI over a prolonged period after administration of the ^{19}F-heptuloses [76]. Additionally, undesirable side effects are minimal because D-mannoheptulose is present in avocados. Collectively, ^{19}F-heptulose is a novel perspective for non-invasive imaging of insulin-producing cells.

If non-invasive imaging is used to estimate pancreatic beta cell mass, it could have a significant impact on clinical diabetes management and islet cell transplantation. Measurement of C-peptide production is currently used to assess beta cell function [77]. However, this has a limited application for *in vivo* detection. Optical imaging using near-infrared (NIR) is a promising high-sensitivity technique that can be utilized for a wide variety of molecules *in vivo,* as NIR light permits relatively deep photon penetration into tissue, minimizes tissue autofluorescence, and has high optical contrast. NIR probes for pancreatic beta cell imaging are based on the beta cell specific ligand streptozotocin (STZ) labeled with the fluorescent dye cyanine-5.5 (Cy5.5) [78]. This probe was designed based on the properties of STZ which interacts with the GLUT2 transporter [79] that is only expressed in beta cells of pancreatic islet [80]. Although the structure of STZ is similar to glucose, it is not recognized by other glucose receptors, such as GLUT1, GLUT3, and GLUT4 [79], because of the nitrosourea moiety of STZ, which is the only part of the molecule that differs from glucose. Moreover, to avoid STZ toxicity in pancreatic beta cells, which primarily arises from the alkylation of DNA and nitric oxide (NO), this probe was constructed using a longer chain to replace the methyl group of STZ since a longer chain has a much lower DNA alkylating ability [81]. After testing the specificity of the synthesized probes, it was shown that the probes labeled almost all INS-1 cells (insulinoma pancreatic beta cells) in a dose-dependent manner, in contrast to other cell lines, such as the CAPAN-2 (pancreatic adenocarcinoma) and HEK 293 (embryonic kidney) cell lines [78]. When the INS-1 cell line was treated with GLUT2 inhibitor, phloretin, for 8 h, the uptake of the probes was also inhibited in a concentration-dependent manner. These results indicated that the cellular uptake of the probe was only mediated by the GLUT2 transporter. Purified human islets that were treated with this probe showed the same intracellular accumulation. Although additional toxicity of longer chain-STZ should be assessed, it is a promising result for pancreas- and/or islet cell transplantation monitoring *in vivo*.

2.7 Vesicular Monoamine Transporter-Targeted Nanoparticles

Vesicular monoamine transporters (VMAT) are categorized into two types, type 1 and 2, on the basis of structure, substrate specificity, and pattern of tissue expression [82]. VMAT2 proteins translocate monoamines into storage vesicles and are expressed by beta cells and monoaminergic neurons. However, VMAT2 is not detectable in exocrine islet cells stained with anti-glucagon or somatostatin [83] and also absent from liver, kidney, and other abdominal organs, which suggests that it might be useful as a biomarker of pancreatic beta cell mass [84]. Several previous studies have identified radioligands for low molecular weight VMAT2, such as dihydrotetrabenazine which is highly selective for VMAT2. Dihydrotetrabenazine and its analogs were developed for positron emission tomography (PET) imaging and have been shown to be 10,000 times more selective for VMAT2 than for VMAT1 [85]. The advantage of dehydrotetrabenazine labelling of VMAT2 is that it is already used clinically for PET imaging of dopaminergic neurons within the central nervous system. Nonetheless, an evaluation of the fidelity with which this molecule reflects beta cell mass and different methods to quantitate VMAT2 in the pancreas is needed [86-88]. ^{11}C-dihydrotetrabenazine materials were used for a PET study of patients with type 1 diabetes to evaluate the feasibility of quantification of VMAT2 binding in the pancreas of healthy controls and patients in the Paul E. Harris group. They hypothesized that the total pancreatic VMAT2 functional binding capacity in long-standing type 1 diabetes was markedly reduced. They also determined a method to measure beta cell functionality: the area under curve (AUC) of C-peptide, with PET-based measurements of VMAT2 binding in the pancreas. In PET studies, ^{11}C-dihydrotetrabenazine uptake in the pancreas was excellent. However, functional binding capacity appears to overestimate the beta-cell mass, as beta cell mass was near-completely depleted in long-standing type 1 diabetes patients. This may be due to higher nonspecific binding in the pancreas than in the renal cortex [87]. Therefore, the overestimation of VMAT2 binding in the pancreas should be addressed in further studies on clinical PET imaging of beta cell mass.

3 Conjugation of Nanoparticles with Islets

In order to monitor islets by MR imaging, it is important to label islets with magnetic nanoparticles which can remain stable after transplantation into an *in vivo* environment [89]. As introduced in section 2, islets have been labeled with magnetic nanoparticles as the contrast agent by culturing islets with concentrated SPIO for 12-48 hours to induce cellular uptake [39]. Although these methods are often effective and simple, labeling efficiency is not high because the intracellular SPIO uptake occurs via random endocytosis without any specific interaction. This problem can be overcome by transfection agents, such as poly-L-lysine, poly(ethylene imine), Lipofectamine, and islet receptor-targeted ligands as mentioned above [18, 90]. These methods can enhance the labeling efficiency of islets, however, there are still some limitations related to availability of agents and toxicity. Iron uptake is potentially toxic to islets, thereby damaging cell viability and insulin secretion function. An alternative strategy is conjugation of nanoparticles to the surface of the islets. Conjugation of nanoparticles onto the islet surface provides a consistent and higher labeling efficiency of islets. Moreover, the potential toxicity can be reduced because nanoparticles are conjugated to an extracellular site. Thus, the conjugation of nanoparticles to the surface of islets may be a good method for islet labeling. In this section, two conjugation methods - (1) Polymer linker-mediated conjugation (2) Specific ligand mediated conjugation - will be introduced (Fig. 2.) [91-93].

3.1 Polymer Linker-Mediated Conjugation of Nanoparticles with the Islets

Polymer linker can be used to conjugate nanoparticles to the surface of islets. One example is a heterofunctional PEG-mediated conjugation method by covalent-linkage [91]. The chemical conjugation was mediated by a heterobifunctional PEG polymer, maleimide-PEG-N-hydroxysuccinimide ester (NHS), which is biocompatible with blood, cell, and tissue. The maleimide group of the PEG polymer can react with the sulfhydryl group of the heparinized superparamagnetic iron oxide (HSPIO) nanoparticle, while the NHS group of the PEG polymer can react with the amine group of the collagen matrix in the islet surface. To this end, the heparin polysaccharide was chemically modified in advance to introduce a sulfhydryl group. The NHS of the PEG polymer reacted with the amine group on the islet surface. Results showed that 1 hour is enough time to conjugate HSPIO-PEG to the surface of the islets. TEM images demonstrated that the HSPIO nanoparticles only existed on the surface of the islets. On the other hand, when islets were treated with control SPIO for 1 h, most islets were not labeled. This result indicated that PEG-mediated conjugation of nanoparticles with the islets could provide stable and efficient labeling with SPIO nanoparticle. Also, HSPIO-conjugated islets retained viability, insulin secretion-function, and morphological integrity compared with the untreated islets, which indicates that the surface conjugation method may not be toxic to the islets. When islets were transplanted to the left renal subcapsular site of STZ-induced diabetic nude mice, there was no significant difference in achieving normal blood glucose levels between the HSPIO-conjugated islets and the control unlabeled islets. MR images showed a marked decrease in signal intensity at the transplanted site for over 30 days, whereas unlabeled islets were not visualized well. Also, histological analysis showed that HSPIO-conjugated islets retained their insulin secretion ability and that HSPIO was stably conjugated on the surface of islets. Collectively, surface conjugation of islets with HSPIO nanoparticles by covalent-linkage can provide accurate quantitative analysis of transplanted islets by MR imaging.

3.2 Ligand-Mediated Conjugation of Nanoparticles with the Islets

Chemical conjugation technology of islets with SPIO nanoparticles is based on DNA hybridization and was developed to image transplanted islets by MR imaging [92]. To introduce deoxy-thymidine (dT) onto islets, oligo-(thymidylic acid)-poly(ethylene glycol)-1,2-dipalmitoyl-3-phosphatidyl ethanolamine ($(dT)_{20}$-PEG-DPPE) was used to modify the surface of the islets. When $(dT)_{20}$-PEG-DPPEs were added to an islet suspension, they spontaneously formed a thin layer on the surface of islets through the hydrophobic interaction between the PEG-DPPE and the lipid bilayer of islets [94]. To prepare deoxy-adenine (dA) carrying SPIO nanoparticles, a $(dA)_{20}$-SH solution was added to the SPIO nanoparticles modified with maleic acid. Then, the thiol group of $(dA)_{20}$ reacted with the maleic acid of the SPIO nanoparticles through the Michael reaction. Finally, SPIO nanoparticles were conjugated with islets through DNA hybridization (dT-dA). ssDNA-modified SPIO nanoparticles fully labeled the surface of islets. Interestingly, SPIO nanoparticles anchored on the islet membranes were subsequently transferred into endosomes inside the islets. This enables visualization of not only the external part of islets, but also the internal part of islets. Thus, SPIO labeling of islets by DNA hybridization can improve hypointensity of dark regions and provides high contrast imaging by MRI.

The multilayered nano-coatings including the SPIO nanoparticles were attached to islet surfaces using layer-by-layer technology [93]. The principle of the layer-by-layer method is electrostatic forces between opposite charged materials, which enable islets to form successive layers. Chemically modified chitosan was used as a positively charged material while heparin or SPIO nanoparticles were used as negatively charged materials. Sequentially, they can associate with the negatively charged islet surfaces. The coated nanofilms were strengthened by further addition of a chitosan/heparin bilayer on the top of the nanofilm layer. The main advantage of this strategy is that it enables different building materials, such as charged SPIO nanoparticles and polyelectrolyte, to be captured between the layers to form the nanofilm composites. In addition, the positively charged groups on chitosan provide a large number of binding sites for the negatively charged SPIO nanoparticles, thereby providing an enhanced MR image. Thus, multilayer coating of SPIO nanoparticles onto the islet surface can be an effective method of labeling islets for MR imaging.

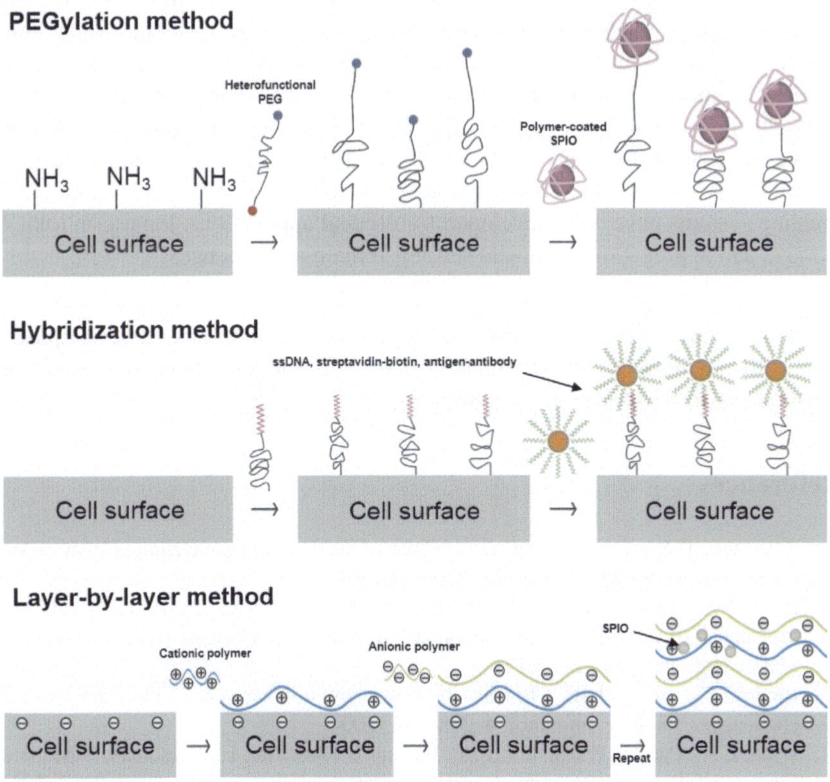

Fig. 2 Methods of chemical conjugation of islets with SPIO nanoparticles

4 Conclusion and Perspective

Non-invasive imaging systems allow us to evaluate the functionality and viability of the islet grafts. They also provide a better understanding of mechanisms related to the graft loss, thereby enabling us to protect islets from host immune rejection via an optimal treatment of immunosuppressive drugs. Among many non-invasive imaging modalities, MRI and PET imaging systems have been most widely used to image transplanted islets. When nanotechnology is applied to MRI and PET modalities, it can provide a more accurate evaluation of transplanted islets. Nanoparticles as contrast agents can enhance the sensitivity of the MR image, and can increase the specificity and half-life of radiolabelled tracers in PET imaging. There are many strategies that can be used to deliver nanoparticles to islet cells. One method is the intracellular uptake of nanoparticles. Although simple co-incubation of islets with nanoparticles can be done, cationic material or antibody-conjugated nanoparticles can enhance the labeling efficiency of islets. Also, ligands of surface receptors (sulfonylurea receptor and glucagon like peptide-1 receptor) and transporters (glucose transporter and vesicular monoamine transporter) can be utilized to specifically target islets. Another method of islet labeling is surface conjugation of nanoparticles with islets. Surface conjugation strategies provide not only stable and efficient labeling of nanoparticles, but also reduce the potential toxicity of nanoparticles because they are conjugated to the extracellular site of islets. Although nanoparticle-based non-invasive molecular imaging systems have been developed for clinical applications, long-term toxicity, storage, and aggregation in clinical settings remain as challenges.

Acknowledgement This book chapter was supported by a Mid-Career Research program grant (NRF-2015R1A2A1A05001832) through the National Research Foundation of Korea (NRF) funded by the Korean government.

References

1. Andersen, D.K.: The practical importance of recognizing pancreatogenic or type 3c diabetes. Diabetes-Metab. Res. **28**, 326–328 (2012)
2. Ewald, N., Kaufmann, C., Raspe, A., Kloer, H.U., Bretzel, R.G., Hardt, P.D.: Prevalence of diabetes mellitus secondary to pancreatic diseases (type 3c). Diabetes-Metab. Res. **28**, 338–342 (2012)
3. Ricordi, C., Strom, T.B.: Clinical islet transplantation: Advances and immunological challenges. Nat. Rev. Immunol. **4**, 258–268 (2004)
4. Ryan, E.A., Paty, B.W., Senior, P.A., Bigam, D., Alfadhli, E., Kneteman, N.M., et al.: Five-year follow-up after clinical islet transplantation. Diabetes **54**, 2060–2069 (2005)
5. Sakata, N., Hayes, P., Tan, A., Chan, N.K., Mace, J., Peverini, R., et al.: MRI Assessment of Ischemic Liver After Intraportal Islet Transplantation. Transplantation **87**, 825–830 (2009)

6. Johansson, H., Lukinius, A., Moberg, L., Lundgren, T., Berne, C., Foss, A., et al.: Tissue factor produced by the endocrine cells of the islets of Langerhans is associated with a negative outcome of clinical islet transplantation. Diabetes **54**, 1755–1762 (2005)
7. Shapiro, A.M.J., Ricordi, C., Hering, B.J., Auchincloss, H., Lindblad, R., Robertson, P., et al.: International trial of the edmonton protocol for islet transplantation. New Engl. J. Med. **355**, 1318–1330 (2006)
8. Shapiro, A.M.J., Lakey, J.R.T., Ryan, E.A., Korbutt, G.S., Toth, E., Warnock, G.L., et al.: Islet transplantation in seven patients with type 1 diabetes mellitus using a glucocorticoid-free immunosuppressive regimen. New Engl. J. Med. **343**, 230–238 (2000)
9. Ayres, R.C.S., Dousset, B., Wixon, S., Buckels, J.A.C., Mcmaster, P., Mayer, A.D.: Peripheral Neurotoxicity with Tacrolimus. Lancet **343**, 862–863 (1994)
10. Berthoux, F.: European best practice guidelines for renal transplantation (Part 2) - Preface. Nephrol. Dial. Transpl. **17**, 2 (2002)
11. Pascual, M., Theruvath, T., Kawai, T., Tolkoff-Rubin, N., Cosimi, A.B.: Medical progress - Strategies to improve long-term outcomes after renal transplantation. New Engl. J. Med. **346**, 580–590 (2002)
12. Radu, R.G., Fujimoto, S., Mukai, E., Takehiro, M., Shimono, D., Nabe, K., et al.: Tacrolimus suppresses glucose-induced insulin release from pancreatic islets by reducing glucokinase activity. Am. J. Physiol-Endoc. M. **288**, E365–E371 (2005)
13. Desai, N.M., Goss, J.A., Deng, S.P., Wolf, B.A., Markmann, E., Palanjian, M., et al.: Elevated portal vein drug levels of sirolimus and tacrolimus in islet transplant recipients: Local immunosuppression or islet toxicity? Transplantation **76**, 1623–1625 (2003)
14. Shivaswamy, V., McClure, M., Passer, J., Frahm, C., Ochsner, L., Erickson, J., et al.: Hyperglycemia induced by tacrolimus and sirolimus is reversible in normal sprague-dawley rats. Endocrine **37**, 489–496 (2010)
15. Lu, Y.X., Dang, H., Middleton, B., Zhang, Z., Washburn, L., Campbell-Thompson, M., et al.: Bioluminescent monitoring of islet graft survival after transplantation. Mol. Ther. **9**, 428–435 (2004)
16. Chen, X.J., Zhang, X.M., Larson, C.S., Baker, M.S., Kaufman, D.B.: In vivo bioluminescence imaging of transplanted islets and early detection of graft rejection. Transplantation **81**, 1421–1427 (2006)
17. Virostko, J., Radhika, A., Poffenberger, G., Chen, Z.Y., Brissova, M., Gilchrist, J., et al.: Bioluminescence Imaging in Mouse Models Quantifies beta Cell Mass in the Pancreas and After Islet Transplantation. Mol. Imaging Biol. **12**, 42–53 (2010)
18. Massoud, T.F., Gambhir, S.S.: Molecular imaging in living subjects: seeing fundamental biological processes in a new light. Gene Dev. **17**, 545–580 (2003)
19. Jirak, D., Kriz, J., Herynek, V., Andersson, B., Girman, P., Burian, M., et al.: MRI of transplanted pancreatic islets. Magnet Reson. Med. **52**, 1228–1233 (2004)
20. Toso, C., Zaidi, H., Morel, P., Armanet, M., Andres, A., Pernin, N., et al.: Positron-emission tomography imaging of early events after transplantation of islets of Langerhans. Transplantation **79**, 353–355 (2005)
21. Lu, Y.X., Dang, H., Middleton, B., Campbell-Thompson, M., Atkinson, M.A., Gambhir, S.S., et al.: Long-term monitoring of transplanted islets using positron emission tomography. Mol. Ther. **14**, 851–856 (2006)

22. Yang, B., Cai, H.L., Qin, W.J., Zhang, B., Zhai, C.X., Jiang, B., et al.: Bcl-2-functionalized ultrasmall superparamagnetic iron oxide nanoparticles coated with amphiphilic polymer enhance the labeling efficiency of islets for detection by magnetic resonance imaging. International Journal of Nanomedicine **8**, 3977–3990 (2013)
23. Evgenov, N.V., Medarova, Z., Dai, G., Bonner-Weir, S., Moore, A.: In vivo imaging of islet transplantation. Nature Medicine **12**, 144–148 (2006)
24. Jiao, Y., Peng, Z.H., Zhang, J.Y., Qin, J., Zhong, C.P.: Liposome-Mediated Transfer Can Improve the Efficacy of Islet Labeling With Superparamagnetic Iron Oxide. Transpl P. **40**, 3615–3618 (2008)
25. Juang, J.H., Wang, J.J., Shen, C.R., Kuo, C.H., Chien, Y.W., Kuo, H.Y., et al.: Magnetic Resonance Imaging of Transplanted Mouse Islets Labeled With Chitosan-Coated Superparamagnetic Iron Oxide Nanoparticles. Transpl. P. **42**, 2104–2108 (2010)
26. Zhang, S.Z., He, H.J., Lu, W.W., Xu, Q.J., Zhou, B.J., Tang, X.D.: Tracking Intrahepatically Transplanted Islets Labeled With Feridex-Polyethyleneimine Complex Using a Clinical 3.0-T Magnetic Resonance Imaging Scanner. Pancreas **38**, 293–302 (2009)
27. Huber, D.L.: Synthesis, properties, and applications of iron nanoparticles. Small **1**, 482–501 (2005)
28. Ito, A., Shinkai, M., Honda, H., Kobayashi, T.: Medical application of functionalized magnetic nanoparticles. Journal of Bioscience and Bioengineering **100**, 1–11 (2005)
29. Elias, A., Tsourkas, A.: Imaging circulating cells and lymphoid tissues with iron oxide nanoparticles. Hematology / the Education Program of the American Society of Hematology American Society of Hematology Education Program, 720–726 (2009)
30. Chemically prepared magnetic nanoparticles. International Materials Reviews **49**, 125–170 (2004)
31. Hwang, Y.H., Lee, D.Y.: Magnetic resonance imaging using heparin-coated superparamagnetic iron oxide nanoparticles for cell tracking in vivo. Quantitative Imaging in Medicine and Surgery **2**, 118–123 (2012)
32. Sun, C., Lee, J.S., Zhang, M.: Magnetic nanoparticles in MR imaging and drug delivery. Adv. Drug Deliv. Rev. **60**, 1252–1265 (2008)
33. McCarthy, J.R., Weissleder, R.: Multifunctional magnetic nanoparticles for targeted imaging and therapy. Adv. Drug Deliv. Rev. **60**, 1241–1251 (2008)
34. Laurent, S., Forge, D., Port, M., Roch, A., Robic, C., Vander Elst, L., et al.: Magnetic iron oxide nanoparticles: synthesis, stabilization, vectorization, physicochemical characterizations, and biological applications. Chemical Reviews **108**, 2064–2110 (2008)
35. Wang, Y.X., Hussain, S.M., Krestin, G.P.: Superparamagnetic iron oxide contrast agents: physicochemical characteristics and applications in MR imaging. European Radiology **11**, 2319–2331 (2001)
36. Bonnemain, B.: Superparamagnetic agents in magnetic resonance imaging: physicochemical characteristics and clinical applications. A Review, Journal of Drug Targeting **6**, 167–174 (1998)
37. Harisinghani, M.G., Barentsz, J., Hahn, P.F., Deserno, W.M., Tabatabaei, S., van de Kaa, C.H., et al.: Noninvasive detection of clinically occult lymph-node metastases in prostate cancer. The New England Journal of Medicine **348**, 2491–2499 (2003)

38. Singh, A., Patel, T., Hertel, J., Bernardo, M., Kausz, A., Brenner, L.: Safety of ferumoxytol in patients with anemia and CKD. American Journal of Kidney Diseases: The Official Journal of the National Kidney Foundation **52**, 907–915 (2008)
39. Ris, F., Lepetit-Coiffe, M., Meda, P., Crowe, L.A., Toso, C., Armanet, M., et al.: Assessment of human islet labeling with clinical grade iron nanoparticles prior to transplantation for graft monitoring by MRI. Cell Transplantation **19**, 1573–1585 (2010)
40. Veranth, J.M., Kaser, E.G., Veranth, M.M., Koch, M., Yost, G.S.: Cytokine responses of human lung cells (BEAS-2B) treated with micron-sized and nanoparticles of metal oxides compared to soil dusts. Particle and Fibre Toxicology **4**, 2 (2007)
41. Hafeli, U.O., Riffle, J.S., Harris-Shekhawat, L., Carmichael-Baranauskas, A., Mark, F., Dailey, J.P., et al.: Cell uptake and in vitro toxicity of magnetic nanoparticles suitable for drug delivery. Molecular Pharmaceutics **6**, 1417–1428 (2009)
42. Jeng, H.A., Swanson, J.: Toxicity of metal oxide nanoparticles in mammalian cells. Journal of Environmental Science and Health Part A, Toxic/Hazardous Substances & Environmental Engineering **41**, 2699–2711 (2006)
43. Hussain, S.M., Hess, K.L., Gearhart, J.M., Geiss, K.T., Schlager, J.J.: In vitro toxicity of nanoparticles in BRL 3A rat liver cells. Toxicology in Vitro: An International Journal Published in Association with BIBRA **19**, 975–983 (2005)
44. Frank, J.A., Miller, B.R., Arbab, A.S., Zywicke, H.A., Jordan, E.K., Lewis, B.K., et al.: Clinically applicable labeling of mammalian and stem cells by combining; Superparamagnetic iron oxides and transfection agents. Radiology **228**, 480–487 (2003)
45. Zhang, Z.L., van den Bos, E.J., Wielopolski, P.A., de Jong-Popijus, M., Duncker, D.J., Krestin, G.P.: High-resolution magnetic resonance imaging of iron-labeled myoblasts using a standard 1.5-T clinical scanner. Magn. Reson. Mater Phy. **17**, 201–209 (2004)
46. Klepka, M.T., Nedelko, N., Greneche, J.M., Lawniczak-Jablonska, K., Demchenko, I.N., Slawska-Waniewska, A., et al.: Local atomic structure and magnetic ordering of iron in Fe-chitosan complexes. Biomacromolecules **9**, 1586–1594 (2008)
47. Laudenslager, M.J., Schiffman, J.D., Schauer, C.L.: Carboxymethyl Chitosan as a Matrix Material for Platinum, Gold, and Silver Nanoparticles. Biomacromolecules **9**, 2682–2685 (2008)
48. Zhang, J., Wang, Q., Wang, L., Wang, A.: Manipulated dispersion of carbon nanotubes with derivatives of chitosan. Carbon **45**, 1917–1920 (2007)
49. Yoksan, R., Akashi, M., Miyata, M., Chirachanchai, S.: Optimal gamma-ray dose and irradiation conditions for producing low-molecular-weight chitosan that retains its chemical structure. Radiat. Res. **161**, 471–480 (2004)
50. Tsai, Z.T., Wang, J.F., Kuo, H.Y., Shen, C.R., Wang, J.J., Yen, T.C.: In situ preparation of high relaxivity iron oxide nanoparticles by coating with chitosan: A potential MRI contrast agent useful for cell tracking. J. Magn. Magn. Mater. **322**, 208–213 (2010)
51. Kim, S.J., Nian, C.L., Widenmaier, S., McIntosh, C.H.S.: Glucose-dependent insulinotropic polypeptide-mediated up-regulation of beta-cell antiapoptotic Bcl-2 gene expression is coordinated by cyclic AMP (cAMP) response element binding protein (CREB) and cAMP-responsive CREB coactivator 2. Mol. Cell Biol. **28**, 1644–1656 (2008)
52. Iwahashi, H., Hanafusa, T., Eguchi, Y., Nakajima, H., Miyagawa, J., Itoh, N., et al.: Cytokine-induced apoptotic cell death in a mouse pancreatic beta-cell line: Inhibition by Bcl-2. Diabetologia **39**, 530–536 (1996)

53. Aguilar-Bryan, L., Clement, J.P., Gonzalez, G., Kunjilwar, K., Babenko, A., Bryan, J.: Toward understanding the assembly and structure of K-ATP channels. Physiol. Rev. **78**, 227–245 (1998)
54. Proks, P., Reimann, F., Green, N., Gribble, F., Ashcroft, F.: Sulfonylurea stimulation of insulin secretion. Diabetes **51**, S368–S376 (2002)
55. Schwanstecher, M., Schwanstecher, C., Dickel, C., Chudziak, F., Moshiri, A., Panten, U.: Location of the Sulfonylurea Receptor at the Cytoplasmic Face of the Beta-Cell Membrane. Brit. J. Pharmacol. **113**, 903–911 (1994)
56. Schneider, S., Feilen, P.J., Schreckenberger, M., Schwanstecher, M., Schwanstecher, C., Buchholz, H.G., et al.: In vitro and in vivo evaluation of novel glibenclamide derivatives as imaging agents for the non-invasive assessment of the pancreatic islet cell mass in animals and humans. Exp. Clin. Endocr. Diab. **113**, 388–395 (2005)
57. Schneider, S., Ueberberg, S., Korobeynikov, A., Schechinger, W., Schwanstecher, C., Schwanstecher, M., et al.: Synthesis and evaluation of a glibenclamide glucose-conjugate: A potential new lead compound for substituted glibenclamide derivatives as islet imaging agents. Regul Peptides **139**, 122–127 (2007)
58. Schottelius, M., Wester, H.J., Reubi, J.C., Senekowitsch-Schmidtke, R., Schwaiger, M.: Improvement of pharmacokinetics of radioiodinated Tyr(3)-octreotide by conjugation with carbohydrates. Bioconjugate Chem. **13**, 1021–1030 (2002)
59. Zhang, Y.F., Chen, W.G.: Radiolabeled glucagon-like peptide-1 analogues: a new pancreatic beta-cell imaging agent. Nucl. Med. Commun. **33**, 223–227 (2012)
60. Holst, J.J., Deacon, C.F., Vilsboll, T., Krarup, T., Madsbad, S.: Glucagon-like peptide-1, glucose homeostasis and diabetes. Trends Mol. Med. **14**, 161–168 (2008)
61. Deacon, C.F., Johnsen, A.H., Holst, J.J.: Degradation of Glucagon-Like Peptide-1 by Human Plasma in-Vitro Yields an N-Terminally Truncated Peptide That Is a Major Endogenous Metabolite in-Vivo. J. Clin. Endocr. Metab. **80**, 952–957 (1995)
62. Wu, Z.H., Liu, S.L., Nair, I., Omori, K., Scott, S., Todorov, I., et al.: Cu-64 Labeled Sarcophagine Exendin-4 for MicroPET Imaging of Glucagon like Peptide-I Receptor Expression. Theranostics **4**, 770–777 (2014)
63. Wu, Z.H., Liu, S.L., Hassink, M., Nair, I., Park, R., Li, L., et al.: Development and Evaluation of F-18-TTCO-Cys(40)-Exendin-4: A PET Probe for Imaging Transplanted Islets. J. Nucl. Med. **54**, 244–251 (2013)
64. Wu, Z., Todorov, I., Li, L., Bading, J.R., Li, Z.B., Nair, I., et al.: In Vivo Imaging of Transplanted Islets with Cu-64-DO3A-VS-Cys(40)-Exendin-4 by Targeting GLP-1 Receptor. Bioconjugate Chem. **22**, 1587–1594 (2011)
65. Brom, M., Woliner-van der Weg, W., Joosten, L., Frielink, C., Bouckenooghe, T., Rijken, P., et al.: Non-invasive quantification of the beta cell mass by SPECT with In-111-labelled exendin. Diabetologia **57**, 950–959 (2014)
66. Brom, M., Joosten, L., Oyen, W.J.G., Gotthardt, M., Boerman, O.C.: Radiolabelled GLP-1 analogues for in vivo targeting of insulinomas. Contrast Media Mol. **7**, 160–166 (2012)
67. Wang, Y., Lim, K., Normandin, M., Zhao, X.J., Cline, G.W., Ding, Y.S.: Synthesis and evaluation of [F-18]exendin (9-39) as a potential biomarker to measure pancreatic beta-cell mass. Nucl. Med. Biol. **39**, 167–176 (2012)
68. Zhang, B., Yang, B., Zhai, C.X., Jiang, B., Wu, Y.L.: The role of exendin-4-conjugated superparamagnetic iron oxide nanoparticles in beta-cell-targeted MRI. Biomaterials **34**, 5843–5852 (2013)

69. Brand, C., Abdel-Atti, D., Zhang, Y.C., Carlin, S., Clardy, S.M., Keliher, E.J., et al.: In Vivo Imaging of GLP-1R with a Targeted Bimodal PET/Fluorescence Imaging Agent. Bioconjugate Chem. **25**, 1323–1330 (2014)
70. Zanzonico, P.: Principles of Nuclear Medicine Imaging: Planar, SPECT, PET, Multi-modality, and Autoradiography Systems. Radiat. Res. **177**, 349–364 (2012)
71. Pittet, M.J., Weissleder, R.: Intravital Imaging. Cell **147**, 983–991 (2011)
72. Malaisse, W.J.: On the track to the beta-cell. Diabetologia **44**, 393–406 (2001)
73. Guillam, M.T., Hummler, E., Schaerer, E., Yeh, J.I., Birnbaum, M.J., Beermann, F., et al.: Early diabetes and abnormal postnatal pancreatic islet development in mice lacking Glut-2. Nature Genetics **17**, 327–330 (1997)
74. Srinivas, M., Heerschap, A., Ahrens, E.T., Figdor, C.G., de Vries, I.J.: (19)F MRI for quantitative in vivo cell tracking. Trends in Biotechnology **28**, 363–370 (2010)
75. Lanza, G.M., Yu, X., Winter, P.M., Abendschein, D.R., Karukstis, K.K., Scott, M.J., et al.: Targeted antiproliferative drug delivery to vascular smooth muscle cells with a magnetic resonance imaging nanoparticle contrast agent: implications for rational therapy of restenosis. Circulation **106**, 2842–2877 (2002)
76. Malaisse, W.J., Zhang, Y., Louchami, K., Sharma, S., Dresselaers, T., Himmelreich, U., et al.: (19)F-heptuloses as tools for the non-invasive imaging of GLUT2-expressing cells. Archives of Biochemistry and Biophysics **517**, 138–143 (2012)
77. Larsen, M.O., Rolin, B., Wilken, M., Carr, R.D., Gotfredsen, C.F.: Measurements of Insulin Secretory Capacity and Glucose Tolerance to Predict Pancreatic β-Cell Mass In Vivo in the Nicotinamide/Streptozotocin Göttingen Minipig, a Model of Moderate Insulin Deficiency and Diabetes. Diabetes **52**, 118–123 (2003)
78. Ran, C., Pantazopoulos, P., Medarova, Z., Moore, A.: Synthesis and testing of beta-cell-specific streptozotocin-derived near-infrared imaging probes. Angewandte Chemie (International ed in English) **46**, 8998–9001 (2007)
79. Wang, Z., Gleichmann, H.: GLUT2 in pancreatic islets: crucial target molecule in diabetes induced with multiple low doses of streptozotocin in mice. Diabetes **47**, 50–56 (1998)
80. Orc, L., Thorens, B., Ravazzola, M., Lodish, H.F.: Localization of the pancreatic beta cell glucose transporter to specific plasma membrane domains. Science **245**, 295–297 (1989)
81. Duncan, R.H., Davies, G.S.: Alkylation of DNA bases by carcinogenic N-nitrosamine metabolites: A theoretical study. International Journal of Quantum Chemistry **35**, 665–677 (1989)
82. Weihe, E., Schafer, M.K., Erickson, J.D., Eiden, L.E.: Localization of vesicular monoamine transporter isoforms (VMAT1 and VMAT2) to endocrine cells and neurons in rat. Journal of Molecular neuroscience: MN **5**, 149–164 (1994)
83. Anlauf, M., Eissele, R., Schafer, M.K., Eiden, L.E., Arnold, R., Pauser, U., et al.: Expression of the two isoforms of the vesicular monoamine transporter (VMAT1 and VMAT2) in the endocrine pancreas and pancreatic endocrine tumors. The Journal of Histochemistry and Cytochemistry: Official Journal of the Histochemistry Society **51**, 1027–1040 (2003)
84. Maffei, A., Liu, Z., Witkowski, P., Moschella, F., Del Pozzo, G., Liu, E., et al.: Identification of tissue-restricted transcripts in human islets. Endocrinology **145**, 4513–4521 (2004)
85. Erickson, J.D., Schafer, M.K., Bonner, T.I., Eiden, L.E., Weihe, E.: Distinct pharmacological properties and distribution in neurons and endocrine cells of two isoforms of the human vesicular monoamine transporter. Proceedings of the National Academy of Sciences of the United States of America **93**, 5166–5171 (1996)

86. Simpson, N.R., Souza, F., Witkowski, P., Maffei, A., Raffo, A., Herron, A., et al.: Visualizing pancreatic beta-cell mass with [11C]DTBZ. Nucl. Med. Biol. **33**, 855–864 (2006)
87. Goland, R., Freeby, M., Parsey, R., Saisho, Y., Kumar, D., Simpson, N., et al.: 11C-dihydrotetrabenazine PET of the pancreas in subjects with long-standing type 1 diabetes and in healthy controls. J. Nucl. Med. **50**, 382–389 (2009)
88. Kung, H.F., Lieberman, B.P., Zhuang, Z.P., Oya, S., Kung, M.P., Choi, S.R., et al.: In vivo imaging of vesicular monoamine transporter 2 in pancreas using an (18)F epoxide derivative of tetrabenazine. Nucl. Med. Biol. **35**, 825–837 (2008)
89. Oishi, K., Miyamoto, Y., Saito, H., Murase, K., Ono, K., Sawada, M., et al.: In vivo imaging of transplanted islets labeled with a novel cationic nanoparticle. PLoS One **8**, e57046 (2013)
90. Toso, C., Valle, J.P., Morel, P., Ris, F., Demuylder-Mischler, S., Lepetit-Coiffe, M., et al.: Clinical magnetic resonance imaging of pancreatic islet grafts after iron nanoparticle labeling. Am. J. Transplant. **8**, 701–706 (2008)
91. Jung, M.J., Lee, S.S., Hwang, Y.H., Jung, H.S., Hwang, J.W., Kim, M.J., et al.: MRI of transplanted surface-labeled pancreatic islets with heparinized superparamagnetic iron oxide nanoparticles. Biomaterials **32**, 9391–9400 (2011)
92. Kitamura, N., Nakai, R., Kohda, H., Furuta-Okamoto, K., Iwata, H.: Labeling of islet cells with iron oxide nanoparticles through DNA hybridization for highly sensitive detection by MRI. Bioorganic & Medicinal Chemistry **21**, 7175–7181 (2013)
93. Wang, Y., Blanco-Andujar, C., Zhi, Z.L., So, P.W., Thanh, N.T., Pickup, J.C.: Multilayered nanocoatings incorporating superparamagnetic nanoparticles for tracking of pancreatic islet transplants with magnetic resonance imaging. Chemical Communications **49**, 7255–7257 (2013)
94. Teramura, Y., Kaneda, Y., Iwata, H.: Islet-encapsulation in ultra-thin layer-by-layer membranes of poly(vinyl alcohol) anchored to poly(ethylene glycol)-lipids in the cell membrane. Biomaterials **28**, 4818–4825 (2007)

Dendritic Nanomaterials for Therapeutic and Diagnostic Applications

Ja Hye Myung, Kevin A. Tam and Seungpyo Hong

Abstract Dendritic nanomaterials have attracted a great deal of scientific interest for use in various biomedical applications primarily due to their well-defined structure and capacity for multifunctionalization. In this chapter, we present a comprehensive overview of the recent advances in various dendrimers and their modified dendritic materials with a particular focus on their applications such as drug delivery, gene delivery, diagnostics and prognostics, and detoxification/ prophylaxis. We also discuss the fundamentals of the biological interactions of dendritic nanomaterials in regard to non-specific cell interactions, multivalent binding, stimuli-responsive processes, and biodistribution. In addition, this chapter highlights the current drawbacks of dendrimers that have hindered their rapid clinical translation and introduce recent approaches to overcoming these challenges.

1 Introduction to Dendritic Nanomaterials

Nanotechnology has impacted a wide range of biomedical applications for therapeutic and diagnostic purposes.[1-4] Nanotechnology is generally defined as a broad range of research and applications to manipulate materials with at least one dimension at the nanoscale (from 1 to 100 nanometers), of which contributes to the unique functions of the materials. In pharmaceuticals, the use of nanotechnology has allowed for efficient and efficacious drug delivery, while reducing side effects.[4,5] The impact of nanomaterials in diagnostic applications has been also significant by dramatically improving detection sensitivity and specificity for a variety of diseases.[6]

J.H. Myung · K.A. Tam · S. Hong(✉)
Department of Biopharmaceutical Sciences, College of Pharmacy,
University of Illinois, Chicago, IL 60612, USA
e-mail: sphong@uic.edu

© Springer International Publishing Switzerland 2016

H. Jo et al. (eds.), *Biomedical Engineering: Frontier Research and Converging Technologies*,
Biosystems & Biorobotics 9,
DOI: 10.1007/978-3-319-21813-7_3

One reason that polymers have been commonly used in nanotechnologies is that their structural diversity confers the ability to fit functional demand within desired physical, chemical, biological, and biodegradable properties.[7] In particular, among the various polymers, dendrimers have been primarily applied to the biomedical field for their unique and well-defined structure, precisely controlled molecular weight distributions, capability for multifunctionalization, and high degree of molecular flexibility.[8-12] In the utilization of the unique features of dendrimers, a variety of dendritic nanomaterials and their hybrids with other natural and synthetic materials have recently emerged in respect to their unique properties tailored to specific biomedical applications.[11-16]

In this chapter, the physiochemical properties of dendrimers, dendritic nanomaterials, and hybrid nanomaterials will be reviewed. We will then discuss the fundamental mechanisms underlying biological behaviors of dendritic nanomaterials in terms of specific and non-specific cell interactions, stimuli-responsive processes, and biodistribution. In the following section, the recent advances in various dendrimers and their modified dendritic nanomaterials that are currently being investigated for drug delivery, gene delivery, diagnostics and prognostics, and detoxification and prophylaxis will be described in a comprehensive manner. In addition, the current drawbacks of dendrimers that have hindered their rapid clinical translation will be discussed, followed by introduction of recent approaches to overcoming those challenges.

1.1 Dendrimers

Dendrimers are regularly hyperbranched, deformable, spherical, and tree-like polymers that display a high density of peripheral functional groups. The name of dendrimers comes from the Greek word "*dendro*", which translates "tree". Depending on the synthetic routes and repeating units, several different types of dendrimers, such as poly(amidoamine) (PAMAM), poly(propylenimine) (PPI), poly(L-lysine) (PLL), and carbosilane (C-Si) dendrimers, have been developed for biomedical applications.[10] The size of dendrimers is generally in a range of 2 to 10 nm depending on their generation and surface modification.[17,10] The unique synthetic procedure of dendrimers produces nano-scale macromolecules with close-to-monodispersed molecular weight distributions.[10] As depicted in Fig. 1a, dendrimers are comprised of three functional domains: the core, branches, and periphery. Because dendrimers possess a large number of surface groups, a multifunctional platform can be achieved via functionalization of dendrimer termini with targeting, therapeutic, and diagnostic moieties.[18,19] Moreover, owing to their large number of surface groups and high degree of molecular flexibility,[20] binding interactions are promoted between the surface conjugated targeting ligands and their corresponding cell surface receptors via multivalent binding effect,[21,22] which makes them one of the ideal platforms for targeted delivery of therapeutics and the use in diagnostics.[18,17]

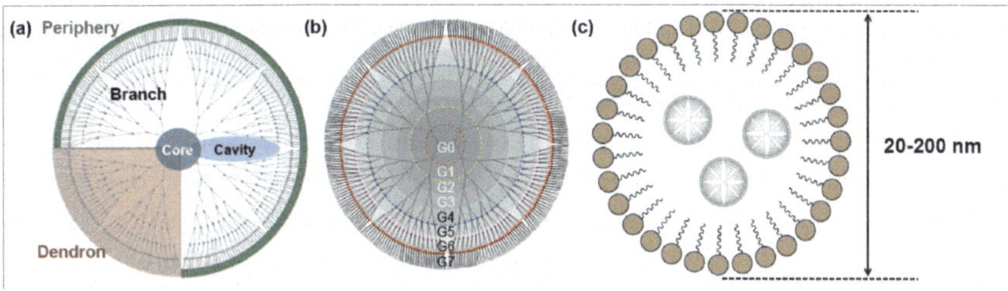

Fig. 1 Structures of dendritic nanomaterials. (a) Dendrimers have a well-defined hierarchical structure consisting of a core, branch, and periphery. (b) Each generation is added to the structure by repeating the reaction sequence of stepwise synthesis, which generates two new branching points. (c) Dendrimers can be encapsulated into relatively large nanoparticles such as liposomes and polymeric nanoparticles (50-200 nm in diameter), which can aim both passive and active targeting strategies.

Among the several types of dendrimers, PAMAM dendrimers are one of the most extensively investigated due to their modifiable peripheral functional groups, biocompatibility, and high deformability, in addition to the characteristic properties that other dendrimers typically have. Different methods of dendrimer synthesis have been described, including convergent, divergent, and click chemistry-based.[10] For example, PAMAM dendrimers can be synthesized using the divergent method from amine-functional cores.[8] Each generational growth step is composed of successive alkylation and amidation reactions, doubling the molecular weight and the number of peripheral functional groups (Fig. 1b). Complete generations of PAMAM dendrimers carry positively charged primary amine groups on the surface, while half-generations possess negatively charged, carboxyl termini.[23] The amine and carboxyl termini offer functional handles to enable facile conjugation with various bioactive moieties. For example, imaging agents, targeting ligands, and therapeutic molecules can be conjugated on PAMAM dendrimers via simple chemical reactions.[19] The peripheral amine and carboxyl termini can be also modified into other end groups such as acetyl, hydroxyl, and methoxy to fit a desired purpose.[21]

1.2 Dendrons

Dendrons are close-to-monodisperse wedge-shaped dendrimer sections with multiple terminal groups and a single reactive functional at the focal point (Fig. 1a). Dendrons, which are the highly branched segments that comprise dendrimers, are synthesized using methods analogous to dendrimers; however, more commonly via convergent synthesis.[24,10] The unique physiochemical and biological properties of dendrons come from those of dendritic structures as well as their focal points.[14,11,12,16] Similar to dendrimers, multiple termini on the wedge-shaped dendritic sections have modular capability to be multifunctionalized with various bioactive molecules. In addition to the multifunctionality of the peripheral groups, their focal points can be modified with other polymeric structures to fit a variety of applications.[14]

Compared to dendrimers, the size (from nanoscale up to microscale), shape (from spherical to rod-like), and structure of dendrons can be engineered more extensively. This structural/morphological flexibility of the materials has allowed for a library of hundreds of dendrons to be synthesized.[14] Some of the unique examples include Janus dendron hybrids and asymmetrically dendronized complexes that have been created via a mix-and-match approach of several dendron scaffolds with different architecture and topology.[14,11,16]

Of particular interest in biomedical engineering are self-assembling dendrons such as linear dendritic-block copolymers (LDBCs) that combine dendrons together with one or multiple linear polymer chains through covalent conjugations.[15,25] Amphiphilic LDBCs are capable of self-assembling into supramolecular structures through a combination of physical forces such as van der Waals, electrostatic, hydrophobic interactions, and hydrogen and coordination bonds, allowing for the encapsulation of small molecules and proteins.[15,26,27] Self-assembled supramolecules of LDBCs are characterized by their larger loading capacity than that of dendrimers while maintaining multivalency.[15,28] One of the key parameters affecting the self-assembly and encapsulation efficiency is hydrophilic-lipophilic balance (HLB) of such amphiphilic copolymers. Compared to linear polymer-based counterparts, dendron-based supramolecules demonstrate that the pre-organized conical dendron architecture facilitates molecular assemblies, as measured by ultra low critical micelle concentrations (CMCs) at high HLBs.[29] The dendron-based self-assembled supramolecules thus have great potential to be engineered to exhibit features amenable for applications in drug and gene delivery.[14,13,15,27]

Besides utilizing polymers to direct the assembly of LDBCs, peptide-based LDBCs offer another class of supramolecules with great biocompatibility. For example, an amphiphilic comb-dendritic block copolymer was developed using poly(gamma-n-dodecyl-L-glutamate) (PDLG), serving as the hydrophobic block, and a hydrophilic polyester dendron modified with PEG chains.[30] The CMC was in the order of 10^{-8} M and approximately 30 wt% of the model drug triclosan could be encapsulated. The tumor targeting efficiencies of polypeptide-based LDBCs were evaluated using a cancer xenograft mouse model and exhibited a four-fold enhancement in efficacy of paclitaxel with reduced toxicity.[31]

1.3 Hybrid Nanoparticles

One of the emerging strategies to overcome the drawbacks of dendrimers and dendritic nanomaterials is their hybridization with other types of nanoparticles, which is most commonly used in targeted cancer drug delivery. The tumor characteristic of enhanced permeability and retention (EPR) effect allows polymeric nanoparticles, micelles, and liposomes of relatively large size (typically 50-200 nm in diameter) to be selectively accumulated in tumor sites.[32] This size-based accumulation in tumor sites, called passive targeting, cannot be easily exploited by dendrimers due to their small size (<10 nm). In contrast, active targeting can be achieved by nanocarriers bearing targeting ligands, and its efficiency is highly dependent upon the tumor penetration of the nanocarriers. Larger nanoparticles typically exhibit limited diffusivity through tumor tissue, compared to the efficient tissue penetration of dendrimers.[12] For this reason, novel hybrid nanoparticle systems have been engineered by combining two different sizes of nanomaterials (Fig. 1c).[33,34]

These hybrid nanoparticles can be prepared via a combination of nanoprecipitation of hydrophobic polymers and subsequent self-assembly of hydrophilic polymers to form the shell of the hybrid. For example, poly(lactic-*co*-glycolic acid) (PLGA) and PEG-conjugated 1,2-distearoyl-*sn*-glycero-3-phosphoethanolamine (DSPE) were used as the polymeric core and shell of the hybrid nanoparticles, respectively.[33] Because the PLGA core stabilized the lipid shell and slowly released the encapsulated drugs, these polymer-lipid hybrid nanoparticles displayed tunable size, high drug loading yield, sustained drug release profile, and good stability in serum. When functionalized with prostate-specific RNA aptamers on the shell, the lipid-polymer hybrid nanoparticles showed cellular targeting ability to prostate cancer cell lines and improved therapeutic effectiveness.[33,35]

Another method to create hybrid nanoparticles is encapsulation of smaller nanoparticles, such as dendrimers, into larger nanoparticles, such as polymeric nanoparticle and liposomes using a double emulsion method.[36] Double emulsions may be formed with either the water-in-oil-in-water (W/O/W) or the oil-in-water-in-oil types (O/W/O). The W/O/W double emulsion method is more commonly used as it stabilizes the particles in the aqueous phase and encapsulates hydrophilic materials into a variety of polymers or copolymers with a controlled particle size.[37] This method allows for the preparation of hybrid nanoparticles that have two-size scale, i.e., polymeric nanoparticles of 50-200 nm and encapsulated dendrimers or other polymer conjugates of <10 nm).[36,38] The encapsulation process is affected by several parameters such as the chemical composition of the outer layers, size and surface charges of dendrimers, and ratios of the outer and inner nanoparticles.[36,25]

A dendrimer-encapsulated hybrid system was introduced by Sunoqrot et al. to engineer a nanocarrier system that met the two different size requirement to exploit the EPR effect and to achieve efficient tumor penetration, as described above.[36,38,39] Folate (FA)-targeted PAMAM dendrimers were encapsulated into PLGA-PEG nanoparticles via double emulsion. The hybrid nanoparticles displayed highly efficient tumor selectivity by both passive targeting based on size control of the PLGA-PEG shell nanoparticles and active targeting through FA-conjugated dendrimers.[39] The controlled release profile of the FA-conjugated dendrimers from the nanohybrids and their targeting kinetics *in vitro* and *in vivo* will be discussed further in section 3.1.3. In addition, the outer layer can protect the inner polymers from rapid blood clearance and accumulation by the reticuloendothelial system (RES, also known as mononuclear phagocyte system or MPS).[40]

1.4 Dendrimersomes

Dendrimersomes have been of recent interest for their stable, size-controllable, and self-assembled unilamellar vesicles of amphiphilic Janus dendrimers in water.[41,13] Compared to liposomes and polymersomes, dendrimersomes display favorable material properties such as suitable stability in various media, low polydispersity, and good mechanical properties.[42]

Unlike liposomes that involve complex compositions and preparation steps, the formation of dendrimersomes can be achieved using simple methods such as the injection of ethanol or other water miscible solvents into water or buffer. Percec et al. synthesized 11 distinct libraries containing a total of 107 amphiphilic Janus dendrimers and described their modular preparation methodology.[13] By tailoring the composition of the hydrophilic and hydrophobic branched segments, a variety of morphologies such as spheres, cubosomes, disks, tubular vesicles, and helical ribbons were observed. Although relatively less investigated than LDBCs, dendrimersomes represent a new class of drug delivery vehicles that combine the properties of liposomes and polymersomes.

2 Fundamental Understanding of Biological Behaviors of Dendritic Nanomaterials

For dendritic nanomaterials to be effective in biomedical applications, a number of parameters need to be considered and optimized to control their biological properties. In this section, we will discuss, with specific examples, the current understanding of physicochemical and biological behaviors of dendritic nanomaterials. Specifically, we will review: a) physical property-dependent non-specific cell interaction; b) cell-specific recognition and multivalent binding effect; c) stimuli-responsive processes for drug delivery; and d) biodistribution and pharmacokinetics of various dendritic nanomaterials. A fundamental understanding is essential for designing and modifying dendritic nanomaterials to exhibit favorable pharmacokinetic profiles, tolerable cytotoxicity, and highly specific biological interactions.

2.1 Physical Property-Dependent Non-specific Cell Interactions

Cell membranes, consisting of lipid bilayers with embedded proteins, separate interior and exterior of cells and are selectively permeable to protect the intracellular environment. The lipid bilayer typically permits diffusion of small and non-polar molecules while preventing diffusion of polar and large molecules. Most of the nanoscale macromolecules are not able to freely diffuse across the cell membranes but can be engineered to internalize into cells through various mechanisms upon contact with the cell membrane. Critical to their use in biomedical applications, the interaction of nanoparticles with cell membranes is frequently determined by their surface properties. The effects of the surface and physiochemical properties of nanomaterials with respect to their interactions with cells need to be understood to design well-controlled nanomaterials for biomedical applications. Multiple properties of nanomaterials, such as particle size, surface charge (positive, neutral, and negative), shape (spherical, rod-like, and irregular), and deformability (flexible versus rigid), can all affect nanomaterial-cell interactions.[43-47]

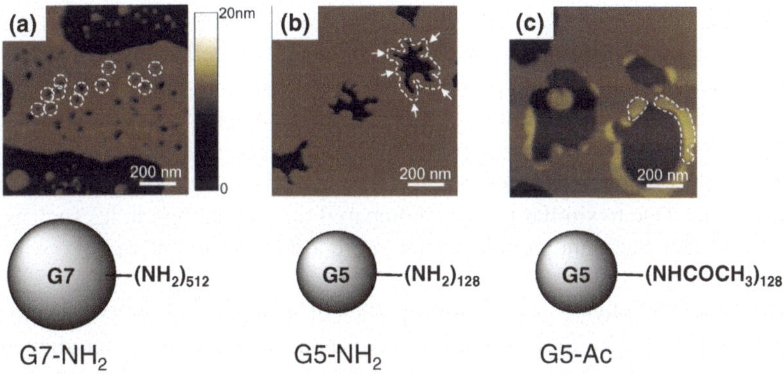

Fig. 2 End group modification of PAMAM dendrimers and their interaction with DMPC bilayer.[44] (a) $G7-NH_2$ initiated the formation of holes with 15–40 nm in diameter on the intact parts of the bilayer, but (b) $G5-NH_2$ affected only the edges of existing bilayer defects. In contrast, charge neutral G5-Ac did not cause disruption of the lipid bilayer. Instead, (c) they adsorbed to the edges of existing bilayer defects. The white dotted lines in the AFM height images highlight the bilayers affected by dendrimers. (Copyright © 2004, American Chemical Society)

The size and generation of dendrimers have been shown to strongly affect their interaction with lipid bilayers and cellular membranes. Hong et al. reported that the interaction of positively charged PAMAM dendrimers destabilized the 1,2-dimyristoyl-sn-glycero-3-phosphocholine (DMPC) lipid bilayers, with G7 dendrimer forming new holes (15-40nm in diameter) and expanding preexisting defects.[44] In cells, these nanoscale holes lead to dendrimer internalization, the outward diffusion of cytosolic proteins, and the two-way diffusion of dye molecules across the cell membranes. However, amine-terminated G5 PAMAM dendrimers were only able to expand existing defects and failed to initiate hole formation at similar concentrations. The size dependency was explained in a way that lipid-bilayer patches interacted with dendrimers and formed dendrimer-filled vesicles, which are more stable with larger dendrimers due to their low curvature.

The Surface charge of dendrimers and dendritic nanoparticles can be chemically modified to be positively, neutral, or negatively charged, which affects their interactions with negatively charged cell membranes. After modifying the amine-termini of G5 PAMAM dendrimers to acetamide, their ability of hole expansion on a DMPC lipid bilayer was impaired (Fig. 2).[44] In agreement with this result obtained using the model membranes, upon surface acetyl or carboxyl modification, the modified PAMAM dendrimers did not show any non-specific interactions and no apparent cytotoxicity.[45] These results indicate that membrane interactions and toxicities of PAMAM dendrimers can be easily controlled by surface modification to charge neutral or negatively charged surface groups. These observation led to the conclusion that the propensity of pore formation is significantly dependent on size and surface chemistry of dendrimers.[43,44] In addition, surface chemistry of dendrimers and dendritic nanoparticles can be also altered by the surrounding environment, such as solvent, biological conditions, pH, and other salts, affecting their cell interactions.[48]

The shape, flexibility, and polydipersity of nanoparticles have been also found to greatly influence their cellular interaction and uptake. Spherical particles of similar size were internalized up to 500% more than rod-shaped particles, which is explained by greater time required for the membrane to endocytose elongated particles.[46] Due to similar reasons, compared to rigid nanoparticles such as gold and silica, flexible nanoparticles were more effective in interacting with cell membrane.[47]

Therefore, the physicochemical properties of nanoparticles must be uniformly controlled for the biological behaviors of nanoparticles to be predictable as polydispersity within one batch of nanoparticles or between different batches of the same nanomaterials could result in significantly different interactions with cell membranes. Considering their spherical, highly deformable, and close-to-monodispersed properties, dendrimers and dendritic nanoparticles offer a great opportunity to precisely control their cellular interactions, which is advantageous for their use in biomedical applications.

2.2 Cell-Specific Recognition and Multivalent Binding Effect

Recognition of cell surface markers is essential in biomedical engineering. Depending on cell type and disease state, expression of cell surface markers varies to a large degree. Targeted drug delivery systems require cell-specific recognition to minimize side effects and maximize targeting efficacy. Diagnostic systems also require cell-specific recognition to minimize false-positive results and maximize detection accuracy.

Cell-specific binding of dendritic nanomaterials is designed to mimic natural phenomena such as binding between cell surface receptors and ligands. Key features of extracellular binding will be discussed in this section because of the importance of the extracellular binding processes in drug delivery and diagnostics. After the docking of the appropriate ligand to its receptor, their interaction, or binding complex, can trigger a biological cascade. These ligand-receptor interactions occur based on various intermolecular forces, such as covalent binding, ionic bonds, hydrogen bonds, and van der Waals forces. With the exception of covalent binding being nonreversible, most intermolecular bindings between ligands and receptors are characterized by binding affinity (net rate of association and dissociation). Given that binding specificity and affinity of ligands on nanomaterials are critical to achieve efficient drug delivery and cell-specific detection, strategies to increase the binding kinetics have been widely investigated.

Multivalent interactions, defined as the simultaneous binding of multiple ligands to multiple receptors, are ubiquitously observed in biological systems and are one of the most commonly used strategies to increase binding kinetics of nanoparticles.[49] For example, multivalent interactions strengthen or permit the binding of pathogens to the cell surface[50,51] and initiate the infection process of host cells.[52-54] During the infection process, the attachment of the influenza virus on the host cell surface is facilitated by multivalent interactions.[55] These interactions enhance the binding strength of monovalent ligand-receptor interactions (affinity) by orders of magnitude through the collective strength of multiple binding pairs (avidity).[49,56,52] Depending on the biological event, these multivalent interactions result in higher binding strength, amplified signals for cell-to-cell communication, enhancements in binding specificity, or prevention of off-target interactions. Because their dendritic structures can localize multiple ligands, dendritic nanomaterials have been considered one of the most efficient mediators of multivalent binding.

The binding of ligands on dendritic nanomaterials to surface-bound receptors can be affected by their distribution on the surface as well as the flexibility of dendritic nanomaterials. Dendrimers are some of the most ideal platforms for exploiting the multivalent interactions because of their many surface terminal groups available for conjugation with multiple ligands and their highly deformable structure, allowing for the optimization of ligand orientations for maximum substrate binding.[20] The high deformability of the actively targeted dendrimers and dendritic nanoparticles compared to the rigid nanoparticles can also potentially facilitate the full utilization of the multivalent binding effect.[57,47] Dendrimers conjugated with 16 mannoside moieties showed significantly stronger binding toward concanavalin A and pea lectins compared to free mannoside.[58,59] When quantitatively compared to free folic acid (FA) using surface plasmon resonance (SPR), G5 PAMAM with different numbers of surface-functionalized FA showed up to a 170,000-fold enhancement in binding avidity to FA-binding protein.[21] However, as the number of conjugated FA increased, the material's properties were changed (increased molecular weight distribution and poorer water solubility) and the increases in cell binding became marginal when the number of FA per dendrimer was above seven. This observation implies that the number of conjugated ligands per nanoparticle must be optimized to maximize the effects of multivalent binding without deteriorating structural regularity.

The introduction of flexible linkers such as PEG onto dendritic nanoparticles could be used to enhance multivalent binding through the facilitation of a more desirable geometric orientation of ligand towards their receptors in a thermodynamically favorable manner.[60-63] The length of the flexible linkers must also be optimized to prevent undesirable back-folding of the conjugated ligands.[60,64,65] However, there is no universal rules in this regard that are applicable for all kinds of dendritic nanoparticles. Due to large variations among dendritic nanoparticles, ligands, and receptors, the optimization processes are required to design the best-performing dendritic multivalent ligands for the individual cases.

2.3 Stimuli-Responsive Processes for Drug Delivery

Environmental changes (stimuli) can trigger the initiation of dynamic reactions in biomolecules through conformational, chemical, or phase changes (responses). In nature, stimuli-responsive reactions are typically fast and reversible upon removal of the stimulus. These biomolecules can undergo several transition cycles before any loss in responsiveness. Their responses to stimuli are mostly non-linear, and the transition point between observing a response and not observing a response to a given stimuli is defined as the threshold.[66] Many external signals are recognized as potential biological stimuli, such as changes in temperature, chemistry, mechanical stress, light irradiation, and exposure to electric or magnetic fields.

Stimuli-responsive dendrimers and dendritic nanomaterials have been used to transport therapeutically and diagnostically active ingredients into the cytosol of cells. During cellular uptake, dendrimers and dendritic nanoparticles need to breach the cell-membrane barriers of endolysosomes to reach the cytosolic machinery or the nuclei of the cells.[67] Nanoparticles responsive to biological stimuli such as pH, temperature, and protein binding can be programmed to escape the endolysosome and translocate into the cytosol or nuclei in a kinetically and spatially controlled manner.[68,69] Using such methods, after encountering or subjecting the nanocarriers to external stimulus, the payload can be also triggered to release.[68,70]

pH sensitivity is one of the most commonly utilized features in drug/gene delivery systems. Various amines of PAMAM dendrimers have been used for pH-responsive applications, given that the pKa values of primary, secondary, and tertiary amines in the macromolecules are 7-9, 5-8, 3-6 respectively.[71] During endocytosis, dendrimers are exposed to different pH, i.e., the physiological environment (pH 7.4), endosomes (pH ~6.0), and lysosomes (pH ~5.0), which affects the protonation behavior of every amine group of the dendrimers. For examples, in lysosomes where pH is low (between pH 4.5 and pH 5.0), the number of positively charged groups increases due to the protonation of tertiary amine groups that are spared at low pH. This behavior, termed "sponge-effect", results in destabilization of the late-endosomal and lysosomal membranes, leading to efficient endosomal escape.[72] Another example is that the increased charge repulsion between positively charged groups at low pH destabilizes the nanocarriers, which triggers the drug release. pH-responsive conformational changes (from dense core to dense shell at low pH) of PAMAM dendrimers in lysosome can also facilitate the release of guest molecules.[72,73] Besides the intracellular pH change, the tumor microenvironment exhibits a relatively acidic pH, compared to normal tissues offering another mechanism to increase the localized release of drug molecules.[74] For instance, pH-sensitive poly(propyleneimine) dendrimers and supramolecular assemblies of linear dendritic copolymers retained encapsulated Nile Red and pyrene at pH 7.4 and released them in acidic conditions via structure disintegration at low pH.[75,69] Other stimuli-responsive dendritic nanoparticles for biomedical applications will be further discussed in section 3.1.4.

2.4 *Biodistribution and Pharmacokinetics*

The pharmacokinetics and biodistribution of dendrimers primarily depend on their size and surface composition. While lower generation dendrimers are more rapidly cleared through renal clearance due to lower molecular weight and smaller size, larger dendrimers induce more effective uptake by the RES and leads to increased RES clearance.[76,77] With an increase of PAMAM dendrimer generation (G2-G9), plasma circulation time increases, peaking at G7.[78,79] Due to the increased uptake of G8 and G9 dendrimers in liver and spleen, their plasma circulation times were reported to be shorter than that of G7.

The size of dendritic supramolecules and hybrid nanoparticles (typically 20-200 nm in diameter) affects their biodistribution *in vivo*. As their hydrodynamic diameters are larger than threshold of kidney (~5.5 nm), their elimination rate from the body is typically slower and circulation time in the body is longer than dendrimers.[80] Another clearance mechanism via RES following intravenous administration is inversely dependent on their size, i.e., more efficient macrophage uptake of ~90 nm nanoparticles than that of ~15 nm nanoparticles.[81]

Surface charge of dendritic nanomaterials is another important parameter that affects their pharmacokinetic profiles. Amine-terminated dendrimers (e.g. PAMAM, PPI, and PLL dendrimers) carry positive surface charges under physiological conditions. These cationic dendrimers were shown to strongly bind to vasculature of highly perfused organs due to non-specific electrostatic interactions, resulting in rapid clearance from the blood circulation and high initial volume of distribution in rats compared to anionic dendrimers.[82,83] Using ^{125}I-labeled dendrimers, Malik et al. revealed that the circulation time of anionic dendrimers in Wistar rats was 10-20 fold longer than that of cationic dendrimers.[82]

The toxicological profiles of dendrimers and dendritic nanoparticles can be also controlled by their size and surface modifications. Importantly, the charge-based non-specific interaction between positively charged dendrimers and cell membranes are typically associated with toxicity.[84,85] As discussed earlier (Section 2.A non-specific cell interaction), the electrostatic interactions of cationic dendrimers with cell membrane induced membrane disruption, which could be a reason for the apparent toxicity observed in animal studies.[86] Surface modification of terminal amine groups to different groups (charge neutral or negatively charged) resulted in a tolerable toxicity profile of dendrimers.[87,85] In addition, peripheral amine groups can be modified by benzene sulfonate, benzene disulfonate, or succinate groups, which exhibit significantly lower vasculature binding and enhanced metabolic stability, leading to reduced toxicity.[76]

The surface conjugation of non-fouling agents, such as PEG, to dendritic nanoparticles is a frequently employed method to improve their pharmacokinetic profiles by providing properties that help avoid opsonization and reduce RES recognition.[88] PEGylation of dendrimer surfaces is a widely adopted method to mask the surface charges.[89] Similar to PEGylated PLL dendrimers,[76] PEGylated PAMAM dendrimers exhibited longer circulation time compared to non-PEGylated dendrimers.[90] The prolonged circulation time can be attributed to the high degree of surface coverage of PEG and the increased molecular weight (size).

3 Biomedical Applications

The preceding sections served as an introduction to the physical properties and biological behaviors of various dendritic nanomaterials and provide the fundamentals necessary for the discussion of designing biomedical systems. In the following sections, we will focus on examples of the use of dendritic nanomaterials in the recent advances in drug delivery, gene delivery, diagnostics and prognostics, and detoxification and prophylaxis with the aim of providing ideas for the future direction of dendritic biomedical system designs.

3.1 Drug Delivery

3.1.1 Dendrimers

The most commonly explored disease for the development of selective drug delivery systems, or nanocarriers, is cancer. Dendrimers are among the most promising nanoscale materials that have been vigorously investigated as nanocarriers for enhanced cancer treatment over the past two decades. Selective drug delivery using dendrimers has been heavily focused on achieving active targeting by conjugating targeting agents to the surfaces of dendrimers, together with drug incorporation either via covalent conjugation or physical entrapment.

PAMAM dendrimers can be multifunctionalized via covalent conjugation with therapeutic, targeting, and imaging (radiolabels or fluorophores) agents. For example, dendrimer-FA conjugates were designed for active targeting to tumor cells through selective binding to the folate receptors (FAR) that are overexpressed by a wide variety of cancer cells.[91,92] Therapeutic agents can be incorporated into dendrimers through chemical conjugation, typically with ester bonds that are easily cleavable by esterase. In a mouse xenograft model using FAR-overexpressing KB cells, the FA-targeted G5 dendrimers conjugated with methotrexate primarily accumulated in the tumor and liver over the course of 4 days, resulting in enhanced therapeutic efficacy, as measured by the reduction in tumor volume and decreased off-target toxicity.[93]

Although relatively unstable, the inner cavity of dendrimers can be used to encapsulate therapeutic payloads, along with surface-conjugated targeting agents. For example, 5-flurouracil (5-FU) was encapsulated within FA-targeted PEGylated PAMAM dendrimers and evaluated for its anti-tumor efficacy.[94] These PEGylated dendrimers showed significantly enhanced targeting efficiency and reduction in tumor volume, compared to free 5-FU in a mouse xenograft model bearing FAR-overexpressing KB cells. PEGylated formulations displayed longer retention times of 5-FU (t_{max}=2.5-5 hrs) compared to non-PEGylated dendrimers (t_{max}=1-2.5 hrs). In addition to FA conjugation, active targeting can be also engineered into dendrimers through the surface conjugation of other targeting agents, such as antibodies, aptamers, and peptide-based ligands that have strong binding affinities to the target tumor tissue.[95,96] A number of other ligand-functionalized dendrimers have shown successful results *in vivo* for anticancer drug delivery, which are comprehensively reviewed elsewhere.[10,11]

Other multifunctional dendrimers include dendritic nanocarriers with three or more functions. For example, pH-sensitive multifunctional dendrimers targeting brain gliomas were prepared by surface conjugation of transferrin for targeting, PEG for shielding, and doxorubicin via a pH sensitive linker.[97] Another chemotherapeutic drug, tamoxifen, was physically entrapped into the interior cavity of G4 PAMAM dendrimers. The multifunctional dendrimers could efficiently transport both anticancer drugs across the blood-brain barrier through the combined actions of transferrin receptor-mediated endocytosis, pH-triggered doxorubicin release, and inhibition of the drug efflux transport by tamoxifen *in vitro*.[97]

3.1.2 Self-Assemblies of Linear Dendritic Block Copolymers (LDBCs)

Self-assembled structures provide inspiration in the development of more complex and thermodynamically stable nanostructures for drug delivery. Dendrons exhibit a hyperbranched, conical molecular structure. This unique morphology facilitates the precise packing of LDBCs to form spherical micelles.[15,25] Their capability to form thermodynamically stable micelles, ability to be multifunctionized, and high drug encapsulation efficiency have made LDBCs one of the emerging drug delivery platforms. Furthermore, the functionalities of the LDBC-based micelles are easily tunable using surface conjugations, charge modifications, and mixed self-assemblies of LDBCs with distinct functions.

Multi-functional micelles self-assembled from LDBCs have been developed for intracellular delivery of antitumor drugs. Self-assembled micelles of amphiphilic LDBCs comprised of semi-polyamidoamine-*b*-poly(D,L-lactic acid) (PALA) were used to deliver docetaxel (DTX) to MCF-7 cells *in vitro*. In rat models, the DTX-loaded PALA micelles showed enhanced plasma concentrations of DTX and prolonged drug clearance in comparison to a clinically used formulation of DTX, or Taxotere®.[98] Another micelle of linear PEG-block-dendritic cholic acid copolymers were able to encapsulate paclitaxel, and their ability to use active targeting to enhance the delivery of paclitaxel (PTX) was investigated.[99,100] The self-assembled nanocarriers were functionalized with a high-affinity peptide targeting ligand "OAO2" (cdG-HoCit-GPQc-Ebes-K-alkyne) that binds selectively to alpha-3 integrin receptors. Compared to non-targeted NPs and Taxotere®, the decoration of the targeted nanocarriers led to increased uptake efficiency in SKOV-3 and ES-2 ovarian cancer cells *in vitro* and improved tumor accumulation, anti-tumor efficacy, and toxicity profiles in SKOV-3 tumor xenografts.

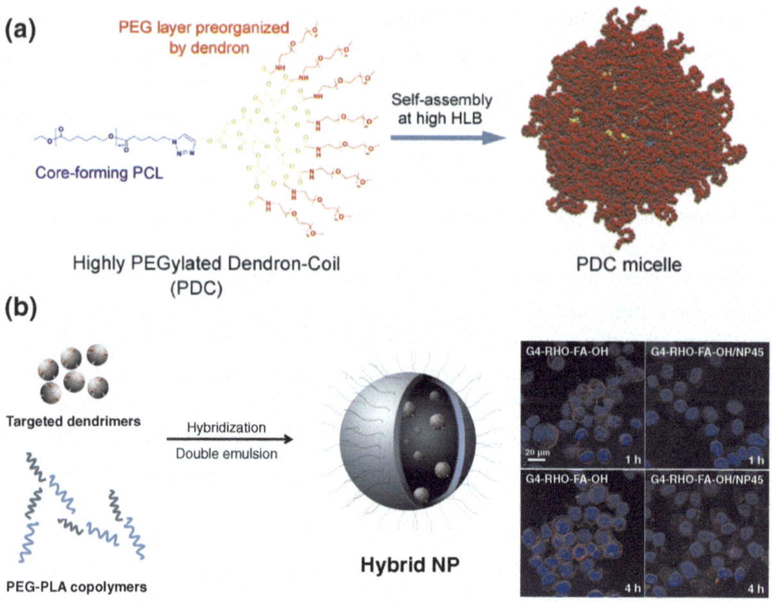

Fig. 3 Novel self-assembled drug delivery systems. (a) Self-assembled nanoscale micelles composed of highly PEGylated dendron coils (PDC). Self-assembly of PDCs is thermodynamically favorable at high HLB and low CMC due to the unique architecture of dendrons.[29] (Copyright © 2011, Royal Society of Chemistry) (b) Formation of nanohybrids through the encapsulation of folate-conjugated dendrimers by self-assembly of PEG-b-PLA block copolymers. Compared to unencapsulated dendrimers (left column of cellular uptake images), the encapsulated dendrimer-nanohybrids (right column) showed the kinetically controlled targeting to FAR-expressed cancer cells. (red, dendrimer conjugates; blue, cell nuclei, scale bar = 20 μm)[38] (Copyright © 2012, American Chemical Society)

Besides high thermodynamic stability and drug encapsulation efficiency, the cellular uptake and tissue penetration of LDBC micelles for drug delivery can be tuned by surface functionalization with targeting agents and changes in surface charge. For instance, a dendron-based nanocarrier platform was developed based on a single linear hydrophobic PCL conjugated with a highly PEGylated dendron, termed PEGylated dendron coil (PDC), forming a micelle that has a densely PEGylated surface (Fig. 3a).[29] The cellular interaction and uptake of PDC micelles depended on the surface charges of PEG outer layers and PEG chain length.[64,65] In brief, the PDC micelles with positively charged amine surface groups readily interacted with cells only if PEG of 600 Da was used. In contrast, the same micelles with amines but with longer PEG chains (2,000 Da) did not exhibit any cell interactions. Charge neutral or negatively charged micelles did not interact with cells regardless of the PEG chain length. Skin penetration efficiency of the PDC micelles with various surface groups (-NH_2, -COOH, or -Ac) was also evaluated after encapsulating endoxifen (EDX).[101,102] Compared to ethanol, a chemical penetration enhancer, and other PDC micelles with amine or acetyl ends, the carboxylated PDC micelles were able to enhance the permeation of EDX across mouse skin up to 20-fold and human skin up to 4-fold, indicating the role of surface groups determining skin penetration of the micelles.

Using a mix-and-match approach of targeted and untargeted LDBCs, a library of LDBC micelles with various binding avidities can be prepared. Hydrophobic poly(benzyl-L-aspartate) (PBLA) was used to form the polypeptide linear block that was coupled to a generation 4 (G4; 16 end groups) polyester dendron to which sixteen PEG (600 Da) chains were attached. The targeting ligand FA was conjugated to the distal ends of the PEG chains to study the effect of ligand clustering on the cellular interactions of the dendritic micelles.[103] The LDBCs with ~3.2 FA molecules per polymer chain that mixed in a 60:40 ratio to untargeted polymers resulted in the highest level of cell uptake, the highest binding avidity, and the greatest *in vivo* tumor localization. As observed using dendrimers, the cluster arrangement of targeting ligands on the surface was attributed to the increased efficacy via multivalent binding effect.

3.1.3 Dendrimer-Linear Polymer Hybrid Nanoparticles

Recently, novel nanoparticle systems that hybridize two different nanoparticles have been introduced in an effort of taking advantage of the both species in an integrated manner. Sunoqrot et al. reported a hybrid nanoparticle system that encapsulated FA-targeted dendrimers into PEG-PLA nanoparticles.[38,36] They revealed that a controlled release of the FA-conjugated dendrimers from the nanohybrids dictated their kinetically controlled specific interactions with FA receptor-overexpressing KB cancer cells, as depicted in Fig. 3b.[38] The release rates of the dendrimers were controlled by the molecular weight of the PLA block in PEG-PLA shell, i.e., longer PLA induced slower release kinetics of the encapsulated dendrimers.[38] Furthermore, biodistribution and plasma circulation time of the hybrid nanoparticles were investigated using BALB/c athymic nude mice bearing an FAR-overexpressing KB xenograft.[39] Compared to the fast clearance of free dendrimers by renal filtration due to their small size (<5 nm in diameter), the hybrid nanoparticles showed a similar behavior of larger polymeric nanoparticle (~100 nm in diameter), as demonstrated by prolonged blood circulation and increased liver/spleen accumulation. In a simulated tissue penetration study using multicellular tumor spheroids, the dendrimers displayed significantly enhanced penetration efficiency upon release from the hybrid nanoparticles whereas the larger polymeric shell showed poor penetration.[104] These results indicated that the hybrid design can be used to exploit the benefits of the both nanocarriers at different size scale, while minimizing the drawbacks of each, presenting a promising strategy to engineer highly effective drug delivery systems.

3.1.4 Environmentally Responsive Dendritic Nanomaterials

Stimuli-responsiveness can be engineered into the design of nanocarriers and is a useful function to be added to drug delivery systems. After encountering or subjecting the nanocarriers to external stimulus, the payload can be triggered to be released. Various stimuli including changes in pH, temperature, and protein binding in biological conditions have been utilized to trigger the responses of dendritic nanomaterials and the release of the their payloads.

Thermo-responsive polymers have been used to release encapsulated drugs in response to temperature changes. These polymers typically undergo a rapid and reversible hydration–dehydration at lower critical solution temperature (LCST, from soluble to insoluble) or upper critical solution temperature (UCST, from insoluble to soluble) upon heating.[71] For example, a thermo-responsive polymer, poly(2-isopropyl- 2-oxazoline) (PiPrOx), was conjugated to pH-responsive poly(benzyl ether)dendrons, which exhibited sharp thermal transitions as well as pH-dependent behaviors (Fig. 4a).[105]

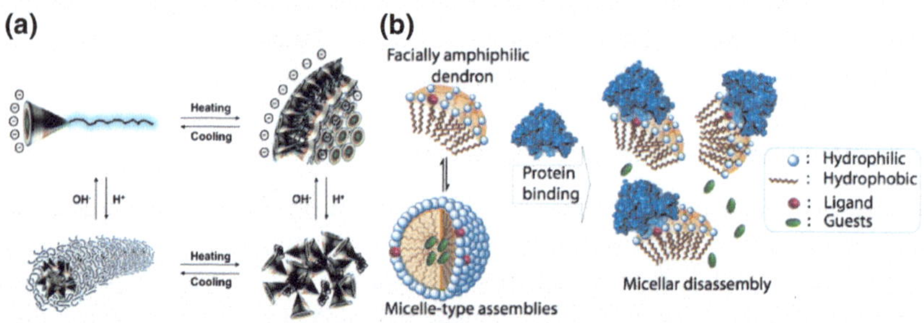

Fig. 4 Novel drug delivery systems using stimuli-responsive polymers. (a) Morphology changes of poly(2-isopropyl- 2-oxazoline) (PiPrOx) and poly(benzyl ether)dendrons dendritic-linear block copolymers conjugates are Temperature- and pH-dependent.[105] (Copyright © 2012, Royal Society of Chemistry) (b) Protein-binding-induced disassembly of dendron-based micelles. This inducible micelle disassembly is selective to the targeted protein and the disassembly can mediate release of the encapsulated therapeutic drugs.[106] (Copyright © 2010, American Chemical Society)

The binding of proteins can induce conformational changes in polymers. Azagarsamy et al. designed a dendritic micellar container that is capable of disassembly upon binding with protein, as depicted in Fig. 4b.[106] By attaching a specific protein binding ligand (e.g. biotin) at the surface of the self-assembled micelle, the micelle could disassemble upon protein binding (e.g. extravidin, a neutral form of avidin with minimal nonspecific interactions) due to the significant alteration of the HLB of the micelle. These stimuli-responsive nanocarriers offer an opportunity to precisely control the release of the payloads upon exposure to specific environments of diseased sites in the body, increasing the overall delivery efficacy.

3.2 Gene Delivery

Viruses are one of the most effective vectors to introduce foreign genetic material into cells. Although viruses do not contain any organelles or cytoplasm (acellular), viruses can survive using resources from the infected host cells, reproduce and replicate either their DNA or RNA within host cells, and mutate. For example, the human immunodeficiency virus (HIV) delivers its RNA payload to host cells. Subsequent replication of the virus leads to the destruction of CD4-positive T cells, resulting in the development of acquired immune deficiency syndrome (AIDS).[107] A number of factors, including the affinity of cell surface receptors to the virus, recognition of viral promoters, ability of the cell to support viral replication, and physical barriers (temperature and pH), all can affect to tissue-dependent tropism of viruses.[108,109]

Viral transfection is the transfer of genetic information (i.e. DNA and RNA) from virus to host cells for the purpose of viral replication. To engineer the transfection process, an efficient delivery system is required. Despite their very high gene delivery efficiency (95-100%) and simplicity of infection, however, viral vectors are not the most preferred delivery systems of choice due to concerns of potential toxicity caused by severe inflammation.

Many cationic polymers have been developed as non-viral gene delivery vectors to overcome safety concerns related to the use of viruses.[110,111] Non-viral vectors are also more cost-effective and less immunogenic and pathogenic than viruses.[112,11] Cationic polymer can complex with negatively charged nucleic acids via electrostatic interactions, protecting the genetic materials from degradation. Typically, the net charge of the complexes (polyplexes) is maintained to be positive, allowing their efficient cell entry for transfection. Furthermore, these polymers can be easily functionalized in a variety of ways to introduce specificity, imaging capability, and improved polyplex stability. However, limitations of the non-viral vectors are highlighted by their significantly lower *in vivo* transfection efficiency, compared to viral vectors. This section will highlight positively charged dendrimers and dendritic nanomaterials that have been used in various gene delivery applications.

3.2.1 Dendrimers

Dendrimers have been investigated as a vector for nucleic acid-based therapies for nearly 20 years, and PAMAM and PPI dendrimers are the two most commonly used kinds that are commercially available.[113-115] Their unique chemical architecture with all primary, secondary, and tertiary amines enabling the proton-sponge effect described above make dendrimers one of the ideal platforms for gene delivery. Furthermore, the close-to-monodispersed chemical structure of dendrimers allows the precise control over their functionalities, which can minimize the unpredictable transfection efficiencies observed in the cases of heterogeneous liposomes and other polymer-based systems.[116]

One of the key parameters determining transfection efficiency is the N/P ratio between the positively charged amines (N) on dendrimers and negatively charged DNA phosphates (P). Both transfection efficiency and cytotoxicity are typically proportional to the N/P ratios. Therefore, it is critical to find an N/P ratio that provides good complexation efficiency while producing minimal cytotoxicity. Another important parameter to consider is size of dendrimers. It was revealed that high generation PAMAM dendrimers more strongly bind to plasmid DNA than lower generations, due to the increased number of surface amine groups and low curvature that facilitates the binding process, but induce higher cytotoxicity.[117,45,44]

Numerous variations of dendrimers have been synthesized and evaluated in an effort to increase the transfection efficiency and decrease the positive charge-related toxicity.[118,119] The cytotoxicity of cationic PAMAM dendrimers was significantly decreased after surface modification using lipid, PEG, peptides, or other charge-neutral chemicals. Jevprasesphant et al prepared G4 PAMAM dendrimers conjugated with L-arginine (Arg).[120]. Using human embryonic kidney 293 cells and mouse albino neuroblastoma Neuro 2A cells, they revealed that the transfection efficiency of the dendrimer-Arg conjugates was significantly higher than those of PEI and unmodified PAMAM dendrimers by more than one order of magnitude.[120] This enhancement was obtained without the cost of cytotoxicity. Interestingly, the IC_{50} value (a higher IC_{50} value indicates a lower toxicity) of the dendrimer-Arg conjugates was measured at 375 μg/mL, which was significantly higher than those of unmodified G4 (150 μg/mL) and PEI (~10 μg/mL).[120] In addition, other modifications including PEGylation and lipid conjugation demonstrated similar improvements.[118,119] These results indicate that surface modifications of dendrimers with functional agents can increase their efficiency without increasing cytotoxicity.

Multifunctional dendrimers have been also used to deliver small interfering RNAs (siRNAs) for RNA interference allowing the sequence-specific suppression of gene expression. For instance, when several passenger strands of siRNA and one targeting ligand per PAMAM dendrimer were covalently conjugated, monodipersity of the dendritic siRNA nanostructures exhibited the effective control over gene expression in neural stem cells and primary neurons.[121] Multifunctionalized PPI dendrimers with a combination of CD44 siRNA, paclitaxel, and a targeting peptide showed a simultaneous therapeutic potential for CD44 suppression, cell death induction, and ovarian cancer-specific targeting in a xenograft mice model.[122]

3.2.2 Dendritic Nanomaterials

Although investigated to a lesser extent compared to dendrimers, LDBCs have been also explored for their gene delivery applications. Wood et al. developed a library of dendritic PAMAM-PEG-(galactose or mannose) of G0 to G6 (1 to 64 surface groups) that could form polyplexes with luciferase DNA.[123] The polyplexes of G4-G6 copolymers displayed higher transfection efficiencies in P388D1 macrophages and HepG2 hepatocytes than those of commercially available polyethylenimine (PEI) systems under serum-containing culture conditions. Linear dendritic cationic vectors were developed for efficient DNA polyplexing and delivery.[124] The pH buffering capacity, size, and zeta potential of methoxy-PEG5K-dendritic polyglycerol-graft-tris(2-aminoethyl)amine (mPEG-DPG-g-TAEA) were characterized. By tailoring the dendritic DPG block and the degree of TAEA substitution, the cationic TAEA was able to bind strongly to DNA with low amine content density. As a result, the mPEG-DPG-g-TAEA complexed with DNA resulted in significantly less cytotoxicity compared to PEI5K, while displaying similar transfection efficiency to that of lipofectamine 2000. The high efficiency with decreased cytotoxicity demonstrates the potential of LDBCs as next generation of gene delivery vectors.

3.3 *Diagnostics and Prognostics*

The unique properties of dendritic nanomaterials, as discussed previously, lend themselves to the design of diagnostic devices.[17,21] Particles or surfaces can be conjugated with imaging agents (e.g. MRI contrast agents or fluorescent dyes) and targeting agents (e.g. antibodies, aptamers, nucleotides, proteins, and small molecules) to perform diagnostic functions.[95,96] These advanced materials can be injected for analysis *in vivo* for direct detection, mixed with specimens *in vitro*, or immobilized on the surface for *ex vivo* analysis. Substrates such as inorganic and organic particles, glass slides, and polymeric matrices are all potential candidates for dendritic nanoparticles-based diagnostic and prognostic detection.

3.3.1 Dendrimer Conjugates with Contrast Agents

While dendrimers are widely investigated as therapeutic delivery agents, they are also employed to enhance the sensitivity of current imaging techniques via conjugation with contrast agents. Among several imaging techniques, dendrimers have been frequently used to modify contrast agents for magnetic resonance imaging (MRI). The multivalency, monodisperse characteristics, and relatively high molecular weight of dendrimers make them well suited for carrying T1 MRI agents such as gadolinium (Gd).[125] Gd(III) complexes, such as Gd(III)DTPA (diethylenetriaminepentaacetic acid) and Gd(III)DOTA (1,4,7,10-tetraazacyclododecane-1,4,7,10-tetraacetic acid), have been proven to increase the T1 relaxation time and enhance the contrast in T1-weighted images. However, due to their low molecular weight, these agents possess high renal clearance and low molar relaxivity properties, which resulted in a loss of specificity and contrast efficiency.

It has been shown by many research groups that conjugating Gd(III) chelators onto dendrimers achieved contrast efficiency enhancement which was attributed to high relaxivities and longer circulation time of the conjugates.[126-128] Interestingly, higher organ specificities were achieved by optimizing the generation of these dendrimer-based MRI contrast agents. The diameters of dendrimers increase with increasing generation, which result in differences in organ or tissue accumulation. For example, smaller contrast agents (3-6 nm in diameter) could be applied to functional renal imaging,[129] while larger ones (7-12 nm) are more suitable for blood pool imaging.[126] Backer et al. reported that FA and DOTA-Gd(III) conjugated G5 PAMAM dendrimers were able to target tumor cells *in vivo*.[130] Carboxyl-terminated FA-PEG-PAMAM dendrimers were reported to form complexes with Gd(III) ions for tumor targeting without using conventional chelators.[131] In addition to Gd(III), dendrimers have been used to improve the colloid stability of iron oxide nanoparticles for T2 MRI.[132]

3.3.2 Dendrimer Conjugates with Targeting Agents

The potential role of dendrimers as interaction amplifiers through multivalent binding effect has resulted in increased attention within the field of biosensing and surface recognition applications. Dendrimers has been leveraged to develop highly sensitive devices for the detection of DNA,[133] toxins,[134] antigens,[135] and cells.[22] Multivalent binding can be used to increase the binding strength of individual ligands through cooperative, localized binding events, and these interactions can be used to enhance the net sensitivity and selectivity of detection assays.[136]

A dendrimer-immobilized substrate was used for ligand immobilization to strongly and securely capture trapped target molecules such as nucleic acids, proteins, and cells.[137,138] For instance, dendrimers functionalized on substrates enhanced the capture of oligomers by 2-fold in nucleic acid hybridization experiments using fluorophore-labeled complementary oligonucleotide targets, compared to the untreated intact substrate.[139] Myung et al. developed a dendrimer-based detection platform for efficient surface capture of tumor cells using a dendrimer-immobilized glass slide functionalized with antibodies against epithelial cell adhesion molecule (anti-EpCAM) (Figs. 5a and 5b).[22] The use of dendrimers enhanced the binding constant of the antibody to EpCAM by remarkable one million fold. Compared to linear PEG-anti-EpCAM functionalized surfaces, the dendrimer-anti-EpCAM surface had up to a 7-fold enhancement of tumor cell detection.[22] The significant enhancement of the dendrimer-coated surfaces was recently expanded to various antibodies and proven effective in capturing tumor cells from clinically relevant blood samples as well.[140] Additional uses of dendrimers have been reported to improve sensitivity of enzyme-linked immunosorbent assays (ELISA)[141] and promote cell adhesion and proliferation.[142,143,138]

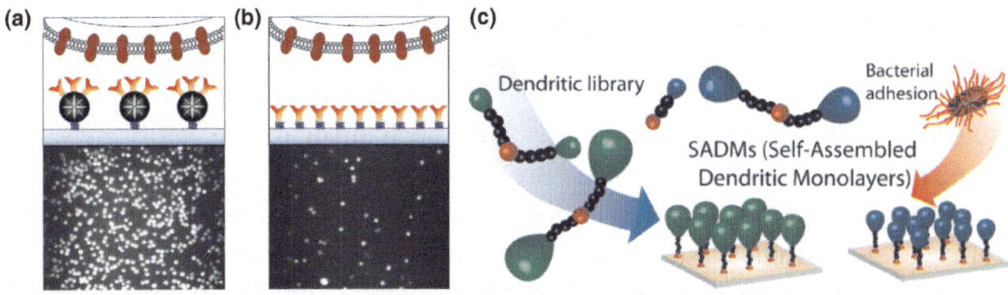

Fig. 5 Dendrimers and dendron-based SAMs in diagnostic platforms. (a-b) Schematic and comparison of dendrimer- and PEG-immobilized surfaces (top row) and fluorescence images of cell adhesion (bottom row). The dendrimer-immobilized surfaces (left, a) showed more immobilization of detecting antibodies and the significant enhancement in target cancer cell capture than the linear polymer, PEG-immobilized ones (right, b).[22] (Copyright © 2011 WILEY-VCH Verlag GmbH & Co. KGaA, Weinheim) (c) Schematic overview of self-assembled dendritic monolayers (SADMs) to provide functional gold sensor surfaces for *E. coli* MS7fim+ bacteria recognition.[147] (Copyright © 2012 American Chemical Society)

3.3.3 Dendrons

Due to their dense presentation of surface functional groups as dendrimers, dendrons and other dendritic nanomaterials have been similarly used to achieve enhanced detection. Despite all the benefits of using dendrimers in the biosensing and surface recognition applications discussed above, the dendrimer-based approaches have drawbacks such as multiple steps for surface functionalization and the inability to control the orientation of functional groups conjugated on the dendrimer surface. Alternatively, dendritic scaffolds can be prepared using dendron-linear block copolymers with an organosulfur core (disulfide or thiol). Thiol-coated scaffolds can recognize the organosulfur core of the dendron-linear block copolymers and undergo spontaneous self-assembly to yield monolayers (SAMs).[144,145] The dendritic structures can provide sufficient surface coverage that increases the monolayer density, depending on several factors, such as type of focal substitution, size of dendron, peripheral end-group functionality, and temperature.[146]

Inorganic scaffolds (Ag, SiO_2, or Fe_2O_3) have been used for dendron-linear block copolymers as well. Inorganic nanoparticles (Ag or Fe_2O_3 nanoparticles) have been encapsulated or incorporated into self-assembling nanostructures (multilayers or shells) of diblock copolymers (e.g. a dendron based on the 2,2-bis(methylol)propionic acid (bis-MPA) monomer).[146,147] These inorganic-dendron assemblies provide diagnostic sensing platforms that can be tailored to detect various biomolecules.[147] For instance, the fabrication of gold surfaces with self-assembled monolayers of G3 dendrons displayed high molecular covering densities of 0.3–2.2 molecules/nm^2, which enhanced the surface recognition for E. coli MS7fim+ bacteria by 2.5 times (Fig. 5c).[147]

3.4 *Detoxification and Prophylaxis*

The anti-pathogenic activities of dendrimer-based nanomaterials have been evaluated on a variety of pathogens. The mechanism of the antibacterial and antifungal activities of dendrimers are primarily related to the formation of electrostatic interactions with the membrane of pathogens, followed by membrane disruption and cytoplasm leakage.[148] However, the lack of selectivity and specificity in the disruption of pathogen membranes and cytotoxicity to normal human cells have prevented the use of dendrimers as an anti-pathogenic agent in clinics. Nevertheless, dendrimers have been used as a carrier of anti-microbial drugs for improved solubility, bioavailability, and activity.[149,150]

3.4.1 Dendrimers

Anti-viral dendrimers mainly prevent the binding of viruses to the host cell surface or inhibit replication of the viral genome. The multivalent interactions between dendrimer-based nanoparticles and the viral surfaces are extremely strong and stable, which effectively prevents the virus from interacting with host cells. Reuter et al. showed SA-functionalized dendrimers enhanced the activity of SA to inhibit the HA caused by influenza by 5×10^4-fold.[151] The SA-functionalized PAMAM dendrimers completely prevented the infection of murine influenza pneumonitis in mice *in vivo*.[152] Besides PAMAM dendrimers, sialyllactose-functionalized C-Si dendrimers and polysulfated galactose-derived PPI dendrimers inhibited the infection of influenza and HIV-1, respectively.[153] Recently, naphthalene disulfonic acid (DNAA)-based dendrimers were also used to inhibit the infection of HIV and herpes simplex virus (HSV), which has been translated into a topical microbicide gel, VivaGel®, to prevent sexually transmitted diseases and bacterial vaginosis.[154,155] VivaGel® is currently under phase III of clinical trials for prevention of the recurrence of bacterial vaginosis.

In addition to inhibiting viral binding, dendrimers have been shown to exhibit the ability to disturb the replication of viruses. Amine-terminated PAMAM dendrimers have been reported to interact with transacting response elements RNA, which affected the viral replication.[156] Moreover, viral protease and reverse transcriptase were inhibited by carboxylated fullerene-based dendrimers and sulfonated dendrimers, respectively.[157]

Interestingly, unmodified PAMAM dendrimers can be used to chelate heavy metal ions such as Cu(II) and remove environmental pollutants such as phenanthrene (Phe). PAMAM dendrimers have demonstrated effective loading and release of agents for the chelation of Cu(II) in drinking water and for the removal of perchlorates, pesticides, and volatile organic compounds from wastewater streams. The binding measurements using extended X-ray absorption fine structure (EXAFS) spectroscopy showed that tertiary amine, primary amine groups, and ethylenediamine core of PAMAM dendrimers are involved in the uptake of Cu(II) in aqueous solutions at pH 7.0. The protonation of these groups controlled the ability of the dendrimers to bind Cu(II).[158,159] In addition, the interaction of PAMAM dendrimers with Phe, a polycyclic aromatic hydrocarbon and a major environmental pollutant, was measured by the energy transfer between the autofluorescent Phe and the fluorescently labeled PAMAM dendrimer in aqueous solutions at pH 8.0. Among mixtures of phenanthrene and dendrimer, the energy transfer efficiency was most efficient in 1:2 molar ratio mixtures due to the strongest binding interactions between phenanthrene and the primary amines of the dendrimer.[160]

4 Challenges in Clinical Translation

Although dendrimer-based technologies have made significant progress in the biomedical field over the past two decades, dendrimers and dendritic nanomaterials have yet to be clinically translated. Several limitations of dendrimers, dendritic supramolecules, and hybrids must be overcome for their clinical translation. First, their synthetic procedures are complex and require multiple steps to achieve well-defined dendritic architecture, which makes the scalability and reproducibility of the promising nanomaterials difficult. Second, while the covalent drug conjugation onto the surfaces is preferred to the physical drug entrapment for dendrimers to stably carry drug molecules, the conjugation method is inherently limited by low drug loading. Moreover, excessive conjugation with drug payload or targeting moieties would likely result in undesirable changes of the material properties, such as increased polydispersity and poor water solubility. Third, similar to other nanocarriers, the development of dendrimers often faces poor or unpredictable correlation between the results from animal studies and clinical trials. The steps taken to address these challenges that virtually all nanocarriers face are comprehensively discussed in the cited review articles.[161,12] Given that these challenges seem inevitable for every emerging technique in the biomedical field, more effort and studies for individual cases are obviously needed to transform promising preliminary results into clinical products.

5 Conclusions and Future Perspectives

In spite of the challenges mentioned above, dendritic nanoparticles are still one of the most promising platforms due to their branched structure, precisely controlled molecular weight distributions, capability for multifunctionalization, and high degree of molecular flexibility. Extensive *in vitro* and *in vivo* studies have demonstrated the advantages of the use of dendritic architectures in biomedical engineering for therapeutic and diagnostic purposes. Various engineering approaches are currently under investigation to overcome the drawbacks of dendritic nanoparticles, such as potential toxicity and low circulation time, particularly by the structural engineering (e.g. PEGylation, terminal group modification, and encapsulation into larger nanoparticles), and the incorporation of functional moieties (e.g. targeting ligands). Their issues with scalability can be overcome, given the rapidly growing field of dendrimer chemistry and chemical engineering. Lastly, the hybridization of dendron and linear polymers creates new platforms with additional benefits that are not seen in either system alone. Although the extensive biological validation of dendritic nanoparticles is still required, the development of the promising nanomaterials is under progress at an unprecedented rate because of the vast array of applications available (e.g. drug delivery, gene delivery, and diagnostic and prognostic applications). With these efforts, breakthroughs in clinical translation and profound impacts on human health are foreseeable in the future of dendritic nanomaterials.

Acknowledgments This work was supported by National Science Foundation (NSF) under grant # CBET-0931472 and DMR-1409161 as well as by the University of Illinois Cancer Center and the Office of Technology Managements of the University of Illinois at Chicago.

References

1. Brannon-Peppas, L., Blanchette, J.O.: Nanoparticle and targeted systems for cancer therapy. Adv. Drug. Deliv. Rev. **56**(11), 1649–1659 (2004). doi:10.1016/j.addr.2004.02.014
2. Mura, S., Couvreur, P.: Nanotheranostics for personalized medicine. Adv. Drug. Deliv. Rev. **64**(13), 1394–1416 (2012). doi:10.1016/j.addr.2012.06.006
3. Chauhan, V.P., Jain, R.K.: Strategies for advancing cancer nanomedicine. Nat. Mater. **12**(11), 958–962 (2013). doi:10.1038/nmat3792
4. Peer, D., Karp, J.M., Hong, S., FaroKHzad, O.C., Margalit, R., Langer, R.: Nanocarriers as an emerging platform for cancer therapy. Nat. Nanotechnol. **2**(12), 751–760 (2007). doi:10.1038/Nnano.2007.387
5. Doane, T.L., Burda, C.: The unique role of nanoparticles in nanomedicine: imaging, drug delivery and therapy. Chemical Society reviews **41**(7), 2885–2911 (2012). doi:10.1039/c2cs15260f
6. Lee, D.E., Koo, H., Sun, I.C., Ryu, J.H., Kim, K., Kwon, I.C.: Multifunctional nanoparticles for multimodal imaging and theragnosis. Chemical Society reviews **41**(7), 2656–2672 (2012). doi:10.1039/c2cs15261d
7. Ulery, B.D., Nair, L.S., Laurencin, C.T.: Biomedical Applications of Biodegradable Polymers. J. Polym. Sci. Pt. B-Polym. Phys. **49**(12), 832–864 (2011). doi:10.1002/Polb.22259
8. Tomalia, D.A., Baker, H., Dewald, J., Hall, M., Kallos, G., Martin, S., Roeck, J., Ryder, J., Smith, P.: A New Class of Polymers - Starburst-Dendritic Macromolecules. Polym. J. **17**(1), 117–132 (1985). doi:10.1295/Polymj.17.117
9. Tomalia, D.A., Naylor, A.M., Goddard, W.A.: Starburst Dendrimers - Molecular-Level Control of Size, Shape, Surface-Chemistry, Topology, and Flexibility from Atoms to Macroscopic Matter. Angew. Chem. Int. Edit. **29**(2), 138–175 (1990). doi:10.1002/Anie.199001381
10. Lee, C.C., MacKay, J.A., Frechet, J.M., Szoka, F.C.: Designing dendrimers for biological applications. Nat. Biotechnol. **23**(12), 1517–1526 (2005). doi:10.1038/nbt1171

11. Pearson, R.M., Sunoqrot, S., Hsu, H.J., Bae, J.W., Hong, S.: Dendritic nanoparticles: the next generation of nanocarriers? Ther. Deliv. **3**(8), 941–959 (2012)
12. Pearson, R.M., Hsu, H.J., Bugno, J., Hong, S.: Understanding nano-bio interactions to improve nanocarriers for drug delivery. MRS Bull. **39**(3), 227–237 (2014). doi:10.1557/Mrs.2014.9
13. Percec, V., Wilson, D.A., Leowanawat, P., Wilson, C.J., Hughes, A.D., Kaucher, M.S., Hammer, D.A., Levine, D.H., Kim, A.J., Bates, F.S., Davis, K.P., Lodge, T.P., Klein, M.L., DeVane, R.H., Aqad, E., Rosen, B.M., Argintaru, A.O., Sienkowska, M.J., Rissanen, K., Nummelin, S., Ropponen, J.: Self-Assembly of Janus Dendrimers into Uniform Dendrimersomes and Other Complex Architectures. Science **328**(5981), 1009–1014 (2010). doi:10.1126/Science.1185547
14. Rosen, B.M., Wilson, C.J., Wilson, D.A., Peterca, M., Imam, M.R., Percec, V.: Dendron-mediated self-assembly, disassembly, and self-organization of complex systems. Chemical reviews **109**(11), 6275–6540 (2009). doi:10.1021/cr900157q
15. Wurm, F., Frey, H.: Linear-dendritic block copolymers: The state of the art and exciting perspectives. Prog. Polym. Sci. **36**(1), 1–52 (2011). doi:10.1016/J.Progpolymsci.2000.07.009
16. Bugno, J., Hsu, H.-j., Hong, S.: Recent advances in targeted drug delivery approaches using dendritic polymers. Biomater. Sci. U.K. (2015). doi:10.1039/C4BM00351A
17. Esfand, R., Tomalia, D.A.: Poly(amidoamine) (PAMAM) dendrimers: from biomimicry to drug delivery and biomedical applications. Drug. Discov. Today **6**(8), 427–436 (2001). doi:10.1016/S1359-6446(01)01757-3
18. Baars, M.W.P.L., Karlsson, A.J., Sorokin, V., de Waal, B.F.W., Meijer, E.W.: Supramolecular modification of the periphery of dendrimers resulting in rigidity and functionality. Angew. Chem. Int. Edit. **39**(23), 4262-+ (2000). doi:10.1002/1521-3773(20001201)39:23<4262:Aid-Anie4262>3.0.Co;2-Y
19. Majoros, I.J., Myc, A., Thomas, T., Mehta, C.B., Baker, J.R.: PAMAM dendrimer-based multifunctional conjugate for cancer therapy: Synthesis, characterization, and functionality. Biomacromolecules **7**(2), 572–579 (2006). doi:10.1021/Bm0506142
20. Mecke, A., Lee, I., Baker Jr, J.R., Holl, M.M., Orr, B.G.: Deformability of poly(amidoamine) dendrimers. The European Physical Journal E., Soft. Matter. **14**(1), 7–16 (2004). doi:10.1140/epje/i2003-10087-5
21. Hong, S., Leroueil, P.R., Majoros, I.J., Orr, B.G., Baker Jr, J.R., Banaszak Holl, M.M.: The binding avidity of a nanoparticle-based multivalent targeted drug delivery platform. Chem. Biol. **14**(1), 107–115 (2007). doi:10.1016/j.chembiol.2006.11.015. S1074-5521(06)00471-6 [Pii]
22. Myung, J.H., Gajjar, K.A., Saric, J., Eddington, D.T., Hong, S.: Dendrimer-mediated multivalent binding for the enhanced capture of tumor cells. Angew. Chem. Int. Edit. **50**(49), 11769–11772 (2011). doi:10.1002/anie.201105508
23. Medina, S.H., El-Sayed, M.E.H.: Dendrimers as Carriers for Delivery of Chemotherapeutic Agents. Chemical reviews **109**(7), 3141–3157 (2009). doi:10.1021/Cr900174j
24. Hawker, C.J., Frechet, J.M.J.: Preparation of Polymers with Controlled Molecular Architecture - a New Convergent Approach to Dendritic Macromolecules. J. Am. Chem. Soc. **112**(21), 7638–7647 (1990). doi:10.1021/Ja00177a027
25. Pearson, R.M., Sunoqrot, S., H-j, H., Bae, J.W., Hong, S.: Dendritic nanoparticles: the next generation of nanocarriers? Ther. Deliv. In Press (2012). doi:10.4155/TDE.12.176

26. Whitesides, G.M., Mathias, J.P., Seto, C.T.: Molecular self-assembly and nanochemistry: a chemical strategy for the synthesis of nanostructures. Science **254**(5036), 1312–1319 (1991)
27. Qiu, L.Y., Bae, Y.H.: Polymer architecture and drug delivery. Pharm. Res. **23**(1), 1–30 (2006). doi:10.1007/S11095-005-9046-2
28. Al-Jamal, K.T., Ramaswamy, C., Florence, A.T.: Supramolecular structures from dendrons and dendrimers. Adv. Drug. Deliv. Rev. **57**(15), 2238–2270 (2005). doi:10.1016/J.Addr.2005.09.015
29. Bae, J.W., Pearson, R.M., Patra, N., Sunoqrot, S., Vukovic, L., Kral, P., Hong, S.: Dendron-mediated self-assembly of highly PEGylated block copolymers: a modular nanocarrier platform. Chem. Commun. **47**(37), 10302–10304 (2011). doi:10.1039/c1cc14331j
30. Tian, L., Hammond, P.T.: Comb-dendritic block copolymers as tree-shaped macromolecular amphiphiles for nanoparticle self-assembly. Chem. Mat. **18**(17), 3976–3984 (2006). doi:10.1021/Cm060232i
31. Poon, Z., Lee, J.A., Huang, S.W., Prevost, R.J., Hammond, P.T.: Highly stable, ligand-clustered "patchy" micelle nanocarriers for systemic tumor targeting. Nanomed. – Nanotechnol. Biol. Med. **7**(2), 201–209 (2011). doi:10.1016/J.Nano.2010.07.008
32. Maeda, H., Bharate, G.Y., Daruwalla, J.: Polymeric drugs for efficient tumor-targeted drug delivery based on EPR-effect. Eur. J. Pharm. Biopharm. **71**(3), 409–419 (2009). doi:10.1016/J.Ejpb.2008.11.010
33. Zhang, L.F., Chan, J.M., Gu, F.X., Rhee, J.W., Wang, A.Z., Radovic-Moreno, A.F., Alexis, F., Langer, R., Farokhzad, O.C.: Self-assembled lipid-polymer hybrid nanoparticles: A robust drug delivery platform. ACS Nano. **2**(8), 1696–1702 (2008). doi:10.1021/Nn800275r
34. Jin, S.E., Bae, J.W., Hong, S.: Multiscale observation of biological interactions of nanocarriers: from nano to macro. Microscopy research and technique **73**(9), 813–823 (2010). doi:10.1002/jemt.20847
35. Hu, C.M., Aryal, S., Zhang, L.: Nanoparticle-assisted combination therapies for effective cancer treatment. Ther. Deliv. **1**(2), 323–334 (2010)
36. Sunoqrot, S., Bae, J.W., Jin, S.E., Ryan, M.P., Liu, Y., Hong, S.: Kinetically controlled cellular interactions of polymer-polymer and polymer-liposome nanohybrid systems. Bioconjugate Chem. **22**(3), 466–474 (2011). doi:10.1021/bc100484t
37. Mora-Huertas, C.E., Fessi, H., Elaissari, A.: Polymer-based nanocapsules for drug delivery. Int. J. Pharm. **385**(1–2), 113–142 (2010). doi:10.1016/j.ijpharm.2009.10.018
38. Sunoqrot, S., Bae, J.W., Pearson, R.M., Shyu, K., Liu, Y., Kim, D.H., Hong, S.: Temporal control over cellular targeting through hybridization of folate-targeted dendrimers and PEG-PLA nanoparticles. Biomacromolecules **13**(4), 1223–1230 (2012). doi:10.1021/bm300316n
39. Sunoqrot, S., Bugno, J., Lantvit, D., Burdette, J.E., Hong, S.: Prolonged blood circulation and enhanced tumor accumulation of folate-targeted dendrimer-polymer hybrid nanoparticles. J. Control Release **191**, 115–122 (2014). doi:10.1016/j.jconrel.2014.05.006
40. Ko, Y.T., Bhattacharya, R., Bickel, U.: Liposome encapsulated polyethylenimine/ODN polyplexes for brain targeting. J. Control Release **133**(3), 230–237 (2009). doi:10.1016/J.Jconrel.2008.10.013

41. Discher, B.M., Won, Y.Y., Ege, D.S., Lee, J.C.M., Bates, F.S., Discher, D.E., Hammer, D.A.: Polymersomes: Tough vesicles made from diblock copolymers. Science **284**(5417), 1143–1146 (1999). doi:10.1126/Science.284.5417.1143
42. Zhang, S.D., Sun, H.J., Hughes, A.D., Draghici, B., Lejnieks, J., Leowanawat, P., Bertin, A., De Leon, L.O., Kulikov, O.V., Chen, Y.C., Pochan, D.J., Heiney, P.A., Percec, V.: "Single-Single" Amphiphilic Janus Dendrimers Self-Assemble into Uniform Dendrimersomes with Predictable Size. ACS Nano. **8**(2), 1554–1565 (2014). doi:10.1021/Nn405790x
43. Leroueil, P.R., Hong, S.Y., Mecke, A., Baker, J.R., Orr, B.G., Holl, M.M.B.: Nanoparticle interaction with biological membranes: Does nanotechnology present a janus face? Accounts Chem. Res. **40**(5), 335–342 (2007). doi:10.1021/Ar600012y
44. Hong, S.P., Bielinska, A.U., Mecke, A., Keszler, B., Beals, J.L., Shi, X.Y., Balogh, L., Orr, B.G., Baker, J.R., Holl, M.M.B.: Interaction of poly(amidoamine) dendrimers with supported lipid bilayers and cells: Hole formation and the relation to transport. Bioconjug. Chem. **15**(4), 774–782 (2004). doi:10.1021/Bc049962b
45. Hong, S., Rattan, R., Majoros, I.J., Mullen, D.G., Peters, J.L., Shi, X., Bielinska, A.U., Blanco, L., Orr, B.G., Baker Jr, J.R., Holl, M.M.: The role of ganglioside GM1 in cellular internalization mechanisms of poly(amidoamine) dendrimers. Bioconjug. Chem. **20**(8), 1503–1513 (2009). doi:10.1021/bc900029k
46. Chithrani, B.D., Ghazani, A.A., Chan, W.C.W.: Determining the size and shape dependence of gold nanoparticle uptake into mammalian cells. Nano. Lett. **6**(4), 662–668 (2006). doi:10.1021/Nl052396o
47. Singh, A.K., Kasinath, B.S., Lewis, E.J.: Interaction of polycations with cell-surface negative charges of epithelial cells. Biochim. Biophys. Acta. **1120**(3), 337–342 (1992)
48. Maiti, P.K., Cagin, T., Lin, S.T., Goddard, W.A.: Effect of solvent and pH on the structure of PAMAM dendrimers. Macromolecules **38**(3), 979–991 (2005). doi:10.1021/Ma0491681
49. Mammen, M., Choi, S.K., Whitesides, G.M.: Polyvalent interactions in biological systems: Implications for design and use of multivalent ligands and inhibitors. Angew. Chem. Int. Edit. **37**(20), 2755–2794 (1998)
50. Mourez, M., Kane, R.S., Mogridge, J., Metallo, S., Deschatelets, P., Sellman, B.R., Whitesides, G.M., Collier, R.J.: Designing a polyvalent inhibitor of anthrax toxin. Nat. Biotechnol. **19**(10), 958–961 (2001)
51. Lee, R.T., Lee, Y.C.: Affinity enhancement by multivalent lectin-carbohydrate interaction. Glycoconjugate J. **17**(7–9), 543–551 (2000)
52. Kiessling, L.L., Gestwicki, J.E., Strong, L.E.: Synthetic multivalent ligands as probes of signal transduction. Angew. Chem. Int. Edit. **45**(15), 2348–2368 (2006). doi:10.1002/Anie.200502794
53. Hlavacek, W.S., Faeder, J.R., Blinov, M.L., Perelson, A.S., Goldstein, B.: The complexity of complexes in signal transduction. Biotechnol. Bioeng. **84**(7), 783–794 (2003). doi:10.1002/Bit.10842
54. Bertozzi, C.R., Kiessling, L.L.: Chemical Glycobiology. Science **291**(5512), 2357–2364 (2001). doi:10.1126/science.1059820
55. Sieben, C., Kappel, C., Zhu, R., Wozniak, A., Rankl, C., Hinterdorfer, P., Grubmller, H., Herrmann, A.: Influenza virus binds its host cell using multiple dynamic interactions. Proc. Natl. Acad. Sci. U.S.A. (2012). doi:10.1073/pnas.1120265109

56. Vance, D., Martin, J., Patke, S., Kane, R.S.: The design of polyvalent scaffolds for targeted delivery. Adv. Drug. Deliv. Rev. **61**(11), 931–939 (2009). doi:10.1016/j.addr.2009.06.002. S0169-409X(09)00219-1 [pii]
57. Pavan, G.M., Mintzer, M.A., Simanek, E.E., Merkel, O.M., Kissel, T., Danani, A.: Computational Insights into the Interactions between DNA and siRNA with "Rigid" and "Flexible" Triazine Dendrimers. Biomacromolecules **11**(3), 721–730 (2010). doi:10.1021/Bm901298t
58. Page, D., Aravind, S., Roy, R.: Synthesis and lectin binding properties of dendritic mannopyranoside. Chem. Commun. **16**, 1913–1914 (1996)
59. Ashton, P.R., Hounsell, E.F., Jayaraman, N., Nilsen, T.M., Spencer, N., Stoddart, J.F., Young, M.: Synthesis and biological evaluation of alpha-D-mannopyranoside-containing dendrimers. J. Org. Chem. **63**(10), 3429–3437 (1998)
60. Rele, S.M., Cui, W.X., Wang, L.C., Hou, S.J., Barr-Zarse, G., Taton, D., Gnanou, Y., Esko, J.D., Chaikof, E.L.: Dendrimer-like PEO glycopolymers exhibit anti-inflammatory properties. J. Am. Chem. Soc. **127**(29), 10132–10133 (2005). doi:10.1021/Ja0511974
61. Shewmake, T.A., Solis, F.J., Gillies, R.J., Caplan, M.R.: Effects of linker length and flexibility on multivalent targeting. Biomacromolecules **9**(11), 3057–3064 (2008). doi:10.1021/bm800529b
62. Kim, Y., Lee, J.H., Ryu, J., Kim, D.J.: Multivalent & multifunctional ligands to beta-amyloid. Current Pharmaceutical Design **15**(6), 637–658 (2009)
63. Kane, R.S.: Thermodynamics of multivalent interactions: influence of the linker. Langmuir **26**(11), 8636–8640 (2010). doi:10.1021/la9047193
64. Pearson, R.M., Patra, N., Hsu, H.J., Uddin, S., Kral, P., Hong, S.: Positively Charged Dendron Micelles Display Negligible Cellular Interactions. ACS Macro Letters **2**(1), 77–81 (2013). doi:10.1021/mz300533w
65. Hsu, H.J., Sen, S., Pearson, R.M., Uddin, S., Kral, P., Hong, S.: Poly(ethylene glycol) Corona Chain Length Controls End-Group-Dependent Cell Interactions of Dendron Micelles. Macromolecules **47**(19), 6911–6918 (2014). doi:10.1021/Ma501258c
66. Burton, G.J.: Evidence for non-linear response processes in the human visual system from measurements on the thresholds of spatial beat frequencies. Vision. Res. **13**(7), 1211–1225 (1973)
67. Conner, S.D., Schmid, S.L.: Regulated portals of entry into the cell. Nature **422**(6927), 37–44 (2003). doi:10.1038/Nature01451
68. Kojima, C.: Design of stimuli-responsive dendrimers. Expert Opin. Drug Deliv. **7**(3), 307–319 (2010). doi:10.1517/17425240903530651
69. Gillies, E.R., Jonsson, T.B., Frechet, J.M.J.: Stimuli-responsive supramolecular assemblies of linear-dendritic copolymers. J. Am. Chem. Soc. **126**(38), 11936–11943 (2004). doi:10.1021/Ja0463738
70. Stuart, M.A.C., Huck, W.T.S., Genzer, J., Muller, M., Ober, C., Stamm, M., Sukhorukov, G.B., Szleifer, I., Tsukruk, V.V., Urban, M., Winnik, F., Zauscher, S., Luzinov, I., Minko, S.: Emerging applications of stimuli-responsive polymer materials. Nat. Mater. **9**(2), 101–113 (2010). doi:10.1038/Nmat2614
71. Schmaljohann, D.: Thermo- and pH-responsive polymers in drug delivery. Adv Drug Deliver Rev **58**(15), 1655–1670 (2006). doi:10.1016/J.Addr.2006.09.020
72. Benjaminsen, R.V., Mattebjerg, M.A., Henriksen, J.R., Moghimi, S.M., Andresen, T.L.: The Possible "Proton Sponge" Effect of Polyethylenimine (PEI) Does Not Include Change in Lysosomal pH. Mol. Ther. **21**(1), 149–157 (2013). doi:10.1038/mt.2012.185

73. Liu, Y., Bryantsev, V.S., Diallo, M.S., Goddard, W.A.: PAMAM Dendrimers Undergo pH Responsive Conformational Changes without Swelling. J. Am. Chem. Soc. - + **131**(8), 2798 (2009). doi:10.1021/Ja8100227
74. Vaupel, P., Kallinowski, F., Okunieff, P.: Blood-Flow, Oxygen and Nutrient Supply, and Metabolic Microenvironment of Human-Tumors - a Review. Cancer Res. **49**(23), 6449–6465 (1989)
75. Pistolis, G., Malliaris, A., Tsiourvas, D., Paleos, C.M.: Poly(propyleneimine) dendrimers as pH-sensitive controlled-release systems. Chem.-Eur. J. **5**(5), 1440–1444 (1999). doi:10.1002/(Sici)1521-3765(19990503)5:5<1440:Aid-Chem1440>3.0.Co;2-M
76. Kaminskas, L.M., Boyd, B.J., Porter, C.J.H.: Dendrimer pharmacokinetics: the effect of size, structure and surface characteristics on ADME properties. Nanomedicine U.k. **6**(6), 1063–1084 (2011). doi:10.2217/Nnm.11.67
77. Wijagkanalan, W., Kawakami, S., Hashida, M.: Designing Dendrimers for Drug Delivery and Imaging: Pharmacokinetic Considerations. Pharm. Res. **28**(7), 1500–1519 (2011). doi:10.1007/S11095-010-0339-8
78. Kobayashi, H., Kawamoto, S., Saga, T., Sato, N., Hiraga, A., Konishi, J., Togashi, K., Brechbiel, M.W.: Micro-MR angiography of normal and intratumoral vessels in mice using dedicated intravascular MR contrast agents with high generation of polyamidoamine dendrimer core: Reference to pharmacokinetic properties of dendrimer-based MR contrast agents. J. Magn. Reson. Imaging **14**(6), 705–713 (2001). doi:10.1002/Jmri.10025
79. Kobayashi, H., Sato, N., Hiraga, A., Saga, T., Nakamoto, Y., Ueda, H., Konishi, J., Togashi, K., Brechbiel, M.W.: 3D-micro-MR angiography of mice using macromolecular MR contrast agents with polyamidoamine dendrimer core with reference to their pharmacokinetic properties. Magn. Reson. Med. **45**(3), 454–460 (2001). doi:10.1002/1522-2594(200103)45:3<454:Aid-Mrm1060>3.0.Co;2-M
80. Soo Choi, H., Liu, W., Misra, P., Tanaka, E., Zimmer, J.P., Itty Ipe, B., Bawendi, M.G., Frangioni, J.V.: Renal clearance of quantum dots. Nat. Biotech. **25**(10), 1165–1170 (2007). doi:10.1038/nbt1340
81. Walkey, C.D., Olsen, J.B., Guo, H., Emili, A., Chan, W.C.: Nanoparticle size and surface chemistry determine serum protein adsorption and macrophage uptake. J. Am. Chem. Soc. **134**(4), 2139–2147 (2012). doi:10.1021/ja2084338
82. Malik, N., Wiwattanapatapee, R., Klopsch, R., Lorenz, K., Frey, H., Weener, J.W., Meijer, E.W., Paulus, W., Duncan, R.: Dendrimers: Relationship between structure and biocompatibility in vitro, and preliminary studies on the biodistribution of I-125-labelled polyamidoamine dendrimers in vivo. J. Control Release **68**(2), 299–302 (2000). doi:10.1016/S0168-3659(00)00283-2
83. Boyd, B.J., Kaminskas, L.M., Karellas, P., Krippner, G., Lessene, R., Porter, C.J.H.: Cationic poly-L-lysine dendrimers: Pharmacokinetics, biodistribution, and evidence for metabolism and bioresorption after intravenous administration to rats. Mol. Pharmaceut. **3**(5), 614–627 (2006). doi:10.1021/Mp060032e
84. Duncan, R., Izzo, L.: Dendrimer biocompatibility and toxicity. Adv. Drug Deliv. Rev. **57**(15), 2215–2237 (2005). doi:10.1016/J.Addr.2004.09.019
85. Satija, J., Gupta, U., Jain, N.K.: Pharmaceutical and biomedical potential of surface engineered dendrimers. Crit. Rev. Ther. Drug. Carr. Syst. **24**(3), 257–306 (2007)

86. Wang, W., Xiong, W., Zhu, Y.H., Xu, H.B., Yang, X.L.: Protective Effect of PEGylation Against Poly(amidoamine) Dendrimer-Induced Hemolysis of Human Red Blood Cells. J. Biomed. Mater. Res. B. **93B**(1), 59–64 (2010). doi:10.1002/Jbm.B.31558
87. Pryor, J.B., Harper, B.J., Harper, S.L.: Comparative toxicological assessment of PAMAM and thiophosphoryl dendrimers using embryonic zebrafish. Int. J. Nanomed. **9**, 1947–1956 (2014). doi:10.2147/Ijn.S60220
88. Owens, D.E., Peppas, N.A.: Opsonization, biodistribution, and pharmacokinetics of polymeric nanoparticles. Int. J. Pharmaceut. **307**(1), 93–102 (2006). doi:10.1016/J.Ijpharm.2005.10.010
89. Jain, K., Kesharwani, P., Gupta, U., Jain, N.K.: Dendrimer toxicity: Let's meet the challenge. Int. J. Pharmaceut. **394**(1–2), 122–142 (2010). doi:10.1016/J.Ijpharm.2010.04.027
90. Kojima, C., Regino, C., Umeda, Y., Kobayashi, H., Kono, K.: Influence of dendrimer generation and polyethylene glycol length on the biodistribution of PEGylated dendrimers. Int. J. Pharmaceut. **383**(1–2), 293–296 (2010). doi:10.1016/J.Ijpharm.2009.09.015
91. Ross, J.F., Chaudhuri, P.K., Ratnam, M.: Differential regulation of folate receptor isoforms in normal and malignant tissues in vivo and in established cell lines. Physiologic. and clinical implications Cancer **73**(9), 2432–2443 (1994). doi:10.1002/1097-0142(19940501)73:9<2432:aid-cncr2820730929>3.0.co;2-s
92. Quintana, A., Raczka, E., Piehler, L., Lee, I., Myc, A., Majoros, I., Patri, A.K., Thomas, T., Mule, J., Baker, J.R.: Design and function of a dendrimer-based therapeutic nanodevice targeted to tumor cells through the folate receptor. Pharm. Res. **19**(9), 1310–1316 (2002). doi:10.1023/A:1020398624602
93. Kukowska-Latallo, J.F., Candido, K.A., Cao, Z.Y., Nigavekar, S.S., Majoros, I.J., Thomas, T.P., Balogh, L.P., Khan, M.K., Baker, J.R.: Nanoparticle targeting of anticancer drug improves therapeutic response in animal model of human epithelial cancer. Cancer Res. **65**(12), 5317–5324 (2005)
94. Singh, P., Gupta, U., Asthana, A., Jain, N.K.: Folate and folate-PEG-PAMAM dendrimers: synthesis, characterization, and targeted anticancer drug delivery potential in tumor bearing mice. Bioconjugate Chem. **19**(11), 2239–2252 (2008). doi:10.1021/bc800125u
95. Lom, B., Healy, K.E., Hockberger, P.E.: A Versatile Technique for Patterning Biomolecules onto Glass Coverslips. J. Neurosci. Methods **50**(3), 385–397 (1993)
96. Jung, H., Kulkarni, R., Collier, C.P.: Dip-pen nanolithography of reactive alkoxysilanes on glass. J. Am. Chem. Soc. **125**(40), 12096–12097 (2003). doi:10.1021/Ja0363720
97. Li, Y., He, H., Jia, X.R., Lu, W.L., Lou, J.N., Wei, Y.: A dual-targeting nanocarrier based on poly(amidoamine) dendrimers conjugated with transferrin and tamoxifen for treating brain gliomas. Biomaterials **33**(15), 3899–3908 (2012). doi:10.1016/J.Biomaterials.2012.02.004
98. Qiao, H.Z., Li, J., Wang, Y., Ping, Q.N., Wang, G.J., Gu, X.C.: Synthesis and characterization of multi-functional linear-dendritic block copolymer for intracellular delivery of antitumor drugs. Int. J. Pharmaceut. **452**(1–2), 363–373 (2013). doi:10.1016/J.Ijpharm.05.003
99. Zhou, Z.Y., D'Emanuele, A., Attwood, D.: Solubility enhancement of paclitaxel using a linear-dendritic block copolymer. Int. J. Pharmaceut. **452**(1–2), 173–179 (2013). doi:10.1016/J.Ijpharm.2013.04.075

100. Xiao, K., Li, Y.P., Lee, J.S., Gonik, A.M., Dong, T., Fung, G., Sanchez, E., Xing, L., Cheng, H.R., Luo, J.T., Lam, K.S.: "OA02" Peptide Facilitates the Precise Targeting of Paclitaxel- Loaded Micellar Nanoparticles to Ovarian Cancer In Vivo. Cancer Res. **72**(8), 2100–2110 (2012). doi:10.1158/0008-5472.Can-11-3883
101. Yang, Y., Pearson, R.M., Lee, O., Lee, C.W., Chatterton, R.T., Khan, S.A., Hong, S.: Dendron-Based Micelles for Topical Delivery of Endoxifen: A Potential Chemo-Preventive Medicine for Breast Cancer. Adv. Funct. Mater. **24**(17), 2442–2449 (2014). doi:10.1002/Adfm.201303253
102. Yang, Y., Bugno, J., Hong, S.: Nanoscale polymeric penetration enhancers in topical drug delivery. Polym. Chem. U.k. **4**(9), 2651–2657 (2013). doi:10.1039/C3py00049d
103. Poon, Z., Chen, S., Engler, A.C., Lee, H.I., Atas, E., von Maltzahn, G., Bhatia, S.N., Hammond, P.T.: Ligand-Clustered "Patchy" Nanoparticles for Modulated Cellular Uptake and In Vivo Tumor Targeting. Angew. Chem. Int. Edit. **49**(40), 7266–7270 (2010). doi:10.1002/Anie.201003445
104. Turturro, S., Sunoqrot, S., Ying, H., Hong, S., Yue, B.Y.: Sustained release of matrix metalloproteinase-3 to trabecular meshwork cells using biodegradable PLGA microparticles. Mol. Pharm. **10**(8), 3023–3032 (2013). doi:10.1021/mp4001052
105. Kim, J.H., Lee, E., Park, J.S., Kataoka, K., Jang, W.D.: Dual stimuli-responsive dendritic-linear block copolymers. Chem. Commun. **48**(30), 3662–3664 (2012). doi:10.1039/C2cc17205d
106. Azagarsamy, M.A., Yesilyurt, V., Thayumanavan, S.: Disassembly of Dendritic Micellar Containers Due to Protein Binding. J. Am. Chem. Soc. **132**(13), 4550 (2010). doi:10.1021/Ja100746d
107. Fauci, A.S.: The human immunodeficiency virus: infectivity and mechanisms of pathogenesis. Science **239**(4840), 617–622 (1988)
108. Kim, S., Ikeuchi, K., Groopman, J., Baltimore, D.: Factors affecting cellular tropism of human immunodeficiency virus. J. Virol. **64**(11), 5600–5604 (1990)
109. Steinhauer, D.A.: Role of Hemagglutinin Cleavage for the Pathogenicity of Influenza Virus. Virology **258**(1), 1–20 (1999). doi:10.1006/viro.1999.9716
110. Glover, D.J., Lipps, H.J., Jans, D.A.: Towards safe, non-viral therapeutic gene expression in humans. Nat. Rev. Genet. **6**(4), 299-U229 (2005). doi:10.1038/Nrg1577
111. Pack, D.W., Hoffman, A.S., Pun, S., Stayton, P.S.: Design and development of polymers for gene delivery. Nat. Rev. Drug. Discov. **4**(7), 581–593 (2005). doi:10.1038/Nrd1775
112. Khalil, I.A., Kogure, K., Akita, H., Harashima, H.: Uptake pathways and subsequent intracellular trafficking in nonviral gene delivery. Pharmacol. Rev. **58**(1), 32–45 (2006). doi:10.1124/Pr.58.1.8
113. Qin, L.H., Pahud, D.R., Ding, Y.Z., Bielinska, A.U., Kukowska-Latallo, J.F., Baker, J.R., Bromberg, J.S.: Efficient transfer of genes into murine cardiac grafts by starburst polyamidoamine dendrimers. Hum. Gene. Ther. **9**(4), 553–560 (1998). doi:10.1089/Hum.1998.9.4-553
114. Bielinska, A.U., KukowskaLatallo, J.F., Baker, J.R.: The interaction of plasmid DNA with polyamidoamine dendrimers: mechanism of complex formation and analysis of alterations induced in nuclease sensitivity and transcriptional activity of the complexed DNA. Biochim. Biophys. Acta.-Gene. Struct. Expression **1353**(2), 180–190 (1997). doi:10.1016/S0167-4781(97)00069-9

115. Haensler, J., Szoka, F.C.: Polyamidoamine Cascade Polymers Mediate Efficient Transfection of Cells in Culture. Bioconjug. Chem. **4**(5), 372–379 (1993). doi:10.1021/Bc00023a012
116. Whitehead, K.A., Langer, R., Anderson, D.G.: Knocking down barriers: advances in siRNA delivery. Nat. Rev. Drug. Discov. **8**(2), 129–138 (2009). doi:10.1038/nrd2742
117. Chen, W., Turro, N.J., Tomalia, D.A.: Using ethidium bromide to probe the interactions between DNA and dendrimers. Langmuir **16**(1), 15–19 (2000). doi:10.1021/La981429v
118. Tang, Y., Li, Y.B., Wang, B., Lin, R.Y., van Dongen, M., Zurcher, D.M., Gu, X.Y., Holl, M.M.B., Liu, G., Qi, R.: Efficient in Vitro siRNA Delivery and Intramuscular Gene Silencing Using PEG-Modified PAMAM Dendrimers. Mol. Pharmaceut. **9**(6), 1812–1821 (2012). doi:10.1021/Mp3001364
119. Patil, M.L., Zhang, M., Minko, T.: Multifunctional Triblock Nanocarrier (PAMAM-PEG-PLL) for the Efficient Intracellular siRNA Delivery and Gene Silencing. ACS Nano. **5**(3), 1877–1887 (2011). doi:10.1021/Nn102711d
120. Jevprasesphant, R., Penny, J., Jalal, R., Attwood, D., McKeown, N.B., D'Emanuele, A.: The influence of surface modification on the cytotoxicity of PAMAM dendrimers. Int. J. Pharm. **252**(1–2), 263–266 (2003)
121. Brunner, K., Harder, J., Halbach, T., Willibald, J., Spada, F., Gnerlich, F., Sparrer, K., Beil, A., Mockl, L., Brauchle, C., Conzelmann, K.K., Carell, T.: Cell-Penetrating and Neurotargeting Dendritic siRNA Nanostructures. Angew. Chem. **54**(6), 1946–1949 (2015). doi:10.1002/anie.201409803
122. Shah, V., Taratula, O., Garbuzenko, O.B., Taratula, O.R., Rodriguez-Rodriguez, L., Minko, T.: Targeted nanomedicine for suppression of CD44 and simultaneous cell death induction in ovarian cancer: an optimal delivery of siRNA and anticancer drug. Clinical Cancer Research : an Official Journal of the American Association for Cancer Research **19**(22), 6193–6204 (2013). doi:10.1158/1078-0432.CCR-13-1536
123. Wood, K.C., Little, S.R., Langer, R., Hammond, P.T.: A family of hierarchically self-assembling linear-dendritic hybrid polymers for highly efficient targeted gene delivery. Angew. Chem. **44**(41), 6704–6708 (2005). doi:10.1002/Anie.200502152
124. Yang, B., Sun, Y.X., Yi, W.J., Yang, J., Liu, C.W., Cheng, H., Feng, J., Zhang, X.Z., Zhuo, R.X.: A linear-dendritic cationic vector for efficient DNA grasp and delivery. Acta. Biomater. **8**(6), 2121–2132 (2012). doi:10.1016/J.Actbio.2012.02.013
125. Langereis, S., Dirksen, A., Hackeng, T.M., van Genderen, M.H.P., Meijer, E.W.: Dendrimers and magnetic resonance imaging. New J. Chem. **31**(7), 1152–1160 (2007). doi:10.1039/B616960k
126. Kobayashi, H., Brechbiel, M.W.: Nano-sized MRI contrast agents with dendrimer cores. Adv. Drug. Deliver Rev. **57**(15), 2271–2286 (2005). doi:10.1016/J.Addr.2005.09.016
127. Venditto, V.J., Regino, C.A.S., Brechbiel, M.W.: PAMAM dendrimer based macromolecules as improved contrast agents. Mo. Pharmaceut. **2**(4), 302–311 (2005). doi:10.1021/Mp050019e
128. Villaraza, A.J.L., Bumb, A., Brechbiel, M.W.: Macromolecules, Dendrimers, and Nanomaterials in Magnetic Resonance Imaging: The Interplay between Size, Function, and Pharmacokinetics. Chem. Rev. **110**(5), 2921–2959 (2010). doi:10.1021/Cr900232t
129. Kobayashi, H., Jo, S.K., Kawamoto, S., Yasuda, H., Hu, X., Knopp, M.V., Brechbiel, M.W., Choyke, P.L., Star, R.A.: Polyamine dendrimer-based MRI contrast agents for functional kidney imaging to diagnose acute renal failure. J. Magn. Reson. Imaging **20**(3), 512–518 (2004). doi:10.1002/jmri.20147

130. Swanson, S.D., Kukowska-Latallo, J.F., Patri, A.K., Chen, C.Y., Ge, S., Cao, Z.Y., Kotlyar, A., East, A.T., Baker, J.R.: Targeted gadolinium-loaded dendrimer nanoparticles for tumor-specific magnetic resonance contrast enhancement. Int. J. Nanomed. **3**(2), 201–210 (2008)
131. Zhang, W.L., Li, N., Huang, J., Yu, J.H., Wang, D.X., Li, Y.P., Liu, S.Y.: Gadolinium-Conjugated FA-PEG-PAMAM-COOH Nanoparticles as Potential Tumor-Targeted Circulation-Prolonged Macromolecular MRI Contrast Agents. J. Appl. Polym. Sci. **118**(3), 1805–1814 (2010). doi:10.1002/App.32494
132. Strable, E., Bulte, J.W.M., Moskowitz, B., Vivekanandan, K., Allen, M., Douglas, T.: Synthesis and characterization of soluble iron oxide-dendrimer composites. Chem Mat. **13**(6), 2201–2209 (2001). doi:10.1021/Cm010125i
133. Stears, R.L., Getts, R.C., Gullans, S.R.: A novel, sensitive detection system for high-density microarrays using dendrimer technology. Physiol. Genomics **3**(2), 93–99 (2000)
134. Song, X.D., Swanson, B.I.: Direct, ultrasensitive, and selective optical detection of protein toxins using multivalent interactions. Anal. Chem. **71**(11), 2097–2107 (1999)
135. Angenendt, P., Glokler, J., Sobek, J., Lehrach, H.: Cahill DJ (2003) Next generation of protein microarray support materials: evaluation for protein and antibody microarray applications. J. Chromatogr. A **1–2**, 97–104 (1009)
136. Haes, A.J., Van Duyne, R.P.: A nanoscale optical blosensor: Sensitivity and selectivity of an approach based on the localized surface plasmon resonance spectroscopy of triangular silver nanoparticles. J. Am. Chem. Soc. **124**(35), 10596–10604 (2002). doi:10.1021/Ja020393x
137. Woller, E.K., Walter, E.D., Morgan, J.R., Singel, D.J., Cloninger, M.J.: Altering the strength of lectin binding interactions and controlling the amount of lectin clustering using mannose/hydroxyl-functionalized dendrimers. J. Am. Chem. Soc. **125**(29), 8820–8826 (2003). doi:10.1021/ja0352496
138. Myung, J.H., Gajjar, K.A., Han, Y.E., Hong, S.P.: The role of polymers in detection and isolation of circulating tumor cells. Polym. Chem. U.k. **3**(9), 2336–2341 (2012). doi:10.1039/C2py20420g
139. Benters, R., Niemeyer, C.M., Wohrle, D.: Dendrimer-activated-solid supports for nucleic acid and protein microarrays. ChemBioChem **2**(9), 686–694 (2001)
140. Myung, J.H., Gajjar, K.A., Chen, J., Molokie, R.E., Hong, S.: Differential detection of tumor cells using a combination of cell rolling, multivalent binding, and multiple antibodies. Anal. Chem. **86**(12), 6088–6094 (2014). doi:10.1021/ac501243a
141. Han, H.J., Kannan, R.M., Wang, S.X., Mao, G.Z., Kusanovic, J.P., Romero, R.: Multifunctional Dendrimer-Templated Antibody Presentation on Biosensor Surfaces for Improved Biomarker Detection. Adv. Funct. Mater. **20**(3), 409–421 (2010). doi:10.1002/Adfm.200901293
142. Lundgren, A., Hed, Y., Oeberg, K., Sellborn, A., Fink, H., Lowenhielm, P., Kelly, J., Malkoch, M., Berglin, M.: Self-Assembled Arrays of Dendrimer-Gold-Nanoparticle Hybrids for Functional Cell Studies. Angew. Chem. Int. Edit. **50**(15), 3450–3453 (2011). doi:10.1002/Anie.201006544
143. Benhabbour, S.R., Sheardown, H., Adronov, A.: Cell adhesion and proliferation on hydrophilic dendritically modified surfaces. Biomaterials **29**(31), 4177–4186 (2008). doi:10.1016/J.Biomaterials.2008.07.016

144. Zhang, L., Zou, B., Dong, D., Huo, F.W., Zhang, X., Chi, L.F., Jiang, L.: Self-assembled monolayers of new dendron-thiols: manipulation of the patterned surface and wetting properties. Chem. Commun. **19**, 1906–1907 (2001). doi:10.1039/B103903m
145. Zhang, L., Huo, F.W., Wang, Z.Q., Wu, L.X., Zhang, X., Hoppener, S., Chi, L.F., Fuchs, H., Zhao, J.W., Niu, L., Dong, S.J.: Investigation into self-assembled monolayers of a polyether dendron thiol: Chemisorption, kinetics, and patterned surface. Langmuir **16**(8), 3813–3817 (2000). doi:10.1021/La990923b
146. Chechik, V., Schonherr, H., Vancso, G.J., Stirling, C.J.M.: Self-assembled monolayers of branched thiols and disulfides on gold: Surface coverage, order and chain orientation. Langmuir **14**(11), 3003–3010 (1998). doi:10.1021/La971090x
147. Oberg, K., Ropponen, J., Kelly, J., Lowenhielm, P., Berglin, M., Malkoch, M.: Templating Gold Surfaces with Function: A Self-Assembled Dendritic Monolayer Methodology Based on Monodisperse Polyester Scaffolds. Langmuir **29**(1), 456–465 (2013). doi:10.1021/La3041314
148. Polcyn, P., Jurczak, M., Rajnisz, A., Solecka, J., Urbanczyk-Lipkowska, Z.: Design of Antimicrobially Active Small Amphiphilic Peptide Dendrimers. Molecules **14**(10), 3881–3905 (2009). doi:10.3390/Molecules14103881
149. Cheng, Y.Y., Qu, H., Ma, M.L., Xu, Z.H., Xu, P., Fang, Y.J.: Xu TW (2007) Polyamidoamine (PAMAM) dendrimers as biocompatible carriers of quinolone antimicrobials: An in vitro study. Eur. J. Med. Chem. **42**(7), 1032–1038 (2006). doi:10.1016/J.Ejmech.12.035
150. Mishra, M.K., Kotta, K., Hali, M., Wykes, S., Gerard, H.C., Hudson, A.P., Whittum-Hudson, J.A., Kannan, R.M.: PAMAM dendrimer-azithromycin conjugate nanodevices for the treatment of Chlamydia trachomatis infections. Nanomed-Nanotechnol Biol Med **7**(6), 935–944 (2011). doi:10.1016/J.Nano.2011.04.008
151. Reuter, J.D., Myc, A., Hayes, M.M., Gan, Z.H., Roy, R., Qin, D.J., Yin, R., Piehler, L.T., Esfand, R., Tomalia, D.A., Baker, J.R.: Inhibition of viral adhesion and infection by sialic-acid-conjugated dendritic polymers. Bioconjugate Chem. **10**(2), 271–278 (1999). doi:10.1021/Bc980099n
152. Landers, J.J., Cao, Z., Lee, I., Piehler, L.T., Myc, P.P., Myc, A., Hamouda, T., Galecki, A.T., Baker Jr, J.R.: Prevention of influenza pneumonitis by sialic Acid-conjugated dendritic polymers. J. Infect. Dis. **186**(9), 1222–1230 (2002). doi:10.1086/344316
153. Kensinger, R.D., Catalone, B.J., Krebs, F.C., Wigdahl, B., Schengrund, C.L.: Novel polysulfated galactose-derivatized dendrimers as binding antagonists of human immunodeficiency virus type 1 infection. Antimicrob. Agents Chemother. **48**(5), 1614–1623 (2004). doi:10.1128/Aac.48.5.1614-1623.2004
154. Rupp, R., Rosenthal, S.L., Stanberry, L.R.: VivaGel (SPL7013 Gel): a candidate dendrimer–microbicide for the prevention of HIV and HSV infection. Int J Nanomedicine **2**(4), 561–566 (2007)
155. Telwatte, S., Moore, K., Johnson, A., Tyssen, D., Sterjovski, J., Aldunate, M., Gorry, P.R., Ramsland, P.A., Lewis, G.R., Paull, J.R., Sonza, S., Tachedjian, G.: Virucidal activity of the dendrimer microbicide SPL7013 against HIV-1. Antiviral Reserch **90**(3), 195–199 (2011). doi:10.1016/j.antiviral.2011.03.186
156. Zhao, H., Li, J.R., Xi, F., Jiang, L.: Polyamidoamine dendrimers inhibit binding of Tat peptide to TAR RNA. FEBS Lett. **563**(1–3), 241–245 (2004). doi:10.1016/S0014-5793(04)00284-4

157. Witvrouw, M., Fikkert, V., Pluymers, W., Matthews, B., Mardel, K., Schols, D., Raff, J., Debyser, Z., De, C.E., Holan, G., Pannecouque, C.: Polyanionic (i.e., polysulfonate) dendrimers can inhibit the replication of human immunodeficiency virus by interfering with both virus adsorption and later steps (Reverse transcriptase/integrase) in the virus replicative cycle. Mol. Pharmacol. **58**(5), 1100–1108 (2000)
158. Diallo, M.S., Christie, S., Swaminathan, P., Balogh, L., Shi, X.Y., Um, W., Papelis, C., Goddard, W.A., Johnson, J.H.: Dendritic chelating agents. 1. Cu(II) binding to ethylene diamine core poly(amidoamine) dendrimers in aqueous solutions. Langmuir **20**(7), 2640–2651 (2004). doi:10.1021/La036108k
159. Theron, J., Walker, J.A., Cloete, T.E.: Nanotechnology and water treatment: Applications and emerging opportunities. Crit. Rev. Microbiol. **34**(1), 43–69 (2008). doi:10.1080/10408410701710442
160. Lard, M., Kim, S.H., Lin, S., Bhattacharya, P., Ke, P.C., Lamm, M.H.: Fluorescence resonance energy transfer between phenanthrene and PAMAM dendrimers. Phys. Chem. Chem. Phys. **12**(32), 9285–9291 (2010). doi:10.1039/B924522g
161. Riehemann, K., Schneider, S.W., Luger, T.A., Godin, B., Ferrari, M., Fuchs, H.: Nanomedicine–challenge and perspectives. Angewandte Chemie **48**(5), 872–897 (2009). doi:10.1002/anie.200802585

Nanomaterials for Diagnostic Imaging of the Brain

Ellen Qin and Hyunjoon Kong

Abstract Various brain diseases including Alzheimer's disease, stroke, and cancer are major causes of death worldwide. Due to the notion that early diagnosis significantly increases success in treatments, several non-invasive bioimaging modalities such as MRI, CT, and PET are increasingly used to locate pathologic sites in the brain. To further enhance the quality of diagnostic imaging, efforts are incrementally made to couple imaging contrasts of interests to macromolecules or nanoparticles designed to cross over the brain-blood barrier and to bind to pathologic tissue. This chapter will therefore review such important emerging technologies for diagnostic imaging of brain and some preclinical and clinical success, so we can ultimately assist efforts to take diagnosis quality to the next level.

Keywords: Brain · Bioimaging · Nanoparticle · Microfabrication · Brain-blood barrier

1 Introduction

Brain diseases including Alzheimer's disease, stroke, tumors, infections, and seizures are among the major causes of death worldwide. For reference, in 2014, it is estimated that 5 million people have Alzheimer's disease in the United States, and it is listed as the sixth-leading cause of death in the United States. (1) Additionally, in 2015, it is estimated that there will be 23,000 new cases of brain tumors in the United States. (2) Early diagnosis is especially important for providing the best treatment options for those diseases. Current non-invasive bioimaging modalities such as magnetic resonance imaging (MRI), computer

E. Qin · H. Kong(✉)
Department of Chemical and Biomolecular Engineering,
University of Illinois at Urbana-Champaign, Champaign, USA
e-mail: hjkong06@illinois.edu

© Springer International Publishing Switzerland 2016
H. Jo et al. (eds.), *Biomedical Engineering: Frontier Research and Converging Technologies*,
Biosystems & Biorobotics 9,
DOI: 10.1007/978-3-319-21813-7_4

tomography (CT), and positron emission topography (PET) provide valuable information on various pathological and physiological conditions in the brain. Regardless of the imaging modality used, the intensity of the signal should be strong enough to differentiate the area of interest from the surrounding tissues.

To this end, a series of contrast agents have been developed to enhance the contrast of tissues of interest. Success in using these agents greatly relies on the ability to cross over the blood brain barrier (BBB), a tightly packed layer of endothelial cells lining the brain capillaries, and to further bind with target tissue. In this context, efforts have been increasingly made to load the agents on or into nanocarriers engineered to direct the transport of the contrast agents into the brain. In this application, nanocarriers should be small enough to pass through BBB; yet, they should be large enough to transport high payload of contrast agents. Additionally, biomolecular moieties such as peptides, antibodies, and nucleic acids that can specifically bind to target tissue should be conjugated to the nanocarriers, so the contrast agents are delivered to the target tissue and monitor disease progression and treatment outcomes over an extended time period.

This chapter will first outline different modalities of the brain imaging and contrast agents developed to complement these modalities. Second, we will review a series of nanocarriers devised for targeted delivery of contrast agents into pathologic area of the brain, and their roles in improving the diagnosis of brain disease.

2 Bioimaging Modalities for Brain Diagnosis

Computed axial tomography (CT), magnetic resonance imaging (MRI), and positron electron tomography (PET) are among the main bioimaging modalities for brain diagnosis. CT and MRI are often used to provide images of tissue anatomy, whereas PET is used to provide images of metabolic functions of the brain. Hybrid systems such as CT-PET and MRI-PET have also been used to combine the unique features of each modality in order to synergistically improve diagnostic capability. Depending on the application, one modality may be preferred to another one for brain imaging. To identify pathologic sites in these images, a series of contrast agents are being used as a beacon. This section will summarize basic principles of various bioimaging modalities for brain diagnosis, as well as contrast agents that are being used to complement these modalities.

2.1 Computerized Axial Tomography (CT)

First, computed axial tomography (CT or CAT scan) is a non-invasive diagnostic imaging modality that combines X-ray with computer technology to produce 3D reconstructions of tissues from horizontal or axial images of the body. CT is based on X-ray attenuation by tissues, and the difference of absorption between tissues allows for clear imaging of electron-dense materials. The resulting images are used for the diagnosis of disease and assessment of treatment. For instance, CT is being used to examine the integrity of the blood-brain barrier. A leaky blood-brain barrier can occur in neurological disorders such as brain tumors, multiple sclerosis, infectious diseases, and stroke. Its clinical availability, short scanning time and high sensitivity make CT one of the most widely used non-invasive clinical imaging modalities. However, CT presents potential concerns from ionizing radiation, which can lead to allergic reactions or higher risks for cancer. (3) It is also difficult to monitor subtle changes of soft tissues such as brain, thus necessitating the use of contrast agents.

CT contrast agents that are currently being used are iodine-based molecules, due to their capability to effectively absorb X-ray radiation. Even though iodinated agents are generally safe in a body (4), the large amount of contrast agents required for CT imaging can cause severe adverse effects and their use is limited by their short circulation time. Gold nanoparticles have also been considered due to low toxicity and high contrast imaging. Intravenously injected gold nanoparticles have been shown to cross the blood-tumor barrier, and localize specifically to brain gliomas. (5)

2.2 Magnetic Resonance Imaging (MRI)

Due to excellent tissue contrast, high penetration, and resolution, magnetic resonance imaging (MRI) is widely used for diagnosis of brain diseases. MRI is based on nuclear magnetic resonance (NMR) spectroscopy, and it provides three-dimensional images of tissues and organs by detecting the nuclear spin reorientation in an applied magnetic field, specifically by measuring the proton relaxation of water. A strong external magnetic field aligns the hydrogen atoms in the body to the direction of the magnetic field, and the aligned nuclei then start to gyroscopically precess along that direction. When a resonant radio frequency transverse pulse is applied perpendicularly to the magnetic field, the hydrogen atom absorbs the energy from the pulse and proceeds to precess towards the perpendicular plane. As soon as the radio frequency is turned off, the hydrogen atom relaxes and realigns with the magnetic field, back to its lowest energy state. The emitted energy received by a computer is used to reconstruct 2-D or 3-D images of the body.

To visualize small changes in tissues with MRI, contrast agents that induce changes in the interaction with biomolecules in the tissue are often used. The contrast agents change the longitudinal and transfer relaxation times referred as T_1 and T_2, respectively. Paramagnetic ions, such as gadolinium ions (Gd^{3+}), magnesium ions (Mg^{2+}), or manganese ions (Mn^{2+}), increase the contrast in T_1 weighted images. Gadolinium diethylenetriaminepentaacetate (Gd-DTPA) is the most widely used T_1 agent, and can be used to detect the breakage of the blood brain barrier and changes in vascularity, perfusion, and flow dynamics. Several of the gadolinium-based contrast agents (e.g. Prohance, Magnevist, and Dotarem) have been approved by the FDA for clinical use in whole body imaging and the central nervous system. Magnesium-enhanced MRI (MEMRI) has also been used as a contrast agent for animals. Mn^{2+} has a unique T_1 contrast that is sensitive to many biological processes, because Mn^{2+} can enter excitable cells via voltage gated calcium channels and accumulate into active areas in the brain (6). Meanwhile, super paramagnetic materials such as superparamagnetic iron oxide nanoparticles (SPION) and ultrasmall paramagnetic iron oxide nanoparticles (USPION) shorten the T_2 relaxation time to produce darker T_2–weighted images. The sensitivity of the MR images can be improved by the combination of both T_1 and T_2 agents.

Various new technologies have also been developed which relies on magnetic resonance technology. For instance, magnetic resonance angiography (MRA) evaluates blood flow through the arteries, and it can be used to detect stenosis, occlusions, intracranial aneurysms and other vascular abnormalities in the brain. (7) Meanwhile, magnetic resonance spectroscopy (MRS) can assess chemical abnormalities in the body including human immunodeficiency virus (HIV) dementia in the brain, hypoxia, stroke, or head injury. (8) The specific location of the brain where a specific function, such as speech or memory, is impaired can be determined with functional magnetic resonance imaging (fMRI). (9)Finally, dynamic susceptibility contrast perfusion MR imaging (DSC-MRI) is useful for predicting brain tumor grade based on relative cerebral blood volume (rCBV). (10)

2.3 *Positron Emission Tomography (PET)*

Positron emission tomography (PET) scan is a non-invasive diagnostic imaging modality that provides images describing various aspects of hemodynamics, proliferation, and metabolic activity of tissues in the brain. A radiopharmaceutical (i.e., radioactive tracer or radioisotope) is first injected into the vein. As the radioactive tracer undergoes positron emission decay, positrons are emitted. The positron travels through tissues until it collides with a nearby electron. Their mutual annihilation produces two gamma rays ~180° with respect from each other, which are detected by a scintillation detector. A computer then analyzes the gamma rays and creates an image map that presents a level of metabolic function. PET is useful as it can detect biochemical changes due to diseases even before CT or MRI can detect anatomical changes. As a consequence, it can be used to diagnose various diseases in the brain related to metabolic processes, such as Alzheimer's, Parkinson's, and Huntington's disease. (11)

3 Biomolecular Carrier of Imaging Contrast Agents

Biomolecules, such as proteins, lipids, and carbohydrates, are naturally present in the body, which make them suitable as carriers of imaging contrast agents. Biomolecules involved in metabolism are labeled with a radioisotope and observed for metabolic changes that mark disease progression. Alternatively, they can self-assemble to encapsulate contrast agents.

3.1 Radiopharmaceutical Analogs

PET scans rely on radiopharmaceuticals as contrast agents, which can be used for imaging of tumor metabolism. The radioactive tracers are formed by labelling cell metabolites (for example, peptides, glucose, or nucleic acids) with positron-emitting isotopes (^{11}C or ^{18}F) (Fig. 1). Early diagnosis can be accomplished by observing radiopharmaceuticals that undertake metabolic processes associated with the diseases.

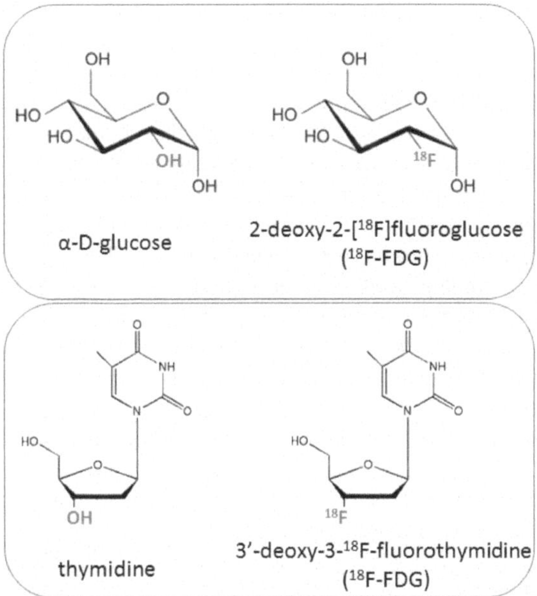

Fig. 1 Examples of radiopharmaceuticals. Radiopharmaceuticals can be formed by labelling cell metabolites such as glucose or thymidine with positron-emitting isotopes (e.g., ^{18}F).

Various PET tracers have been developed for tumors, so they present different aspects of metabolism in tumor. The most widely used radiopharmaceutical in the clinic to detect for tumor is the glucose analog 2-[^{18}F]fluoro-2-deoxy-d-glucose (^{18}F-FDG). (12) Like glucose, ^{18}F-FDG is transported into cells via the glucose transporters and phosphorylated by the enzyme hexokinase. Tumors cells have increased expression of glucose transporters and hexokinase, which in turn, increases ^{18}F-FDG uptake and phosphorylation. Once the ^{18}F-FDG is phosphorylated, the resulting FDG 6-phosphate accumulates in the cells where it can be visualized using PET. The main advantages of ^{18}F-FDG include its relatively long half-life, ease of synthesis, and difference in uptake level between tumor and normal cells; however, FDG has low sensitivity for slowly growing and recurrent tumors, and encounters difficulties in differentiating tumor recurrence from radiation necrosis. (13) Many other tracers have been developed to address certain limitations of FDG in early diagnosis of brain tumors. (14)

Some brain tumors show hypo- or isometabolism on FDG PET, which makes it hard to detect for primary or recurrent tumors. One PET tracer that was investigated for detecting primary or recurrent tumors in PET imaging is ^{11}C-methionine (MET), an amino acid. (15) Unlike FDG, MET was shown to have high sensitivity and good contrast when evaluating the tumors. However, the clinical applications of MET are limited because MET has a shorter half-life, is less available, and is more difficult to synthesize than ^{18}F-FDG. 3,4-dihydroxy-6-18F-fluoro-L-phenylalanine (^{18}F-FDOPA) was also reported to have similar tumor uptake to (MET). (16) The advantage of ^{18}F-FDOPA is that in contrast to ^{18}F-FDG, ^{18}F-FDOPA can visualize both high-grade and low-grade tumors, evaluate recurrent tumors, and differentiate tumor recurrence from radiation necrosis. (17)

The thymidine analog 3'-deoxy-3-^{18}F-fluorothymidine (^{18}F-FLT) has also been developed to image cell proliferation and is also useful for detection of brain tumor. (18) ^{18}F-FLT is taken up by the cells and phosphorylated by thymidine kinase 1 (TK_1), an enzyme expressed during the DNA synthesis phase of the cell cycle. (19) The negatively charged FLT monophosphate that is formed is also trapped within the cell, and thereby results in an accumulation of radioactivity. This makes it useful for distinguishing tumor from normal cells as TK_1 activity is three to four times higher in malignant cells than in benign cells. (20) ^{18}F-FLT is also more sensitive than ^{18}F-FDG for evaluating recurrent high grade gliomas due to the fact that there is less uptake of ^{18}F-FLT than ^{18}F-FDG in normal brain cells. (21)

Not only have radiopharmaceuticals been developed for the diagnosis of tumors, radiopharmaceuticals have also been developed for the detection of Alzheimer's disease. Thioflavin-T, an amyloid binding histological dye, has been modified with a radioisotope for MRI imaging, (22) and the thioflavin-T analog known as Pittsburgh Compound-B (PIB) was the lead compound for the first human trial with amyloid-imaging agents. (23)

4 Nanoparticle Carrier of Imaging Contrast Agents

Nanoparticles functionalized with targeting ligands can also carry imaging agents to tissues of interests while minimally stimulating immune system and also extending half-life of payloads. They can even deliver both diagnostic and therapeutic agents, thus enabling image-guided therapies, termed as theranostics. (24) Such functionality is highly valuable for diagnosis and treatment of brain diseases. In many instances, a wide array of peptides, lipids, and natural or synthetic polymers have been used as a building block to modulate solubility, reduce toxicity, prevent magnetic saturation, and also improve circulation time of the nanoparticle. Contrast agents are either covalently conjugated with the polymer or encapsulated within the nanoparticle core. Peptides and antibodies are further conjugated onto the nanocarriers to increase target specificity.

4.1 Liposomal Carriers

Liposomes are vesicles with a phospholipid bilayer and can be used to encapsulate imaging agents or drugs. Saito *et al.* demonstrated that convection-enhanced delivery of liposome encapsulated with Gadoteridol (a gadolinium chelate) was could track distribution in various regions of the brain using MRI in real time (Fig. 2). (25) The liposomes were prepared with a phospholipid, cholesterol, and a pegylated phospholipid. Gadolinium (Gd) by itself has a short blood half-life, and requires repeated injections (26) at high dosages, which increases the risk of adverse effects. The retention time of the Gd post-infusion increased when the Gd was in the liposome, and despite the enhanced distribution, the toxicity of the liposomal Gd was almost negligible. The results showed that liposomes were a promising carrier for gadolinium based contrast agents for MRI, and can be used to monitor liposomal drug delivery to tumor tissues in real time.

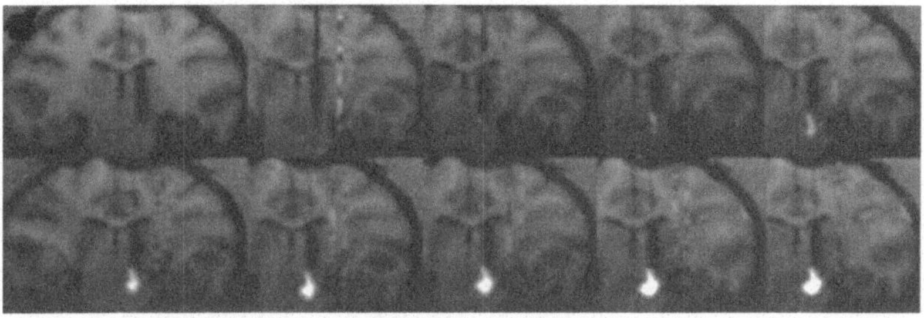

Fig. 2 Real time MRI monitoring of liposomal Gd at the brain stem. The T_1 weighted MR images were taken in ten minute intervals. (25)

4.2 Nanoparticles Enabling Targeting and Theranostics

The surface of nanoparticles that are used as a contrast agent of MR imaging are often modified with a biocompatible material (e.g. poly(ethylene glycol)) and targeting ligands (e.g. antibody, peptide, or glycoprotein). (27) For example, pathogenic amyloid beta oligomers (AβOs) had been shown to be responsible for memory loss and to appear early in Alzheimer's disease. (28) Thus, AβO-specific antibodies were coupled with magnetic nanoparticles to create a molecular MRI contrast probe that can detect Alzheimer's disease. (29) Mono-dispersed, nitro-dopamine (nDOPA) was used as a surface anchoring group and poly(ethylene glycol) (PEG) was used as a spacer, and the antibody was attached via carboxyl-amine conjugation. The results of the study showed that the magnetic nanoparticle was able to bind specifically to AβO in the brain and to give off MRI signals (Fig. 3).

The theranostic property of magnetic nanoparticles has also been explored as a way to diagnose and to treat brain tumors. To this end, nanocarriers were devised to transport both therapeutic and contrast agents to brain tumor. Liu *et al.* have utilized polymer-coated magnetic nanoparticles to deliver anticancer drug to the brain and simultaneously provide MRI contrast agent for real time detection (Fig. 4). (30) The polymer [poly[aniline-co-N-(1-one-butyric acid)] aniline (SPAnH) was used to encapsulate superparamagnetic iron oxide nanoparticles (SPIONs). Epirubicin, a cytotoxic anticancer agent used to treat malignant tumors, was then immobilized covalently on the SPAnh/SPION nanoparticle. Their study combined focused ultrasound (FUS) with magnetic targeting (MT) to deliver nanoparticles across the BBB. FUS locally disrupted the integrity of the BBB for several hours, and allow the nanoparticles to passively transport across the barrier, and a magnetic field was also applied to actively enhance their accumulation at the target site.

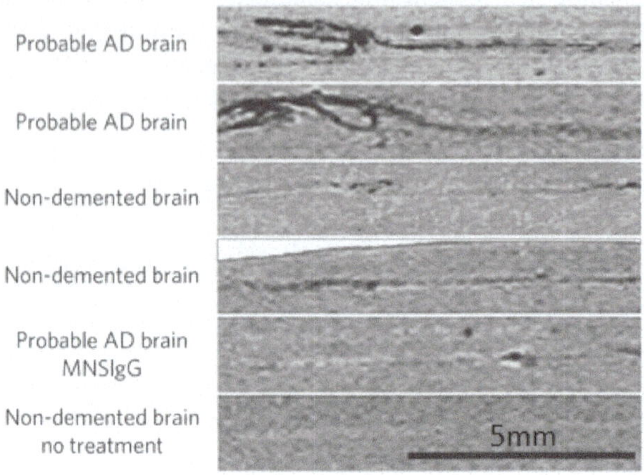

Fig. 3 MRI of the brain injected with AβO-specific magnetic nanoparticles. A strong signal was detected in brain with Alzheimer's (top two rows), and the signal was greatly reduced in age-matched patients without Alzheimer's (next two rows). No signal was found with magnetic nanoparticles conjugated with non-specific IgG and without nanoparticles (last two rows). (29)

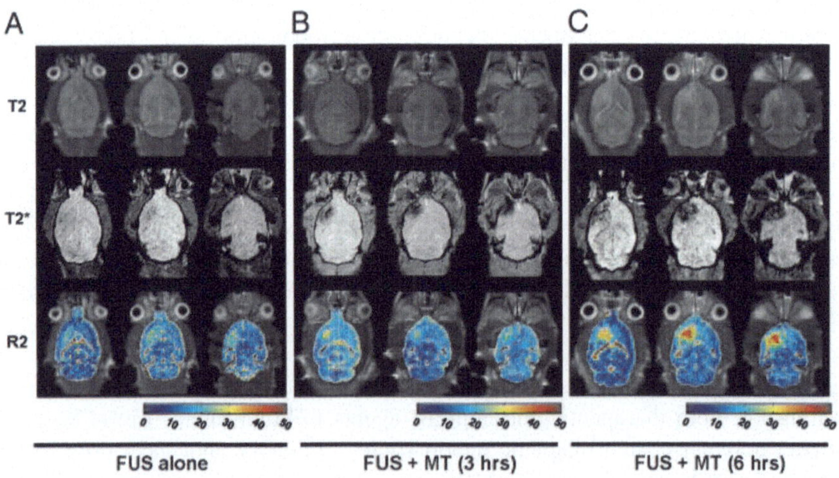

Fig. 4 In vivo imaging of SPIONs in the brain using **(a)** FUS alone **(b)** FUS and MT after 3 hours and **(c)** FUS and MT after 6 hours. Top row(T2) shows T2 weighted MR images, middle row (T2*) shows T2* weighted images, and the bottom row(R) shows T2 and T2* combined. (30)

Theranostic nanocarriers were further modified to enable active targeting. Transactivator of Transcription (TAT) peptides were attached to gold nanoparticles via PEG to enhance endothelial cell membrane interaction and improve penetration into brain tumors. (31) The study showed that doxorubicin (Dox) can be conjugated to the gold nanoparticles through an acid-labile hydrazone linker, or gadolinium (Gd) can be chelated to the surface of gold nanoparticles coated with thiol-modified diethylenetriaminepentaacetic acid (DTPA). In both instances, the gold nanoparticles were able to cross the BBB and improve brain imaging and treatment (Fig. 5).

In another instance, nanoparticle surface was modified with F3 peptides that can target tumor vasculature, while the core was used to encapsulate photodynamic therapy drugs (e.g., Photofrin) and MRI contrast agent (e.g., SPIONs). (32) The nanoparticle core was synthesized with polyacrylamide, which was embedded with Photofrin and SPIONs, and a PEG linker was used to attach the F3 peptide to the core matrix. As a consequence, the therapeutic agent could be selectively delivered to glioma cells, and the SPIONs allowed the glioma cells in mice to be imaged with MRI. When activated by light irradiation, Photofrin produced a cytotoxic, short-lived species that treated glioma.

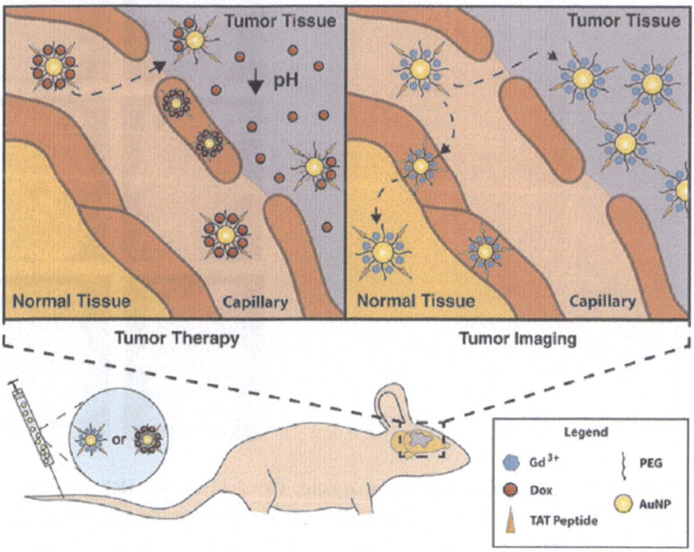

Fig. 5 Scheme of gold nanoparticles functionalized to deliver therapeutic and contrast agents. Dox, a chemotherapeutic drug, was conjugated onto the nanoparticle via a pH sensitive linker (top left), and the contrast agent Gd was chelated onto the nanoparticle surface (top right). The gold nanoparticles circulate in the blood, pass through the BBB and accumulate in tumors via the enhanced permeation retention (EPR) effect. (31)

4.3 Nanoparticles Enabling Multimodal Imaging

Nanoparticles were also assembled to enable multimodal imaging by loading different imaging contrast agents. For example, MRI-fluorescence imaging was extensively studied because it allowed for disease localization, identification, and surgical procedure planning. (33,34) To support this dual imaging, a multifunctional probe that can target glioma cells and can be detected by both MRI and fluorescence microscopy was synthesized. (34) The probe was synthesized by chemically conjugating glioma tumor-targeting molecule (e.g., chlorotoxin (Cltx)) and the near-infrared fluorescing (NIRF) peptide (e.g., Cy5.5) to the surface of SPIONs covalently bounded with PEG. When 9L glioma cells and noncancerous rat cardiomyocyte cells were both incubated with the nanoparticles, the MR image showed that nanoparticles had a higher specificity for the glioma cells than the normal cells. In another study, Cy5.5 was covalently attached to SPIONs coated with cross-linked dextran. (35) *In vitro* studies showed that the nanoparticle was internally sequestered in both brain tumors and activated microglia/macrophages, and it can be used intraoperatively to delineate brain tumor.

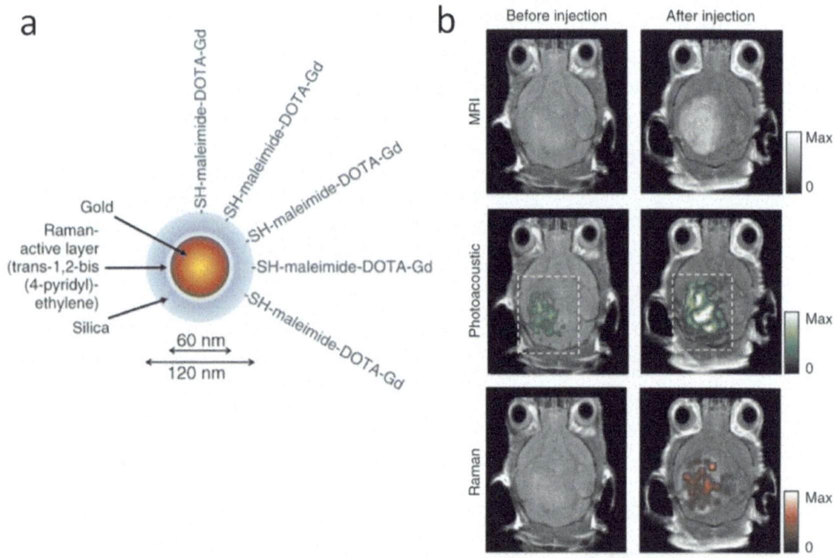

Fig. 6 **(a)** Scheme of triple-imaging nanoparticle. **(b)** 2-D axial MRI, photoacoustic, and Ramen images of tumor in brain before and after injection of nanoparticle. Tumor margin is clearly defined for all three modalities. (36)

Separately, nanoparticles enabling triple imaging including MR imaging, photacoustic imaging, and Raman imaging were developed to accurately delineate brain tumors margins in living mice before and during surgery. (36) The nanoparticles were synthesized by covering a gold nanoparticle with a thin Raman-active outer layer containing the molecular tag *trans*-1,2-bis(4-pyridyl)-ethylene, and was further protected by a silica coating (Fig. 6). Maleimide-DOTA-Gd was then conjugated to the thiol (SH) group on the silica to form the resulting triple-modality nanoparticle. The resulting nanoparticles were intravenously injected into a mouse with glioblastoma and exclusively accumulated in the tumor. Without specific targeting ligands, the nanoparticles relied on the enhanced permeability retention (EPR) effects, which allowed them to diffuse across the disrupted BBB and be retained by the tumor. Minimal fraction of nanoparticles was found in the surrounding healthy tissue. According to *in vivo* kinetic studies, the resulting nanoparticles sustainably stayed in the brain tumor, unlike free Gd-DOTA that rapidly diffused through the renal clearance.

5 Conclusion

In summary, this chapter described a series of biomolecular and nanocarriers of various imaging modalities that have been used for the diagnosis of brain diseases including brain cancer and Alzheimer's disease. Many of these carriers take into the advantage of peptides and antibodies to target various biomarkers that are overexpressed or present for certain diseases in the brain. These carriers served to extend half-life of imaging contrast agents and also retention time in target tissue, such that clinicians do not only detect pathologic brain tissue but also use the images to enhance surgical outcomes. The reduced toxicity and host response were additional advantages attained with nanoparticles. Moreover, these nanoparticles were functionalized to carriers enabling theranostic and multimodal imaging. Such materials will substantially serve to improve diagnosis of other brain diseases including stroke, infection, and ultimately the quality of clinical treatments.

Acknowledgement This work was supported by the National Institutes of Health (1R01 HL109192 to H.J.K.) and the National Science Foundation (IGERT-CMMB to E.Q. and H.J.K).

References

1. Alzheimer's Disease Facts and Figures **10**(2) (2014)
2. Siegel, R., Ma, J., Zou, Z., Jemel, A.: Cancer statistics. CA-Cancer J. Clin. (2014)
3. Brenner, D., Elliston, C., Hall, E., Berdon, W.: Estimated Risks of Radiation - Induced Fatal Cancer from Pediatric CT. AJR. **176**, 289–296 (2000)
4. Wang, C., Cohan, R., Ellis, H., Adusumilli, S., Dunnick, N.: Frequency, Management, and Outcome of Extravasation of Nonionic Iodinated Contrast Medium in 69 657 Intravenous Injections **243**(1), 80-87 (2007)
5. Hainfeld, J., Smilowitz, H., O'Connor, M., Dilmanian, F., Slatkin, D.: Gold nanoparticle imaging and radiotherapy of brain tumors in mice. Nanomedicine **8**(10), 1601–1609 (2013)
6. Silva, A., Lee, J., Aoki, I., Korestsky, A.: Manganese-enhanced magnetic resonance imaging (MEMRI): methodological and practical considerations. NMR Biomed. **17**, 532–543 (2004)

7. Husson, B., Rodesch, G., Lasjuanias, P., Tardieu, M., Sebire, G.: Magnetic resonance angiography in childhood arterial brain infarcts a comparative study with contrast angiography. Stroke. **33**, 1280–1285 (2002)
8. Lin, A., Ross, B., Harris, K., Wong, W.: Efficacy of Proton Magnetic Resonance Spectroscopy in Neurological Diagnosis and Neurotherapeutic Decision Making. NeuroRx. **2**(2), 197–214 (2005)
9. Binder, J., Frost, J., Hammeke, T., Cox, R., Rao, S., Prieto, T.: Human Brain Language Areas Identified by Functional Magnetic Resonance Imaging. J. Neurosci. 353-362 (1997)
10. Law, M., Yang, S., Babb, J., Knopp, E., Golfinos, J., Zagzag, D., et al.: Comparison of cerebral blood volume and vascular permeability from dynamic susceptibility contrast-enhanced perfusion MR imaging with glioma grade. AJNR Am. J. Neuroradiol. **25**(5), 746–755 (2004)
11. Herholz, K.: PET studies in dementia. Ann. Nucl. Med. **17**(2), 79–89 (2003)
12. Brindle, K.: New approaches for imaging tumor responses to treatment. Nature **8** (2008)
13. Ricci, P., Karis, J., Heiserman, J., Fram, E., Bice, A., Drayer, B.: Differentiating recurrent tumor from radiation necrosis: time for re-evaluation of positron emission tomography? Am. J. Neuroradiol. **19**, 407–413 (1998)
14. Chen, W.: Clinical Applications of PET in Brain Tumors. J. Nucl. Med. **48**(9), 1468–1481 (2007)
15. Chung, J., Kim, Y., Kim, S., Lee, Y., Paek, S., Yeo, J., et al.: Usefulness of 11C-methionine PET in the evaluation of brain lesions that are hypo- or isometabolic on 18F-FDG PET. Eur. J. Nucl. Med. **29**, 176–182 (2002)
16. Becherer, A., Karanikas, G., Szabó, M., Zettinig, G., Asenbaum, S., Marosi, C., et al.: Brain tumour imaging with PET: a comparison between [18F]fluorodopa and [11C]methionine. Eur. J. Nucl. Med. Mol. Imaging **30**(11), 1561–1567 (2003)
17. Chen, W., Silverman, D.H.S., Delaloye, S., Czernin, J., Kamdar, N., Pope, W., et al.: 18F-FDOPA PET Imaging of Brain Tumors: Comparison Study with 18F-FDG PET and Evaluation of Diagnostic Accuracy. J. Nucl. Med. **47**(6), 904–911 (2006)
18. Shields, A., Grierson, J., Dohmen, B., Machulla, H., Stayanoff, J., Lawhorn-Crews, J., et al.: Imaging proliferation in vivo with [F-18]FLT and positron emission tomography. Nat. Med. **4**(11), 1334–1336 (1998)
19. Been, L., Suurmeijer, A.J., Cobben, D., Jager, P., Hoekstra, H., Elsinga, P.: [18F]FLT-PET in oncology: current status and opportunities. Eur. J. Nucl. Med. Mol. Imaging **31**(12), 1659–1672 (2004)
20. Boothman, D., Davis, T., Sahijdak, W.: Enhanced expression of thymidine kinase in human cells following ionizing radiation. Int. J. Radiat. Oncol. Bio. Phys. **30**(2), 391–398 (1994)
21. Chen, W., Cloughesy, T., Kamdar, N., Satyamurthy, N., Bergsneider, M., Liau, L., et al.: Imaging Proliferation in Brain Tumors with 18F-FLT PET: Comparison with 18F-FDG. J. Nucl. Med. **46**, 945–952 (2005)
22. Mathis, C., Bacskai, B., Kajdasz, S., McLellan, M., Frosch, M., Hyman, B., et al.: A lipophilic thioflavin-T derivative for positron emission tomography (PET) imaging of amyloid in brain. Bioorg. Med. Chem. Lett. **12**(3), 295–298 (2002)
23. Klunk, W., Engler, H., Nordberg, A., Wang, Y., Blomqvist, G., Holt, P., et al.: Imaging Brain Amyloid in Alzheimer's Disease with Pittsburgh Compound-B. Ann. Neurol. **55**, 306–319 (2004)

24. Bhojani, M.S., Van Dort, M., Rehemtulla, A., Ross, B.D.: Targeted Imaging and Therapy of Brain Cancer Using Theranostic Nanoparticles. Mol. Pharmaceutics **7**(6), 1921–1929 (2010)
25. Saito, R., Krauze, M., Bringas, J., Noble, C., McKnight, T., Jackson, P., et al.: Gadolinium-loaded liposomes allow for real-time magnetic resonance imaging of convection-enhanced delivery in the primate brain. Exp. Neurol. **196**, 381–389 (2005)
26. Ludermann, L., Hamm, B., Zimmer, C.: Pharmacokinetic analysis of glioma compartments with dynamic Gd-DTPA-enhanced magnetic resonance imaging. Magn. Reson. Imaging **18**(10), 1201–1214 (2000)
27. Gupta, A., Gupta, M.: Synthesis and surface engineering of iron oxide nanoparticles for biomedical applications. Biomaterials **26**(18), 3995–4021 (2005)
28. Lacor, P., Buniel, M., Chang, L., Fernandez, S., Gong, Y., Viola, K., et al.: Synaptic targeting by Alzheimer's-related amyloid beta oligomers. J. Neurosci. **45**, 10191–10200 (2004)
29. Viola, K., Sbarboro, J., Sureka, R., De, M., Bicca, M., Wang, J., et al.: Towards non-invasive diagnostic imaging of early-stage Alzheimer's disease. Nat. Nanotechnol. **10**, 91–98 (2015)
30. Liu, H., Hua, M., Yang, H., Huang, C., Chu, P., Wu, J., et al.: Magnetic resonance monitoring of focused ultrasound/magnetic nanoparticle targeting delivery. PNAS. **107**(34), 15205–15210 (2010)
31. Cheng, Y., Dai, Q., Morshed, R., Fan, X., Wegscheid, M., Wainwright, D., et al.: Blood-brain barrier permeable gold nanoparticles: An efficient delivery platform for enhanced malignant glioma therapy and imaging. Small. **10**(24), 5137–5150 (2014)
32. Reddy, G., Bhojani, M., McConville, P., Moody, J., Moffat, B., Hall, D., et al.: Vascular Targeted Nanoparticles for Imaging and Treatment of Brain Tumors. Clin. Cancer. Res. **12**(22), 6677–6686 (2006)
33. Kircher, M., Mahmood, U., King, R., Weissleder, R., Josephson, L.: A Multimodal Nanoparticle for Preoperative Magnetic Resonance Imaging and Intraoperative Optical Brain Tumor Delineation. Cancer Res. **63**, 8122–8125 (2003)
34. Veiseh, O., Sun, C., Gunn, J., Kohler, N., Gabikian, P., Lee, D., et al.: Optical and MRI Multifunctional Nanoprobe for Targeting Gliomas. Nano Lett. **5**(6), 1003–1008 (2005)
35. Trehin, R., Figueiredo, J., Pettet, M., Weissleder, R., Josephson, L., Mahmood, U.: Fluorescent Nanoparticle Uptake for Brain Tumor Visualization. Neoplasia. **8**(4), 302–311 (2006)
36. Kircher, M., de la Zerda, A., Jokerst, J., Zavaleta, C., Kempen, P., Mittra, E., et al.: A brain tumor molecular imaging strategy using a new triple-modality MRI-photoacoustic-Ramen nanoparticle. Nat. Med. **18**(5), 829–834 (2012)

Part II
Regenerative Medicine

Cellular Interactions with Self-assembled Biomaterials and Composites for Bone Tissue Engineering

Jeremy B. Vines, Grant C. Alexander, Lily Deng, Patrick T J Hwang, Dhruv Patel and Ho-Wook Jun

Abstract Bone is an extremely important tissue with great versatility that provides support, movement, and protection for the organs of the body. Various bone related pathologies such as osteoporosis have the potential to create large defects that do not have the capability to heal with only basic clinical intervention. In these circumstances, the principles of tissue engineering are utilized in order to provide additional material to either replace or repair the damaged bone regions. Tissue engineering involves the utilization of various combinations of growth factors, cells, and biomaterials to recreate organs or lost tissue regions. Currently, the most prominent materials for this purpose are limited by issues such as availability for autografts, possible antigenicity for allografts, or insufficient of bioactivity as is the case with many synthetic materials. To address these concerns, research has shifted focus towards development of synthetic materials that faithfully mimic native bone's natural structure and function. Bone formation and regeneration are dependent upon the dynamic interactions between bone cells and extracellular matrix components. Therefore, it is important that the native extracellular matrix components are recapitulated to the greatest extent possible. Specifically, this new paradigm shift requires the development of scaffolds that recreate both the structural and functional qualities of bone tissue. These materials are intended to enhance the interactions of bone forming cells and biomaterials. Adopting this strategy, many groups have developed self-assembling materials to

---

J.B. Vines · G.C. Alexander · L. Deng · P. T J Hwang ·
D. Patel · H.-W. Jun(✉)
Department of Biomedical Engineering, University of Alabama School of Medicine,
University of Alabama at Birmingham, Birmingham, AL 35294, USA
e-mail: hwjun@uab.edu

J.B. Vines
NuTech Medical, Birmingham, AL 35216, USA

© Springer International Publishing Switzerland 2016

H. Jo et al. (eds.), *Biomedical Engineering: Frontier Research and Converging Technologies*,
Biosystems & Biorobotics 9,
DOI: 10.1007/978-3-319-21813-7_5

serve as functionalized scaffolds for enhancing the dynamic interactions between the bone cells and the biomaterials for subsequent bone tissue regeneration. In this chapter, the underlying principles of bone tissue engineering, bone cell interactions, various bioactive, self-assembling biomaterial technologies, and the means by which they improve interactions with bone forming cells are discussed.

1 Introduction

Bone is a dense connective tissue comprised of a multi-component extracellular matrix. Bone is an extremely important tissue with great versatility that provides support, movement, and protection for the organs of the body [1]. The extracellular matrix (ECM) is comprised of organic (collagen I, bone sialoprotein, osteopontin, and vitronectin, etc.), cellular (osteoblasts, osteoclasts, etc.) and inorganic hydroxyapatite (HA) components [2-4]. The term ECM refers to the extracellular part of animal tissue that usually provides structural support to the cells in addition to performing various other important functions. The ECM can be thought of as a structural scaffold made up of fibrous and connective components that give the cells of the body something to hold on to. Bone nanostructure is made from the mineralization of collagen fibrils with HA nanocrystals. These mineralized fibrils then bundle together into structural units to make lamellar bone (Figure 1).

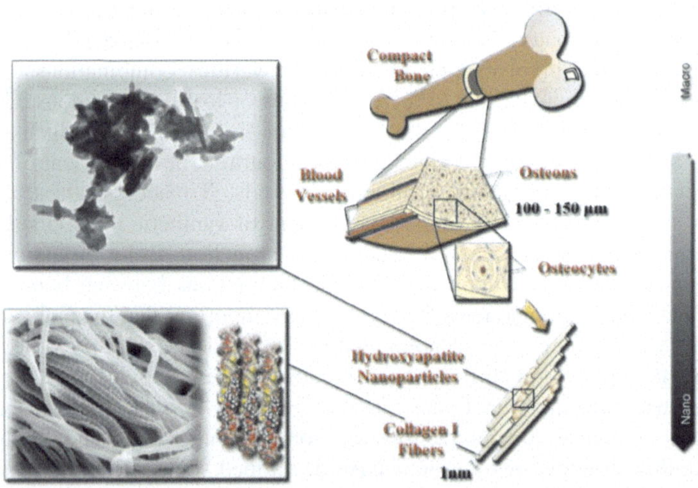

Fig. 1 *Illustration of Macroscopic Bone Composition and Structure.* Adapted from "Bonzani IC, George JH, Stevens MM. Novel materials for bone and cartilage regeneration. *Current opinion in chemical biology*. 2006;10:568-575." Copyright © 2006 Elsevier Ltd. All rights reserved.

The organic component of the extracellular matrix (making up ~20% of bone by weight[4]) is a system of polysaccharides and proteins that directly regulates cellular behaviors through direct interactions between the cell and the extracellular matrix [5]. These behaviors will be discussed in detail within a later section. The main organic component of the bone extracellular matrix is Collagen I.

The inorganic extracellular matrix component is made of up hydroxyapatite $[Ca_{10}(PO_4)_6(OH)_2]$. Hydroxyapatite is the main component of bone (making up ~70% of the native ECM [2-4] by weight) and serves as its load bearing constituent, providing structural support that is comparable to steel by weight [6].

The third main component of bone is cells, which inhabit the bone and maintain its structural integrity. In this regard, bone is populated by three main cell types: osteoblasts, osteoclasts, and osteocytes [7]. Osteoblasts are mononucleated cells that are responsible for bone formation. They are a specialized form of fibroblasts and are responsible for producing the main organic (collagen, bone sialoprotein) and inorganic (hydroxyapatite) components of the bone matrix. These are essentially bone "forming" cells. The second type, osteoclasts, are a type of bone cell that removes bone tissue by removing its mineralized matrix and breaking up the organic bone in a process known as bone resorption. These cells can be identified by the fact that they contain multiple nuclei. The third cell type, osteocytes, are quiescent cells responsible for mediating mineral production. It is thought that they control mineralization in response to mechanical stimuli. In this regard, they become mechanosensory detectors and modulate bone formation and breakdown by osteoblasts and osteoclasts. Hence, if bones are not submitted to weight bearing loads, osteocytes may signal osteoclasts to become more active and osteoblasts to become less active [8].

In addition to the three main cell types above, there are two main adult stem cell types from which the main bone cell types are derived. These are known as mesenchymal stem cells, and hematopoietic stem cells (HSCs).

Mesenchymal stem cells (MSCs) are multipotent cells that can differentiate into a variety of cell types, including: osteoblasts, chondrocytes (cartilage cells), and adipocytes (fat cells), etc. [9]. These progenitor cells can be found lining the exterior of the bone surface underneath the periosteum and within the bone marrow. They can also be found in other connective tissues within the body such as fat. Hematopoietic stem cells (HSCs) also spelled haematopoietic stem cells, are multipotent stem cells that give rise to all the blood cell types from the myeloid (monocytes, macrophages, neutrophils, basophils, eosinophils, erythrocytes, and megakaryocytes/platelets, dendritic cells), and lymphoid lineages (T-cells, B-cells, NK-cells) [10]. Additionally, these cells are responsible for differentiating into osteoclasts. When exposed to the proper environment, these stem cells can differentiate into the main cell types in bone. In an osteogenic environment, MSCs become osteoblasts and HSCs become osteoclasts.

Bone macrostructure (Figure 2) consists of a dense exterior called compact bone with the less dense interior segment called trabecular or spongy bone [11]. Bone derives its nutritional sustenance from an extensive vascular network in which blood vessels pervade the bone through Haversian and Volkmann's canals. Compact bone is made up of osteons, which serve as its basic functional unit. Osteons are long, cylindrical structures oriented parallel to the long axis of the bone and to the main compression stresses. Functionally, osteons can be viewed as miniature weight bearing pillars made up of layers of concentric tubes known as lamella. In-between each layer of tubes are lacunae, where osteocytes can be found.

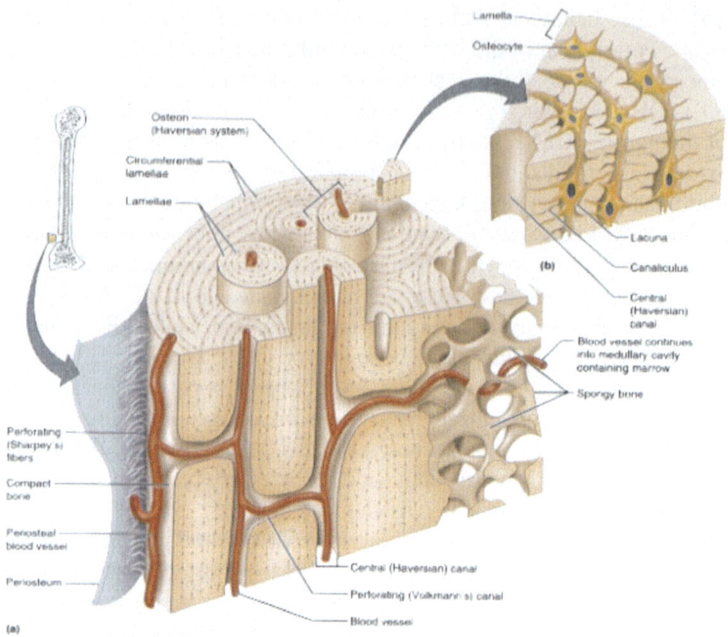

Fig. 2 *Illustration of Macroscopic Bone Composition and Structure.* From : Marieb, E. N., J. Mallatt, et al. (2013). Human Anatomy 7th ed. San Francisco, California, Pearson Benjamin Cummings. Copyright © 2013 Pearson Benjamin Cummings. All rights reserved.

During development, bone is formed via one of two pathways. These pathways are known as endochondral ossification and intramembranous ossification [12]. Endochondral ossification primarily involves the development, growth, and repair of long bones while intramembranous ossification is reserved for the development, growth, and repair of bones of the head. The main difference between the two processes is that during endochondral ossification, osteoprogenitors will first differentiate into chondrocytes and deposit cartilage before becoming hypertrophic, recruiting osteoblasts, and subsequently mineralizing the matrix. In contrast, intramembranous ossification does not involve this intermediate step and directly forms mineralized matrix without prior chondrogenic differentiation.

Physiologically, when bone is fractured, the surrounding vascular tissue is ruptured, causing the tissues to bleed. The blood will then coagulate to form a large clot around the fracture within the time frame of a week. This large clot is known as a fracture hematoma. Anywhere from two to three weeks following the fracture, blood vessels will penetrate the matrix of the hematoma, recruiting phagocytes to remove non-viable material; furthermore, recruited fibroblasts will remodel the ECM and form collagen fibers, resulting in a rubbery callus consisting of hyaline cartilage. Around the beginning of week four, osteoblasts are recruited to mineralize this collagen matrix with HA nanocrystals, leading to the formation of immature bone within 16 weeks. Over time, this immature bone matrix is remodeled to create mature lamellar bone [13-17].

The remodeling of bone structure results from the activities of osteoblasts and osteoclasts, which provide constant deposition and re-adsorption of bone matrix, respectively. This occurs through a phenomenon known as Wolff's Law: in response to the application of weight bearing loads, often through daily routines such as walking or exercising, osteoblasts and osteoclasts will remodel bone structure [18,19].

2 Cellular Interactions

All cells in general interact with their environment through either direct contact, or the detection of soluble molecules secreted by other cells in the body. These interactions guide multiple cellular behaviors such as mitosis, migration, and differentiation [20]. No matter what the communication type, the overall concept is the same. On the outer membrane of cells are molecules known as cell receptors. Think of a cell receptor like a lock that can only be unlocked with a specific key. The keys that unlock these cell receptors will be called signaling molecules. These signaling molecules can come in many forms. They can be a free floating molecule (soluble signal), or attached to something else such as the surface of a cell or an extracellular matrix protein such as collagen. Each signaling molecule is specific to certain receptors built into the cell membrane, just as a key is specific to a certain lock [21].

Every cell response has a specific pattern. A signaling molecule first binds to a cell receptor. This then unlocks a cascade which alters gene expression, resulting in a change in cell behavior. As mentioned earlier, there are two main classifications of cellular communication. The first is direct contact, and the second is by soluble signals.

For cellular communication mediated by direct contact, there are two sub classifications. These are cell to extracellular matrix and cell to cell communication. For cell to extracellular matrix signaling, cells interact with the extracellular matrix via various signaling mechanisms in which cells recognize and bind to signaling molecules transcribed into organic extracellular matrix proteins via integrins [22,23]. An integrin is a trans-membrane cell receptor capable of binding to signaling molecules built into extracellular matrix proteins outside of the cell. Cellular adhesive ligands are signaling molecules built into various extracellular matrix proteins and are designed to bind to the integrins, leading to various cellular responses [24]. Cellular adhesive ligands will be discussed in more detail later (See "Principles of Bone Tissue Engineering" section). Integrins also serve as anchoring molecules, allowing the cells to attach to fixate themselves to the extracellular matrix. This is shown in Figure 3 [25].

In instances of cell to cell signaling, molecules built onto the surface of one cell bind to a cell receptor built onto the surface of another cell. For soluble factor mediated signaling, there are three sub-classifications. These sub-classifications are paracrine, autocrine, and endocrine soluble signaling pathways [26]. A paracrine factor is a signaling molecule that is secreted by a cell to act on other cells within its immediate vicinity. If a signaling molecule is released by the same cell that the molecule acts on, it is called an autocrine signal. The overall concept of endocrine signaling is the same as that of paracrine signaling. The main distinction, however, is the distance over which the signaling molecule travels. Paracrine signaling is described as signaling that takes place over a short distance. Endocrine signaling is a signal that travels from one part of the body to the other.

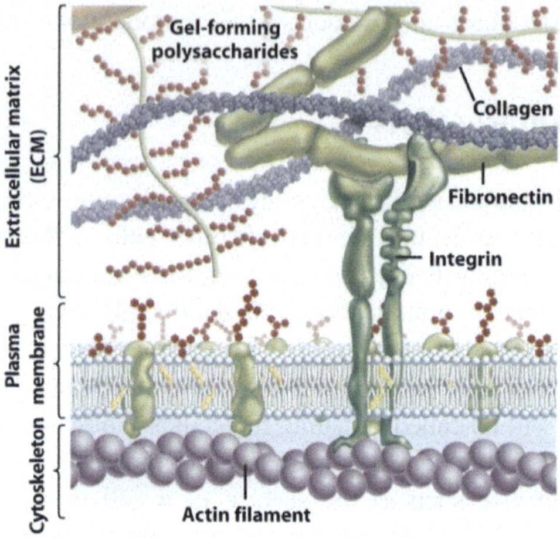

Fig. 3 *Illustration of Cell to Extracellular Matrix Signaling.* From: Freeman, S. (2005). Biological Science, 2nd Ed. New Jersey, Prentice Hall. Copyright © 2005 Pearson Prentice Hall. All rights reserved.

Paracrine, Autocrine, Endocrine, and Cell to Cell signaling pathways are demonstrated in Figure 4 [27].

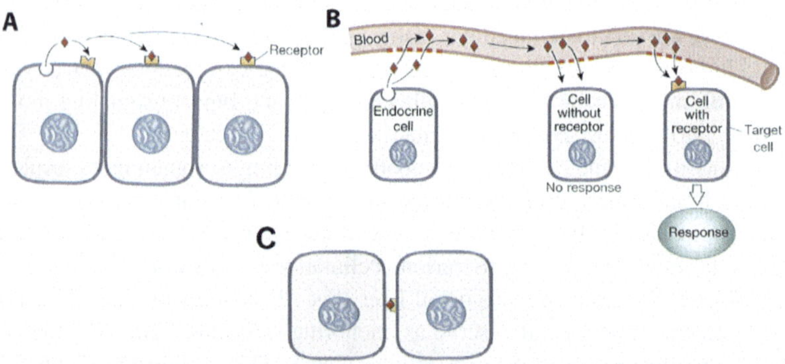

Fig. 4 *Cellular Signaling.* A) Autocrine signals act on the same cell that secreted them while paracrine signals act on adjacent cells in the local vicinity. B) Endocrine signaling takes places over longer distances than paracrine or autocrine signaling. Like all signaling modalities, only cells with receptors specific to certain signaling molecules will respond. C) Direct Contact Cell to Cell Signaling requires interactions between membrane molecules on two different cells. The signaling molecule is depicted as a red square with the cell receptor depicted in yellow. From: Silverthorn, D. U. (2007). Human Physiology: An Integrated Approach, 4th Ed. San Francisco, CA, Pearson Benjamin Cummings. Copyright © 2007 Pearson Benjamin Cummings. All rights reserved.

3 Current Biomaterials for Bone Grafting: Strengths and Weaknesses

Depending on fracture size and severity, bone may heal in different ways. If the defect is merely small or acute, bone tissue is usually able to regenerate, regaining the full function of its state prior to injury. If, however, the fracture results in a defect above a critical size, the tissue may be incapable of filling the segmental void. This can result in a permanent loss of function. Therefore, in order to provide proper healing across a large defect, some sort of filler material that is able to bridge the gap between the disjoined bone segments must be added. This material allows osteogenic progenitor cells to infiltrate the defect space and regenerate the lost bone. There are currently three categories of filler materials utilized in segmental defects: allografts, autografts, and synthetics.

Autografts are bone tissue harvested from a donor site within the same individual (typically from the iliac crest of the pelvis) and transplanted to the fracture site. The innately autologous nature of these scaffolds combined with their highly osteoinductive properties has led them to be considered the "gold standard". However, while they are useful for providing a filler material that is conducive to bone regeneration without any risk of rejection, material is limited and the removal of bone from the donor site often results in residual pain and morbidity [28].

There are currently three main autologous graft types used as adjuncts for bone repair. These types are as following: traditional bone grafts, bone marrow aspirate concentrate (BMAC) and platelet rich plasma (PRP). Traditional bone grafts are typically harvested from the iliac crest or another de-localized region within the body and then placed at the site of the defect. BMAC involves a procedure in which a physician aspirates (removes by suction) a small amount of the patient's bone marrow [29]. Following this, the bone marrow is placed into a centrifuge, which then concentrates stem cells present within the bone marrow. This concentrate is then combined with another material such as a synthetic or an allograft and implanted into the surgical site of interest. PRP is blood plasma that has been enriched with platelets. As a concentrated source of autologous platelets, PRP contains (and releases through degranulation) several different growth factors and other cytokines that stimulate healing of bone and soft tissue [30,31].

If there is an insufficient amount of material at the donor site, an alternative treatment must be adopted. In allografting, bone is harvested from a cadaver, and sterilized before implantation into the individual. Unfortunately, sterilizing the bone can cause it to lose much of its biological signaling capabilities. Furthermore, these sterilization measures are not always successful and rejection can still occur [32].

The most common allogeneic graft types include demineralized bone matrices, and other allogeneic bone matrices of various configurations including both whole and ground cancellous/trabecular and cortical/compact bone grafts. Demineralized bone matrix (DBM) is allograft bone that has had the inorganic mineral removed, leaving behind the organic collagen matrix. It was first discovered by Marshall

Urist in 1965 that the removal of the bone mineral exposes more biologically active bone morphogenetic proteins [33]. These growth factors modulate the differentiation of progenitor cells into osteoprogenitor cells, which are responsible for bone and cartilage formation. As a result of the demineralization process, DBM is more biologically active than non-demineralized bone grafts; conversely the mechanical properties are significantly diminished. Ground cancellous matrix is an allograft containing both the organic and inorganic facets of bone. Cancellous bone is utilized due to its improved handling characteristics and the fact that it has bioactivity similar to demineralized bone matrix [34]. Synthetic materials are materials which have been created in a laboratory or industrial environment and not harvested from another biological source. To bypass some of the difficulties associated with allografts and autografts, synthetic materials can be utilized instead.

The main synthetic materials currently used for bone tissue engineering purposes are calcium phosphates such as hydroxyapatite and tri-calcium phosphate [34]. Hydroxyapatite (HA), is a naturally occurring mineral form of calcium apatite with the formula $Ca_5(PO_4)_3(OH)$, but is usually written $Ca_{10}(PO_4)_6(OH)_2$ to denote that the crystal unit cell comprises two entities. As mentioned earlier, it is structurally similar to the mineral component of bone. HA has a slow degradation profile with degradation occurring on the order of months or years. It has long been determined to be osteoconductive. Tricalcium phosphate is a calcium salt of phosphoric acid with the chemical formula $Ca_3(PO_4)_2$. It is also known as tribasic calcium phosphate and bone phosphate of lime (BPL). Calcium phosphate is one of the main combustion products of bone. In general cases, calcium phosphate is also commonly derived from inorganic sources such as mineral rock. This material has a faster degradation profile than HA. It is generally combined with hydroxyapatite or some other material in order to increase the speed of degradation to provide space for tissue regeneration.

Traditional synthetic materials, such as titanium or cobalt chromium, are tailored to replace the structural support provided by native bone. These materials are designed to remain inert in order to minimize undesired interactions with surrounding tissue. In addition, these materials are mechanically mismatched with native bone tissues. Most traditional synthetic materials are mechanically stronger than bone, lessening the load on endogenous bone and resulting in a loss of bone density. This is often referred to as stress shielding [35]. This loss in bone density makes surrounding bone susceptible to further bone fracture, leading to the eventual failure of the implant.

Due to the inherent problems associated with allografts, autografts, and inert synthetic materials, research has moved towards the development of synthetic bioactive materials that mimic the natural structure and function of native bone tissue. In order to develop an ideal synthetic bone scaffold, the following attributes must be considered (Fig. 5) [36]:

Characteristics	General remarks
Biocompatible	Biologically compatible to host tissue i.e. should not provoke any rejection, inflammation, and immune responses.
Bioactive	To facilitate a direct biochemical bonding to host tissue.
Biodegradable	The rate of biodegradation has to be adjusted to match the rate of bone tissue formation.
Mode of degradation	Bulk or surface erosion.
Osteoconductive	Capable of supporting in-growth of sprouting capillaries, perivascular mesenchymal tissues, and osteoprogenitor cells from the recipient host into the three dimensional structure of a graft that act as a scaffold.
Vascular supportive	Should provide channels for blood supply for fast and healthy bone regeneration.
Porous structure	To maximize the space for cellular adhesion, growth, ECM secretion, revascularization, adequate nutrition and oxygen supply.
Three dimensional structure	For the assistance of cellular in-growth and transportation of nutrition and oxygen.
Adequate mechanical strength Sterilizable	To withstand *in-vivo* stimuli during bone formation. To avoid toxic contamination.
Cost-effective	Affordable to all.

Fig. 5 *Desired Properties of an Ideal Synthetic Scaffold. From:* Murugan R, Kumar TSS, Ramakrishna S. Scaffolds for bone tissue restoration from biological apatite. *Trends Biomater. Artif. Organs*. 2006;20:35-39.

Currently, the only materials considered to be capable of meeting these criteria are autografts. This is because they are harvested from viable, native bone. However, as mentioned earlier, they are not always an acceptable choice due to the lack of material availability. To address these concerns, research has shifted focus towards development of synthetic materials that faithfully mimic native bone's natural structure and function. In particular, the new aim is to develop a scaffold that recapitulates both the structure and function of bone ECM as a means to meet as many of the above criteria as possible.

4 Principles of Bone Tissue Engineering

In order to regenerate the human body, it has become apparent that progressively intricate solutions that endeavor to replicate both the biological and structural characteristics of native tissue are required. This need drives the continual development of novel biomedical technologies. The complexity of this problem is beyond the scope of any single scientific discipline; therefore, expertise from a wide variety of areas must be utilized in this effort. The result of this conjunction is the emergence in the last 20 years of the highly interdisciplinary field dubbed "tissue engineering." Under the umbrella of tissue engineering, the fields of engineering and life sciences intertwine to develop novel therapeutic technologies to maintain, restore, enhance, or replace tissue functions and whole organs [37,38].

To achieve successful tissue regeneration, the three main tissue components: cells, biomaterials, and biochemical cues, must dynamically interact and work together to produce a fully functional tissue construct. Of these, the oftentimes inert synthetic biomaterials are functionalized through the addition of cellular adhesive ligands, with the aim of heightening the material's bioactivity. Cellular adhesive ligands, small molecular parts of various extracellular matrix proteins, function in signaling and in binding to integrins, modulating a variety of cell responses [24]. Integrins, a type of anchoring molecule, permit cellular binding and anchoring to the extracellular matrix. Two specific cellular adhesive ligand sequences have emerged for bone tissue engineering due to their ability to induce considerable osteogenic differentiation via integrin-mediated binding: Arg-Gly-Asp (RGD) and Asp-Gly-Glu-Ala (DGEA).

The peptide sequence RGD is well-known as a prominent cell attachment cue [39,40]. RGD is found in many different bone extracellular matrix proteins, including fibronectin, vitronectin, bone sialoprotein, and osteopontin. Through integrins such as alpha5-beta1, alpha V-beta3, and alphaIIb-beta3, RGD can interact with cells [41]. It has been shown that the integrins that bind to RGD can mediate a variety of cell behaviors, including morphology, proliferation, differentiation, and gene expression [42]. Furthermore, the RGD sequence is particularly abundant in fibronectin; previous studies found that the combination of fibronectin and several HA surfaces heightened osteoblastic adhesion, proliferation, and differentiation [43,44]. In a similar manner, the peptide sequence DGEA is a cellular adhesive ligand derived from type I collagen. DGEA functions through interaction with the alpha2-beta1 integrin receptor, which is known to be involved in eliciting osteogenic differentiation of bone marrow cells [45].

5 Self-assembling Biomaterials

Bottom-up self-assembly is utilized to construct the structural constituents of bone at the nanoscale level. Therefore, in order to develop a biomimetic composite effectively, this nanoscale environment must be replicated. Nano-architectural scaffolds have exhibited enhanced performance when assessed against comparable scaffolds composed of microscale architectures. For example, nanoscale scaffolds have demonstrated more natural cell adhesion and spreading, which demonstrates their higher suitability to promote cell-ECM interactions. In regard to HA, nanoscale sizes result in enhanced hardness, strength, and super plasticity [46-48]. Furthermore, scaffolds incorporating nanoscale HA have demonstrated enhanced human mesenchymal stem cell (hMSC) attachment and osteogenic differentiation when compared to scaffolds of HA produced with microscale architecture (Figure 6) [49].

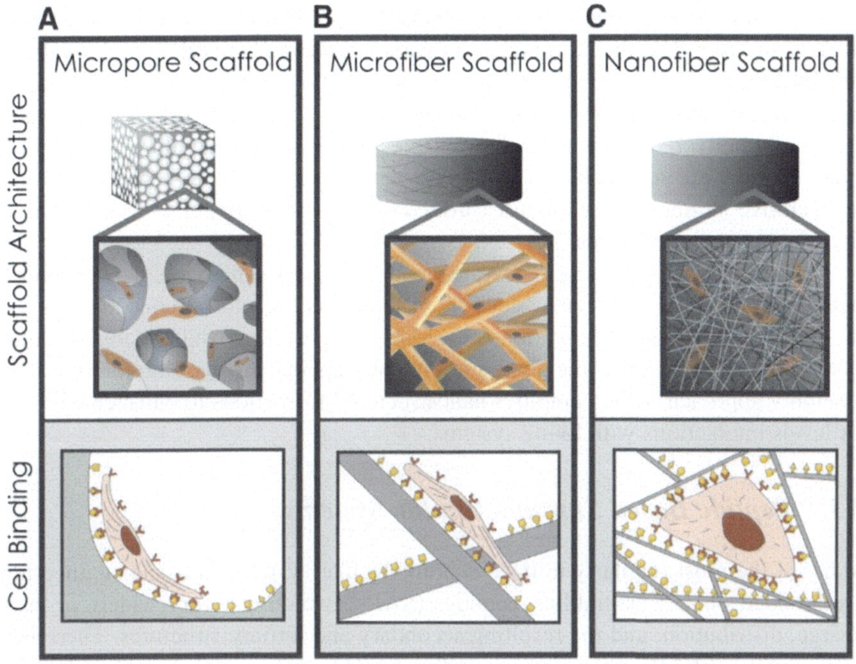

Fig. 6 *Scaffold Architecture Versus Cell Binding:* The inherent benefit of utilizing a nanofibrous scaffold architecture (C) is increased surface area for protein absorption as well as enhanced cell-ECM interactions via cell-membrane receptors. Microscale architecture (A,B) is not as effective in this sense. From "Exploring and Engineering the Cell Surface Interface" by M.M. Stevens and J.H. George, 2005. *Science*, *310*, p. 1135. Copyright 2005 by AAAS.

The ideal biomimetic material would consist of a bottom-up fabrication method that recreates the nanostructure of bone by merging an organically mimetic ECM component with an inorganic ECM component. The material would then be supplemented with a biological component to facilitate ECM remodeling and bone regeneration. The goal is to provide an osteoconducive and osteoinductive environment similar to that found in native bone tissue as a means by which to encourage subsequent bone tissue regeneration.

As mentioned, a major challenge in tissue engineering is the creation of an optimized balance between structural and biological properties, since any potential therapies must be able to both interact with cells on the molecular level as well as provide the structured arrangement and biomechanical cues of native tissue. To this end, a promising solution to this challenge is supramolecular systems, which can reproduce the structure of native biological systems. Using tissue engineered supramolecular systems, researchers can therefore synthetically recreate the microenvironments in which tissues and cells properly function. This bottom-up self-assembly approach affords these tissue engineered supramolecular scaffolds with the ability to encapsulate progenitor cells for a variety of purposes, including healing, promotion of intracellular contacts, the provision of nanostructured formations that simulate the morphology of native tissue, and stimulation of localized cell infiltration and replacement by host tissue. This multitude of capabilities is present within these physiologically biocompatible supramolecular systems, which are simply delivered via implantation or injection. Furthermore, these supramolecular systems are applicable to a variety of tissue engineering applications and regenerative strategies, such as for cardiovascular, dental, bone, cartilage, neural, liver, and pancreatic tissues.

Although many materials types are available, the main focus is on peptide-based supramolecular systems, since these materials are most frequently utilized in tissue engineering applications. Overall, it is evident that supramolecular systems have great potential in tissue engineering due to their bottom-up self-assembly approach, which employs nanostructured formations to enhance molecular levels interactions with living systems.

5.1 Peptide Based Supramolecular Systems

Many of the most regularly utilized supramolecular systems in tissue engineering are founded on self-assembling peptides (SAPs), which can vary widely in their charge, distribution, and the resulting secondary and tertiary structures. There is a wide variety of SAPs available, which Zhang and colleagues have classified under four major categories [50]. A brief description of each SAP type is discussed below.

The first is Type I SAPs, which form β-sheet structures in aqueous solutions. This formation is enabled by the molecule's structural nature, which is composed of an alternating hydrophobic and hydrophilic composite face. The hydrophilic surfaces allow the formation of complimentary ionic bonds with recurrent peptide blocks [51]; this property gave rise to these SAPs being dubbed "molecular Lego" structures, since they contain structural "pegs" and "holes". Moreover, the size of the "pegs" and "holes" can be varied through alteration of the peptide sequences employed, as exemplified by sequences such as RARADADA and RARARADADADA.

Physiological buffers, including cell culture media and monovalent alkaline ions (e.g., sodium), can cause these SAPs to self-assemble into nanofibers. The alternating charges within the SAPs direct this self-assembly process and the subsequently formed self-assemble nanofibers, which have typical diameters of between 10-20 nanometers and lengths of up to several microns. The resultant self-assembled peptide nanofiber network remains stable over a wide range of pH and temperatures. Within the nanofiber network are pores between 50-200 nanometers [51-53]. Additionally, the nanofiber network density can be controlled through alteration of the peptide solution concentration. Therefore, highly tunable Type I SAPs can provide the requisite physiological size scales and porosity that numerous vital biomolecules require.

Second are Type II SAPs, which can be depicted as "molecular switches" due to their ability to rapidly alter their secondary molecular structure to form α-helices or β-sheets in response to pH or temperature changes. Two contrasting features of the molecule contribute to this dual structural assembly. On the one hand, Type II SAPs contain the distinctive hydrophobic and hydrophilic faces that facilitate assembly of β-sheets, as in Type I SAPs. On the other hand, Type II SAPs also contain a group of negatively charged amino acid residues near the amino terminus, which is opposed by positively charged amino acid residues clustered at the carboxy terminus. The resulting polarized charge distribution promotes the formation of the α-helices.

As mentioned, the transition from one Type II SAP secondary structure to another is abrupt and elicited by either pH or temperature. An example of this phenomenon is the 16 residue self-complimentary oligopeptide DAR16-IV developed by Zhang *et al.*, which exhibits a 5 nm long β-sheet structure at room temperature, but when heated to 60°C it experiences a rapid structural transition to a stable 2.3 nm long α-helix [54,55]. This transition temperature is governed by the peptide sequence: those sequences that have more stable β-sheet structures must be heated to higher temperatures for the transition to α-helices to occur. Once this transition occurs, however, it is extremely stable, taking weeks for the α-helices to revert to the initial β-sheet structures. In a similar manner, depending on the constituent amino acid charge, pH can be adjusted to induce structure transformation.

Next are Type III SAPs, which are also known as molecular paint or molecular Velcro as a result of their ability to self-assemble onto surfaces rather than merely self-assembling amongst themselves. This property facilitates their use in creation of monolayers on many different typically inert surfaces, thus biologically functionalizing the surfaces to encourage cellular attachment [56]. Usually, Type III SAPs usually encompass the following three components: a ligand, an anchor group to foster binding to a variety of different surfaces, and a linker to connect the ligand and the anchor. These components can be modified depending on the desired application. Various ligands can be utilized to tailor to the target cell type. The peptide anchor can be altered with specific chemical groups designed to create self-assembled coating and to react with the requisite surface. As for the linker, not only does it serve as a connector between the ligand and the anchor, but it also supplies versatile mechanical control through modifications in the amino acid choice that can alter the stiffness and flexibility of the molecule. For instance, incorporation of a glycine backbone into the linker yields a more flexible molecule; in contrast, a valine sequence creates a stiff connection. Thus, Type III SAPs can be employed to address many different questions over specific cell-cell and cell-tissue interactions. Moreover, by utilizing the cell-responsive ligand as a molecular hook, surface molecule detecting "intelligent" diagnostic devices can be developed.

Finally, there are Type IV SAPs, which mimic characteristics of polymeric and lipid surfactant molecules. Type IV SAPs are by nature amphiphilic: the leading head group contains at least one hydrophilic amino acid, which is trailed by a sequence of six identical hydrophobic amino acids to create a hydrophobic tail. It is known that both cationic and anionic amphiphiles will self-assemble into tubular structures at neutral pH [57]. First, the peptides assemble into bilayers in order to isolate the hydrophobic tail sequence from the aqueous phase, which is then followed by hydrogen bonding between the adjacent units. This process facilitates the formation of higher order structures, such as vesicular or tubular morphologies. Through incorporation of molecular recognition sites, these vesicles and tubes may be exploited for drug delivery to specific cell types with few off-target effects.

As explained by Zhang and colleagues, these are the four major categories of SAPs employed for tissue engineering purposes. The tissue engineering applications of other supramolecular systems will be expounded upon in the sections to follow in specific relation to bone tissue. Supramolecular system application to some of the major challenges and unique requirements of each system, especially with respect to specific tissue regeneration, will be discussed in the following sections.

6 Self-assembling Biomaterials for Scaffold Based Bone Tissue Regeneration

Approaching this new strategy, many different supramolecular self-assembling peptides have been developed and investigated by various research groups as potential functionalized scaffolds for stimulating bone tissue regeneration.

6.1 Surface Modification of Devices

Currently, synthetic materials such as cobalt chromium or titanium are used and implanted in conventional treatment for replacement of bone lost through trauma or degeneration [58]. In order to prevent detrimental interactions between the bone and the implant, such as adverse immune responses, these synthetic materials are purposefully designed to remain biologically inert [59,60]. Although these implants typically have no adverse effects on native bone aside form stress shielding, host cells will endeavor to interact with the implants through non-specific processes, which can lead to desorption, degradation, or molecular dissolution on the implant surface [61]. These adverse processes can cause deterioration of the bond between the implant and the endogenous bone, which may ultimately lead to implant failure at the implant-bone interface. In order to address this challenge and prevent implant failure as a result of these adverse interactions, many groups have researched methods to functionalize implants using supramolecular scaffold systems, thus improving cellular interactions between the implants and native tissue.

Peptide Amphiphiles (PAs) present a promising solution for bone tissue engineering as they are able to be utilized in surface modification of devices for enhanced efficacy. For example, Sargeant *et al.* proposed employing PAs to functionalize ordinarily inert nickel titanium surfaces. Such modification can enhance native cellular attachment, minimize any adverse non-specific interactions, and improve stability through prevention of desorption, degradation, and molecular dissolution at the implant surface [62]. It was found that through the treatment of nickel titanium surfaces with aminosilanes to enhance titanium-PA binding and the incorporation of the RGDS sequence into the PAs, the functionalized nickel titanium surfaces exhibited enhanced cell-implant interactions. This furthermore minimized the undesirable effects, such as long term implant failure and fibrotic depositions at the tissue-implant junction. Retention and proliferation of MC3T3-E1 cells were assessed on various nickel titanium surfaces, with and without aminosilane treatment and the addition of self-assembling PAs. Results from these experiments showed that surfaces treated with aminosilane and functionalized with PAs enhanced cellular proliferation and diminished levels of cell detachment when compared to the non-treated controls. These results indicate that PAs are promising materials for surface modification of titanium-based implants to enhance cell-implant interactions.

As mentioned previously, implant lifetime is constrained by loosening at the implant-bone interface, which leads to implant failure and bone density loss due to implant stress shielding. This challenge has created a high demand for biomaterials that facilitate implant stabilization through ingrowth of bone. One method is through increasing the biomaterial's porosity, which lowers the elastic modulus and consequently reduces bone density loss from stress shielding and bone re-adsorption [63,64]. Although decreasing stress shielding through porosity is known to be beneficial, the biological inertness of standard implant materials must be tackled as well. Therefore, novel supramolecular tissue engineered solutions that unite both porosity and bioactivity have emerged as material complexity and improved biomimetic properties for bone regeneration have developed. An example of this is the combination of inert biomaterials such as titanium with synthetic bioactive peptides, which can promote cell attachment, proliferation, differentiation, and subsequent bone tissue in-growth. For these investigations, the Stupp group combined the bioactive self-assembling PAs with an incorporated RGDS sequence with porous titanium implants in order to promote bone growth [65]. This bone regenerative scaffold with dual functionality elicited direct biological responses. In particular, after MC3T3-E1 cells were cultured in a rotating bioreactor on PA-Ti hybrid scaffolds and visualized by scanning electron microscope (SEM) imaging, cell filopodia extending out onto the underlying PA nanofibers were pictured, indicative of cell spreading. This SEM data confirmed the ability of the PAs to self-assemble into cylindrical nanofibers, as well as the cellular capacity to recognize the functional ligand sequences. Additionally, evaluation of long-term osteogenic markers revealed that alkaline phosphatase and osteocalcin levels, both significant indicators of osteogenic differentiation, exhibited increased expression after 14 and 28 days respectively. These results indicated that these self-assembled PAs were able to stimulate osteogenic differentiation with a typical expression profile. Therefore, by using PAs to functionalize the porous titanium scaffolds, both osteogenic differentiation and bone mineralization levels could be enhanced.

Guidance of progenitor cell fate to regenerate native tissue underlies many tissue engineering applications; therefore, molecules devised to explicitly direct stem cell behavior are often integrated within tissue engineered scaffolds. Through functionalization of peptide-based supramolecular biomaterials by incorporation of cell adhesion ligands, many groups endeavor to mimic cell-ECM interactions in order to direct cell behaviors and guide stem cell fate for particular bone tissue engineering applications. For example, in order to evaluate the ability of human mesenchymal stem cells (hMSCs) to recognize and bind to cell adhesion sequences and thus adjust their behavior for osteogenic applications, Anderson *et al.* cultured hMSCs on various biologically functionalized PA coatings for up to 35 days. These PA coatings separately incorporated cells adhesive ligand sequences isolated from various sources, including collagen (PA-DGEA), fibronectin (PA-RGDS), or proteoglycans (PA-KRSR), and exhibited the capability to self-assemble on various material surfaces through a solvent evaporation method, underscoring the PAs' ability to functionalize diverse biomaterial surfaces. PAs with scrambled or entirely lacking cell adhesion sequences (PA-RGES and PA-S, respectively) were utilized as controls. The aim of this research was to accomplish osteogenic differentiation without the use of osteogenic supplements, demonstrating that the cell adhesive ligands within the functionalized PA coatings could single-handedly direct osteoinduction [66]. First of all, self-assembly of the PAs via the solvent evaporation technique was confirmed through transmission electron microscopy, which demonstrated that all of the functionalized peptides formed cylindrical nanofibers with diameters of between 8 and 10 nm on the surface of the substrate. Next, hMSCs were seeded on the different PA surfaces and analyzed at 1 and 4 hours via a DNA quantification assay in order to evaluate the cells' ability to recognize and bind to the various cell adhesive ligand sequences. Results displayed heightened levels of cellular attachment to PA-RGDS and PA-DGEA functionalized surfaces, which indicated that the cells could recognize and interact with the cell adhesive ligand sequences within the PAs. Following this, osteogenic differentiation was studied through quantification of ALP levels up to 21 days and Von Kossa staining was utilized to visually evaluate mineralization. The functionalized peptides PA-RGDS and PA-DGEA were the most favorable for the promotion of osteogenic differentiation without the aid of osteogenic supplementation, demonstrating that cell-ECM signaling can be utilized to direct cellular behaviors. Therefore, functionalized PAs demonstrate great potential for various biomaterial modification and osteogenic induction of progenitor cells via cell-ligand interactions.

6.2 *Scaffold Based Tissue Regeneration*

Recently, bone tissue engineering approaches have shifted towards bioactive and degradable synthetic biomaterials and away from the traditional bioinert materials. The favorable properties of the bioactive materials enable natural tissue to replace them, completely repairing the tissue and re-establishing its former functions. In addition, a frequent strategy to enhance the rate of bone regeneration even more is to encapsulate osteogenic progenitor cells in the biomimetic scaffold, enhancing the cellular behaviors such as migration, proliferation, and differentiation via signaling sequences incorporated within the scaffold.

Therefore, in order to recreate the natural bone environment for the purposes of bone regeneration, many groups have developed various self-assembling materials that can serve as scaffolds intended to mimic the organic facets of bone extracellular matrix by incorporating adhesive ligands derived from specific extracellular matrix proteins. SAPs, an example of new bone tissue engineering technology, have been developed by Zhang *et al.* in order to address osteogenic regeneration. SAPs, which are from the same family of biomaterials as the ones described previously, are modified for bone tissue engineering applications by incorporating ligand sequences found in native bone ECM. Consequently, the SAPs can be utilized as an assembled platform of many molecules which contains defined cellular ligands to use for cell signaling. In order to investigate their usefulness for bone tissue regeneration, non-functional RADA16 SAPs have been combined with functional SAPs with sequences such as ALK (from osteogenic growth peptide), DGR (from osteopontin's cell adhesion sequence), and PRG (2-unit RGD sequences) [42]. Each of these sequences have been demonstrated to play a role in bone morphogenesis. Osteogenic growth peptide (OGP) sequences are essential in the bone regeneration process, increasing during osteogenic reactions and promoting osteogenesis and hematopoiesis. For these reasons, those sequences were incorporated into the SAPs. Similar to OGPs, it has been demonstrated that osteopontin, which is phosphorylated and synthesized by both osteoblasts and osteoclasts, can regulate cellular behaviors, such as migration, adhesion, survival, and calcium deposition, acting as an important mediator in the bone formation process. Additionally, the RGD peptide, which is found in fibronectin, has promoted osteogenic differentiation and cell adhesion in some experiments by mimicking the natural cell ECM.

For all of these SAP sequences, atomic force microscope (AFM) was used to assess their self-assembly into nanofiber-like structures. The images showed that ALK and PRG sequences were not able to form nanofibers when they were used alone, so they needed to be integrated into the RADA16 peptide in order to obtain increased structural support. The PRG and DGR peptides were also combined with the RADA16 sequence, in order to maintain consistent results and scaffold characteristics. After the scaffold stability was obtained by combining peptides, the osteogenic potential of the composite SAPs was evaluated. For these studies, MC3T3-E1 pre-osteoblastic cells were seeded onto scaffold with the different signaling sequences. It was shown that the functional PRG scaffold promoted the highest amount of alkaline phosphatase (ALP) and osteocalcin activity, thus enhancing osteogenic differentiation and calcium deposition the most. Additionally, all of the other peptide sequences showed higher amounts of ALP and osteocalcin activity than just the RADA16 alone, showing that incorporation of functionalized peptides further enhances the levels of osteogenic differentiation.

Alternative methods to functionalize RADA16 SAPs for bone tissue engineering applications, like culturing with platelet-rich plasma (PRP) [67], have been explored by other groups. For example, Ueda *et al.* investigated the effect of adding PRP to Purmatrix (PM and dog mesenchymal stem cells (dMSCs) on osteogenic regeneration. In the experiment, 10 mm mandibular defects were created in adult hybrid dogs, and different graft materials, PM, PM/PRP, PM/dMSCs, and PM/PRP/dMSCs, were implanted at the defect site. The control was just the defect only. SEM was used to image the fibrous structures of PM and PM/dMSCs, showing fibers about 10-20 nm in diameter, which confirmed that the nanofibers had self-assembled to form matrices. The scaffolds were implanted and retrieved at 2, 4, and 8 weeks, and histological, histomorphometric, and radiographic assessment were used to evaluate the levels of osteogenic regeneration. At 8 weeks, radiographic analysis demonstrated that none of the scaffolds had caused infection and that they had been absorbed into the body; the defect in which the PM/dMSCs/PRP scaffold was inserted showed bone regeneration equivalent to almost the original, natural state. After 2, 4, and 8 weeks after the endogenous bone tissue was removed, histomorphometric analysis, utilizing image analysis software, quantified the medullary and cortical bone of the grafts that were retrieved. From this analysis, it was determined that PM/dHMSCs/PRP scaffolds promoted a significantly higher volume of bone regeneration than PM/dHMSCs, PM/PRP, PM, and the control groups. Therefore, it was shown that functional peptide scaffolds can successfully transport therapeutic osteogenic progenitor cells in order to promote bone regeneration.

Multiple other studies have investigated functionalized SAP ligands with a variety of signaling sequences in order to enhance bone tissue engineering applications by mimicking the natural ECM environment. For instance, a novel peptide scaffold made of RGDA16 (Ac-RADARRGDARADARGDA-CONH2) and RADA16 was created by Xie *et al.* This RGDA mix includes the RGD cell adhesion sequence, which has been previously demonstrated to enhance cellular attachment, migration, differentiation, and attachment, and the RADA16 peptide, which mimics the native bone ECM; thus, seeking to mimic native bone ECM environment. In order to assess cellular spreading, attachment, and proliferation on the scaffolds, MC3T3-E1 cells were seeded on the RGDA mix coating and the RADA 16 coating, as a control group. The results showed that the RGDA mix had a quicker rate of proliferation, more spreading, and more attachment than the control scaffold. This suggested that the scaffold was recognized by and influenced cellular behavior through its signaling motifs. The results of this experiment show that it is feasible to design a native bone ECM mimicking scaffold using self-assembling peptides, thus customizing those biomaterials for different bone tissue engineering applications.

Besides SAPs, Hartgerink *et. al.* have developed peptide amphiphiles (PAs) [68], a new class of self-assembling peptide-based biomaterials with potential for bone tissue regeneration. A PA is comprised of three peptide regions, each with different functions, and a hydrophobic aliphatic tail. These components interact with each other, self-assembling to form a cylinder (Figure 7). Particularly, the structure of a PA contains four cysteine amino acids that are adjacent to a hydrophilic aliphatic core. This allows for disulfide bonding, which stabilizes the PA structure when it self-assembles. Furthermore, phosphoserine, which is copious in non-collagenous bone and is believed to interact with HA, elicits biomineralization when it is included in the PA structure. RGD, a cell-adhesion sequence shown to promote cell attachment, proliferation, and osteogenic differentiation, was likewise incorporated into the peptide region. This peptide design, combined with HA crystals that were aligned along the long fiberaxis, confirmed by cryo-TEM imagining, successfully induced biomineralization. Aligning the peptide nanofibers with the HA crystals mimics the environment of native bone, exhibiting its potential as a material for biomimetic scaffolds.

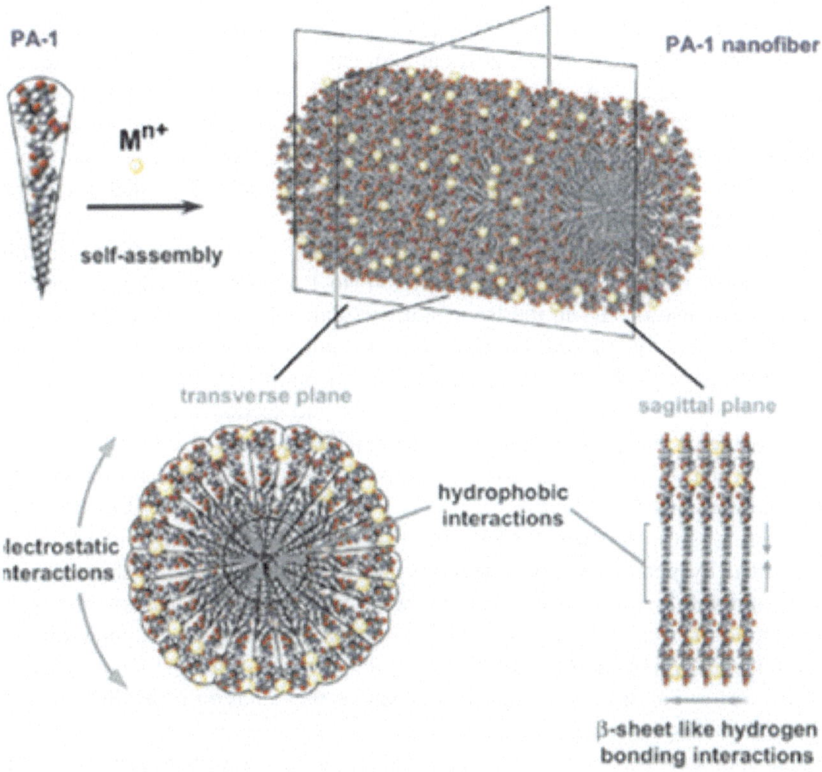

Fig. 7 *Peptide Amphiphile Structure and Self-Assembly.* From "Intermolecular Forces in the Self-Assembly of Peptide Amphiphile Nanofibers" by J.C. Stendahl, M.S. Rao, M.O Guler, and S.I. Stupp. Advanced Functional Materials, 6(4), p. 499. Copyright 2006 by Wiley-VCH Verlag GmbH & Co. KGaA.

While self-assembling peptides functionalized with various cellular adhesion motifs are capable to enhancing cellular interactions via ligand and receptor binding mechanisms for bone tissue engineering, they do not fully encompass the natural bone structure, which is biphasic in nature. As mentioned earlier, bone is a biphasic composite consisting of organic ECM proteins that have been mineralized with hydroxyapatite. Even though a PA nanofiber matrix can enhance cellular behaviors and osteogenic differentiation of hMSCs [66], it can only provide limited osteogenic differentiation because it cannot mimic the biphasic nature of native bone ECM. Currently, synthetic scaffolds are composed of materials that only mimic the inorganic part of bone ECM. Usually, HA tends to be used in those scaffolds because of its calcium phosphate composition, which is similar to native bone. Furthermore, the osteoinductive potential of HA and its effect of enhancing osteogenic differentiation have been proposed in studies [69], although this is controversial and the exact means by which HA induces osteogenic differentiation is unclear.

Although HA provides structural support and possibly promotes osteogenic differentiation, its mechanical properties are not equivalent to native bone tissue, and it is slowly invaded by host tissues when used alone as a bioactive scaffold [70-72]. To address these issues, HA nanoparticles (HANPs) can be combined with the PA nanofiber matrix, providing both the inorganic and organic components of native bone ECM, thus providing cellular behaviors that favor bone regeneration.

In order to enhance the osteogenic capability of PA nanofiber scaffolds, HANPs were incorporated at varying HANP to PA ratios in order to determine how the presence of HA in the PA nanofiber scaffold would influence the osteogenic differentiation of hMSCs [73]. For the purposes of the study, PAs were functionalized with the RGDS cellular adhesion motif (PA-RGDS). Two dimensional coatings were created for each condition and seeded with hMSCs in order to assess their osteogenic response. TEM and SEM imaging confirmed the successful formation of biphasic coatings consisting of HANPs encapsulated within PA nanofiber matrices (Figure 8). It was found that as the HANP to PA-RGDS ratio increases (up to 66% HANP), the osteogenic differentiation of hMSCs, as determined by gene expression analysis and quantification of alkaline phosphatase, improves in a step-wise fashion, with 50% HANP enhancing osteogenic differentiation over 33% HANP, and 66% HANP enhancing osteogenic differentiation over 50% HANP. Overall, the study demonstrated that the incorporation of HANPs to form a biphasic composites enhances the behavior of hMSCs in regard to their capability to regenerate bone.

This approach was then further explored using a three dimensional culture approach, in which these biphasic composites were induced to form three dimensional hydrogels, in which, hMSCs were encapsulated and assessed over time using both in-vitro, and in-vivo study approaches [74]. For the in-vitro approach, hMSCs were encapsulated within biphasic composites consisting of 50% HANPs and PAs functionalized with various cell adhesive ligands (PA-RGDS, PA-DGEA) and a non-functionalized placeholder (PA-S). It was demonstrated that PA-RGDS enhanced osteogenic differentiation over PAs functionalized with DGEA and that HANP incorporation further enhanced osteogenic differentiation over those without. Following this, using a rat critical femoral defect model, it was demonstrated that these PA nanomatrix composites were capable to enhancing bone tissue regeneration as demonstrated by histology and x-ray imaging at 6 weeks.

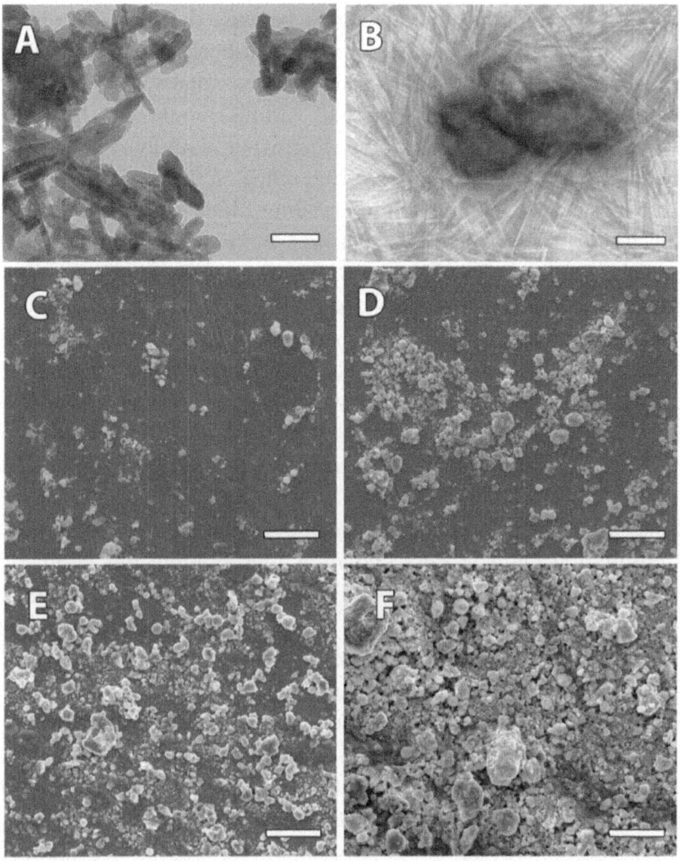

Fig. 8 Characterization of nanomatrix composites. TEM images of (A) HANPs and (B) PA-RGDS (16% HANP) at 42000×. SEM Images of (C) PA-RGDS (16% HANP) (D) PA-RGDS (33% HANP) (E) PA-RGDS (50% HANP) and (F) PA-RGDS (66% HANP) at 1000×. Scale bars represent 100 nm for TEM and 50μm for SEM respectively. Considered together, TEM and SEM images demonstrate aggregates of varying size and distribution made up of HANPs enmeshed within PA nanofibers. Copyright © 2012 Acta Materialia Inc. Published by Elsevier Ltd. All rights reserved.

7 Limitations of Current Approaches

An ideal bone scaffold should be designed to provide structural support and mimic the natural biological characteristics of bone tissue in order to enhance cellular interactions. Other advantageous strategies include the incorporation of different growth factors in order to enhance osteogenesis and angiogenesis, and carefully tailoring degradation to a rate that is more appropriate for the mending of bone. Self-assembling peptide-based biomaterials can provide a useful matrix that can influence cellular behavior, but they do not provide enough mechanical support for applications in bone tissue engineering because they usually self-assemble as hydrogels. Thus, there is still a need to create a bone tissue engineering scaffold that is capable of providing both a functionalized matrix and structural support. Possible solutions for this need include various bioglasses, hydroxyapatite, or nano crystalline hydroxyapatites that is electrospun with PGA or PCL [75-78].

Bone is a highly vascularized tissue that relies on the close spatial and temporal relationship between bone cells and pervading blood vessels to maintain skeletal integrity [79]. The interdependence between the two tissue types necessitates the need for angiogenesis in coordination with new bone formation for fracture repair. Bone tissue engineering approaches, especially engineered scaffolds with seeded cells for implantation, have been extensively researched and developed for promoting fracture healing. However, such scaffold implants commonly rely on vascularization exclusively from the ingrowth of surrounding host blood vessels [80]. Local blood vessel infiltration is initiated upon scaffold implantation, albeit at an insufficient rate of less than one millimeter per day, not enough to vascularize tissues of clinically relevant size [81,82]. Additionally, if successful vascularization within the engineered construct is achieved, the vessels must still connect to the host circulation via a process known as inosculation; this does not occur immediately and has been shown to take up to eight days, possibly leading to ischemia [83]. These are all major concerns to the clinical efficacy of implanted bone scaffolds, especially since tissue regeneration over 100-200 μm exceeds the capacity of nutrient supply and waste removal by diffusion [84,85]. Specifically, a lack of sufficient vascularization for implanted cellular grafts severely limits the capacity to uptake vital substrate molecules (e.g. oxygen, glucose, amino acids) and clear metabolic byproducts (e.g. CO_2, lactate, urea) , which can result in necrosis at the center of engineered bone if an adequate vascular supply is not quickly established [86,87]. Thus, the lack of functional vasculature integrated with the host blood supply has been identified as a main culprit for implant failure and acknowledged as one of the major challenges in bone tissue engineering [82,88,89].

It is essential that multifaceted scaffolds that promote both osteogenesis and angiogenesis continue to be developed and improved for bone tissue engineering, as the integration of vascular networks is vital to ensuring long-term viability of thick three-dimensional tissue constructs. Several strategies for inducing vascularization within bone tissue engineered implants have been investigated, including the release of soluble factors and co-culture of target osteogenic cells with endothelial cells [90]. Each angiogenic strategy has its own merits and has been prevalently researched. The release of soluble factors (e.g. VEGF, bFGF) from scaffolds has been the most commonly investigated, as this approach has an established history for enhancing vascularity [79]. However, exogenous growth factor release for angiogenesis continues to be limited by the short half-life of the proteins, rapid diffusion out of target tissue area, frequent doses required to achieve significant biological response, high cost of multiple cytokine delivery, and potential toxicity issues [91,92]. To this point, the side effects from high levels of angiogenic factors can include permeable vessels, stimulation of tumor growth, abnormal vascular function, and hypervascularity [91]. Consequently, issues of self-sustainability exist that we believe can be better addressed through the use of a co-culture system for promoting angiogenesis. By combining osteogenic and endothelial cells, the interaction between the two populations recreates a physical and chemical environment favorable for the formation of vascular structures, negating the need for exogenous supply of angiogenic stimuli [87].

Many studies have found that cell-based therapies using endothelial cells accelerate vascularization of tissue engineered implants, as the transplanted cells interact with host endothelium and vasculature to quickly establish a blood supply [88,91,93,94]. Concurrently, the co-culture of endothelial cells with target osteoprogenitors has been shown to positively affect osteogenic differentiation and bone formation, both at ectopic and orthotopic locations [95-98]. It is believed that a synergistic relationship occurs because the co-culture of heterogeneous cell types more closely recreates an in vivo microenvironment compared to monotypic cell cultures [90]. Besides the immediate effect of providing a vascular supply, endothelial cells in co-culture participate in many facets of osteogenesis, such as osteoinduction via release of soluble factors, controlling transcription factors of bone cell differentiation, and facilitating tissue mineralization [99]. The osteoprogenitor cells reciprocate in kind, as the presence of non-endothelial cells (e.g. mesenchymal stem cells) can improve the survival, proliferation, and formation of vessel structures by endothelial cells in co-culture [100-102]. Potential limitations for co-culture therapies do exists, though, attributed to the low availability and proliferation of endothelial cells, variable biological response based on phenotypic drift and tissue isolation, and suboptimal life span of endothelial-derived blood vessels [87,94]. However, osteogenic cells have been demonstrated to work in support of endothelial-derived vasculature, counteracting deficient vessel formation by serving as a potential source of stabilizing perivascular cells; for example, in a study by Au et al., vessels formed from co-culture of endothelial and osteogenic cells were stable and remained functional for more than 130 days in vivo [103]. Thus, the co-culture approach does not rely on the ingrowth of host endothelial cells throughout the entire scaffold implant, only the outer regions, which may prompt faster vascularization and aid bone formation [101].

As alluded to earlier, the communication between osteogenic and endothelial cells is of most importance to bone formation. The specific mechanisms for crosstalk between the two cell populations occurs by both indirect cell contact via the release of soluble factors to paracrine and autocrine effect and direct cell-cell contact mediated by gap junctions [104,105]. The intercellular communication is believed to be more heavily influenced by gap junction signaling; predominantly, the connexin43 gap junction has been pinpointed as the main signaling mechanism, up-regulating expression of osteogenic markers [106,107]. The reciprocal relationship between the two cell types has been demonstrated in numerous co-culture studies. Mediated by intercellular communication, osteogenic cells in co-culture release higher amounts of VEGF than in monoculture, driving cellular response of endothelial cells [102,108]. At the same time, endothelial cells can produce BMP growth factors, promoting osteogenic proliferation and differentiation [109,110]. Hence, the intercellular communication between endothelial and osteogenic cell populations can be used to great effect for the development bone tissue engineering solutions.

Examples of co-culture tissue engineered scaffolds for bone regeneration are emerging in the literature and have begun to be reviewed at length [87,107]. Common scaffold materials for co-culture systems include poly-lactide-co-glycolide (PLGA), polycaprolactone (PCL), and collagen gel [89,96,103,111,112]. In one of the first co-culture studies, Villars et al. showed that the intercellular communication between bone marrow stromal cells (bMSCs) and human umbilical vein endothelial cells (HUVECs) is most influenced by direct cell-cell contact, as alkaline phosphate activity (an osteogenic marker) was enhanced based on connexin43 signaling [105]. Progressing to more clinically relevant studies, recent research has shown that the co-culture of osteogenic and endothelial cells for bone tissue engineering can stimulate blood vessel formation, leading to increased bone regeneration. For example, Kaigler et al. co-transplanted endothelial cells with bMSCs using a PLGA scaffold into calvarial rat defects [95]. After six weeks, the transplanted endothelial cells significantly improved bone volume formation and mineralization at the orthotopic site. Furthermore, Grellier et al. investigated bone regeneration for long bone defects in a mouse model, co-transplanting human osteoprogenitors with HUVECs immobilized in alginate microspheres [99]. The co-culture alginate scaffolds were found to increase mineralization after six weeks compared to the monoculture implants. Therefore, a potential strategy for future bone tissue engineering would be to develop a biphasic nanoscale, self-assembling scaffold to serve as a three-dimensional construct for co-cultured endothelial and osteogenic cells. This would perhaps create a self-sustainable bone tissue engineering solution, capable of promoting bone formation and vascularization for improved skeletal repair.

8 Conclusion

Bone is a highly important tissue for providing support, movement, and protecting bodily organs as well as serving as a reservoir for important minerals necessary to sustain bodily function such as calcium and phosphate. Damage to this vital organ can have drastic consequences to an individual's health and quality of life. Fortunately, most insults are capable of healing without further intervention, however, insults and pathologies, such as osteoporosis that result in defects above a certain size and scale do not heal without further clinical intervention. In these instances, tissue engineering is used to replace and repair the damaged bone regions. This emerging field takes advantage of growth factors, cells, and biomaterials in order to recreate organs or lost tissue regions. For the purposes of bone tissue engineering, many of the current materials used for this purpose such as autografts, allografts, and traditional synthetic materials have many shortcomings that prevent them from being the most ideal adjuncts for bone tissue engineering. In order to remedy this problem, current research has moved towards the development of synthetic materials that better mimic the natural structure and function of endogenous bone. In this chapter, the importance of creating self-assembling materials that mimic the native structure of endogenous bone so that the dynamic interactions between bone cells and extracellular matrix components can be improved was discussed. Furthermore, the mechanisms by which cells interact with scaffolds, as well as the self-assembling biomaterial technologies utilized to enhance these interactions, and special considerations that must be made with regard to future development of these technologies for bone tissue engineering were also discussed in depth.

Acknowledgement This was supported by National Science Foundation Career Award (CBET 0952974) and National Institute of Health (1R01HL125391 and 1R03EB017344-01) for HWJ.

References

1. Meghji, S.: Bone remodelling. Br. Dent. J. **172**, 235–242 (1992)
2. Robinson, R.A.: Bone tissue: composition and function. Johns Hopkins Med. J. **145**, 10–24 (1979)
3. Eastoe, J.E.: The Chemical Composition of Bone and Tooth. Adv. Fluorine. Res. **21**, 5–17 (1965)
4. Ashman, R.F., Buckwalter, J.A., Devane, P., Dobbs, M.B., Ferguson, P.J., Flatow, E.L., France, J.C., Green, N.E., Guyton, G.P., Horne, G.: Turek's Orthopaedics: Principles and Their Application, 6th edn. Lippincott Williams and Wilkins, Iowa City (2005)
5. Couchman, J.R., Austria, M.R., Woods, A.: Fibronectin-cell interactions. J. Invest. Dermatol. **94**, 7S–14S (1990)
6. Fung, Y.-C.: Biomechanics: Mechanical properties of living tissues, 2nd edn. Springer, New York (1993)
7. Nijweide, P.J., Burger, E.H., Feyen, J.H.: Cells of bone: proliferation, differentiation, and hormonal regulation. Physiol. Rev. **66**, 855–886 (1986)
8. Heino, T.J., Kurata, K., Higaki, H., Vaananen, H.K.: Evidence for the role of osteocytes in the initiation of targeted remodeling. Technol. Health Care **17**, 49–56 (2009)
9. Pittenger, M.F., Mackay, A.M., Beck, S.C., Jaiswal, R.K., Douglas, R., Mosca, J.D., Moorman, M.A., Simonetti, D.W., Craig, S., Marshak, D.R.: Multilineage potential of adult human mesenchymal stem cells. Science **284**, 143–147 (1999)
10. Muller-Sieburg, C.E., Cho, R.H., Thoman, M., Adkins, B., Sieburg, H.B.: Deterministic regulation of hematopoietic stem cell self-renewal and differentiation. Blood **100**, 1302–1309 (2002)
11. Marieb, E.N., Hoehn, K.: Human anatomy & physiology, 7th edn. Pearson Benjamin Cummings, San Francisco (2007)
12. Gilbert, S.F.: Paraxial and intermediate mesoderm Developmental biology, 6th edn., vol. xviii, 749 p. Sinauer Associates, Sunderland (2000)
13. Cruess, R.L., Dumont, J.: Fracture healing. Can. J. Surg. **18**, 403–413 (1975)
14. Ketenjian, A.Y., Jafri, A.M., Arsenis, C.: Studies on the mechanism of callus cartilage differentiation and calcification during fracture healing. Orthop. Clin. North Am. **9**, 43–65 (1978)

15. Tsunoda, M., Mizuno, K., Matsubara, T.: The osteogenic potential of fracture hematoma and its mechanism on bone formation–through fracture hematoma culture and transplantation of freeze-dried hematoma. Kobe J. Med. Sci. **39**, 35–50 (1993)
16. Macmahon, P., Eustace, S.J.: General principles. Semin. Musculoskelet. Radiol. **10**, 243–248 (2006)
17. Sfeir, C., Ho, L., Doll, B.A., Azari, K., Hollinger, J.O.: Fracture repair. In: Lieberman, J.R., Friedlander, G.E. (eds.) Bone Regeneration and Repair: Biology and Clinical Applications. Humana Press Inc., Totowa (2005)
18. Greer, R.B.: 3rd. Wolff's Law. Orthopaedic Review **22**, 1087–1088 (1993)
19. Cowin, S.C.: Wolff's law of trabecular architecture at remodeling equilibrium. J. Biomech. Eng. **108**, 83–88 (1986)
20. Alberts, B., Alexander, J., Lewis, J., Raff, M., Roberts, K., Walter, P.: Cell Communication Molecular biology of the cell, 4th edn., vol. xxxiv, 1548 p. Garland Science, New York (2002)
21. Cooper, G.M., Hausman, R.E.: Cell Signaling The cell : a molecular approach, 5th edn., vol. xix, 820 p. ASM Press/Sinauer Associates, Washington, D.C./Sunderland (2009)
22. Lafrenie, R.M., Yamada, K.M.: Integrin-dependent signal transduction. J. Cell. Biochem. **61**, 543–553 (1996)
23. Martin, G.R., Kleinman, H.K., Terranova, V.P., Ledbetter, S., Hassell, J.R.: The regulation of basement membrane formation and cell-matrix interactions by defined supramolecular complexes. Ciba Found. Symp. **108**, 197–212 (1984)
24. Keselowsky, B.G., Collard, D.M., Garcia, A.J.: Integrin binding specificity regulates biomaterial surface chemistry effects on cell differentiation. Proc. Natl. Acad. Sci. U.S.A. **102**, 5953–5957 (2005)
25. Freeman, S., Hamilton, H.: Biological science / Scott Freeman; contributors, Healy Hamilton … [et al.], 2nd edn. Pearson Prentice Hall, Upper Saddle River (2005)
26. Lodish, H.F., Berk, A., Zipursky, S.L.: Cell-to-Cell Signaling: Hormones and Receptors Molecular cell biology, 4th edn. W.H. Freeman, New York (2000)
27. Silverthorn, D.U.: Human physiology : an integrated approach, 3rd edn. Pearson/Benjamin Cummings, San Francisco (2004)
28. Arrington, E.D., Smith, W.J., Chambers, H.G., Bucknell, A.L., Davino, N.A.: Complications of iliac crest bone graft harvesting. Clin. Orthop. Relat. Res., 300–309 (1996)
29. Jager, M., Herten, M., Fochtmann, U., Fischer, J., Hernigou, P., Zilkens, C., Hendrich, C., Krauspe, R.: Bridging the gap: bone marrow aspiration concentrate reduces autologous bone grafting in osseous defects. J. Orthop. Res. **29**, 173–180 (2011)
30. Arora, N.S., Ramanayake, T., Ren, Y.F., Romanos, G.E.: Platelet-rich plasma: a literature review. Implant Dent. **18**, 303–310 (2009)
31. Griffin, X.L., Smith, C.M., Costa, M.L.: The clinical use of platelet-rich plasma in the promotion of bone healing: a systematic review. Injury **40**, 158–162 (2009)
32. Cowan, C.M., Soo, C., Ting, K., Wu, B.: Evolving concepts in bone tissue engineering. Curr. Top. Dev. Biol. **66**, 239–285 (2005)
33. Urist, M.R.: Bone: formation by autoinduction. Clin. Orthop. Relat. Res. **2002**, 4–10 (1965)
34. Giannoudis, P.V., Dinopoulos, H., Tsiridis, E.: Bone substitutes: an update. Injury **36**(Suppl. 3), S20–S27 (2005)

35. Kennady, M.C., Tucker, M.R., Lester, G.E., Buckley, M.J.: Stress shielding effect of rigid internal fixation plates on mandibular bone grafts. A photon absorption densitometry and quantitative computerized tomographic evaluation. Int. J. Oral Maxillofac. Surg. **18**, 307–310 (1989)
36. Murugan, R., Kumar, T.S.S., Ramakrishna, S.: Scaffolds for Bone Tissue Restoration from Biological Apatite. Trends Biomater. Artif. Organs **20**, 35–39 (2006)
37. Langer, R., Vacanti, J.P.: Tissue engineering. Science **260**, 920–926 (1993)
38. MacArthur, B.D., Oreffo, R.O.: Bridging the gap. Nature **433**, 19 (2005)
39. Ruoslahti, E., Pierschbacher, M.D.: New perspectives in cell adhesion: RGD and integrins. Science **238**, 491–497 (1987)
40. Rezania, A., Healy, K.E.: Integrin subunits responsible for adhesion of human osteoblast-like cells to biomimetic peptide surfaces. J. Orthop. Res. **17**, 615–623 (1999)
41. Hersel, U., Dahmen, C., Kessler, H.: RGD modified polymers: biomaterials for stimulated cell adhesion and beyond. Biomaterials **24**, 4385–4415 (2003)
42. Horii, A., Wang, X., Gelain, F., Zhang, S.: Biological designer self-assembling Peptide nanofiber scaffolds significantly enhance osteoblast proliferation, differentiation and 3-D migration. PLoS One **2**, e190 (2007)
43. Schonmeyr, B.H., Wong, A.K., Li, S., Gewalli, F., Cordiero, P.G., Mehrara, B.J.: Treatment of hydroxyapatite scaffolds with fibronectin and fetal calf serum increases osteoblast adhesion and proliferation in vitro. Plast. Reconstr. Surg. **121**, 751–762 (2008)
44. Sogo, Y., Ito, A., Matsuno, T., Oyane, A., Tamazawa, G., Satoh, T., Yamazaki, A., Uchimura, E., Ohno, T.: Fibronectin-calcium phosphate composite layer on hydroxyapatite to enhance adhesion, cell spread and osteogenic differentiation of human mesenchymal stem cells in vitro. Biomed. Mater. **2**, 116–123 (2007)
45. Mizuno, M., Fujisawa, R., Kuboki, Y.: Type I collagen-induced osteoblastic differentiation of bone-marrow cells mediated by collagen-alpha2beta1 integrin interaction. J. Cell. Physiol. **184**, 207–213 (2000)
46. Priya, A., Nath, S., Biswas, K., Basu, B.: In vitro dissolution of calcium phosphate-mullite composite in simulated body fluid. J. Mater. Sci. Mater. Med. **21**, 1817–1828 (2010)
47. Kong, Y.M., Bae, C.J., Lee, S.H., Kim, H.W., Kim, H.E.: Improvement in biocompatibility of ZrO2-Al2O3 nano-composite by addition of HA. Biomaterials **26**, 509–517 (2005)
48. Kong, Y.M., Kim, D.H., Kim, H.E., Heo, S.J., Koak, J.Y.: Hydroxyapatite-based composite for dental implants: an in vivo removal torque experiment. J. Biomed. Mater. Res. **63**, 714–721 (2002)
49. Dulgar-Tulloch, A.J., Bizios, R., Siegel, R.W.: Human mesenchymal stem cell adhesion and proliferation in response to ceramic chemistry and nanoscale topography. J. Biomed. Mater. Res. A **90**, 586–594 (2009)
50. Zhang, S.: Emerging biological materials through molecular self-assembly. Biotechnol. Adv. **20**, 321–339 (2002)
51. Zhang, S., Holmes, T., Lockshin, C., Rich, A.: Spontaneous assembly of a self-complementary oligopeptide to form a stable macroscopic membrane. Proc. Natl. Acad. Sci. U.S.A. **90**, 3334–3338 (1993)
52. Zhang, S., Holmes, T.C., DiPersio, C.M., Hynes, R.O., Su, X., Rich, A.: Self-complementary oligopeptide matrices support mammalian cell attachment. Biomaterials **16**, 1385–1393 (1995)

53. Holmes, T.C., de Lacalle, S., Su, X., Liu, G., Rich, A., Zhang, S.: Extensive neurite outgrowth and active synapse formation on self-assembling peptide scaffolds. Proceedings of the National Academy of Sciences of the United States of America **97**, 6728–6733 (2000)
54. Zhang, S., Rich, A.: Direct conversion of an oligopeptide from a beta-sheet to an alpha-helix: a model for amyloid formation. Proc. Natl. Acad. Sci. U.S.A. **94**, 23–28 (1997)
55. Altman, M., Lee, P., Rich, A., Zhang, S.: Conformational behavior of ionic self-complementary peptides. Protein Sci. **9**, 1095–1105 (2000)
56. Zhang, S., Yan, L., Altman, M., Lassle, M., Nugent, H., Frankel, F., Lauffenburger, D.A., Whitesides, G.M., Rich, A.: Biological surface engineering: a simple system for cell pattern formation. Biomaterials **20**, 1213–1220 (1999)
57. Vauthey, S., Santoso, S., Gong, H., Watson, N., Zhang, S.: Molecular self-assembly of surfactant-like peptides to form nanotubes and nanovesicles. Proc. Natl. Acad. Sci. U.S.A. **99**, 5355–5360 (2002)
58. Tarasevicius, S., Tarasevicius, R., Zegunis, V., Smailys, A., Kalesinskas, R.J.: Metal type of the femoral stem in total hip arthroplasty. Medicina (Kaunas) **41**, 932–935 (2005)
59. Endres, S., Bartsch, I., Sturz, S., Kratz, M., Wilke, A.: Polyethylene and cobalt-chromium molybdenium particles elicit a different immune response in vitro. J. Mater. Sci. Mater. Med. **19**, 1209–1214 (2008)
60. Gosheger, G., Goetze, C., Hardes, J., Joosten, U., Winkelmann, W., von Eiff, C.: The influence of the alloy of megaprostheses on infection rate. J. Arthroplasty **23**, 916–920 (2008)
61. Collier, J.P., Mayor, M.B., Jensen, R.E., Surprenant, V.A., Surprenant, H.P., McNamar, J.L., Belec, L.: Mechanisms of failure of modular prostheses. Clin. Orthop. Relat. Res., 129–139 (1992)
62. Sargeant, T.D., Rao, M.S., Koh, C.Y., Stupp, S.I.: Covalent functionalization of NiTi surfaces with bioactive peptide amphiphile nanofibers. Biomaterials **29**, 1085–1098 (2008)
63. Thieme, M., Wieters, K.P., Bergner, F., Scharnweber, D., Worch, H., Ndop, J., Kim, T.J., Grill, W.: Titanium powder sintering for preparation of a porous functionally graded material destined for orthopaedic implants. J. Mater. Sci. Mater. Med. **12**, 225–231 (2001)
64. Bandyopadhyay, A., Espana, F., Balla, V.K., Bose, S., Ohgami, Y., Davies, N.M.: Influence of porosity on mechanical properties and in vivo response of Ti6Al4 V implants. Acta Biomater. **6**, 1640–1648 (2010)
65. Sargeant, T.D., Oppenheimer, S.M., Dunand, D.C., Stupp, S.I.: Titanium foam-bioactive nanofiber hybrids for bone regeneration. J. Tissue Eng. Regen. Med. **2**, 455–462 (2008)
66. Anderson, J.M., Kushwaha, M., Tambralli, A., Bellis, S.L., Camata, R.P., Jun, H.W.: Osteogenic differentiation of human mesenchymal stem cells directed by extracellular matrix-mimicking ligands in a biomimetic self-assembled peptide amphiphile nanomatrix. Biomacromolecules **10**, 2935–2944 (2009)
67. Yoshimi, R., Yamada, Y., Ito, K., Nakamura, S., Abe, A., Nagasaka, T., Okabe, K., Kohgo, T., Baba, S., Ueda, M.: Self-assembling peptide nanofiber scaffolds, platelet-rich plasma, and mesenchymal stem cells for injectable bone regeneration with tissue engineering. J. Craniofac. Surg. **20**, 1523–1530 (2009)

68. Hartgerink, J.D., Beniash, E., Stupp, S.I.: Self-assembly and mineralization of peptide-amphiphile nanofibers. Science **294**, 1684–1688 (2001)
69. Muller, P., Bulnheim, U., Diener, A., Luthen, F., Teller, M., Klinkenberg, E.D., Neumann, H.G., Nebe, B., Liebold, A., Steinhoff, G., Rychly, J.: Calcium phosphate surfaces promote osteogenic differentiation of mesenchymal stem cells. J. Cell. Mol. Med. **12**, 281–291 (2008)
70. Matsumine, A., Myoui, A., Kusuzaki, K., Araki, N., Seto, M., Yoshikawa, H., Uchida, A.: Calcium hydroxyapatite ceramic implants in bone tumour surgery. A long-term follow-up study. J. Bone Joint Surg. Br. **86**, 719–725 (2004)
71. White, E., Shors, E.C.: Biomaterial aspects of Interpore-200 porous hydroxyapatite. Dent. Clin. North Am. **30**, 49–67 (1986)
72. Vuola, J., Taurio, R., Goransson, H., Asko-Seljavaara, S.: Compressive strength of calcium carbonate and hydroxyapatite implants after bone-marrow-induced osteogenesis. Biomaterials **19**, 223–227 (1998)
73. Vines, J.B., Lim, D.J., Anderson, J.M., Jun, H.W.: Hydroxyapatite nanoparticle reinforced peptide amphiphile nanomatrix enhances the osteogenic differentiation of mesenchymal stem cells by compositional ratios. Acta Biomater. **8**, 4053–4063 (2012)
74. Anderson, J.M., Patterson, J.L., Vines, J.B., Javed, A., Gilbert, S.R., Jun, H.W.: Biphasic peptide amphiphile nanomatrix embedded with hydroxyapatite nanoparticles for stimulated osteoinductive response. ACS Nano **5**, 9463–9479 (2011)
75. Phan, P.V., Grzanna, M., Chu, J., Polotsky, A., el-Ghannam, A., Van Heerden, D., Hungerford, D.S., Frondoza, C.G.: The effect of silica-containing calcium-phosphate particles on human osteoblasts in vitro. J. Biomed. Mater. Res. A **67**, 1001–1008 (2003)
76. Tyagi, P., Catledge, S.A., Stanishevsky, A., Thomas, V., Vohra, Y.K.: Nanomechanical properties of electrospun composite scaffolds based on polycaprolactone and hydroxyapatite. J. Nanosci. Nanotechnol. **9**, 4839–4845 (2009)
77. Catledge, S.A., Clem, W.C., Shrikishen, N., Chowdhury, S., Stanishevsky, A.V., Koopman, M., Vohra, Y.K.: An electrospun triphasic nanofibrous scaffold for bone tissue engineering. Biomed. Mater. **2**, 142–150 (2007)
78. Thomas, V., Jagani, S., Johnson, K., Jose, M.V., Dean, D.R., Vohra, Y.K., Nyairo, E.: Electrospun bioactive nanocomposite scaffolds of polycaprolactone and nanohydroxyapatite for bone tissue engineering. J. Nanosci. Nanotechnol. **6**, 487–493 (2006)
79. Kanczler, J.M., Oreffo, R.O.: Osteogenesis and angiogenesis: the potential for engineering bone. Eur. Cell. Mater. **15**, 100–114 (2008)
80. Moon, J.J., West, J.L.: Vascularization of engineered tissues: approaches to promote angio-genesis in biomaterials. Curr. Top. Med. Chem. **8**, 300–310 (2008)
81. Mikos, A.G., Sarakinos, G., Lyman, M.D., Ingber, D.E., Vacanti, J.P., Langer, R.: Prevascularization of porous biodegradable polymers. Biotechnol. Bioeng. **42**, 716–723 (1993)
82. Nomi, M., Atala, A., Coppi, P.D., Soker, S.: Principals of neovascularization for tissue engineering. Mol. Aspects Med. **23**, 463–483 (2002)
83. Lokmic, Z., Mitchell, G.M.: Engineering the microcirculation. Tissue Eng. Part B Rev. **14**, 87–103 (2008)
84. Jain, R.K.: Normalization of tumor vasculature: an emerging concept in antiangiogenic therapy. Science **307**, 58–62 (2005)
85. Koike, Y., Uzuki, M., Kokubun, S., Sawai, T.: Angiogenesis and inflammatory cell infiltration in lumbar disc herniation. Spine (Phila Pa 1976) **28**, 1928–1933 (2003)

86. Ochoa, E.R., Vacanti, J.P.: An overview of the pathology and approaches to tissue engineering. Ann. N.Y. Acad. Sci. **979**, 10–26 (2002). discussion 35-8
87. Santos, M.I., Reis, R.L.: Vascularization in bone tissue engineering: physiology, current strategies, major hurdles and future challenges. Macromol. Biosci. **10**, 12–27 (2010)
88. Koike, N., Fukumura, D., Gralla, O., Au, P., Schechner, J.S., Jain, R.K.: Tissue engineering: creation of long-lasting blood vessels. Nature **428**, 138–139 (2004)
89. Yu, H., VandeVord, P.J., Mao, L., Matthew, H.W., Wooley, P.H., Yang, S.Y.: Improved tissue-engineered bone regeneration by endothelial cell mediated vascularization. Biomaterials **30**, 508–517 (2009)
90. Kaully, T., Kaufman-Francis, K., Lesman, A., Levenberg, S.: Vascularization–the conduit to viable engineered tissues. Tissue Eng. Part B Rev. **15**, 159–169 (2009)
91. Epstein, S.E., Kornowski, R., Fuchs, S., Dvorak, H.F.: Angiogenesis therapy: amidst the hype, the neglected potential for serious side effects. Circulation **104**, 115–119 (2001)
92. Moioli, E.K., Clark, P.A., Chen, M., Dennis, J.E., Erickson, H.P., Gerson, S.L., Mao, J.J.: Synergistic actions of hematopoietic and mesenchymal stem/progenitor cells in vascularizing bioengineered tissues. PLoS One **3**, e3922 (2008)
93. Holder, W.D., Gruber, H.E., Roland, W.D., Moore, A.L., Culberson, C.R., Loebsack, A.B., Burg, K.J.L., Mooney, D.J.: Increased Vascularization and Heterogeneity of Vascular Structures Occurring in Polyglycolide Matrices Containing Aortic Endothelial Cells Implanted in the Rat. Tissue Eng. **3**, 149–160 (1997)
94. Schechner, J.S., Nath, A.K., Zheng, L., Kluger, M.S., Hughes, C.C., Sierra-Honigmann, M.R., Lorber, M.I., Tellides, G., Kashgarian, M., Bothwell, A.L., Pober, J.S.: In vivo formation of complex microvessels lined by human endothelial cells in an immunodeficient mouse. Proc. Natl. Acad. Sci. U.S.A. **97**, 9191–9196 (2000)
95. Kaigler, D., Krebsbach, P.H., Wang, Z., West, E.R., Horger, K., Mooney, D.J.: Transplanted endothelial cells enhance orthotopic bone regeneration. J. Dent. Res. **85**, 633–637 (2006)
96. Kaigler, D., Krebsbach, P.H., West, E.R., Horger, K., Huang, Y.C., Mooney, D.J.: Endothelial cell modulation of bone marrow stromal cell osteogenic potential. Faseb J. **19**, 665–667 (2005)
97. Stahl, A., Wenger, A., Weber, H., Stark, G.B., Augustin, H.G., Finkenzeller, G.: Bi-directional cell contact-dependent regulation of gene expression between endothelial cells and osteoblasts in a three-dimensional spheroidal coculture model. Biochem. Biophys. Res. Commun. **322**, 684–692 (2004)
98. Wenger, A., Stahl, A., Weber, H., Finkenzeller, G., Augustin, H.G., Stark, G.B., Kneser, U.: Modulation of in vitro angiogenesis in a three-dimensional spheroidal coculture model for bone tissue engineering. Tissue Eng. **10**, 1536–1547 (2004)
99. Grellier, M., Granja, P.L., Fricain, J.C., Bidarra, S.J., Renard, M., Bareille, R., Bourget, C., Amedee, J., Barbosa, M.A.: The effect of the co-immobilization of human osteoprogenitors and endothelial cells within alginate microspheres on mineralization in a bone defect. Biomaterials **30**, 3271–3278 (2009)
100. Ko, H.C., Milthorpe, B.K., McFarland, C.D.: Engineering thick tissues–the vascularisation problem. Eur. Cell. Mater. **14**, 1–18 (2007). discussion 18-9
101. Rouwkema, J., de Boer, J., Van Blitterswijk, C.A.: Endothelial cells assemble into a 3-dimensional prevascular network in a bone tissue engineering construct. Tissue Eng. **12**, 2685–2693 (2006)

102. Unger, R.E., Sartoris, A., Peters, K., Motta, A., Migliaresi, C., Kunkel, M., Bulnheim, U., Rychly, J., Kirkpatrick, C.J.: Tissue-like self-assembly in cocultures of endothelial cells and osteoblasts and the formation of microcapillary-like structures on three-dimensional porous biomaterials. Biomaterials **28**, 3965–3976 (2007)
103. Au, P., Tam, J., Fukumura, D., Jain, R.K.: Bone marrow-derived mesenchymal stem cells facilitate engineering of long-lasting functional vasculature. Blood **111**, 4551–4558 (2008)
104. Guillotin, B., Bourget, C., Remy-Zolgadri, M., Bareille, R., Fernandez, P., Conrad, V., Amedee-Vilamitjana, J.: Human primary endothelial cells stimulate human osteoprogenitor cell differentiation. Cell. Physiol. Biochem. **14**, 325–332 (2004)
105. Villars, F., Guillotin, B., Amedee, T., Dutoya, S., Bordenave, L., Bareille, R., Amedee, J.: Effect of HUVEC on human osteoprogenitor cell differentiation needs heterotypic gap junction communication. Am. J. Physiol. Cell Physiol. **282**, C775–C785 (2002)
106. Gomes, M.E., Sikavitsas, V.I., Behravesh, E., Reis, R.L., Mikos, A.G.: Effect of flow perfusion on the osteogenic differentiation of bone marrow stromal cells cultured on starch-based three-dimensional scaffolds. J. Biomed. Mater. Res. A **67**, 87–95 (2003)
107. Grellier, M., Bordenave, L., Amedee, J.: Cell-to-cell communication between osteogenic and endothelial lineages: implications for tissue engineering. Trends Biotechnol. **27**, 562–571 (2009)
108. Fuchs, S., Hofmann, A., Kirkpatrick, C.J.: Microvessel-like structures from outgrowth endothelial cells from human peripheral blood in 2-dimensional and 3-dimensional co-cultures with osteoblastic lineage cells. Tissue Eng. **13**, 2577–2588 (2007)
109. Bouletreau, P.J., Warren, S.M., Spector, J.A., Peled, Z.M., Gerrets, R.P., Greenwald, J.A., Longaker, M.T.: Hypoxia and VEGF up-regulate BMP-2 mRNA and protein expression in microvascular endothelial cells: implications for fracture healing. Plast. Reconstr. Surg. **109**, 2384–2397 (2002)
110. Yamaguchi, A., Ishizuya, T., Kintou, N., Wada, Y., Katagiri, T., Wozney, J.M., Rosen, V., Yoshiki, S.: Effects of BMP-2, BMP-4, and BMP-6 on osteoblastic differentiation of bone marrow-derived stromal cell lines, ST2 and MC3T3-G2/PA6. Biochem. Biophys. Res. Commun. **220**, 366–371 (1996)
111. Santos, M.I., Unger, R.E., Sousa, R.A., Reis, R.L., Kirkpatrick, C.J.: Crosstalk between osteoblasts and endothelial cells co-cultured on a polycaprolactone-starch scaffold and the in vitro development of vascularization. Biomaterials **30**, 4407–4415 (2009)
112. Sun, H., Qu, Z., Guo, Y., Zang, G., Yang, B.: In vitro and in vivo effects of rat kidney vascular endothelial cells on osteogenesis of rat bone marrow mesenchymal stem cells growing on polylactide-glycoli acid (PLGA) scaffolds. Biomed. Eng. Online **6**, 41 (2007)

Engineering Integrative Stem Cell and Biomaterial Therapies for Peripheral Artery Disease

Daniel A. Balikov, Sue Hyun Lee, Timothy C. Boire, Jung-bok Lee, Angela L. Zachman and Hak-Joon Sung

Abstract Because of their potential to regenerate tissues and organs, stem cells garner extensive interest worldwide for treating a wide range of degenerative diseases. In numerous efforts, a variety of stem cell formats have been considered for therapeutic purposes to combat such pathologies, one being vascular-related diseases. In particular, the prevalence of peripheral artery disease (PAD) has steadily increased with the growth of the aging population in many first-world countries. Considering this disturbing trend as well as the obesity epidemic and the ballooning population growth in third-world nations, the burden of PAD is expected to increase worldwide to alarming levels in the coming decades. The advent of stem cell treatments could stymie this burden by alleviating the complications and improving upon the less-than-satisfactory outcomes from current standard-of-care surgical/pharmacological interventions for PAD. This chapter reviews the relevant, cutting-edge clinical and animal model research efforts in the field and explores the remaining questions as they pertain to convergence technologies that may provide the potential for stem cells to reverse PAD-induced tissue damage. Harnessing and translating this potential to create more viable and efficacious PAD treatments should be of paramount global and public health concern.

1 Introduction

Stem cells are principally characterized as cells that have the ability to self-replicate and provide a progenitor base from which specialized cell types arise and form specific tissues and organs. The lineage potential of stem cells is categorized

D.A. Balikov · S.H. Lee · T.C. Boire · J.-b. Lee · A.L. Zachman · H.-J. Sung(✉)
Department of Biomedical Engineering, Vanderbilt University, Nashville, USA
e-mail: hak-joon.sung@vanderbilt.edu

© Springer International Publishing Switzerland 2016
H. Jo et al. (eds.), *Biomedical Engineering: Frontier Research and Converging Technologies*, Biosystems & Biorobotics 9,
DOI: 10.1007/978-3-319-21813-7_6

into the following classes: totipotent, pluripotent, multipotent, oligopotent, and unipotent. Totipotent stem cells are created from the fusion of an egg and sperm cells, and together they produce a complete, viable organism via differentiation into embryonic and extraembryonic lineages. During the early stages of embryonic development, pluripotent stem cells, also known as embryonic stem cells (ESCs), reside within the blastocyst and can differentiate into all three primary germ layers (i.e. ectoderm, mesoderm, endoderm). The strict regulation and scant availability of these cell lines, however, limits their therapeutic potential. Multipotent, oligopotent and unipotent stem cells typically refer to stem cell niche populations within adult tissues that give rise to more specialized cells that comprise the tissue from which the stem cells are housed. Many of these stem cell types are extensively studied for their ability to produce vascular, chondrocytic, adipogenic, osteogenic, and neurogenic lineages, among many others that can be useful for tissue regeneration applications.

The most readily available sources of stem cells are multipotent cells that reside within the bone marrow (e.g. mesenchymal and hematopoietic stem cells), adipose tissue (e.g. adipose-derived stem cells), or blood (e.g. endothelial progenitor cells). An improving grasp of the mechanisms and conditions responsible for stem cell physiology, as well as the evolving genetic engineering techniques that manipulate stem cell phenotypes, has led to the development of a new laboratory-derived stem cell: induced pluripotent stem cells (iPSCs). iPSCs are pluripotent stem cells that are dedifferentiated from specialized cell types, such as fibroblasts, and can be forced down virtually any lineage path given the correct *in vitro* culture conditions. These pluripotent cell lines are being investigated as treatments for a plethora of human pathologies including diabetes, Alzheimer's disease, cardiovascular disease, and wound healing following severe burns [1,2]. Yet, few stem cell-based therapies have thus far entered the clinic, which highlights the urgent need to address the key limitations in their translation into the clinic. In particular, the retention and differentiation of stem cell therapies at the disease site will need to be optimized in a consistent, controlled manner to effectively replace the diseased tissue of interest. In order to achieve this and ensure clinical safety and efficacy, the most appropriate stem cell sources and means for delivery and expression control will need to be determined for each pathology being addressed by these therapies.

For the purposes of this chapter, the primary focus is on the past, current and future directions related to the therapeutic potential of adult stem cells on one specific disease, peripheral artery disease (PAD). PAD is commonly characterized by atherosclerosis, with plaque buildup in peripheral arteries limiting blood flow to the extremities. This can ultimately result in leg pain and critical limb ischemia (CLI). Blockages can develop in the arteries of the leg including the femoral, popliteal, posterior tibial, and dorsalis pedis arteries. As the level of perfusion to the limbs is reduced, leg tissue begins to become hypoxic and eventually dies leading to gangrene, thereby necessitating the need for limb amputation to prevent the spread of gangrenous tissue and risk of sepsis.

Due to the increase in life expectancy in first-world countries and the rising obesity epidemic, the prevalence of PAD is likely to surge and induce an unprecedented cost burden on society [3]. PAD is a cause of major clinical and financial burdens throughout the world. In the United States alone, approximately 12% of the adult population has some form of PAD. Furthermore, a study exploring the costs of treating PAD in 2001 totaled $4.37 billion dollars alone [4], costs that can go above $10,000 per patient over a two year period [5]. This prevalence rises to 20% in the elderly and diabetic populations [6], while smokers, hypertensive, and high-cholesterol patients are also at an increased risk [7,8]. Existing treatments (e.g. pharmacological and surgical approaches to be discussed later) have demonstrated marginal improvements in PAD outcomes, but the progression of the disease is rarely, if ever, permanently abrogated. Multiple amputations are often unavoidable and only temporarily halt its progression. New methods of treating PAD are therefore requisite to reduce the expected rise in PAD patients. Innovative regenerative medicine strategies employing stem cells are being explored to meet this challenge. We will discuss past, existing and future work within the theme of biomaterials engineering- and stem cell-based therapies in this chapter in order to highlight the active role that research and clinical communities are playing to remedy the destructive toll of PAD on the adult population.

2 Background on Bone Marrow Mononuclear Stem Cells and Peripheral Blood Mononuclear Stem Cells

Before discussing the application of adult stem cells in therapeutic applications, it is important to articulate some of the different adult cell types and elaborate on their location in the body. Adult stem cells are typically stem cells found within the body and maintain their presence in post-natal humans. Several different adult stem cell niches exist such as bone marrow, intestinal, musculoskeletal and neural stem cells [9]. Historically, bone marrow stem cells have been the most heavily studied and most widely used in PAD-based therapies. The bone marrow is, generally speaking, an amalgamation of several cell types and matrix materials that reside in the center of large and long bones in most vertebrate organisms. Within these spaces, many different cell types are created, the primary two types being hematopoietic and mesenchymal stromal lineages. Hematopoietic lineages are involved in creating all the blood cell types that are found in the body while mesenchymal stromal lineages mainly give rise to cell types within connective tissue matrices and general soft tissues. Both umbrella lineages described are mononuclear stem cells at the most upstream lineage state that inevitably give rise to a whole host of cell lineages, some of which can become multinuclear (e.g. megakaryocytes) and further differentiate into more specific cell types [10].

Despite hematopoietic stem cells being a major cell population within the bone marrow, stem cell therapies have focused on the mesenchymal stromal lineages for regenerative medicine potential in patients [11]. This is primarily because the hematopoietic cells found in the bone marrow only give rise to blood cells, both red and white, as well as platelets, none of which can serve to replace lost functional cells in organs or the connective tissue. The specific cell in the non-hematopoietic stem class most often studied is the mesenchymal stromal cell (MSC). The first evidence of their existence was described decades ago by Friedenstein *et al.* as spindle-shaped cells with multi-lineage differentiation potential (Figure 1A) and adherent behavior to culture plastic [12]. While many surface profiling studies have been conducted on mesenchymal stromal cells, it is estimated that they represent an extremely small fraction (0.01 to 0.001%) of total cells found in the bone marrow [13]. As such, the dominant role they appear to play in directing the lineage fate of whole bone marrow mononuclear cell aspirates remains to be seen as the literature is slowly painting a picture of the mechanisms that MSCs execute to direct stem cell differentiation within the bone marrow.

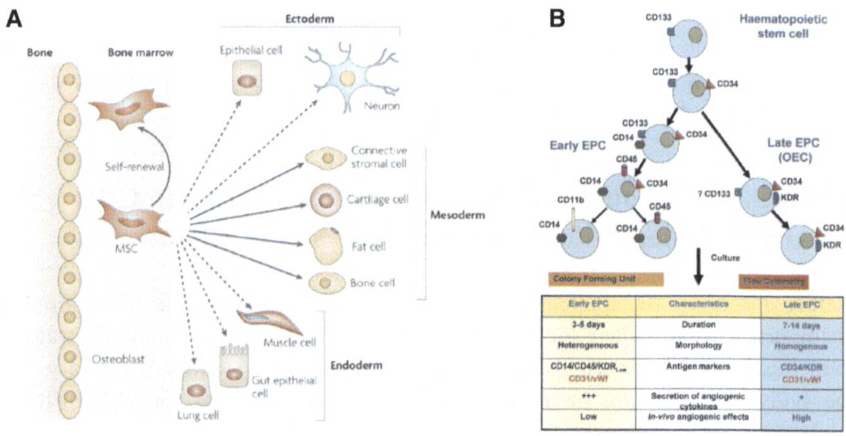

Fig. 1 Bone marrow and peripheral blood stem cells. (A) Example of a bone marrow-derived MSC and its potential to differentiate into various lineages. Although not depicted in this figure, MSCs also have the potential to differentiate into vascular endothelial cells and promote angiogenesis for vascular repair [19,20]. (B) Differentiation pathway of CD34$^+$ peripheral blood stem cells from a hematopoietic stem cell lineage, which can serve to become endothelial cells within the vasculature and possess greater pro-angiogenic effects. Both figures reprinted from [21] and [22], respectively, with permission.

Peripheral blood mononuclear stem cells have also been investigated for their therapeutic potential in PAD, but are more related to the hematopoietic stem cell niche than the stromal cell niche. Depending on the surface profile used, peripheral blood mononuclear stem cells are also referred to as endothelial progenitor cells given that they are easily recruited and differentiated into endothelial cells [14,15]. Within the progenitor cell purview, the vascular biology community has identified that progenitor cells (Figure 1B) that are CD133$^+$/CD34$^+$ produce endothelial cells in the vasculature setting and can prove critical in restoring normal blood flow to ischemic tissue [16,17]. It remains to be seen if further sorting the $CD34^+$ fraction for therapeutic use has significant advantages over unfractionated peripheral blood mononuclear stem cells, as CD34+ cells are known to be more involved in mediating the attachment to the vascular architecture [18].

3 Definition of Peripheral Artery Disease

PAD is most commonly characterized by plaque buildup within the artery walls of the extremities (Figure 2), whereby blood flow is severely compromised and leads to critical hypoxia necessitating amputation. At the molecular level, the low levels of oxygen within the plaque prevents the degradation of hypoxia inducible factor 1-alpha (HIF-1α) and stimulates the increased production of HIF-1α by both endothelial cells and macrophages in a feed-forward fashion [23,24] (Figure 3A). This event is what accelerates the development of PAD and allows it to spread within the vasculature uninhibited, eventually leading to limb loss. As reviewed by Gao *et al*, HIF-1α production is increased by high levels of reactive oxygen species (ROS), TNF-α, and IL-1β produced by inflammatory cells. In turn, HIF-1α increases the accumulation of low density lipoproteins (LDL) by macrophages to form foam cells, increasing the size of the plaque and decreasing the levels of oxygen within the plaque [23,25] (Figure 3B). As the plaque grows, smooth muscle cells also migrate into the intima and deposit ECM fibers [8]. The macrophages, smooth muscle cells, and endothelial cells in the plaque then undergo apoptosis due to the depletion of oxygen and nutrients. As a result, the plaque becomes highly susceptible to rupturing or dislodging from the blood vessel, causing the formation of a thrombus [23]. A thrombus can also form by secretion of matrix metalloproteinase (MMP) via endothelial cells, which degrade the basement membrane surrounding the plaque [8]. The thrombus can then break free of the artery wall and travel through the vasculature to occlude smaller vessels downstream of the original blockage, further contributing to the risk of subsequent thrombosis in the downstream microvessels.

However, the body can attempt to repair itself as PAD is developing. The production of ROS and HIF-α inside the plaque stimulates the formation of microvessels to supply nutrients to the cells in the plaque [26]. HIF-α also induces expression of vascular endothelial growth factor receptors 1 and 2 (VEGFR1 and VEGFR2) that promote angiogenesis to restore blood flow to tissue distal from the blockage [23]. Yet, depending on how advanced the disease is, this response mechanism of creating collateral vessels can be too little and too late, begetting the PAD disease cycle to continue and amplify its effect on the local tissue.

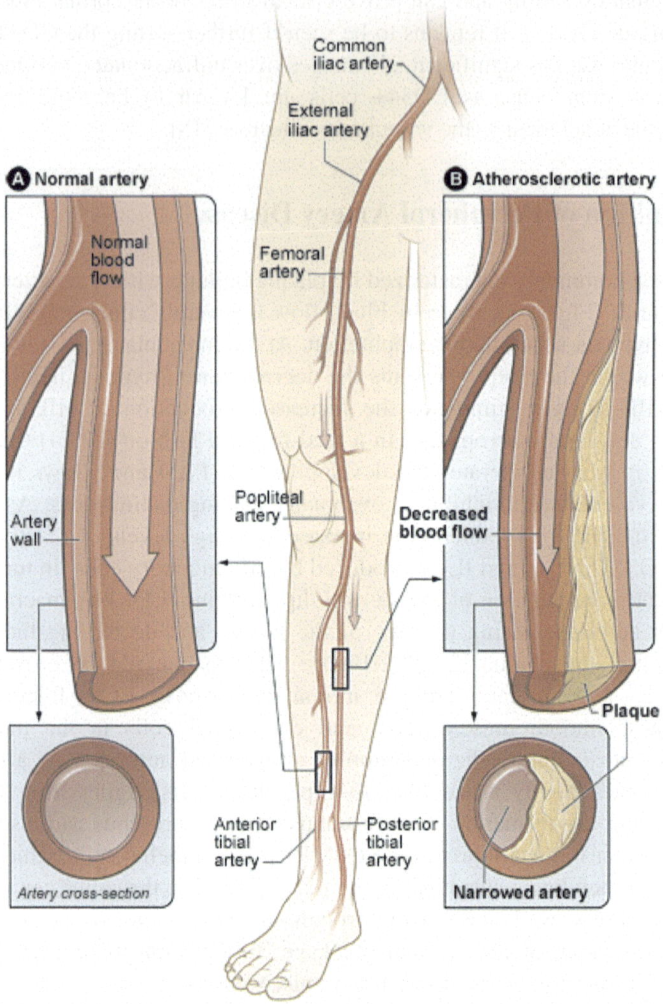

Fig. 2 Anatomical illustration of common sites for PAD occurrence. Reprinted with permission from the National Heart, Lung, and Blood Institute; National Institutes of Health; U.S. Department of Health and Human Services.

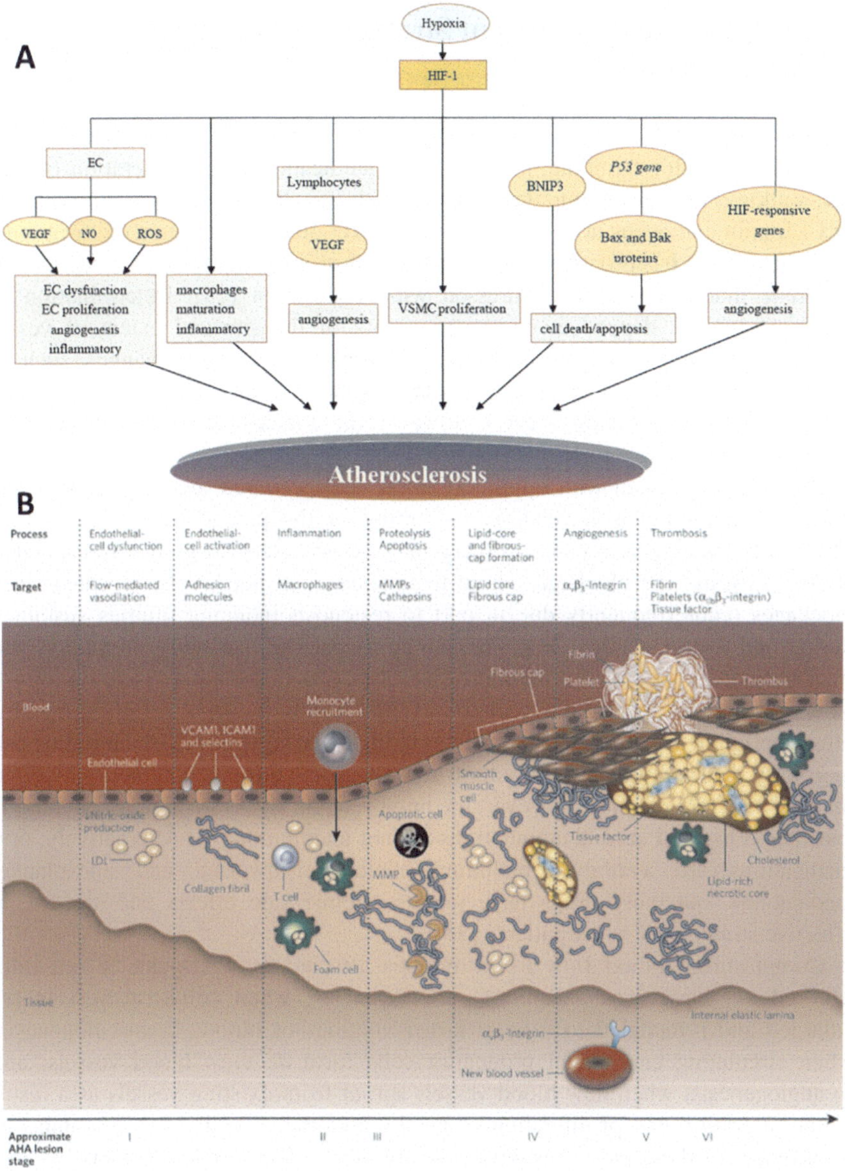

Fig. 3 Progression of atherosclerosis. Reprinted with permission from A) Gao et al [27] and B) Sanz et al [25]. BNIP3, Bcl-2/E1B interacting protein.

4 Symptoms and Treatment Strategies for Peripheral Artery Disease

Patients with PAD often present with impaired walking, known as claudication, the Latin word for "limping." Classically, claudication causes alteration of gait (i.e. walking) due to a cramp that subsides with rest. When the disease progresses to CLI, pain can occur at rest. Even worse, PAD often has little or no symptoms until the disease has progressed to CLI. Without proper treatment, CLI can lead to gangrene that can only be addressed with amputation [28]. Determination of treatment type is dependent on the proximity of the site and extent of vessel occlusion, as well as the surgical risk to the patient. Lifestyle changes, such as cessation of smoking, increased exercise, weight loss, and careful nutrition can slow down the progression of PAD. However, these measures are not adequate for patients with moderate to severe ischemia [8].

PAD can be treated with minimally invasive procedures such as percutaneous transluminal angioplasty, stenting, or cutting of the plaque (i.e., atherectomy) [29-31]. In more severe cases, surgery may be necessary to graft a transplanted artery to bypass the blockage. These treatments often need to be repeated since blockages recur frequently due in part to restenosis-inducing injuries sustained during and following their implementation, increasing morbidity, mortality, and financial burdens [8,31]. Furthermore, such procedures are not a viable option for patients with widespread atherosclerosis [32]. Diabetic patients with CLI are also often ineligible for invasive treatments and have a higher risk of mortality and morbidity than non-diabetic patients with CLI. Amputation is commonly required for diabetic PAD patients; however, the three year survival rate after amputation is less than 50%, and second amputations are not uncommon [8]. Alternatively, oral medications are available, including cholesterol-lowering and anti-platelet medications, but these drugs have many systemic side effects and are not an effective strategy for many patients [33].

Compromised blood flow to extremities affected by PAD can be naturally restored via three mechanisms: 1) arteriogenesis, when collateral vessels are enlarged and remodeled to support larger volumes of blood; 2) vasculogenesis, when circulating endothelial progenitor cells form *de novo* blood vessels; and 3) angiogenesis, when new blood vessels sprout from existing vessels as a result of the migration and proliferation of nearby endothelial cells [34]. Spontaneous occurrences of these processes, however, are insufficient in PAD patients [32,35]. There is therefore a need to develop and implement therapies that promote arteriogenesis, vasculogenesis, and angiogenesis; stem cells provide a potential solution towards this aim.

5 Clinical Trials Treating PAD with Stem Cells

5.1 *Initial Trials with Bone Marrow Mononuclear Stem Cell Injections*

The earliest pilot studies using stem cells in patients for treating PAD started in the late 1990s. First published in 2002, Tateishi-Yuyama *et al.* compared injections of bone marrow-mononuclear cells and endothelial progenitor cells in the gastrocnemius muscles of two patients groups [36]. Much of the preclinical work preceding the trial had numerous *in vitro* and *in vivo* animal studies exploring how both stem cell populations could combat the detrimental effects of PAD, where it was found that implantation of those cells rescued ischemic limb damage via collateral vessel formation. The pilot study consisted of two test groups (47 patients total). The first group was injected with bone marrow-mononuclear cells in one ischemic limb while the second limb was injected with saline; the second group would be injected with bone marrow-mononuclear cells in one leg and endothelial progenitor cells in the other. After 4 weeks post-injection, the second group exhibited lower ankle-brachial index (ABI, where lower values indicate appropriate anatomical blood pressures) responses in the leg receiving the bone marrow cells compared to the endothelial progenitor cells. Parallel improvements were also noticed with respect to transcutaneous oxygen pressure (TcO_2), resting pain and pain-free walking duration. The first group also had similar response profiles in the bone marrow cell-injected leg compared to the saline controls. Bone marrow-mononuclear cells were even effective in some extreme PAD cases, aiding in the healing of leg ulcers and gangrene progression. This first-of-its-kind study supports the notion that bone marrow-mononuclear cells are safe and effective in treating PAD.

Unlike the Tateishi-Yuyama *et al.* trial, a handful of the other studies conducted had significantly less patients in the test groups. In some cases, certain trials would be unblinded or limited in nested conditions. However, given that PAD can progress unrelentingly, their findings are informative in ascertaining a baseline understanding of the effect of stem cells on the disease. These studies are especially informative given that patients suffering from refractory PAD (i.e. unresponsive to standard-of-care treatments) were not excluded from these clinical trials. Two small-scale studies by Esato *et al.* [37] and Miyamoto *et al.* [38] assessed the safety and efficacy of administering autologous bone marrow mononuclear stem cells to treat pathological tissues for patients who did not exhibit positive responses to standard-of-care therapy. Esato *et al* designed their study to look at 8 refractory PAD patients. Interestingly, the research group used autologous bone marrow mononuclear stem cells instead of allogenic stem cells for the injections into the calf, foot or hand regions afflicted by PAD. Collateral vessel formation was seen and ulcer healing occurred for most of the ulcer site injections. Subjective symptoms were also improved in 7 of the 8 patients.

Unfortunately, no investigation was made into the inflammatory response to highlight the advantage of using an autologous stem cell source.

Miyamoto *et al.* were able to demonstrate the short-term safety of injecting autologous bone marrow mononuclear stem cells for PAD patients with thromboangiitis obliterans that were refractory towards conventional treatments [38]. Thromboangiitis obliterans, or Buerger's disease, is a condition in which the arteries and veins of the hands and feet acutely and chronically inflame and thrombose, commonly resulting in pain and the formation of ulcers in the affected areas. After 4 weeks, decreased pain in the injection limbs were reported and ulcers were either healed or extensively in the process of healing. However, long-term adverse events were reported in 50% of the test subjects (4 of 8), bringing into question the long-term safety and/or efficacy of this therapy for CLI patients with Buerger's disease. The events included death nearly 20-months post-implantation, incomplete ulcer healing that progressively worsened after 4 months, worsening of resting state pain after 8 months, and development of an arteriovenus shunt that spontaneously regressed after 1 year. The authors concluded that careful long-term monitoring would be required as there is little control over stem cell behavior at the injection site, and even if the stem cells were active at the injury site, this does not necessarily translate to cessation or reversal of the natural progression of PAD.

5.2 Further Trials Solely Using Bone Marrow Mononuclear Stem Cell Injections

As these aforementioned initial clinical trials were being published, the baseline information gathered from those studies allowed for more focused trials. A larger clinical trial was able to more convincingly demonstrate the benefit of bone marrow mononuclear stem cell treatments in PAD patients with Buerger's disease refractory towards conventional treatments [39]. With a study population of 67 patients and 90 limbs, the affected leg was injected with autologous bone marrow mononuclear cells in the calf area while the control leg, if a patient had an unaffected limb, was given a saline injection. With a mean follow-up time of 16.6 months, 50% of patients had improved ABI values in addition to rest pain score, peak walking time and patient reported quality of life. 83% of patients also saw healing of ischemic ulcers without the need for narcotic analgesics. Collateral vessel network formations were also confirmed in 78.5% of the study group. Hence, the clinical trial authors could conclude that at least one specific disease under the PAD umbrella term exhibited very favorable responses to this stem cell treatment.

The power of these stem cell treatments was further corroborated by a study that explored if such medical interventions could promote amputation-free survival, a metric not previously reported on in prior studies. Murphy *et al* recruited 29 patients with CLI, injected autologous bone marrow mononuclear cells, and conducted, at the time, some of the most thorough endpoint measurements that had been reported in the literature [40]. Most importantly, the

1-year amputation-free survival rate was 86.3%. Additionally, significant increases were reported in first toe pressure (FTP), toe-brachial index (TBI), ABI, and perfusion index by positron emission tomography-computed tomography (PET-CT) with intra-arterial water infusion.

Bone Marrow Mononuclear Stem Cell Injections: Are Mesenchymal Stem Cells or Mononuclear Cells Driving this Therapeutic Effect?

Despite these initially promising results, some studies began to bring up questions regarding which cells from the bone marrow stromal population are better suited for injection, given that this could affect the consistency of therapeutic outcomes and may be important from a regulatory perspective as well [41]. Lu *et al* asked whether bone marrow mesenchymal stem cells or bone marrow mononuclear cells provided the best healing performance in diabetic critical limb ischemia and foot ulcers [42]. The double-blind, randomized control trial consisting of 41 subjects documented that mesenchymal stem cells provided a higher ulcer healing rate than bone marrow mononuclear cells at the same time point. Other measures like ABI, TcO_2 and magnetic resonance angiography (MRA) had similarly significant trends. The most interesting fact, though, was that significantly less volume (30mL versus 300mL) was needed for mesenchymal stem cell injection compared to bone marrow mononuclear cells, bringing to light that the mesenchymal stem cell-purified populations would be better suited for future stem cell therapies treating PAD, similar to adipose-derived mesenchymal stem cells in other PAD trials [43,44]. Yet, other studies that were conducted concurrently or shortly after the publication by Lu *et al* challenged the findings based on low performance measures [45].

5.3 *Bone Marrow Mononuclear Stem Cell Injections: The Influence of Delivery Methods on Therapeutic Effects*

Other studies began investigating the role that delivery methods play on therapeutic effects of bone marrow mononuclear stem cells in PAD. One of the most recent trial studies using bone marrow mononuclear cells was verifying if intra-arterial injections had any advantages in affected PAD [46]. The 40-patient phase II double-blind randomized trial had patients separated into two groups: stem cell injection or placebo by intra-arterial route, or placebo injection by intra-arterial route followed by stem cell injection 3 months later, also by an intra-arterial administration. ABI, the primary endpoint reading that has high clinical prognosis relevance, did not exhibit a significant increase compared to the placebo. (Of note, the ABI finding had since been refuted in another study using the same injection method, indicating further study into this injection methodology is necessary [47,48].) Despite the fact that no improvement in ABI was observed in this study, the intra-arterial delivery route was found to improve ulcer healing and reduce rest pain. Limb amputation was not reduced with this method compared to controls. While it was concluded that intra-arterial delivery of stem cells was shown to be safe, the comparison against intramuscular delivery was not tested in this trial. Therefore, larger, more controlled studies are required to determine if the route of stem cell injection to the site of injury does in fact have a significant impact on patient prognosis.

5.4 Peripheral Blood Mononuclear Stem Cell Injections: A Potentially Attractive Alternative Stem Cell Source

As Tateishi-Yuyama *et al* were able to demonstrate that endothelial progenitor cells could aid in the healing of limbs afflicted with PAD, other groups decided to focus on peripheral blood mononuclear cells to further explore their performance against PAD. The first follow-up study conducted by Huang *et al* used granulocyte colony-stimulating factor (G-CSF)-mobilized peripheral blood mononuclear cells as the injection cell for 28 diabetic patients with critical limb ischemia [49]. G-CSF was injected over 5 days to help collect the progenitor cells; after collection, the cells were injected intramuscularly at the site of ischemia. Three months after injection, limb pain and ulcer healing were greatly improved, and Doppler blood perfusion measurements confirmed increased perfusion in the injection limb compared to controls. No adverse reactions were reported in the injection patients and limb amputation was absent in the trial population. Interestingly, for these diabetic patients, injection of the G-CSF-mobilized peripheral blood mononuclear cells improved blood glucose metabolism, an unexpected benefit from the treatment. These results demonstrate the efficacy of endothelial progenitors. Moreover, these progenitor cells may be preferable to their bone marrow counterparts because they may be easier to obtain and are believed to be involved in the recruitment of other cells that aid in the healing process.

A similarly designed study by Lenk *et al* also looked at G-CSF-stimulated circulating progenitor cells to salvage PAD-damaged limbs [50]. Despite only investigating seven patients, the study subjects had severe PAD. Thus, if any changes were seen towards halting and reversing disease progression, this could be indicative of a therapeutic effect of these circulating cells. Like the prior endothelial progenitor cells studied, G-CSF was given to the patients and no adverse events were reported. The circulating progenitor cells were collected, injected and then the injured limbs were evaluated after 3 months. Across the board, many indicators like ABI, TcO_2, flow-dependent vasodilation and response to adenosine were significantly improved. While some critics may note that the findings have limited ramifications without the control group, the credibility of these results are strengthened by the fact that measured factors such as ABI and TcO_2 are difficult to influence subjectively.

Amputation outcomes were also explored for $CD34^+$ peripheral blood mononuclear stem cell injections in a randomized controlled pilot study conducted by Losordo *et al.* [51]. 28 patients received the CD34+ cell injections (control, low dose and high dose) intramuscularly and were monitored over one year for any amputation events. At 6 months, two-thirds of the control subjects had experienced a major or minor amputation. While fewer patients treated with low or high doses of these cells necessitated amputation at 6 months, these results were not statistically significant to the control until 12 months. At 12 months, 75% of control patients had undergone an amputation while there was no change over the 6-to-12 month follow-up interval for the treated groups (43% and 22% of low dose and high dose patients, respectively). An expanded trial was later conducted by Kinoshita *et al* utilizing $CD34^+$ cell injections for treating critically ischemic limbs over 208 weeks of observation [52]. The first patient evaluations conducted at weeks 52 and 104 found that 88.2% of the injection subjects were ischemia-free. This number fluctuated to 92.3% at week 156 and 84.6% at week 208. TBI and TcO_2 were both elevated significantly compared to paired baselines, and decreased pain, ulcer size and increased exercise tolerance were also significantly improved. The study group was also large enough to specifically study a patient subgroup with Buerger's disease, which reflected the same trends as the larger cohort.

5.5 A Direct Comparison between Peripheral Blood Mononuclear Stem Cell and Bone Marrow Mononuclear Stem Cell Injections

Two years later, a 150-patient study using peripheral blood mononuclear cells was published [53]. The patient population was split into two groups. One group received the peripheral blood stem cells while the second groups received bone marrow mononuclear stem cells. To much surprise, the group having received the peripheral blood stem cells performed significantly better on all major indices than the bone marrow mononuclear stem cells. ABI, rest pain and skin temperature were dramatically influenced towards recovery from the disease. However, no significant difference was seen with respect to pain-free walking distance, TcO_2, ulcer healing or rate of limb amputation, all of which are contradictory to the Tateishi-Yuyama *et al.* findings. Comparatively, both stem cell treatment groups had improvement for the patients, but the results were clear that peripheral blood stem cells were better suited for combating PAD, as well as being more easily obtained and prepared for the treatment routines.

5.6 Therapeutic Effects of Purified Peripheral Blood Mononuclear Stem Cell Injections

Research groups interested in endothelial progenitor cells obtained $CD34^+$ stem cells from their patient donors using similar purification steps to those used for separating out mesenchymal stem cells from the whole bone marrow mononuclear stem cell population. In a phase I/IIa, single-blinded, dose-escalation trial, investigators injected the purified stem cell population into the legs (intramuscularly) of patients with severe ischemia resulting from Buerger's disease [54]. Following 12 weeks of healing time after the injection, patients were noted to have increased TBI and TcO_2, as well as decreased pain and ulcer size. Of note, while adverse events from such stem cell transplants of this nature are rare, mild reactions to G-CSF and leukapheresis events were present in a small number of the patients. As the authors of the study noted, at this point, no other clinical trial had explored the use of purified endothelial progenitor cells to aid in combating PAD with chronic ischemia. Like the prior study mentioned above that isolated and concentrated mesenchymal stem cells for intramuscular injection in PAD-affected limbs, a similar isolation of endothelial progenitor cells appears to act more efficaciously to retard the disease progression. Overall, these overlapping findings suggest that only a small fraction of bone marrow mononuclear cells truly contribute the most therapeutic benefit.

5.7 Effect of Non-mobilized Peripheral Blood Mononuclear Stem Cell Injections

A curious question asked by one group of clinicians was what the response of ischemic limbs would be to non-mobilized peripheral blood mononuclear stem cells. Unlike all previous clinical trials with stem cells, Mutirangura *et al.* neglected the use of G-CSF when harvesting the peripheral stem cells for patient use [55], which would have otherwise primed the blood circulation with peripheral blood stem cells to maximize their chance of affecting a healing process at the PAD injury site. When injected into the limb with critical ischemia, 83.3% of test subjects showed local improvement in circulation within the distal limb, while one patient unfortunately had spreading of granulation tissue from a severe foot infection that preceded the clinical trial. A total of 6 patients were used, which admittedly is a limitation to interpreting the results; making firm conclusions to tease out the effects of non-mobilized cells on PAD outcomes requires extensive follow-up in a more robust, expansive, randomized clinical trial. This was indirectly addressed again by a clinical trial using VesCell therapy to extract non-mobilized peripheral mononuclear stem cells from 20 patients and found a positive correlation with the peripheral stem cell-injected legs and clinical measures such as ABI, circulation, wound healing and decreased pain [56].

5.8 Conclusion from PAD Clinical Trials with Adult Stem Cells

Provided all the past and ongoing clinical trials using stem cells, and in some select cases growth factors to combat PAD [57], the performance measure results for PAD regression and prevention of amputation are and still remain somewhat mixed. In many stem cell therapy settings, the most common attributable cause for failure of efficacy across a host of testable conditions is that of low cell engraftment to the surrounding tissue. This becomes even more apparent when soluble growth factors alone are injected into the pathological site of interest, as they diffuse into the tissue and are not bound to proteins or engulfed by cells the clinicians intended to target. It is clear that a new vehicle is needed to more specifically deliver stem cells and growth factors to the injury site. One viable candidate to overcome these challenges is a biomaterial scaffold [58]. Aside from their potential to more effectively engineer stem cell behavior and locality to facilitate a more robust, faster healing responses than the stem cell-only injections demonstrated, many biomaterials systems can also be administered to the site of injury in a minimally-invasive manner. This capability reduces the risk of surgical complications that would otherwise be incurred from implantation of tradition scaffolds. PAD could stand to greatly benefit from such a strategy and the results from trials that used biomaterials would likely have better patient outcomes than those just described.

6 Biomaterials as a Means to Improve Stem Cell Delivery

The particular biomaterials that would serve best for stem cell delivery for PAD are hydrogels. Hydrogels have the potential to effectively encapsulate and retain the stem cells at the injury site while maintaining sufficient nutrient and oxygen diffusion to the cells to maintain cell viability. They can also be delivered to the site of injury with the encapsulated cells in a minimally-invasive manner via injection. Hydrogels can be split into two major families: natural (i.e. those being derived of proteins formed in nature) and synthetic (i.e. constructed in a laboratory setting and made of polymers not found in nature) (Figure 4). Three natural polymers that are commonly studied include alginate, collagen and gelatin.

6.1 Alginate Hydrogels

Alginate has a long history of use given its availability from brown seaweed. An anionic polysaccharide, alginate can create crosslinked networks upon addition of divalent cations, typically calcium [59]. In commercially provided alginate, the polymer is a block copolymer consisting of L-guluronate and D-mannuronate [60,61]. Upon addition of calcium, repeated L-guluronate residues mediate crosslinking. Because of its extensive use in healing applications and cell encapsulation, alginate is generally regarded as a biocompatible, minimally cytotoxic and economic biomaterial [62,63]. Additionally, because humans lack alginase, an enzyme that degrades alginate, slow degradation by ion exchange within the extracellular environment has been observed in animal and human studies *in vivo*.

There are four methods of gelation that can be employed to generate crosslinked hydrogel formats. The first method, which was alluded to earlier, is ionic crosslinking induced by divalent cations like calcium. In the presence of calcium, the multiple guluronate units in the alignate polymer chain form junctions with each other as well as adjacent polymer chains, thereby creating a network of polymer chains with a three-dimensional structure. Depending on the source of calcium (e.g. $CaCl_2$, $CaCO_3$ or $CaSO_4$), the rate of gelation can be tuned [64].

The second method of alginate gelation is covalent crosslinking. To do this, alginate must be modified with other reactive groups in order to form the covalent bonds creating the polymer networks. Addition of poly(ethylene glycol)-diamines to alginate can tune stiffness properties that are not influenced over time by the *in vivo* environment [65], and addition of methacrylate groups followed by ultraviolet light irradiation allows for *in situ* formation of degradation-resistant gels to form *in vivo* structural contours [66].

The third method of gelation involves alginate gel formation dependent on temperature. An example of this includes control of alginate gel concentration in a specified area when thermoresponsive polymers such as N-isopropylacrylamide and poly(ethylene glycol)-*co*-poly(ε-caprolactone) (PEG-PCL) serve as interpenetrating networks and prevent alginate gelation until certain temperatures are achieved [67].

Finally, alginate gels can also be mediated by cell-based crosslinking, where peptides are grafted onto the alginate polymers to provide cell anchoring points that dictate alginate position in space [68].

6.2 Collagen and Gelatin

Collagen is the most abundant protein in humans as well as other mammals [69]. Given its superior strength and material robustness, collagen is found in many fibrous tissues like tendons, ligaments and skin [70]. Because it is so common, collagen can be isolated and purified in large batches and formed into many physical formats for virtually any application [71]. Of the many collagen types that exist in various tissues (28 have been identified and described), scar tissue formed in the healing process is primarily dominated by types I and III collagens [72]. The general structure of collagen is rod-shaped and serves to interact with itself to form fibrils. Compositionally, these fibrils consist of three collagen chains forming a helical structure. Three of these structures then form larger coils that are stabilized by hydrogen bonds and van de Waals forces, although other groups have found some covalent bond structures [73] or quaternary structures that have unusual properties compared to the majority of larger fibrils [74]. At the basic level of a single collagen fiber, a particular sequence homology across many types of collagen exist: a repeat of three amino acids that starts with glycine and follows with proline and any other amino acid. In many cases, proline can become hydroxproline and remain in the second position of the trimer amino acid sequence [75].

On the other hand, gelatin is a product of thermally-denatured insoluble collagen that is not uniform in size or isoionicity (i.e. the pH at which a zwitterion has an equal number of positive and negative charges). The size and charge character depend on the original type of collagen and how it was degraded [76]. Collagen I is the most common collagen source of gelatin. Gelatin is typically manufactured from a variety of sources including cows, pigs, and fish bones and skin. Importantly, the source of collagen also alters the amino acid compositions of the gelatin and is different from that found in humans [77]. In general, two types of gelatin exist: type A gelatin, which contains 18.5% nitrogen, and type B gelatin, which contains fewer amide groups and is comprised of 18.0% nitrogen. The amino acid components of either type of gelatin are highly variable [78]. For both collagen and gelatin, five matrix metalloproteinases (MMP) in humans have been found to degrade their structures. These include MMP1 (fibroblast collagenase), MMP2 (gelatinase-A), MMP3 (stromelysine), MMP8 (neutrophil collagenase), and MMP9 (gelatinase B) [79]. Importantly, gelatin breaks down much easier in the presence of proteases than collagen and is therefore quicker to degrade *in vivo*.

Despite most of the collagen and gelatin structure being conserved across species, use of animal proteins in humans does raise concerns for antigenicity. Three properties of these proteins, helical recognition, central sequence recognition and terminal peptide recognition, generally dictate whether use of collagen or gelatin can be safely administered in humans [80]. The scientific community has considered using recombinant human collagen, which is more expensive [81], but the small associated risk of antigenicity with animal collagen has been deemed acceptable hence allowing their continued widespread use in a variety of commercial and medical arenas [69]. Gelatins, on the other hand, have not been known to be immunogenic, although this research is limited and was conducted almost a century ago [82,83]. It was not until 2008 that a scientific reasoning behind this non-immunogenicity was discovered, pointing to the lack of aromatic radicals within the gelatin structure [84]. Furthermore, the ultra-heat treatment of collagen that is required to derive gelatin helps aid in sterilization, along with other ion exchanges and chemical treatments [85].

Unlike alginate gels, collagen gels require innovative strategies for crosslinking given that the digestion process needed to extract pure collagen damages part of its secondary and tertiary structures and cannot be recovered. As a result, collagen properties, namely mechanical strength and thermal stability, are irreversibly and adversely affected [86]. Many different crosslinking strategies have been explored to overcome this and generate the three-dimensional structure requisite for their use in biomaterials [87,88]. Gelatin gels are different in that they are prone to create many hydrogen bonds with increased weight-to-volume ratios provided the solvent is appropriate. Once above 30°C, the gelatin molecules have induced conformational changes to form a triple helix, creating a network of gelatin fibers that are stabilized by hydrogen bonding [89,90]. Furthermore, both collagen and gelatin have employed shared crosslinking strategies that employ chemical, enzymatic, or physical means. Chemical crosslinking includes, but is not limited to, using aldehydes; dialdehyde starches; diphenylphosphorylazide (DPPA); carbodiimides like EDC chemistry; and phenolic compounds. Enzymatic crosslinking primarily uses transglutaminase and tyrosinase. Finally, physical crosslinking utilizes dehydrothermal treatment, ultraviolet radiation and gamma radiation. All of these examples can be further read in [78].

6.3 Synthetic Hydrogels

As polymer chemistry has made significant advancements over the past century, a plethora of materials for industrial, consumer and biomedical applications has changed the face of society. Since the field of biomaterials is relatively young compared to polymer chemistry as a whole, great strides have been made in tuning polymer products into useful devices with medical purposes. Within biomedical science, three-dimensional cultures of cells for basic and translational research have transformed scientists' understanding on how to recreate human tissues or fight debilitating diseases [91,92]. Such synthetic polymers have distinct advantages over their natural polymer counterparts that span a huge range of properties. For example, degradation rates [93], degree of cell adhesion [94], matrix mechanics [95], surface topography [96] and growth factor release [97] can be exquisitely and robustly controlled, in contrast to natural polymers. This ability to tune numerous material properties fosters the efficacy of biomaterials as a tool in a variety of medical settings.

For applications similar to those of PAD, the operational function of synthetic polymer hydrogels should allow for easy modification of any component of the polymer itself in order to tune its function and favorably impact therapeutic efficacy. Two classic polymer examples within this category are poly(N-isopropylacrylamide) (PNIPAAm) and poly(ethylene glycol) (PEG), because their lower critical solution temperatures nears that of normal body temperature [98]. PEG is particularly attractive because its configuration is highly adaptable to form hydrogels through a host of different chemical groups [99,100]. Methacrylate [101], thiol [102] and Michael-type addition groups have been commonly employed as they provide a means to crosslink hydrogel components *in situ*. PEG can also be grafted to other polymers like PNIPAAm, poly(gycolic acid) (PGA), poly(lactic acid) (PLA) and poly(lactic-co-glycolic acid) (PLGA) in order to create thermally responsive and hydrophobic domain hydrogels, respectively. PNIPAAm [103] and PLGA [104] grafts have been shown to allow for cell encapsulation as the temperature of the copolymer solution increases. Moreover, elastic moduli and swelling of PEG-diacrylate gels have been shown to be modified by the addition of polysaccharide methacrylates, allowing for increased stiffness in the gels while maintaining cell viability, fluid diffusion and gel geometry [105].

Synthetic hydrogels have also been improved to allow for the best interface with cells they are delivering to sites of interest. One common strategy employed is adding binding epitopes that cells can preferentially interact with. In many cases, peptides with key binding motifs from fibronectin, collagen, vitronectin and laminin are employed. For PEG gels, a highly utilized sequence is the arginine-glycine-aspartic acid (RGD), which has been shown in several studies to enhance cell viability, growth and differentiation into preferred cell types [101,106]. Further adjustments in the quantity of RGD peptide were shown to influence cell sensitivity to matrix stiffness [107], and spatially organized RGD (via altered PEG molecular weight and fabrication methods) could organize the encapsulated cells into microstructures more representative of the tissue they are recreating [108]. This method has been extended to include whole recombinant protein oligopeptides and biodegradable units (portions of the gel that can be degraded by MMPs) [109].

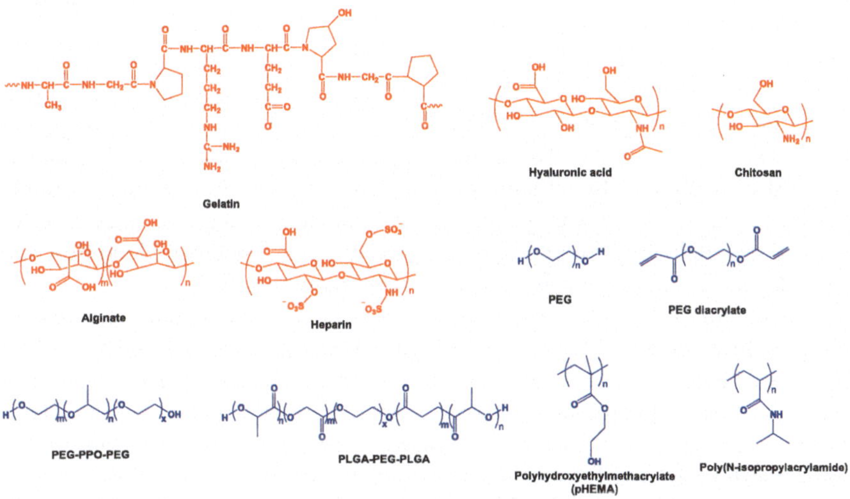

Fig. 4 Chemical structures of natural and synthetic polymers used in biomaterials. Red-colored structures indicate natural polymers formed in nature. Blue-colored structures indicated lab-synthesized polymers.

7 Biomaterials-Based Treatments for Peripheral Artery Disease in Animal Trials

Because biomaterials have become an attractive vehicle for cells and growth factors alike, researchers have begun experimenting with methods to manipulate biomaterials to treat PAD in mouse models. Over the past several years, the most popular biomaterial approaches for reversing tissue damage induced by PAD have been to provide localized, sustained delivery of growth factors known to stimulate a healing response or to improve retainal of stem cells to the diseased site.

7.1 Alginate and Gelatin Hydrogels Utilization for Growth Factor and Stem Cell Delivery

Some of the promising initial animal studies began by loading certain growth factors in gelatin or alginate hydrogels for the purpose of recruiting cells to the damaged tissue areas and stimulating angiogenesis to restore blood flow. For a study using gelatin, Layman *et al.* desired to allow for sustained release of fibroblast growth factor-2 (FGF2) and G-CSF in order to gain long term recovery in ischemic hind limbs of mice affected by PAD [110]. FGF2 would serve to promote endothelial cell proliferation while G-CSF, like the human clinical trials, would increase the presence of peripheral blood mononuclear stem cells to the injury area and aid in mediating blood vessel formation. Because gelatin can degrade in environments with high MMPs, which is present in wound healing sites or ongoing pathologic tissue, they sought to prevent degradation of such growth factors within the hydrogel. To do this, they synthesized gelatin-poly(L-glutamic acid) and loaded the hydrogel with the two growth factors. In as little as 8 weeks after injection, compared with a saline control, the delivered growth factors in the modified gelatin hydrogel outperformed expectations as functional blood flow was restored and signs of previously PAD were severely diminished.

Alginate gels were also used in a similar fashion to the gelatin hydrogels to deliver a different set of growth factors. In one study, hepatocyte growth factor (HGF) was linked with alginate sulfate, a form of alginate that provides high affinity to bind heparin-binding molecules such as HGF [111]. *In vivo*, HGF serves to combat fibrosis that replaces dead tissue within the body, but equally provides cytoprotection for existing healthy cells and plays a significant role in angiogenesis. Evaluation of the HGF-alginate sulfate performance *in vitro* showed sustained release of HGF 3-fold higher than free HGF in the environment, and the bioactivity of HGF was also maintained as extracellular receptor kinase 1/2 (ERK1/2) was maintained in rat neonatal cardiomyocytes (an accepted *in vitro* test model). 9 days post-injection, the hydrogel treatment demonstrated 10-fold enhanced recovery of tissue perfusion and vessel network construction in a murine hindlimb ischemia model, which was also verified by $CD31^{+}$ staining.

One other group used alginate hydrogels to co-delivery vascular endothelial growth factor (VEGF) and stromal derived factor (SDF) to PAD injury sites [112]. Although these growth factors were not previously used in the human clinical trials, it was hypothesized that local and sustained delivery of the two growth factors would better recruit and mobilize circulating angiogenic cells to the injury site. Instead of forming whole gels, Anderson *et al.* created small microbeads filled with the growth factors and injected them into murine ischemic hind limbs intramuscularly. In addition to the microbead injection, they also delivered circulating angiogenic cells (via intracardiac injection to avoid first-pass lung retention) to replicate the stem cell delivery strategies used in human clinical trials. As expected, the localization of the growth factors in the ischemic limbs with the circulating angiogenic cells provided the greatest improvement perfusion recovery, including endothelial cell injection as a control. The circulating angiogenic cells were incredibly responsive to the growth factors and also controlling invading inflammatory cells *in vivo*.

7.2 Efficacy of Modified Synthetic Hydrogels That Accentuate Healing Responses as well as Stem Cell and Growth Factor Integration

As the wound environment in PAD contains enzymes that remodel the tissue, Phelps *et al* designed PEG-based hydrogels that contain cell adhesion motifs, soluble growth factors and protease-degradable sites to provide a dynamic physiological tissue replacement to promote angiogenesis [113]. Such a hydrogel would also be designed for increased size as the synthesis and assembly of the PEG hydrogel could easily be scaled to human anatomical dimensions. Using PEG-DA polymer, the group demonstrated that degradation of the gel only occurred in the presence of cells that excreted MMPs for remodeling purposes and that cells could easily attach throughout the gel surface due to the cell binding motifs (RGD). Vascular endothelial growth factor (VEGF) also demonstrated sustained release that lasted over 12 days before release kinetics were lost, therefore matching levels of soluble VEGF by pure injection which immediately washed out over a short time interval. When injected intramuscularly into a mouse ischemic hind leg, the gel polymerized *in situ* and had 1.5-fold higher perfusion than saline control as evaluated by Doppler imaging after one week. Vessel density continued to increase over the course of one month with little evidence of residual hydrogel after 2 weeks, thereby implying the robust new tissue formation in the leg.

Recently, our research group (Hak-Joon Sung laboratory) took a similar approach using a PEL-PCL hydrogel that contained both pro-angiogenic and anti-inflammatory peptide sequences to modulate nearby cells for healing of PAD in mouse models without the need of cell injections [114]. Pro-angiogenic C16 peptide and anti-inflammatory Ac-SDKP peptides in PEG-PCL was injected into the adductor muscle of an ischemic leg of a mouse following ligation surgery, with injection volumes as little as 1uL. Both peptides were hypothesized to work synchronously to prevent normal tissue fibrosis and scarring in the healing process in order to allow angiogenic cells to form new collateral vessels uninhibited by inflammatory cells. After two weeks post-injection, compared to empty gel controls, the peptide-loaded hydrogel displayed 1.4-fold increase in perfusion by Doppler imaging, 3-fold higher fluorescent reading in microangiography and 0.1-fold reading compared to controls in macrophage presence in the local tissue. Another group used a similar strategy by grafting substance P, a small peptide known to aid in healing by mobilizing stromal cells that helps tissue reconstruction and inflammation control, which had sustained release for nearly one month and demonstrated similar significant improvement in limb perfusion [115].

Biomolecules other than growth factors have also been delivered to promote vascularization in PAD. In one study, reactive oxygen species (ROS)-responsive microspheres (polypropylene sulfide) loaded with curcumin were delivered to mouse ischemic limbs via intramuscular injection to the gastrocnemius and adductor muscles [116]. Curcumin has been known to exhibit strong antioxidant and anti-inflammatory properties to keep inflammatory cells at bay within wound sites. Upon injection, the group reported 1.8-fold increased oxygen saturation in the foot of an ischemic limb, reduces ROS levels and more collateral blood flow compared to unloaded microspheres alone. In a separate study, nitric oxide (NO) delivery has been to improve endothelial cell proliferation while restraining smooth muscle cell proliferation, and Domanchuk *et al.* grafted glycosylated NONOate (NO donor) to chitosan and delivered it to a type I diabetic mouse model exhibiting PAD [117]. Again, an increase (1.5-fold) in microcapillary formation was seen in the treated limb compared to saline control.

8 Future Directions and Considerations

In light of all the research being conducted into treating PAD in animal models and clinical trials, there remains much to be explored in terms of adapting different strategies of cell and growth factor delivery to sites of PAD. As was seen in the clinical trials, the addition of stem cells and some growth factors did show modest improvement in outcomes for patients with severe PAD. However, the few studies that showed negative or non-significant impact on patients bring into question the existing methods of stem cell use for patient therapies. While this may at first appear concerning, researchers and clinicians alike should not despair because there are always new ways to build upon existing techniques to promote better efficacy in PAD treatment.

With respect to pure cell injections, some researchers have considered intravenous delivery of endothelial colony-forming cells with or without mesenchymal stem cells [118]. These endothelial colony units can provide better incorporation into the disturbed vasculature as they have a higher cell density than single cell suspensions, and the mesenchymal stem cells would provide paracrine signaling to reduce inflammation but also recruit pericytes and smooth muscles cells to further develop the new collateral vessels. Human neural stem cells have also been considered as an alternative stem cell source that could fight PAD [119]. *In vivo*, human neural stem cells have been shown to exhibit a paracrine effect that prevents inflammation and allows for angiogenesis in stroke models. In this pilot study, a dose-dependent positive correlation was seen with limb reperfusion and the cell mass injected into the adductor muscle of a PAD mouse model, but especially pronounced in diabetic mice. Innovative approaches based on tissue transfer-printing [120] of myoblast monolayers as well as three-dimensional aggregates of human umbilical vein endothelial cells and cord-blood mesenchymal stem cells [121] have demonstrated promise towards hastening tube formation.

Material-based stem cell delivery strategies have also been further developed either to emphasize encapsulating cells for factor secretion and not integration, and support of iPSC-derived endothelial cells. With respect to the former, two research groups have explored isolated mesenchymal stem cells within alginate gels to take advantage of the paracrine effects on surrounding tissue. In one study, alginate microbeads with mesenchymal stem cells were injected in the thigh region of a PAD mouse and left *in vivo* for one week [122]. The encapsulated cells were then removed and were found to secrete angiogenic factors like VEGF, HGF and angiogenin at significant levels and correlated with 20% improved perfusion levels and 70% increase in vessel density in the thigh muscle bed. iPSC-derived endothelial cells were also injected into PAD-affected limbs demonstrating great promise for customized stem cell therapy. Intramuscular injection of iPSC-dervied endothelial cells in Matrigel improved capillary density by 20% and maintained cell viability at the two-week time point as measured by bioluminescence [123]. Similar findings, although not quantitative, were reported by Mulyasasmita *et al* where proline-rich multi-armed PEG hydrogels, plus RGD hydrogels, delivered VEGF and iPSC-derived endothelial cells and had significant visual improvement of affected limbs [124].

Finally, gene therapy has also garnered attention in the biomedical community to tailor treatments using cells that specifically upregulate expression of certain proteins that have therapeutic potential. This method uses viral vectors, micelles, or plasmids to deliver genetic materials (DNA or RNA) to alter gene expression in a target cell type, and potential genes of interest to treat PAD include HIF-1α, FGF-2, angiopoietin-1 (Ang-1), VEGF, ephrin-B2, and tissue kallikrein [125]. Gene transfer provides several advantages, including cell-type specificity and long-lasting expression once induced, and requires only a single injection of genetic material in a delivery vehicle. However, gene transfer has low-efficiency, can cause a severe inflammatory response, and may result in uncontrollable levels of gene expression [32,125,126]. As an alternative to delivery of genetic material, delivering proteins or peptides to a target tissue allows for dosage control and results in little immune response. Protein/peptide treatments are limited as they have a short half-life in the circulatory system, being quickly cleared by the kidneys. Therefore, the use of peptides and proteins requires multiple systemic injections or controlled local release [127]. Without sustained release of proteins, little if any improvement has been observed in randomized clinical trials as was reported in a study using intramuscular injections of recombinant FGF-2 [35]. Some groups have already begun advanced animal models to validate the potential of these methods including upregulated VEGF-expressing mesenchymal stem cells transported within biodegradable poly-b-aminoester nanoparticles [128], glucagon-like peptide 1-expressing mesenchymal stem cells in alginate microbeads [129], and naked plasmid delivery of hypoxia-regulated HO-1 that mitigates cellular injury in surround cells at the wound site [130].

9 Conclusion

As highlighted in the multiple sections above, it is increasingly becoming clear that the current strategies used to combat PAD show promise in changing the outcomes for many patients. Stem cell therapy does appear to have a place in the treatment regimens and the ease of obtaining autologous stem cells to reduce immunogenicity is of minimal concern. Importantly, regenerative medicine tools such as biomaterials and similar technologies may be able to improve the already significantly improved outcomes of combating PAD with stem and progenitor cells by improving the retention and behavior of stem cells. Yet, even with the cutting edge studies further controlling the delivery of therapeutic agents and cells, a considerable amount of investment is required to fully translate the bench-top work into clinical practice. Moreover, further understanding of the mechanisms that the growth factors and injected cells execute to induce collateral vessel formation and reperfusion remains to be explored. Knowledge of such mechanisms could feed back into the design and implementation of regenerative medicine therapies. Considering the enthusiastic response to the clinical trials and mouse model results in the cited literature, there is significant promise that more collaborative efforts will eventually lead to clinically and economically efficacious solutions to make PAD not a disease of high morbidity and mortality with unrelenting progression, but rather one that can be addressed in outpatient settings with relative ease.

References

1. Botman, O., Wyns, C.: Induced pluripotent stem cell potential in medicine, specifically focused on reproductive medicine. Frontiers in Surgery **1**, 5 (2014)
2. Singh, V.K., Kalsan, M., Kumar, N., Saini, A., Chandra, R.: Induced pluripotent stem cells: Applications in regenerative medicine, disease modeling, and drug discovery. Frontiers in Cell and Developmental Biology **3**, 2 (2015)
3. Fowkes, F.G., Rudan, D., Rudan, I., Aboyans, V., Denenberg, J.O., McDermott, M.M., Norman, P.E., Sampson, U.K., Williams, L.J., Mensah, G.A., Criqui, M.H.: Comparison of global estimates of prevalence and risk factors for peripheral artery disease in 2000 and 2010: A systematic review and analysis. Lancet **382**, 1329–1340 (2013)
4. Hirsch, A.T., Hartman, L., Town, R.J., Virnig, B.A.: National health care costs of peripheral arterial disease in the medicare population. Vasc. Med. **13**, 209–215 (2008)
5. Mahoney, E.M., Wang, K., Keo, H.H., Duval, S., Smolderen, K.G., Cohen, D.J., Steg, G., Bhatt, D.L., Hirsch, A.T.: Reduction of Atherothrombosis for Continued Health Registry I. Vascular hospitalization rates and costs in patients with peripheral artery disease in the united states. Circulation. Cardiovascular Quality and Outcomes **3**, 642–651 (2010)
6. Olin, J.W., Sealove, B.A.: Peripheral artery disease: Current insight into the disease and its diagnosis and management. Mayo Clin. Proc. **85**, 678–692 (2010)

7. Mueller-Schweinitzer, E., Muller, S.E., Reineke, D.C., Kern, T., Carrel, T.P., Eckstein, F.S., Grapow, M.T.R.: Reactive oxygen species mediate functional differences in human radial and internal thoracic arteries from smokers. J. Vasc. Surg. **51**, 438–444 (2010)
8. Jude, E.B., Eleftheriadou, I., Tentolouris, N.: Peripheral arterial disease in diabetes–a review. Diabet. Med. **27**, 4–14 (2010)
9. Boyette, L.B., Tuan, R.S.: Adult stem cells and diseases of aging. Journal of Clinical Medicine **3**, 88–134 (2014)
10. Challen, G.A., Little, M.H.: A side order of stem cells: The sp phenotype. Stem Cells **24**, 3–12 (2006)
11. Salem, H.K., Thiemermann, C.: Mesenchymal stromal cells: Current understanding and clinical status. Stem Cells **28**, 585–596 (2010)
12. Friedenstein, A.J., Petrakova, K.V., Kurolesova, A.I., Frolova, G.P.: Heterotopic of bone marrow. Analysis of precursor cells for osteogenic and hematopoietic tissues. Transplantation **6**, 230–247 (1968)
13. Pittenger, M.F., Mackay, A.M., Beck, S.C., Jaiswal, R.K., Douglas, R., Mosca, J.D., Moorman, M.A., Simonetti, D.W., Craig, S., Marshak, D.R.: Multilineage potential of adult human mesenchymal stem cells. Science **284**, 143–147 (1999)
14. Asahara, T., Masuda, H., Takahashi, T., Kalka, C., Pastore, C., Silver, M., Kearne, M., Magner, M., Isner, J.M.: Bone marrow origin of endothelial progenitor cells responsible for postnatal vasculogenesis in physiological and pathological neovascularization. Circulation Research **85**, 221–228 (1999)
15. Asahara, T., Murohara, T., Sullivan, A., Silver, M., van der Zee, R., Li, T., Witzenbichler, B., Schatteman, G., Isner, J.M.: Isolation of Putative Progenitor Endothelial Cells for Angiogenesis. Science **275**, 964–967 (1997)
16. Kalka, C., Masuda, H., Takahashi, T., Kalka-Moll, W.M., Silver, M., Kearney, M., Li, T., Isner, J.M., Asahara, T.: Transplantation of ex vivo expanded endothelial progenitor cells for therapeutic neovascularization. Proc. Natl. Acad Sci. USA **97**, 3422–3427 (2000)
17. Rehman, J., Li, J., Orschell, C.M., March, K.L.: Peripheral blood "endothelial progenitor cells" are derived from monocyte/macrophages and secrete angiogenic growth factors. Circulation **107**, 1164–1169 (2003)
18. Caballero, S., Sengupta, N., Afzal, A., Chang, K.H., Li Calzi, S., Guberski, D.L., Kern, T.S., Grant, M.B.: Ischemic vascular damage can be repaired by healthy, but not diabetic, endothelial progenitor cells. Diabetes **56**, 960–967 (2007)
19. Oswald, J., Boxberger, S., Jørgensen, B., Feldmann, S., Ehninger, G., Bornhäuser, M., Werner, C.: Mesenchymal stem cells can be differentiated into endothelial cells in vitro. Stem Cells **22**, 377–384 (2004)
20. Silva, G.V., Litovsky, S., Assad, J.A., Sousa, A.L., Martin, B.J., Vela, D., Coulter, S.C., Lin, J., Ober, J., Vaughn, W.K.: Mesenchymal stem cells differentiate into an endothelial phenotype, enhance vascular density, and improve heart function in a canine chronic ischemia model. Circulation **111**, 150–156 (2005)
21. Uccelli, A., Moretta, L., Pistoia, V.: Mesenchymal stem cells in health and disease. Nature Reviews. Immunology **8**, 726–736 (2008)
22. Shantsila, E., Watson, T., Tse, H.F., Lip, G.Y.: New insights on endothelial progenitor cell subpopulations and their angiogenic properties. J. Am. Coll. Cardiol. **51**, 669–671 (2008)
23. Gao, L., Chen, Q., Zhou, X., Fan, L.: The role of hypoxia-inducible factor 1 in atherosclerosis. Journal of Clinical Pathology (2012)

24. Selvin, E., Erlinger, T.P.: Prevalence of and risk factors for peripheral arterial disease in the united states: Results from the national health and nutrition examination survey, 1999-2000. Circulation **110**, 738–743 (2004)
25. Sanz, J., Fayad, Z.A.: Imaging of atherosclerotic cardiovascular disease. Nature **451**, 953–957 (2008)
26. Sluimer, J.C., Daemen, M.J.: Novel concepts in atherogenesis: Angiogenesis and hypoxia in atherosclerosis. J. Pathol. **218**, 7–29 (2009)
27. Gao, L., Chen, Q., Zhou, X., Fan, L.: The role of hypoxia-inducible factor 1 in atherosclerosis. Journal of Clinical Pathology **65**, 872–876 (2012)
28. Golomb, B.A., Dang, T.T., Criqui, M.H.: Peripheral arterial disease: Morbidity and mortality implications. Circulation **114**, 688–699 (2006)
29. Bhatt, D.L., Flather, M.D., Hacke, W., Berger, P.B., Black, H.R., Boden, W.E., Cacoub, P., Cohen, E.A., Creager, M.A., Easton, J.D., Hamm, C.W., Hankey, G.J., Johnston, S.C., Mak, K.H., Mas, J.L., Montalescot, G., Pearson, T.A., Steg, P.G., Steinhubl, S.R., Weber, M.A., Fabry-Ribaudo, L., Hu, T., Topol, E.J., Fox, K.A.: Patients with prior myocardial infarction, stroke, or symptomatic peripheral arterial disease in the charisma trial. J. Am. Coll. Cardiol. **49**, 1982–1988 (2007)
30. Mrc/bhf heart protection study of cholesterol lowering with simvastatin in 20,536 high-risk individuals: A randomised placebo-controlled trial. Lancet **360,** 7–22 (2002)
31. Garcia, L.A., Lyden, S.P.: Atherectomy for infrainguinal peripheral artery disease. J. Endovasc. Ther. **16**(II), 105–115 (2009)
32. Collinson, D.J., Donnelly, R.: Therapeutic angiogenesis in peripheral arterial disease: Can biotechnology produce an effective collateral circulation? European Journal of Vascular and Endovascular Surgery: the Official Journal of the European Society for Vascular Surgery **28**, 9–23 (2004)
33. Berger, J.S., Krantz, M.J., Kittelson, J.M., Hiatt, W.R.: Aspirin for the prevention of cardiovascular events in patients with peripheral artery disease: A meta-analysis of randomized trials. Jama **301**, 1909–1919 (2009)
34. Kalka, C., Baumgartner, I.: Gene and stem cell therapy in peripheral arterial occlusive disease. Vasc. Med. **13**, 157–172 (2008)
35. Pacilli, A., Faggioli, G., Stella, A., Pasquinelli, G.: An update on therapeutic angiogenesis for peripheral vascular disease. Ann. Vasc. Surg. **24**, 258–268 (2010)
36. Tateishi-Yuyama, E., Matsubara, H., Murohara, T., Ikeda, U., Shintani, S., Masaki, H., Amano, K., Kishimoto, Y., Yoshimoto, K., Akashi, H., Shimada, K., Iwasaka, T., Imaizumi, T.: Therapeutic Angiogenesis using Cell Transplantation Study I. Therapeutic angiogenesis for patients with limb ischaemia by autologous transplantation of bone-marrow cells: A pilot study and a randomised controlled trial. Lancet **360**, 427–435 (2002)
37. Esato, K., Hamano, K., Li, T.S., Furutani, A., Seyama, A., Takenaka, H., Zempo, N.: Neovascularization induced by autologous bone marrow cell implantation in peripheral arterial disease. Cell Transplantation **11**, 747–752 (2002)
38. Miyamoto, K., Nishigami, K., Nagaya, N., Akutsu, K., Chiku, M., Kamei, M., Soma, T., Miyata, S., Higashi, M., Tanaka, R., Nakatani, T., Nonogi, H., Takeshita, S.: Unblinded pilot study of autologous transplantation of bone marrow mononuclear cells in patients with thromboangiitis obliterans. Circulation **114**, 2679–2684 (2006)
39. Durdu, S., Akar, A.R., Arat, M., Sancak, T., Eren, N.T., Ozyurda, U.: Autologous bone-marrow mononuclear cell implantation for patients with rutherford grade ii-iii thromboangiitis obliterans. J. Vasc. Surg. **44**, 732–739 (2006)

40. Murphy, M.P., Lawson, J.H., Rapp, B.M., Dalsing, M.C., Klein, J., Wilson, M.G., Hutchins, G.D., March, K.L.: Autologous bone marrow mononuclear cell therapy is safe and promotes amputation-free survival in patients with critical limb ischemia. J. Vasc. Surg. **53**, 1565–1574 (2011). e1561
41. Patel, M.R., Conte, M.S., Cutlip, D.E., Dib, N., Geraghty, P., Gray, W., Hiatt, W.R., Ho, M., Ikeda, K., Ikeno, F., Jaff, M.R., Jones, W.S., Kawahara, M., Lookstein, R.A., Mehran, R., Misra, S., Norgren, L., Olin, J.W., Povsic, T.J., Rosenfield, K., Rundback, J., Shamoun, F., Tcheng, J., Tsai, T.T., Suzuki, Y., Vranckx, P., Wiechmann, B.N., White, C.J., Yokoi, H., Krucoff, M.W.: Evaluation and treatment of patients with lower extremity peripheral artery disease: Consensus definitions from peripheral academic research consortium (parc). J. Am. Coll. Cardiol. **65**, 931–941 (2015)
42. Lu, D., Chen, B., Liang, Z., Deng, W., Jiang, Y., Li, S., Xu, J., Wu, Q., Zhang, Z., Xie, B., Chen, S.: Comparison of bone marrow mesenchymal stem cells with bone marrow-derived mononuclear cells for treatment of diabetic critical limb ischemia and foot ulcer: A double-blind, randomized, controlled trial. Diabetes Research And Clinical Practice **92**, 26–36 (2011)
43. Bura, A., Planat-Benard, V., Bourin, P., Silvestre, J.S., Gross, F., Grolleau, J.L., Saint-Lebese, B., Peyrafitte, J.A., Fleury, S., Gadelorge, M., Taurand, M., Dupuis-Coronas, S., Leobon, B., Casteilla, L.: Phase i trial: The use of autologous cultured adipose-derived stroma/stem cells to treat patients with non-revascularizable critical limb ischemia. Cytotherapy **16**, 245–257 (2014)
44. Marino, G., Moraci, M., Armenia, E., Orabona, C., Sergio, R., De Sena, G., Capuozzo, V., Barbarisi, M., Rosso, F., Giordano, G., Iovino, F., Barbarisi, A.: Therapy with autologous adipose-derived regenerative cells for the care of chronic ulcer of lower limbs in patients with peripheral arterial disease. The Journal of Surgical Research **185**, 36–44 (2013)
45. Gupta, P.K., Chullikana, A., Parakh, R., Desai, S., Das, A., Gottipamula, S., Krishnamurthy, S., Anthony, N., Pherwani, A., Majumdar, A.S.: A double blind randomized placebo controlled phase i/ii study assessing the safety and efficacy of allogeneic bone marrow derived mesenchymal stem cell in critical limb ischemia. Journal of Translational Medicine **11**, 143 (2013)
46. Walter, D.H., Krankenberg, H., Balzer, J.O., Kalka, C., Baumgartner, I., Schluter, M., Tonn, T., Seeger, F., Dimmeler, S., Lindhoff-Last, E., Zeiher, A.M., Investigators, P.: Intraarterial administration of bone marrow mononuclear cells in patients with critical limb ischemia: A randomized-start, placebo-controlled pilot trial (provasa). Circulation. Cardiovascular Interventions **4**, 26–37 (2011)
47. Bartsch, T., Brehm, M., Zeus, T., Strauer, B.E.: Autologous mononuclear stem cell transplantation in patients with peripheral occlusive arterial disease. The Journal of Cardiovascular Nursing **21**, 430–432 (2006)
48. Bartsch, T., Brehm, M., Zeus, T., Kogler, G., Wernet, P., Strauer, B.E.: Transplantation of autologous mononuclear bone marrow stem cells in patients with peripheral arterial disease (the tam-pad study). Clinical Research in Cardiology: Official Journal of the German Cardiac Society **96**, 891–899 (2007)
49. Huang, P., Li, S., Han, M., Xiao, Z., Yang, R., Han, Z.C.: Autologous transplantation of granulocyte colony-stimulating factor-mobilized peripheral blood mononuclear cells improves critical limb ischemia in diabetes. Diabetes Care **28**, 2155–2160 (2005)

50. Lenk, K., Adams, V., Lurz, P., Erbs, S., Linke, A., Gielen, S., Schmidt, A., Scheinert, D., Biamino, G., Emmrich, F., Schuler, G., Hambrecht, R.: Therapeutical potential of blood-derived progenitor cells in patients with peripheral arterial occlusive disease and critical limb ischaemia. European Heart Journal **26**, 1903–1909 (2005)
51. Losordo, D.W., Kibbe, M.R., Mendelsohn, F., Marston, W., Driver, V.R., Sharafuddin, M., Teodorescu, V., Wiechmann, B.N., Thompson, C., Kraiss, L., Carman, T., Dohad, S., Huang, P., Junge, C.E., Story, K., Weistroffer, T., Thorne, T.M., Millay, M., Runyon, J.P., Schainfeld, R.: Autologous CDCTfCLII. A randomized, controlled pilot study of autologous cd34 + cell therapy for critical limb ischemia. Circulation Cardiovascular Interventions **5**, 821–830 (2012)
52. Kinoshita, M., Fujita, Y., Katayama, M., Baba, R., Shibakawa, M., Yoshikawa, K., Katakami, N., Furukawa, Y., Tsukie, T., Nagano, T., Kurimoto, Y., Yamasaki, K., Handa, N., Okada, Y., Kuronaka, K., Nagata, Y., Matsubara, Y., Fukushima, M., Asahara, T., Kawamoto, A.: Long-term clinical outcome after intramuscular transplantation of granulocyte colony stimulating factor-mobilized cd34 positive cells in patients with critical limb ischemia. Atherosclerosis **224**, 440–445 (2012)
53. Huang, P.P., Yang, X.F., Li, S.Z., Wen, J.C., Zhang, Y., Han, Z.C.: Randomised comparison of g-csf-mobilized peripheral blood mononuclear cells versus bone marrow-mononuclear cells for the treatment of patients with lower limb arteriosclerosis obliterans. Thrombosis and Haemostasis **98**, 1335–1342 (2007)
54. Kawamoto, A., Katayama, M., Handa, N., Kinoshita, M., Takano, H., Horii, M., Sadamoto, K., Yokoyama, A., Yamanaka, T., Onodera, R., Kuroda, A., Baba, R., Kaneko, Y., Tsukie, T., Kurimoto, Y., Okada, Y., Kihara, Y., Morioka, S., Fukushima, M., Asahara, T.: Intramuscular transplantation of g-csf-mobilized cd34(+) cells in patients with critical limb ischemia: A phase i/iia, multicenter, single-blinded, dose-escalation clinical trial. Stem Cells **27**, 2857–2864 (2009)
55. Mutirangura, P., Ruangsetakit, C., Wongwanit, C., Chinsakchai, K., Porat, Y., Belleli, A., Czeiger, D.: Enhancing limb salvage by non-mobilized peripheral blood angiogenic cell precursors therapy in patients with critical limb ischemia. Journal of the Medical Association of Thailand = Chotmaihet Thangphaet **92**, 320–327 (2009)
56. Szabo, G.V., Kovesd, Z., Cserepes, J., Daroczy, J., Belkin, M., Acsady, G.: Peripheral blood-derived autologous stem cell therapy for the treatment of patients with late-stage peripheral artery disease-results of the short- and long-term follow-up. Cytotherapy **15**, 1245–1252 (2013)
57. Ungerleider, J.L., Christman, K.L.: Concise review: Injectable biomaterials for the treatment of myocardial infarction and peripheral artery disease: Translational challenges and progress. Stem Cells Translational Medicine **3**, 1090–1099 (2014)
58. Sun, H., Wang, X., Hu, X., Yu, W., You, C., Hu, H., Han, C.: Promotion of angiogenesis by sustained release of rhgm-csf from heparinized collagen/chitosan scaffolds. J. Biomed. Mater. Res. B Appl. Biomater. **100**, 788–798 (2012)
59. Wee, S., Gombotz, W.R.: Protein release from alginate matrices. Adv. Drug Deliv. Rev. **31**, 267–285 (1998)
60. Fischer, F.G., Dorfel, H.: [polyuronic acids in brown algae]. Hoppe-Seyler's Zeitschrift fur physiologische Chemie **302**, 186–203 (1955)
61. Haug, A.: Fractionation of alginic acid. Acta. Chem. Scand. **13**, 601–613 (1959)
62. Lee, J., Lee, K.Y.: Local and sustained vascular endothelial growth factor delivery for angiogenesis using an injectable system. Pharmaceutical Research **26**, 1739–1744 (2009)

63. Orive, G., Ponce, S., Hernandez, R.M., Gascon, A.R., Igartua, M., Pedraz, J.L.: Biocompatibility of microcapsules for cell immobilization elaborated with different type of alginates. Biomaterials **23**, 3825–3831 (2002)
64. Crow, B.B., Nelson, K.D.: Release of bovine serum albumin from a hydrogel-cored biodegradable polymer fiber. Biopolymers **81**, 419–427 (2006)
65. Eiselt, P., Lee, K.Y., Mooney, D.J.: Rigidity of two-component hydrogels prepared from alginate and poly(ethylene glycol)-diamines. Macromolecules **32**, 5561–5566 (1999)
66. Smeds, K.A., Pfister-Serres, A., Miki, D., Dastgheib, K., Inoue, M., Hatchell, D.L., Grinstaff, M.W.: Photocrosslinkable polysaccharides for in situ hydrogel formation. Journal of Biomedical Materials Research **54**, 115–121 (2001)
67. Zhao, S., Cao, M., Li, H., Li, L., Xu, W.: Synthesis and characterization of thermo-sensitive semi-ipn hydrogels based on poly(ethylene glycol)-co-poly(epsilon-caprolactone) macromer, n-isopropylacrylamide, and sodium alginate. Carbohydrate Research **345**, 425–431 (2010)
68. Lee, K., Hong, H., Larson, R., Mooney, D.J.: Hydrogel formation via cell cross-linking. Advanced Materials **15**, 1828–1832 (2003)
69. Friess, W.: Collagen- biomaterial for drug delivery. European Journal of Pharmaceutics and Bioppharmaceutics **45**, 113–136 (1998)
70. Brinckmann, J., Notbohm, H., Muller, P.K.: Collagen: Primer in structure, processing and assembly. Springer-Verlag, Netherlands (2005)
71. Sinha, V.R., Trehan, A.: Biodegradable microspheres for protein delivery. Journal of Controlled Release **90**, 261–280 (2003)
72. Cheng, W., Yan-hua, R., Fang-gang, N., Guo-an, Z.: The content and ration of type i and type iii collagen in skin differ with age and injury. African Journal of Biotechnology **10** (2011)
73. Harkness, R.D.: Collagen. Science Progress **54**, 257–274 (1966)
74. He, L., Mu, C., Shi, J., Zhang, Q., Shi, B., Lin, W.: Modification of collagen with a natural cross-linker, procyanidin. International Journal of Biological Macromolecules **48**, 354–359 (2011)
75. Gorham, S.D.: Collagen as a biomaterial. In: Byron, D. (ed.) Biomaterials, pp. 55–122. Stockton Press, New York (1991)
76. Gomez-Guillen, M.C., Perez-Mateos, M., Gomez-Estaca, J., Lopez-Caballero, E., Gimenez, B., Montero, P.: Fish gelatin: A renewable materials for developing active biodegradable films. Trends in Food Science & Technology **20**, 3–16 (2009)
77. Zhang, F., Xu, S., Wang, Z.: Pre-treatment optimization and properties of gelatin from freshwater fish scales. Food and Bioproducts Processing (2010)
78. Gorgieva, S., Kokol, V.: Collagen- vs. Gelatine-based biomaterials and their biocompatibility: Review and perspectives. Interchopen (2011)
79. Netzel-Arnett, S., Mallya, S.K., Nagase, H., Birkedal-Hansen, H., Van Wart, H.E.: Continuously recording fluorescent assays optimized for five human matrix metalloproteinases. Analytical Biochemistry **195**, 86–92 (1991)
80. Lynn, A.K., Yannas, L.V., Bonfield, W.: Antigenicity of immunogenicity of collagen. Journal of Biomedical Materials Research Part B: Applied Materials **71B**, 343–354 (2004)
81. Olsen, D., Yang, C., Bodo, M., Chang, R., Leigh, S., Baez, J., Carmichael, D., Perala, M., Hamalainen, E.R., Jarvinen, M., Polarek, J.: Recombinant collagen and gelatin for drug delivery. Adv. Drug Deliv. Rev. **55**, 1547–1567 (2003)

82. Hopkins, S.J., Wormall, A.: Pehyl isocyanate protein compounds and their immunological properties. Biophysical Journal **27**, 740–753 (1933)
83. Starin, W.A.: The antigenic properties of gelain. The Journal of Infectious Diseases **23**, 139–158 (1918)
84. Kokare, C.R.: Pharmaceutical microbiology- principles and applications, Pune, India (2008)
85. Taylor, P.M., Cass, A.E.G., Yacoub, M.H.: Extracellular matrix scaffolds for tissue engineering heart valves. Process in Pediatric Cardiology **21**, 219–225 (2006)
86. Sisson, K., Zhang, C., Farach-Carson, M.C., Chase, D.B., Rabolt, J.F.: Evaluation of cross-linking methods for electrospun gelatin on cell growth and viability. Biomacromolecules **10**, 1675–1680 (2009)
87. Yannas, I.V.: Tissue regeneration by use of collagen-glycosaminoglycan co-polymers. Clinical Materials **9**, 179–184 (1992)
88. Tefft, S., Bentz, J., Estridge, T.D.: Collagen and heparin matrices for growth factor delivery. Journal of Controlled Release **48**, 29–33 (1997)
89. Mohanty, B., Bohidar, H.B.: Systematic of alcohol-induced simple coacervation in aqueous gelatin solutions. Biomacromolecules **4**, 1080–1086 (2003)
90. Mohanty, B., Bohidar, H.B.: Microscopic structure of gelatin coacervates. International Journal of Biological Macromolecules **36**, 39–46 (2005)
91. Drury, J.L., Mooney, D.J.: Hydrogels for tissue engineering: Scaffold design variables and applications. Biomaterials **24**, 4337–4351 (2003)
92. Lee, K.Y., Mooney, D.J.: Hydrogels for tissue engineering. Chemical Reviews **101**, 1869–1879 (2001)
93. Kloxin, A.M., Kasko, A.M., Salinas, C.N., Anseth, K.S.: Photodegradable hydrogels for dynamic tuning of physical and chemical properties. Science **324**, 59–63 (2009)
94. Hern, D.L., Hubbell, J.A.: Incorporation of adhesion peptides into nonadhesive hydrogels useful for tissue resurfacing. Journal of Biomedical Materials Research **39**, 266–276 (1998)
95. Huebsch, N., Arany, P.R., Mao, A.S., Shvartsman, D., Ali, O.A., Bencherif, S.A., Rivera-Feliciano, J., Mooney, D.J.: Harnessing traction-mediated manipulation of the cell/matrix interface to control stem-cell fate. Nature Materials **9**, 518–526 (2010)
96. Sundararaghavan, H.G., Burdick, J.A.: Gradients with depth in electrospun fibrous scaffolds for directed cell behavior. Biomacromolecules **12**, 2344–2350 (2011)
97. Sukarto, A., Yu, C., Flynn, L.E., Amsden, B.G.: Co-delivery of adipose-derived stem cells and growth factor-loaded microspheres in rgd-grafted n-methacrylate glycol chitosan gels for focal chondral repair. Biomacromolecules **13**, 2490–2502 (2012)
98. Kim, J.K., Kim, H.J., Chung, J.Y., Lee, J.H., Young, S.B., Kim, Y.H.: Natural and synthetic biomaterials for controlled drug delivery. Archives of Pharmacal Research **37**, 60–68 (2014)
99. Keys, K., Andreopoulos, F.M., Peppas, N.A.: Poly(ethylene glycol) star polymer hydrogels. Macromolecules **31**, 8149–8156 (1998)
100. Liu, V.A., Bhatia, S.N.: Three-dimensional photopatterning of hydrogels containing living cells. Biomedical Devices **4**, 257–266 (2002)
101. Chan, V., Zorlutuna, P., Jeong, J.H., Kong, H., Bashir, R.: Three-dimensional photopatterning of hydrogels using stereolithography for long-term cell encapsulation. Lab on a Chip **10**, 2062–2070 (2010)
102. Kade, M., Burke, D.J., Hawker, C.: The power of thiol-ene chemistry. Journal of Polymer Science Part A **48**, 743–750 (2010)

103. Park, K.H., Na, K., Kim, S.W., Jung, S.Y., Park, K.H., Chung, H.M.: Phenotype of hepatocyte spheroids behavior within thermo-sensitive poly(nipaam-co-peg-g-grgds) hydrogel as a cell delivery vehicle. Biotechnology Letters **27**, 1081–1086 (2005)
104. Jeong, B., Bae, Y.H., Kim, S.W.: Drug release from biodegradable injectable thermosensitive hydrogel of peg–plga–peg triblock copolymers. Journal of Controlled Release **63**, 155–163 (2000)
105. Jeong, J.H., Chan, V., Cha, C., Zorlutuna, P., Dyck, C., Hsia, K.J., Bashir, R., Kong, H.: "Living" microvascular stamp for patterning of functional neovessels; orchestrated control of matrix property and geometry. Adv. Mater. **24**(58–63), 51 (2012)
106. Peyton, S.R., Raub, C.B., Keschrumrus, V.P., Putnam, A.J.: The use of poly(ethylene glycol) hydrogels to investigate the impact of ecm chemistry and mechanics on smooth muscle cells. Biomaterials **27**, 4881–4893 (2006)
107. Kong, H.J., Boontheekul, T., Mooney, D.J.: Quantifying the relation between adhesion ligand-receptor bond formation and cell phenotype. Proc. Natl. Acad. Sci. USA **103**, 18534–18539 (2006)
108. Kong, H.J., Hsiong, S., Mooney, D.J.: Nanoscale cell adhesion ligand presentation regulates nonviral gene delivery and expression. Nano Letters **7**, 161–166 (2007)
109. Rizzi, S.C., Hubbell, J.A.: Recombinant protein-co-peg networks as cell-adhesive and proteolytically degradable hydrogel matrixes. Part i: Development and physicochemical characteristics. Biomacromolecules **6**, 1226–1238 (2005)
110. Layman, H., Sacasa, M., Murphy, A.E., Murphy, A.M., Pham, S.M., Andreopoulos, F.M.: Co-delivery of fgf-2 and g-csf from gelatin-based hydrogels as angiogenic therapy in a murine critical limb ischemic model. Acta Biomaterialia **5**, 230–239 (2009)
111. Ruvinov, E., Leor, J., Cohen, S.: The effects of controlled hgf delivery from an affinity-binding alginate biomaterial on angiogenesis and blood perfusion in a hindlimb ischemia model. Biomaterials **31**, 4573–4582 (2010)
112. Anderson, E.M., Kwee, B.J., Lewin, S.A., Raimondo, T., Mehta, M., Mooney, D.J.: Local delivery of vegf and sdf enhances endothelial progenitor cell recruitment and resultant recovery from ischemia. Tissue engineering. Part A (2014)
113. Phelps, E.A., Landázuri, N., Thulé, P.M., Taylor, W.R., García, A.J.: Bioartificial matrices for therapeutic vascularization. Proceedings of the National Academy of Sciences **107**, 3323–3328 (2010)
114. Zachman, A.L., Wang, X., Tucker-Schwartz, J.M., Fitzpatrick, S.T., Lee, S.H., Guelcher, S.A., Skala, M.C., Sung, H.-J.: Uncoupling angiogenesis and inflammation in peripheral artery disease with therapeutic peptide-loaded microgels. Biomaterials **35**, 9635–9648 (2014)
115. Kim, J.H., Jung, Y., Kim, B.-S., Kim, S.H.: Stem cell recruitment and angiogenesis of neuropeptide substance p coupled with self-assembling peptide nanofiber in a mouse hind limb ischemia model. Biomaterials **34**, 1657–1668 (2013)
116. Poole, K.M., Nelson, C.E., Joshi, R.V., Martin, J.R., Gupta, M.K., Haws, S.C., Kavanaugh, T.E., Skala, M.C., Duvall, C.L.: Ros-responsive microspheres for on demand antioxidant therapy in a model of diabetic peripheral arterial disease. Biomaterials **41**, 166–175 (2015)
117. Domanchuk, K., Ferrucci, L., Guralnik, J.M., Criqui, M.H., Tian, L., Liu, K., Losordo, D., Stein, J., Green, D., Kibbe, M., Zhao, L., Annex, B., Perlman, H., Lloyd-Jones, D., Pearce, W., Taylor, D., McDermott, M.M.: Progenitor cell release plus exercise to improve functional performance in peripheral artery disease: The propel study. Contemporary Clinical Trials **36**, 502–509 (2013)

118. Schwarz, T.M., Leicht, S.F., Radic, T., Rodriguez-Araboalaza, I., Hermann, P.C., Berger, F., Saif, J., Böcker, W., Ellwart, J.W., Aicher, A.: Vascular incorporation of endothelial colony-forming cells is essential for functional recovery of murine ischemic tissue following cell therapy. Arteriosclerosis, Thrombosis, And Vascular Biology **32**, e13–e21 (2012)
119. Katare, R., Stroemer, P., Hicks, C., Stevanato, L., Patel, S., Corteling, R., Miljan, E., Vishnubhatla, I., Sinden, J., Madeddu, P.: Clinical-grade human neural stem cells promote reparative neovascularization in mouse models of hindlimb ischemia. Arteriosclerosis, Thrombosis, And Vascular Biology **34**, 408–418 (2014)
120. Kim, D.W., Jun, I., Lee, T.-J., hye Lee, J., Lee, Y.J., Jang, H.-K., Kang, S., Park, K.D., Cho, S.-W., Kim, B.-S.: Therapeutic angiogenesis by a myoblast layer harvested by tissue transfer printing from cell-adhesive, thermosensitive hydrogels. Biomaterials **34**, 8258–8268 (2013)
121. Chen, D.-Y., Wei, H.-J., Lin, K.-J., Huang, C.-C., Wang, C.-C., Wu, C.-T., Chao, K.-T., Chen, K.-J., Chang, Y., Sung, H.-W.: Three-dimensional cell aggregates composed of huvecs and cbmscs for therapeutic neovascularization in a mouse model of hindlimb ischemia. Biomaterials **34**, 1995–2004 (2013)
122. Landazuri, N., Levit, R.D., Joseph, G., Ortega-Legaspi, J.M., Flores, C.A., Weiss, D., Sambanis, A., Weber, C.J., Safley, S.A., Taylor, W.R.: Alginate microencapsulation of human mesenchymal stem cells as a strategy to enhance paracrine-mediated vascular recovery after hindlimb ischaemia. Journal of Tissue Engineering And Regenerative Medicine (2012)
123. Rufaihah, A.J., Huang, N.F., Jamé, S., Lee, J.C., Nguyen, H.N., Byers, B., De, A., Okogbaa, J., Rollins, M., Reijo-Pera, R.: Endothelial cells derived from human ipscs increase capillary density and improve perfusion in a mouse model of peripheral arterial disease. Arteriosclerosis, Thrombosis And Vascular Biology **31**, e72–e79 (2011)
124. Mulyasasmita, W., Cai, L., Dewi, R.E., Jha, A., Ullmann, S.D., Luong, R.H., Huang, N.F., Heilshorn, S.C.: Avidity-controlled hydrogels for injectable co-delivery of induced pluripotent stem cell-derived endothelial cells and growth factors. Journal of Controlled Release : Official Journal of the Controlled Release Society **191**, 71–81 (2014)
125. Grochot-Przeczek, A., Dulak, J., Jozkowicz, A.: Therapeutic angiogenesis for revascularization in peripheral artery disease. Gene **525**, 220–228 (2013)
126. Sedighiani, F., Nikol, S.: Gene therapy in vascular disease. Surgeon **9**, 326–335 (2011)
127. Khan, T.A., Sellke, F.W., Laham, R.J.: Gene therapy progress and prospects: Therapeutic angiogenesis for limb and myocardial ischemia. Gene Ther. **10**, 285–291 (2003)
128. Yang, F., Cho, S.-W., Son, S.M., Bogatyrev, S.R., Singh, D., Green, J.J., Mei, Y., Park, S., Bhang, S.H., Kim, B.-S.: Genetic engineering of human stem cells for enhanced angiogenesis using biodegradable polymeric nanoparticles. Proceedings of the National Academy of Sciences **107**, 3317–3322 (2010)
129. Katare, R., Riu, F., Rowlinson, J., Lewis, A., Holden, R., Meloni, M., Reni, C., Wallrapp, C., Emanueli, C., Madeddu, P.: Perivascular delivery of encapsulated mesenchymal stem cells improves postischemic angiogenesis via paracrine activation of vegf-a. Arteriosclerosis, Thrombosis And Vascular Biology **33**, 1872–1880 (2013)
130. Jazwa, A., Stepniewski, J., Zamykal, M., Jagodzinska, J., Meloni, M., Emanueli, C., Jozkowicz, A., Dulak, J.: Pre-emptive hypoxia-regulated ho-1 gene therapy improves post-ischaemic limb perfusion and tissue regeneration in mice. Cardiovascular Research **97**, 115–124 (2013)

Modulation of Innate Immune Response for Tissue Engineering

Min-Ho Kim

Abstract The number of biomaterial scaffolds for tissue engineering applications continues to rise and holds promise for regenerative medicine. However, the complexity of the immune response poses a challenging environment for implanted biomaterial scaffolds for tissue repair. Specifically, the innate immune responses characterized by tissue infiltrating neutrophils and macrophages have been shown to govern either pro-inflammatory or tissue reparative microenvironments at the local site of tissue injury depending on their activation and phenotypic status. Thus, a selective strategy for developing immune-modulatory biomaterial scaffolds to improve the modulation of innate immune reactions may offer attractive features for tissue regeneration. The focus of this chapter is to discuss recent progress in the development of biomaterial scaffolds for modulating immune responses and their potential application for tissue repair. Specifically, important design variables for fabricating immuno-modulatory biomaterial scaffolds are highlighted.

1 Introduction

The principle of tissue engineering is to employ biomaterial scaffolds to restore the functional structure of damaged tissue by providing proper micro-environmental cues to promote the desired cellular functions towards regeneration. However, the tissue engineering approach is substantially influenced by the inflammatory phase of healing in response to tissue injury [1], which has been shown to be mediated by cellular components of the innate immune system such as neutrophils and macrophages. These cells are evolved to recognize and respond to tissue injury and infection by initiating and modulating inflammatory responses and provide immediate protection against pathogenic infection [2].

M.-H. Kim(✉)
Department of Biological Sciences, Kent State University, Kent, OH 44242, USA
e-mail: mkim15@kent.edu

© Springer International Publishing Switzerland 2016
H. Jo et al. (eds.), *Biomedical Engineering: Frontier Research and Converging Technologies*, Biosystems & Biorobotics 9,
DOI: 10.1007/978-3-319-21813-7_7

Biomaterial scaffolds implanted at the local site of tissue injury may intensify the inflammatory response by inducing an immunogenic response [3], which is characterized by the recruitment and persistent activation of neutrophils and macrophages that trigger prolonged inflammation. However, many studies also support the concept that tightly regulated inflammation can be beneficial for successful tissue repair by producing potent signaling molecules that are essential for tissue repair [4]. Thus, the precise understanding of the mechanisms by which cells of the innate immune system interact with foreign biomaterials will enable us to better identify targets to control the inflammatory response. As such, a strategy to enhance the beneficial aspects of the inflammatory response while suppressing its deleterious effects by means of tuning characteristics of biomaterial scaffold may facilitate successful tissue regeneration. Despite extensive studies on the role of innate immune responses in the process of wound healing, relatively little attention has been devoted to the potential impact of biomaterial scaffolds on the innate immune response, along with the development of tissue engineered scaffold to tune inflammatory response towards tissue regeneration.

The focus of this chapter is to discuss recent progress in developing tissue engineered scaffolds for modulating innate immune responses for tissue repair. Specifically, important design variables for biomaterial scaffolds that influence the fate of neutrophils and macrophages are highlighted.

2 Perspective on the Role of Innate Immune Response in Wound Healing and Tissue Repair

The wound healing process is a tightly-regulated sequence of events involving inflammatory, proliferative, and maturation phases [5]. The inflammatory response is initiated by rapid recruitment of neutrophils and macrophages from the circulation to the site of injury [5]. Although, this is critical step for providing a host defense against invading pathogens, pathological conditions that elicit either excessive or impaired recruitment of innate immune cells lead to aberrant wound healing by prolonging the release of proteolytic enzymes and oxygen-free radicals [6,7]. Indeed, chronic wound recurrence associated with dysregulated recruitment of innate immune cell has been observed in peripheral vascular disease [8], chronic stress [7] and diabetes [9]. A large body of evidence supports the concept that tightly regulated inflammatory response at the early phase of tissue injury is critical for successful tissue repair [4]. The precise understanding on the critical role of the inflammatory response mediated by innate immune cells to tissue injury will provide strategies to cure diseases associated with various chronic inflammatory disease states. Specifically, this section is focused on discussing our current understanding on the role of neutrophil and macrophage-mediated inflammatory responses during the process of tissue healing.

2.1 *Role of Neutrophils Survival and Apoptosis in Wound Healing*

Neutrophils constitute the largest proportion of leukocytes in human blood and are the first responders of the immune system that are recruited to the site of tissue injury and infection [10]. In response to tissue injury, a large number of neutrophils in the blood circulation transmigrate across the endothelial cells of post-capillary venules to the site of injury and accumulate in numbers in proportional to the extent of tissue insults [10,11]. The tightly regulated recruitment of neutrophils at the early inflammatory phase of tissue injury is essential for bacterial clearance and the removal of cellular debris. The recruitment of neutrophils to the site of injury or infection is triggered by pro-inflammatory cytokines and chemokines produced by tissue resident macrophages, which are activated on the recognition of pathogen-associated molecular patterns (PAMPs) or damage-associated molecular patterns (DAMPs) [11] (Figure 1). It has been generally accepted that neutrophil influx reaches a maximum within the initial 1-2 days after tissue injury and their numbers return to baseline at days 3-4 post-wounding in sterile inflammation [5]. However, the presence of exuberant infection at the local site of wounding can substantially increase the recruitment of neutrophils by triggering their rapid mobilization from the bone marrow compartment, which prolongs the inflammatory response [6,12].

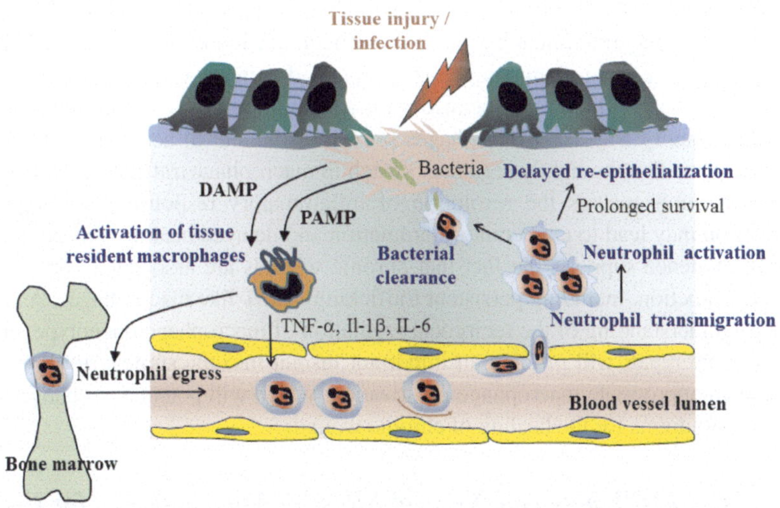

Fig. 1 Schematic on the role of the recruitment of neutrophils to the site of injury and their activation in inflammation and bacterial clearance in response to tissue injury and infection

The tightly regulated neutrophil apoptosis and subsequent clearance by macrophages are essential for the resolution of inflammation in the absence of gross infection [13]. A prolonged neutrophil survival could amplify the inflammatory response, contributing to the chronic inflammatory state in non-infectious chronic diseases such as rheumatoid arthritis [14]. However, it should be noted that, in the presence of bacterial challenge, a functional ability to delay neutrophil apoptosis is critical for defense against infection [12]. As such, a strategy for achieving appropriate recruitment and tightly regulated survival/apoptosis of neutrophils at the local site of tissue injury can be effective immuno-modulatory approaches.

2.2 Macrophages Phenotypes on Inflammation and Tissue Repair

The mediators that are released either by PAMPs or DAMPs act as exogenous or endogenous signals to initiate an inflammatory response orchestrated largely by macrophages [15,16]. Macrophages exhibit plasticity in response to environmental cues and thereby adjust their phenotype and activation status [17,18]. It has been well demonstrated that classically activated macrophages (M1) mediate host defense against invading pathogens, while alternatively activated macrophages (M2) mediate tissue repair [19]. M1 polarization can be induced by IFN-γ or lipopolysaccharides (LPS) and M1 polarized macrophages have been shown to exhibit anti-bacterial activity and secrete pro-inflammatory cytokines including TNF-α, IL-1β, and IL-6, but exhibit decreased capability to phagocytize apoptotic cells. In contrast, M2 polarization is induced by IL-4 or IL-13 and M2 polarized macrophages constitutively express scavenger and mannose receptors (CD163 and CD206) and release anti-inflammatory factors including IL-10 and TGF-β [18].

During the early inflammatory phase of tissue injury induced by either PAMPs or DAMPs, M1-like pro-inflammatory macrophages dominate, whereas at the late inflammatory phase, which is associated with accumulation of apoptotic neutrophils, anti-inflammatory M2-like macrophages prevail at the site of acute wounds. Many studies support that the tightly regulated switch in macrophage trafficking from M1 to M2 polarization ensures the resolution of inflammatory response. Prolonged M1 polarization may lead to exuberant inflammation and delayed healing. Indeed, a large body of evidence supports the fact that chronic wounds are associated with macrophage dysfunction, including persistent trafficking of M1-like macrophages. As such, a precise understanding of the reciprocal regulation of macrophage phenotype and its role in tissue repair will provide an important insight into the strategy to direct and promote an appropriate macrophage polarization, which will potentially change regeneration paradigms for the healing of chronic wounds.

2.3 Cross-Talk between Neutrophils and Macrophages for the Resolution of Inflammation and Tissue Repair

It has been well characterized that the functional interaction between macrophages and neutrophils is critical for mediating inflammatory response to tissue injury (Figure 2). Tissue macrophages express a wide array of receptors for the recognition of PAMPs and DAMPs, which include Toll-like receptors, NOD-like receptors (NLRs), and scavenger receptors [20]. After initial recognition of a microbial challenge, tissue resident macrophages promote the influx of neutrophils, followed by recruitment of monocytes, which subsequently differentiate into inflammatory macrophages. These monocyte-derived macrophages have been shown to exhibit a M1-like phenotype and rapidly dominate the inflammatory site, becoming the majority of all macrophages during the early phase of tissue injury. Any uncontrolled inflammatory environment, as frequently observed in many chronic diseases, triggers persistent trafficking of M1-like macrophages, which might act as a key source of inflammation that drives increased neutrophil trafficking and their excessive activation [15].

The clearance of apoptotic neutrophils by macrophages is a critical step for the resolution of inflammation and tissue repair. This phagocytic clearance provides a critical mechanism by which injured tissues are protected from the exposure of harmful inflammatory contents of dying cells [21]. In addition, the uptake of apoptotic cells by macrophages induces phenotypic switches and activates signaling pathways in macrophages that resemble M2 polarization [22]. Subsequently, M2 polarized macrophages further promote the phagocytic clearance of apoptotic neutrophils, whereas M1 macrophages exhibit reduced uptake of apoptotic cells. It has been shown that increased TNF-α production by M1-like macrophages may inhibit the macrophage phagocytosis of apoptotic neutrophils [23]. Indeed, an elevated number of apoptotic neutrophils has been observed in diabetic wounds, which have been associated with increased trafficking of pro-inflammatory M1-like macrophages and impaired wound healing [24]. If neutrophils die by necrosis by releasing their contents, this may exacerbate the local inflammatory response and impair the resolution of inflammation. Since aberrantly activated neutrophils contribute to the prolonged inflammatory response, the process involving macrophage phagocytosis of apoptotic neutrophils plays an important role in the resolution of inflammation, which in turn leads to the successful tissue repair. The immunosuppressive effects have been shown to be mediated through the increased production of anti-inflammatory cytokines such as TGF-β1 and PGE_2 and a suppression in the release of pro-inflammatory mediators such as IL-8 and TNF-α from macrophages [25-28].

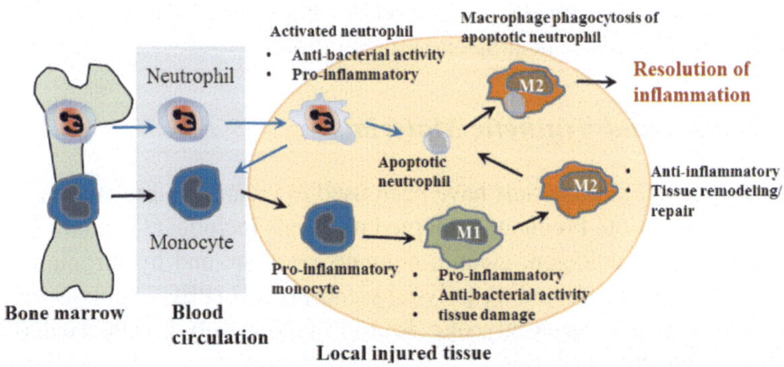

Fig. 2 Functional interactions between neutrophils and macrophages and its role in the resolution of inflammation

3 Design Considerations of Biomaterial Scaffolds for Immunomodulation

Biomaterials have been widely used to restore or augment the physiological function of diseased or damaged tissues via replacement or regeneration of tissue [29-31]. Biomaterial scaffolds can maintain a suitable environment for tissue growth by providing biomechanical and biochemical cues to support cell adhesion, migration, and appropriate activation. Innate immune cells are the first to respond to the implantation of a biomaterial in vascularized tissue [3]. Although, the implantation of most biomaterials can initiate an immune response, the extent of this response can be modulated by tuning properties of scaffold. Appropriate control of micro-environmental cues presented by biomaterials can play a crucial role in modulating the response of cells. For example, physical properties such as substrate stiffness, topography, pore size and chemical properties such as surface chemistry and ligand incorporation of biomaterial scaffolds have been shown to influence the behavior of cells that interact with those scaffolds [29,30]. Since the relative activation and phenotypic status of neutrophils and macrophages can significantly influence the outcome of the immune response, which ranges from the resolution of inflammatory response to functional regeneration of damaged tissue, the strategy to engineer specific immune processes mediated by innate immune cells may improve and facilitate the outcome of tissue healing. In light of this fact, there have been many studies employing biomaterial-based approaches for immunomodulation for tissue engineering.

Engineering the proper cellular niche can provide a suitable environment for tuning host responses by immune cells. Since an immune response is inevitable, a strategy to harness the beneficial aspects of neutrophils and macrophages will be attractive approach for designing biomaterials for modulating immune responses. In this section, we will discuss some important design variables for fabricating suitable biomaterial scaffolds for modulating host innate immune responses, specifically in the context on the role of neutrophils and macrophages. A better understanding of how different type of biomaterial scaffolds interact with innate immune cells is critical for the design and development of tissue engineered constructs that promote tissue healing and regeneration.

3.1 Natural and Synthetic Materials

Numerous types of biomaterials have been used to create a microenvironment that promotes regeneration. Frequently adopted materials include: (1) naturally-derived polymers including collagen, gelatin, alginate, agarose, and hyaluronic acid, and (2) synthetic polymers including poly(ethylene glycol) (PEG), polyglycolic acid (PGA) and poly-dl-lactic-co-glycolic acid (PLGA), poly-l-lactic acid (PLLA), and other engineered peptides [32]. Biodegradability is one of most important design factors for scaffolds for tissue engineering. Naturally-derived polymers have been used due to their inherent biodegradability and biocompatibility.

However, the use of naturally-derived polymers has been limited by batch-to-batch variability, along with limitation in conferring controlled biodegradability and mechanical properties [33]. Compared with naturally-derived polymers, synthetic polymers can confer tunable and reproducible physical and chemical properties, but tend to have lower biological activity. In most tissue engineering applications, a polymer network is required to mimic the three-dimensional microenvironments of living tissues while maintaining proper interactions between the extracellular matrix (ECM) and cells. For this reason, natural polymers such as collagen and gelatin have an advantage due to their intrinsic motifs for cellular attachment, such as the Arg-Gly-Asp (RGD) sequence. Recently, RGD-conjugated synthetic polymers have been used as well. When synthetic polymers are used, the careful design of biocompatibility and biodegradability is necessary for in vivo applications.

3.2 Mechanical Stiffness

It has been thoroughly shown that mechanical signals induced by the stiffness of biomaterials can significantly alter the behavior of cells. The elastic modulus of living tissue varies in the range from a few pascal (Pa) to multiple killopascals (kPa). Many studies highlight the importance of determining cellular functions in mechanical environments that can closely mimic those of in vivo conditions. In view of this, several, recently published studies have examined the extent to which neutrophils and macrophages respond to biomaterial scaffolds with varying degrees of elastic stiffness.

During the process of neutrophil recruitment from the blood stream to the local site of inflammation upon tissue injury or infection, neutrophils experience different physical environment, ranging from the viscous fluid in the blood vessel to the to the variable degree of stiffness at sites of inflammation [34]. These physical environments may provide different mechanical cues, which can differentially affect a neutrophil's function and behavior. It has been shown that the extent of neutrophil adhesion and transmigration are highly dependent on the mechanical properties of vascular endothelial cells [35,36]. Using fibronectin-coated polyacrylamide substrates across a physiologically relevant range of stiffness, Oakes et al. [34] demonstrated that the elastic properties of the substrate could dictate the ability that neutrophils are efficiently adhere and migrate. They revealed that neutrophils sense mechanical signals imparted by substrate stiffness via the phosphatidylinositol 3-kinase (PI3K) activity. On the softer gels (~10 kPa), neutrophils exhibited paths characteristic of random walks, while as the stiffness of the substrate was increased to ~100 kPa, the migration paths of the neutrophils become more directed and persistent. However, the cells moved slower. In a similar study, Stroka et al. [37] further revealed that the impact of substrate stiffness on neutrophil behavior patterns can be influenced by the presence of extracellular matrix proteins coated on the substrate. They revealed that the optimum stiffness for neutrophil motility was dependent on the concentration of fibronectin on the polyacrylamide substrate. For example, for high density of fibronectin (100 μg/mL), the optimum stiffness for neutrophil motility was measured at 4 kPa, however, the optimum stiffness increased to 7 kPa for relatively low density of fibronectin (10 μg/mL). This implies that optimal motility of neutrophils on scaffold can be achieved as functions of substrate stiffness and extracellular matrix coating.

Macrophages also exhibited similar adherent characteristics to neutrophils in response to substrate stiffness. Ferreol et al. [38] examined the effect of substrate stiffness on the sensitivity of alveolar macrophages (AMs), in which cells were cultured on substrates of various stiffness ranging from low and high intermediate stiffness (respectively 40 kPa and 160 kPa) to a highly rigid surface of plastic or glass (respectively 3 MPa and 70 MPa). Results showed that macrophage shape dramatically changed from rounded to flattened as substrate stiffness increased from soft to rigid. In a similar in vitro study using the RAW 264.7 macrophage cell line, Patel et al. [39] revealed that the cells exhibited organized actin filaments and filopodial projections on stiff (150 kPa) polyacrylamide substrates (Figure 3). It was further revealed that the response was mediated via a Rho-GTPase dependent manner since treatment with a Rho-GTPase inhibitor could significantly reduce the response to a level that was observed in softer (1.2 kPa) substrates. However, the extent of macrophage phagocytic ability and pro-inflammatory activation appears to increase with increasing material stiffness. Increasing the stiffness of three-dimensional polyethylene glycol-RGD (PEG-RGD) hydrogels (130, 240, and 840 kPa) increased foreign body reaction and the expression of TNF-α, IL-1β, and IL-6 [40]. In contrast, hydrogels with lower stiffness resulted in reduced macrophage activation.

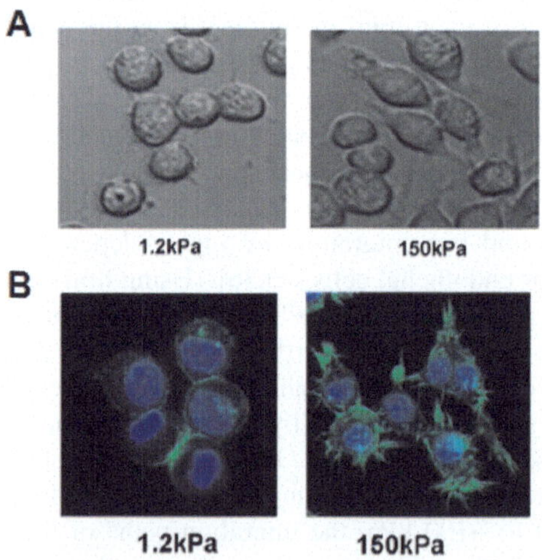

Fig. 3 Cell spreading of RAW 264.7 macrophages cultured on polyacrylamide gel surfaces with varying stiffness. Differential interference contrast microscopy images (A) and confocal images for actin staining (B) from RAW 264.7 macrophages on a less rigid (1.2 kPa) versus rigid (150 kPa) polyacrylamide gel coated with poly-L-lysine. Green color represents actin staining (phalloidin) with blue stained nuclei (Hoechst). Adapted from [39] with permission.

Taken together, these results support the fact that both neutrophils and macrophages are highly sensitive to biomechanical environment dictated by scaffold stiffness and a design strategy for interfering with scaffold stiffness can be an effective approach to modulate innate immune response.

3.3 Surface Topography

The ECM can play important roles in the functional behavior of cells by conferring the surrounding tissue with topographic features. For instance, it can provide cells with a suitable adhesion substratum and can present cells with not only a variety of embedded chemical cues, but also mechanical cues, such as the degree of the matrix rigidity, size, orientation, deformability and local density of the ECM fibers [41]. Growing bodies of evidence strongly support the fact that most mammalian cell types are exquisitely sensitive to the mechanical features of the adhesion substratum that can actively signal to the cells, influencing their phenotypes significantly [42]. Indeed, a broad range of cells including osteoblasts, fibroblasts, endothelial, epithelial, and smooth muscle cells have been shown to differently react to topographic features by changing their responsiveness of cell adhesion, morphology, orientation, or activity [43].

Surface topographic features on the micron scale were shown to influence the fate of neutrophils by activating either non-apoptotic, or apoptotic cell death. Shive et al. [44] revealed that neutrophils underwent a rapid apoptosis on a polyetherurethane urea film even when they were exposed to physiological level of shear stress. Chang et al. [45] have shown that neutrophils cultured on roughened polytetrafluoroethylene (ePTFE) resulted in non-apoptotic cell death by the release of reactive oxygen species, while neutrophils adhered to smooth polystyrene remain viable.

Modification of surface topography at the level of either nano or micron scale has been shown to modulate the morphology and activation status of macrophages [46-48]. The RAW 264.7 macrophage cultured on 2 μm PDMS gratings exhibited an elongated morphology in the direction of the gratings, while cell on the planar substrate maintained its native round morphology [48] (Figure 4A). In poly-L-lactic acid (PLLA) electrospun scaffolds incorporating controlled topographic features including fiber diameter and alignment, a greater number of viable macrophages were observed for aligned and micron-meter scale fibers than those with either randomly oriented or nanometer scale (Figure 4B and 4C) [47]. It was shown that surface topography could elicit expression of pro-inflammatory cytokines in macrophages by triggering NFκB-dependent signaling pathway [49]. Interestingly, increased foreign body reaction by macrophages was observed in scaffold with nanometer range of electrospun fibers associated with high surface-to-volume ratios in poly(ethylene terephthalate) (PET) compared with one with micron scale fibers [50].

Given that chronic, non-healing wounds are associated with extensive ECM damage due to cytotoxic enzymes and free radicals mediated by activated phagocytes, the biomaterial based strategy for re-establishing a functional ECM topography can normalize the appropriate activation and phenotype of infiltrating immune cells. Design considerations for incorporating topographic features can be determined depending on the specific application. For example, it may be reasonable to apply micro-scale topographic patterns in the scaffold to promote the tissue reparative phenotype of macrophages. In contrast, the use of nanostructured biomaterials may be desirable to promote the pro-inflammatory and anti-bacterial phenotype of neutrophils and macrophages, which is necessary for the resolution of wound infection.

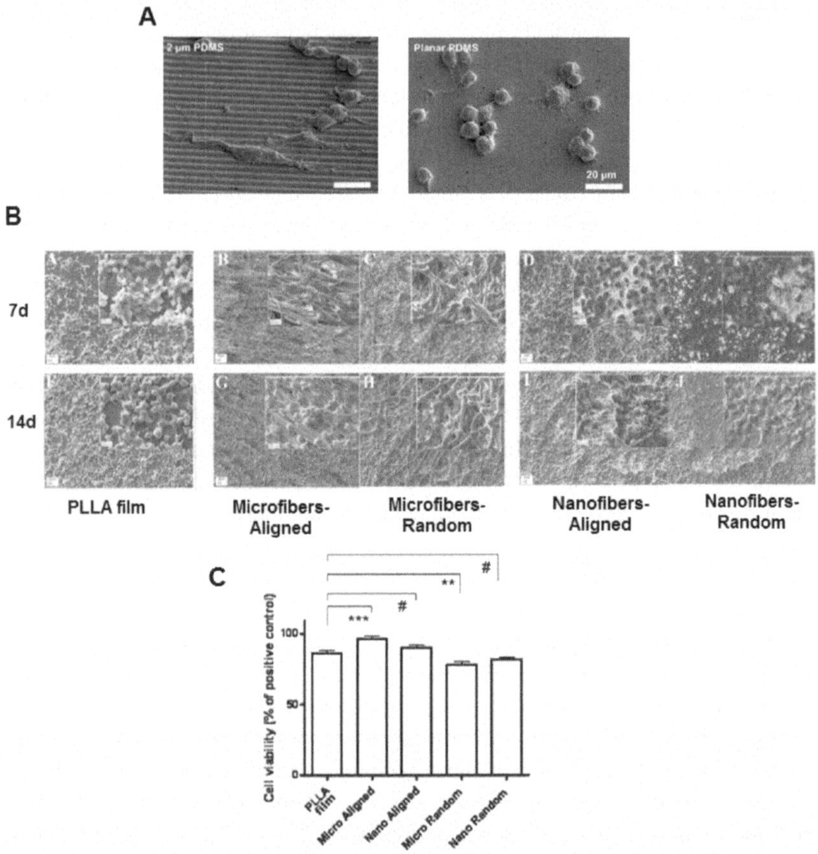

Fig. 4 A. The morphology of RAW 264.7 cells cultured for 48 hours on PDMS with 2µm topological gratings (left) compared to planar control (right). Adapted from [48] with permission. B and C. Representative SEM images (200× magnification and inserts at 1000×) of RAW 264.7 cells cultured for 24 hours and 7 days on PLLA film with microfibrous aligned, microfibrous random, nanofibrous aligned and nanofibrous random scaffolds (B) and the quantification for cell viability (C). Adapted from [47] with permission.

3.4 Scaffold Pore Size

Cell infiltration into the scaffold and its proper activation is critical for successful tissue regeneration. Importantly, three-dimensionally modeled biomaterial scaffolds can mimic the complex microenvironment found within in-vivo tissue. Thus, it is important to better understand the extent of cell infiltration into three-dimensional scaffolds. Many recent studies highlight the importance of pore size in three-dimensional scaffolds as an important consideration factor for modulating the activation status of infiltrating leukocytes. Studies have been performed to examine the effect of scaffold pore size on macrophage activation and behavior, by controlling pore dimension at either nanoscale or microscale. Ferraz et al. [51] examined the effect of alumina nanoporosity on monocyte/macrophage behavior. In this study, human mononuclear cells were cultured on the alumina with pore diameters of 20 and 200 nm and the release of pro-inflammatory cytokines was evaluated. The study demonstrated an increased number of activated cells on the 200 nm porous alumina, while the cells adhering on the 20 nm porous substrate exhibited a reduced pro-inflammatory activity. This implies that the porosity at the nanometer scale can be tuned for controlling the inflammatory response induced by implanted scaffolds.

The effect of micro-scale porosity in biomaterials on tissue healing has also been studied using a series of porous biomaterials made from poly(2-hydroxyethyl methacrylate) (pHEMA) hydrogel, a biodegradable pHEMA-co-caprolactone (PCL), silicone elastomer, fibrin, alginate and polyurethane, which all have pore sizes ranging from 10 to 160 μm [52-57]. In this study, it was observed that porous scaffolds with 30–40 μm diameter pore size exhibited the most efficient healing. However, the mechanism that leads to successful tissue healing in that pore size is not clear. It was proposed that the macrophages that had infiltrated into the scaffold seemed to be directed toward a M2-like tissue healing phenotype. In another study employing a transcutaneous implantation of porous scaffold, Fukano et al. [58] have made similar observations. In that study, scaffolds with a 35-μm pore size lead to successful tissue healing associated with reconstruction of the epidermis and vascularization of the dermis. Recently, Sussman et al. [59] further examined the effect of scaffold porosity on the macrophage polarization, in which either 34 μm porous or 160 μm porous scaffold was implanted in subcutaneous mouse tissue. Interestingly, they observed that macrophages in 34 μm porous implants resulted in increased expression of M1 markers by 63%, along with reduction in M2 marker expression by 85%, which were associated with reduced fibrosis and increased vascularization. In contrast, cells outside the porous structure were associated with an increased infiltration of M2 phenotypic cells. This study implicates that a strategy to achieve a balanced activation of M1 and M2 macrophages may be desirable for successful tissue repair.

3.5 *Incorporation of Signaling Molecules*

During the course of the inflammatory phase of tissue healing, various signaling factors orchestrate the functional behaviors of immune cells. Thus, the extent of immune response can be modulated locally by incorporating appropriate bioactive signaling molecules into the biomaterial scaffold. These biochemical factors can direct cellular adhesion, chemotactic migration, or functional phenotypes.

Henry et al. [60] investigated the effect of the density of adhesive ligands on the neutrophil motility. The study showed that neutrophils on moderate densities of adhesive ligands formed a well-spread morphology without a uropod, along with slow motility with high persistence. However, the cellular morphology changed to an amoeboid shape with increasing ligand density, along with quick migration with low persistence. These results suggest that the extent of neutrophil motility as well as morphology can be tuned by incorporating a proper adhesive environment.

Macrophages interact with implanted biomaterials through a layer of proteins adsorbed on the biomaterial surface. For example, macrophage integrin receptors bind to a milieu of biomaterial-adsorbed adhesive proteins [61,62]. Binding of integrin molecules by macrophages initiates signaling pathways that lead to macrophage activation and the release of inflammatory cytokines [63,64]. Blocking the receptor class of RGD-binding integrins through the controlled release of echistatin, a high affinity RGD peptide, resulted in a 45% reduction in fibrous capsule thickness formed around implants. The dual delivery of MCP-1 and VEGF using alginate particles that achieve distinct release kinetics was investigated in an attempt to promote not only early monocyte recruitment, but also vascular angiogenesis [65]. The results revealed that the endothelial cell therapy combined with delivery of VEGF and MCP-1 could result in more enhanced therapeutic effects by increasing the formation of functional blood vessels than did endothelial cell therapy alone. Interestingly, despite the potential role of MCP-1 in inflammation, macrophage polarization trended towards the development of a M2 phenotype in a time- and dose-dependent manner [65]. In another study using a murine subcutaneous implantation model, scaffolds coated with LPS were infiltrated and degraded by M1 macrophages, while porous collagen scaffolds without LPS coatings were surrounded by a fibrous capsule, coincident with the high expression of M2 macrophage markers, [66]. In a recent study using a murine subcutaneous implantation model, Spiller et al. [67] designed biomaterial scaffolds that can promote the sequential M1 and M2 polarization of host macrophages by incorporating IFN-γ and IL-4 and inducing their sequential release (Figure 5A). The scaffolds with physically adsorbed IFN-gamma are expected to cause initial polarization of macrophages to the M1 phenotype and thereby achieve the release of angiogenic factors from macrophages and initiate the process of angiogenesis. The release of IFN-gamma was followed by the release of IL-4, which could convert M1 macrophages to the M2 phenotype, which further could stabilize the growing vasculature by secreting PDGF-BB. The immunohistochemical analysis of macrophage phenotype after 2 weeks of subcutaneous implantation in mice confirmed an increased vascularization although there were no differences in macrophage phenotype (Figure 5B). Gower et al. [68] examined the effect of IL-10 delivery on the extent of leukocyte infiltration to implanted PLG scaffolds within intraperitoneal fat. The study showed that IL-10 delivery could decrease the infiltration of leukocytes (neutrophils and macrophages) by 50% relative to controls, which was correlated with increased IL-10 expression and decreased IFN-γ expression in macrophages.

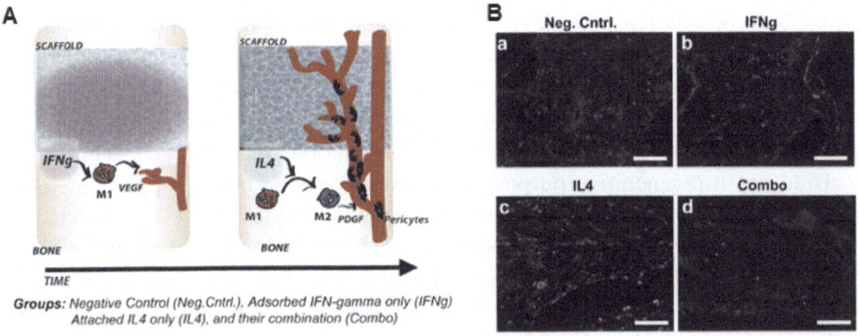

Fig. 5 A. The schematic on the design of scaffolds that can promote the sequential M1 and M2 polarization by achieving the sequential release of IFN-γ and IL-4. B. Immunohistochemical analysis of macrophage phenotype after 2 weeks of subcutaneous implantation in mice. Samples were stained for the M1 marker iNOS (green), the M2 marker Arg1 (blue) and the pan-macrophage marker F480 (red). (a) Negative control, (b) IFN-γ, (c) IL4, (d) Combined treatment of IFN-γ and IL-4. Scale bars represent 100 μm. Adapted from [67] with permission.

Taken together, these studies support the concept that the appropriate incorporation of signaling molecules intended to regulate macrophage activation and polarization onto biomaterial scaffolds can be effective approaches to either modulate foreign body reaction or promote tissue repair processes.

4 Summary and Future Outlook

Inflammation mediated by cells of the innate immune system is like a "double-edged sword" for tissue repair. The inflammatory response is critical in removing deleterious substances and in repairing damaged tissues. Despite its functional importance in host defense and wound healing, dis-regulated or prolonged inflammation can lead to impaired tissue repair. Indeed, the complexity of the immune response can be a challenging environment for the successful application of tissue engineered constructs for tissue repair. Many recent studies support that the extent of morphology and phenotypic activations of innate immune cells can be significantly influenced by not only biomechanical (e.g. stiffness, topography, and porosity), but also biochemical (e.g. presentation of signaling molecules) features of biomaterial scaffolds. As such, an appropriate incorporation of those variables in designing scaffolds will realize the development of novel immuno-modulatory biomaterial scaffolds that can achieve the successful resolution of inflammation by modulating phenotypes and activation status of neutrophils and macrophages (Figure 6).

The specific design criteria of biomaterial scaffolds for immunomodulation can be adapted depending on the specific purpose and application [69]. For example, if the implanted scaffold needs to be integrated permanently in the tissue, the scaffold should be tailored to achieve a predominant M2 polarization of macrophages upon interaction with the scaffold while reducing neutrophil activation. However, it should be noted that the transient presence of M1 macrophages has been shown to be beneficial for angiogenesis and vascularization [67], whereas the prolonged presence of M2 macrophages can result in foreign body reaction [70]. Thus, if the scaffold has a regenerative purpose intended to provide initial support for tissue remodeling and growth, immuno-modulatory strategies need be directed to achieve a more balanced M1/M2 phenotype, rather than achieving a predominant polarization toward a M2 phenotype. In addition, it is also equally important to modulate neutrophil functions, since uncontrolled and excessive neutrophil activation can exacerbate local inflammatory responses, which in turn triggers the trafficking of pro-inflammatory monocytes/macrophages in a positive feedback manner.

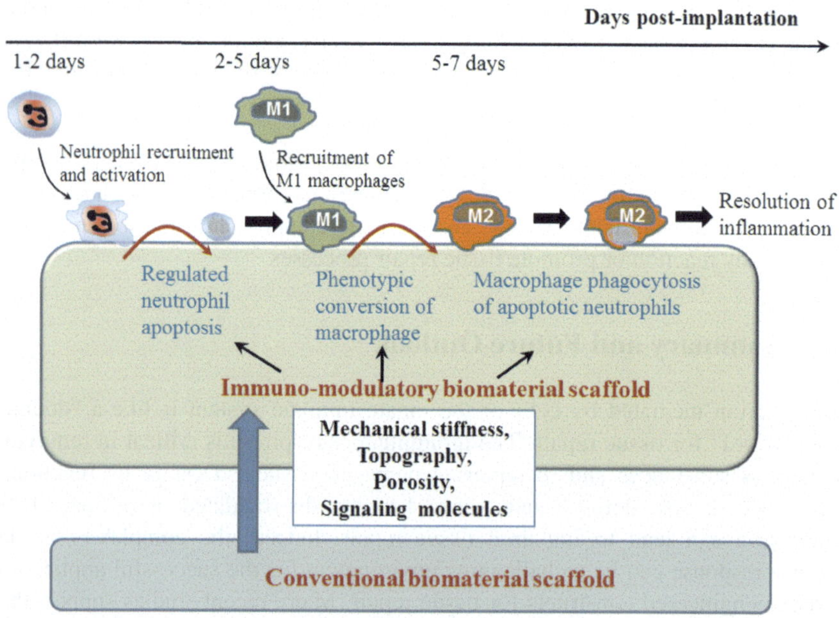

Fig. 6 Schematic on the design of immuno-modulatory biomaterial scaffold that modulates phenotypes and activation status of neutrophils and macrophages

In conclusion, a more detailed and systemic understanding of the interactions between neutrophils and macrophages within biomaterial scaffolds with different design conditions (e.g. mechanical stiffness, pore size, topography, and ligand incorporation) will provide a novel strategy to harness inflammatory responses to improve tissue regeneration, while at the same time, limiting unwanted pro-inflammatory responses.

Acknowledgments The author would like to thank you for the support from Kent State University through a startup fund.

References

1. Bryers, J.D., Giachelli, C.M., Ratner, B.D.: Engineering biomaterials to integrate and heal: the biocompatibility paradigm shifts. Biotechnol. Bioeng. **109**, 1898–1911 (2012)
2. Miller, L.S., O'Connell, R.M., Gutierrez, M.A., Pietras, E.M., Shahangian, A., et al.: MyD88 mediates neutrophil recruitment initiated by IL-1R but not TLR2 activation in immunity against Staphylococcus aureus. Immunity **24**, 79–91 (2006)
3. Ekdahl, K.N., Lambris, J.D., Elwing, H., Ricklin, D., Nilsson, P.H., et al.: Innate immunity activation on biomaterial surfaces: a mechanistic model and coping strategies. Adv. Drug Deliv. Rev. **63**, 1042–1050 (2011)
4. Eming, S.A., Krieg, T., Davidson, J.M.: Inflammation in wound repair: molecular and cellular mechanisms. J. Invest. Dermatol. **127**, 514–525 (2007)
5. Singer, A.J., Clark, R.A.: Cutaneous wound healing. N. Engl. J. Med. **341**, 738–746 (1999)
6. Kim, M.H., Liu, W., Borjesson, D.L., Curry, F.R., Miller, L.S., et al.: Dynamics of neutrophil infiltration during cutaneous wound healing and infection using fluorescence imaging. J. Invest. Dermatol. **128**, 1812–1820 (2008)
7. Kim, M.H., Gorouhi, F., Ramirez, S., Granick, J.L., Byrne, B.A., et al.: Catecholamine stress alters neutrophil trafficking and impairs wound healing by beta2-adrenergic receptor-mediated upregulation of IL-6. J. Invest. Dermatol. **134**, 809–817 (2014)
8. Haumer, M., Amighi, J., Exner, M., Mlekusch, W., Sabeti, S., et al.: Association of neutrophils and future cardiovascular events in patients with peripheral artery disease. J. Vasc. Surg. **41**, 610–617 (2005)
9. Pierce, G.F.: Inflammation in nonhealing diabetic wounds: the space-time continuum does matter. Am. J. Pathol. **159**, 399–403 (2001)
10. Manz, M.G., Boettcher, S.: Emergency granulopoiesis. Nat. Rev. Immunol. **14**, 302–314 (2014)
11. Kolaczkowska, E., Kubes, P.: Neutrophil recruitment and function in health and inflammation. Nat. Rev. Immunol. **13**, 159–175 (2013)
12. Kim, M.H., Granick, J.L., Kwok, C., Walker, N.J., Borjesson, D.L., et al.: Neutrophil survival and c-kit(+)-progenitor proliferation in Staphylococcus aureus-infected skin wounds promote resolution. Blood **117**, 3343–3352 (2011)
13. Sindrilaru, A., Scharffetter-Kochanek, K.: Disclosure of the Culprits: Macrophages-Versatile Regulators of Wound Healing. Adv. Wound Care (New Rochelle) **2**, 357–368 (2013)
14. Parsonage, G., Filer, A., Bik, M., Hardie, D., Lax, S., et al.: Prolonged, granulocyte-macrophage colony-stimulating factor-dependent, neutrophil survival following rheumatoid synovial fibroblast activation by IL-17 and TNFalpha. Arthritis Res. Ther. **10**, R47 (2008)

15. Mosser, D.M., Edwards, J.P.: Exploring the full spectrum of macrophage activation. Nat. Rev. Immunol. **8**, 958–969 (2008)
16. Ricardo, S.D., van Goor, H., Eddy, A.A.: Macrophage diversity in renal injury and repair. J. Clin. Invest. **118**, 3522–3530 (2008)
17. Novak, M.L., Koh, T.J.: Macrophage phenotypes during tissue repair. J. Leukoc. Biol. **93**, 875–881 (2013)
18. Sica, A., Mantovani, A.: Macrophage plasticity and polarization: in vivo veritas. J. Clin. Invest. **122**, 787–795 (2012)
19. Murray, P.J., Wynn, T.A.: Protective and pathogenic functions of macrophage subsets. Nat. Rev. Immunol. **11**, 723–737 (2011)
20. Davies, L.C., Jenkins, S.J., Allen, J.E., Taylor, P.R.: Tissue-resident macrophages. Nat. Immunol. **14**, 986–995 (2013)
21. Michlewska, S., Dransfield, I., Megson, I.L., Rossi, A.G.: Macrophage phagocytosis of apoptotic neutrophils is critically regulated by the opposing actions of pro-inflammatory and anti-inflammatory agents: key role for TNF-alpha. FASEB J. **23**, 844–854 (2009)
22. Ariel, A., Serhan, C.N.: New Lives Given by Cell Death: Macrophage Differentiation Following Their Encounter with Apoptotic Leukocytes during the Resolution of Inflammation. Front. Immunol. **3**, 4 (2012)
23. Feng, X., Deng, T., Zhang, Y., Su, S., Wei, C., et al.: Lipopolysaccharide inhibits macrophage phagocytosis of apoptotic neutrophils by regulating the production of tumour necrosis factor alpha and growth arrest-specific gene 6. Immunology **132**, 287–295 (2011)
24. Khanna, S., Biswas, S., Shang, Y., Collard, E., Azad, A., et al.: Macrophage dysfunction impairs resolution of inflammation in the wounds of diabetic mice. PLoS One **5**, e9539 (2010)
25. Manfredi, A.A., Iannacone, M., D'Auria, F., Rovere-Querini, P.: The disposal of dying cells in living tissues. Apoptosis **7**, 153–161 (2002)
26. Ren, Y., Savill, J.: Apoptosis: the importance of being eaten. Cell Death Differ. **5**, 563–568 (1998)
27. Savill, J., Dransfield, I., Gregory, C., Haslett, C.: A blast from the past: clearance of apoptotic cells regulates immune responses. Nat. Rev. Immunol. **2**, 965–975 (2002)
28. Voll, R.E., Herrmann, M., Roth, E.A., Stach, C., Kalden, J.R., et al.: Immunosuppressive effects of apoptotic cells. Nature **390**, 350–351 (1997)
29. Furth, M.E., Atala, A., Van Dyke, M.E.: Smart biomaterials design for tissue engineering and regenerative medicine. Biomaterials **28**, 5068–5073 (2007)
30. Pashuck, E.T., Stevens, M.M.: Designing regenerative biomaterial therapies for the clinic. Sci. Transl. Med. **4**, 160–164 (2012)
31. Langer, R., Tirrell, D.A.: Designing materials for biology and medicine. Nature **428**, 487–492 (2004)
32. O'Brien, F.J.: Biomaterials & scaffolds for tissue engineering. Materials Today **14**, 88–95 (2011)
33. Nicodemus, G.D., Bryant, S.J.: Cell encapsulation in biodegradable hydrogels for tissue engineering applications. Tissue Eng. Part B Rev. **14**, 149–165 (2008)
34. Oakes, P.W., Patel, D.C., Morin, N.A., Zitterbart, D.P., Fabry, B., et al.: Neutrophil morphology and migration are affected by substrate elasticity. Blood **114**, 1387–1395 (2009)
35. Yeung, T., Georges, P.C., Flanagan, L.A., Marg, B., Ortiz, M., et al.: Effects of substrate stiffness on cell morphology, cytoskeletal structure, and adhesion. Cell Motil. Cytoskeleton **60**, 24–34 (2005)

36. Stroka, K.M., Aranda-Espinoza, H.: Endothelial cell substrate stiffness influences neutrophil transmigration via myosin light chain kinase-dependent cell contraction. Blood **118**, 1632–1640 (2011)
37. Stroka, K.M., Aranda-Espinoza, H.: Neutrophils display biphasic relationship between migration and substrate stiffness. Cell Motil. Cytoskeleton **66**, 328–341 (2009)
38. Fereol, S., Fodil, R., Labat, B., Galiacy, S., Laurent, V.M., et al.: Sensitivity of alveolar macrophages to substrate mechanical and adhesive properties. Cell Motil. Cytoskeleton **63**, 321–340 (2006)
39. Patel, N.R., Bole, M., Chen, C., Hardin, C.C., Kho, A.T., et al.: Cell elasticity determines macrophage function. PLoS One **7**, e41024 (2012)
40. Blakney, A.K., Swartzlander, M.D., Bryant, S.J.: The effects of substrate stiffness on the in vitro activation of macrophages and in vivo host response to poly(ethylene glycol)-based hydrogels. J. Biomed. Mater. Res. A **100**, 1375–1386 (2012)
41. Kim, D.H., Han, K., Gupta, K., Kwon, K.W., Suh, K.Y., et al.: Mechanosensitivity of fibroblast cell shape and movement to anisotropic substratum topography gradients. Biomaterials **30**, 5433–5444 (2009)
42. Bershadsky, A., Kozlov, M., Geiger, B.: Adhesion-mediated mechanosensitivity: a time to experiment, and a time to theorize. Curr. Opin. Cell Biol. **18**, 472–481 (2006)
43. Kim, D.H., Provenzano, P.P., Smith, C.L., Levchenko, A.: Matrix nanotopography as a regulator of cell function. J. Cell Biol. **197**, 351–360 (2012)
44. Shive, M.S., Salloum, M.L., Anderson, J.M.: Shear stress-induced apoptosis of adherent neutrophils: a mechanism for persistence of cardiovascular device infections. Proc. Natl. Acad. Sci. U.S.A. **97**, 6710–6715 (2000)
45. Chang, S., Popowich, Y., Greco, R.S., Haimovich, B.: Neutrophil survival on biomaterials is determined by surface topography. J. Vasc. Surg. **37**, 1082–1090 (2003)
46. Kwon, K.W., Park, H., Song, K.H., Choi, J.C., Ahn, H., et al.: Nanotopography-guided migration of T cells. J. Immunol. **189**, 2266–2273 (2012)
47. Saino, E., Focarete, M.L., Gualandi, C., Emanuele, E., Cornaglia, A.I., et al.: Effect of electrospun fiber diameter and alignment on macrophage activation and secretion of proinflammatory cytokines and chemokines. Biomacromolecules **12**, 1900–1911 (2011)
48. Chen, S., Jones, J.A., Xu, Y., Low, H.Y., Anderson, J.M., et al.: Characterization of topographical effects on macrophage behavior in a foreign body response model. Biomaterials **31**, 3479–3491 (2010)
49. Waterfield, J.D., Ali, T.A., Nahid, F., Kusano, K., Brunette, D.M.: The effect of surface topography on early NFkappaB signaling in macrophages. J. Biomed. Mater. Res. A **95**, 837–847 (2010)
50. Veleirinho, B., Coelho, D.S., Dias, P.F., Maraschin, M., Pinto, R., et al.: Foreign body reaction associated with PET and PET/chitosan electrospun nanofibrous abdominal meshes. PLoS One **9**, e95293 (2014)
51. Ferraz, N., Hong, J., Santin, M., Karlsson Ott, M.: Nanoporosity of alumina surfaces induces different patterns of activation in adhering monocytes/macrophages. Int. J. Biomater. **2010**, 402715 (2010)
52. Beckstead, B.L., Tung, J.C., Liang, K.J., Tavakkol, Z., Usui, M.L., et al.: Methods to promote Notch signaling at the biomaterial interface and evaluation in a rafted organ culture model. J. Biomed. Mater. Res. A **91**, 436–446 (2009)
53. Linnes, M.P., Ratner, B.D., Giachelli, C.M.: A fibrinogen-based precision microporous scaffold for tissue engineering. Biomaterials **28**, 5298–5306 (2007)

54. Madden, L.R., Mortisen, D.J., Sussman, E.M., Dupras, S.K., Fugate, J.A., et al.: Proangiogenic scaffolds as functional templates for cardiac tissue engineering. Proc. Natl. Acad. Sci. U.S.A. **107**, 15211–15216 (2010)
55. Osathanon, T., Giachelli, C.M., Somerman, M.J.: Immobilization of alkaline phosphatase on microporous nanofibrous fibrin scaffolds for bone tissue engineering. Biomaterials **30**, 4513–4521 (2009)
56. Osathanon, T., Linnes, M.L., Rajachar, R.M., Ratner, B.D., Somerman, M.J., et al.: Microporous nanofibrous fibrin-based scaffolds for bone tissue engineering. Biomaterials **29**, 4091–4099 (2008)
57. Saul, J.M., Linnes, M.P., Ratner, B.D., Giachelli, C.M., Pun, S.H.: Delivery of non-viral gene carriers from sphere-templated fibrin scaffolds for sustained transgene expression. Biomaterials **28**, 4705–4716 (2007)
58. Fukano, Y., Knowles, N.G., Usui, M.L., Underwood, R.A., Hauch, K.D., et al.: Characterization of an in vitro model for evaluating the interface between skin and percutaneous biomaterials. Wound Repair Regen. **14**, 484–491 (2006)
59. Sussman, E.M., Halpin, M.C., Muster, J., Moon, R.T., Ratner, B.D.: Porous implants modulate healing and induce shifts in local macrophage polarization in the foreign body reaction. Ann. Biomed. Eng. **42**, 1508–1516 (2014)
60. Henry, S.J., Crocker, J.C., Hammer, D.A.: Ligand density elicits a phenotypic switch in human neutrophils. Integr. Biol. (Camb) **6**, 348–356 (2014)
61. Ratner, B.D., Bryant, S.J.: Biomaterials: where we have been and where we are going. Annu. Rev. Biomed. Eng. **6**, 41–75 (2004)
62. Sun, D.H., Trindade, M.C., Nakashima, Y., Maloney, W.J., Goodman, S.B., et al.: Human serum opsonization of orthopedic biomaterial particles: protein-binding and monocyte/macrophage activation in vitro. J. Biomed. Mater. Res. A **65**, 290–298 (2003)
63. Ingham, E., Fisher, J.: The role of macrophages in osteolysis of total joint replacement. Biomaterials **26**, 1271–1286 (2005)
64. Merkel, K.D., Erdmann, J.M., McHugh, K.P., Abu-Amer, Y., Ross, F.P., et al.: Tumor necrosis factor-alpha mediates orthopedic implant osteolysis. Am. J. Pathol. **154**, 203–210 (1999)
65. Jay, S.M., Shepherd, B.R., Andrejecsk, J.W., Kyriakides, T.R., Pober, J.S., et al.: Dual delivery of VEGF and MCP-1 to support endothelial cell transplantation for therapeutic vascularization. Biomaterials **31**, 3054–3062 (2010)
66. Spiller, K.L., Anfang, R.R., Spiller, K.J., Ng, J., Nakazawa, K.R., et al.: The role of macrophage phenotype in vascularization of tissue engineering scaffolds. Biomaterials **35**, 4477–4488 (2014)
67. Spiller, K.L., Nassiri, S., Witherel, C.E., Anfang, R.R., Ng, J., et al.: Sequential delivery of immunomodulatory cytokines to facilitate the M1-to-M2 transition of macrophages and enhance vascularization of bone scaffolds. Biomaterials **37**, 194–207 (2015)
68. Gower, R.M., Boehler, R.M., Azarin, S.M., Ricci, C.F., Leonard, J.N., et al.: Modulation of leukocyte infiltration and phenotype in microporous tissue engineering scaffolds via vector induced IL-10 expression. Biomaterials **35**, 2024–2031 (2014)
69. Sridharan, R., Cameron, A.R., Kelly, D.J., Kearney, C., O'Brien, F.J.: Biomaterial based modulation of macrophage polarization: a review and suggested design principles. Materials Today (2015). doi:10.1016/j.mattod.2015.01.019
70. Anderson, J.M., Rodriguez, A., Chang, D.T.: Foreign body reaction to biomaterials. Semin. Immunol. **20**, 86–100 (2008)

Part III
Organ-on-Chip

Lipid-Based Nanoparticles for Vaccine Applications

Rui Kuai, Lukasz J. Ochyl, Anna Schwendeman and James J. Moon

Abstract Currently available vaccine adjuvants are ineffective against a wide range of infectious pathogens as well as cancers. Therefore, there is a critical demand for new vaccine strategies that can elicit potent cellular and humoral immune responses. Liposomes have been widely examined as vaccine delivery systems because of their safety, low toxicity, and ease of scale-up. However, successful clinical translation of liposomal vaccines has been hampered by their limited potency to induce strong T and B cell responses. In this chapter, we will present two classes of lipid-based nanoparticle systems designed to address limitations of liposomal vaccines and discuss their potential as vaccine delivery systems. The first class of lipid-based nanoparticles presented in this chapter is termed interbilayer-crosslinked multilamellar vesicles. These novel vaccine nanoparticles are stable vehicles that can effectively deliver antigens and adjuvant molecules to antigen-presenting cells in lymphoid tissues and induce robust T and B cell immune responses in vivo. The second class of vaccine nanoparticles is lipoproteins composed of endogenous proteins and lipids. Applications of lipoproteins for vaccine delivery have recently gained much attention due to their safety and multi-faceted functions as endogenous drug delivery vehicles. We

R. Kuai · L.J. Ochyl · J.J. Moon(✉)
Department of Pharmaceutical Sciences, University of Michigan, Ann Arbor, MI 48109, USA
e-mail: moonjj@med.umich.edu

R. Kuai · L.J. Ochyl · A. Schwendeman · J.J. Moon
Biointerfaces Institute, University of Michigan, Ann Arbor, MI 48109, USA

A. Schwendeman
Department of Medicinal Chemistry, University of Michigan, Ann Arbor, MI 48109, USA
e-mail: annaschw@umich.edu

J.J. Moon
Department of Biomedical Engineering, University of Michigan, Ann Arbor, MI 48109, USA

© Springer International Publishing Switzerland 2016

H. Jo et al. (eds.), *Biomedical Engineering: Frontier Research and Converging Technologies*, Biosystems & Biorobotics 9,
DOI: 10.1007/978-3-319-21813-7_8

provide an overview on the latest advances in this rapidly evolving interdisciplinary area of research, and we discuss biomaterial-based innovations enabled by nanotechnology for improving vaccine design and development.

Keywords: Biomaterials · Vaccine · Nanoparticles · Nanotechnology · Lipid · Antigen · Adjuvant

List of Abbreviations

α-GC - alpha-galactosyl ceramide; APCs - antigen-presenting cells; CM – chylomicrons; CpG - oligonucleotides with unmethylated CpG motifs; CTL - cytotoxic CD8+ T lymphocyte; DC - dendritic cell; ER - endoplasmic reticulum; FPLC - fast protein liquid chromatography; GGC - galactosyl(α 1-2)galactosyl ceramide; HCV - Hepatitis C virus; HDL - high density lipoproteins; HIV - human immunodeficiency virus; HPV - human papilloma virus; ICMVs - interbilayer-crosslinked multilamellar vesicles; IFNγ - Interferon gamma; LDL - low density lipoproteins; LDLR - low density lipoprotein receptor; MHC - major histocompatibility complex; MPLA - monophosphoryl lipid A; NiNLPs - nickel-chelating nanolipoprotein particles; NKT - natural killer T cells; NLPs - nanolipoprotein particles; OVA – ovalbumin; PAMPs - pathogen-associated molecular patterns; PBAE - poly(β-amino ester); PEG - polyethylene glycol; PLGA - poly(lactide-co-glycolide) acid; PRRs - pathogen recognition receptors; TLR - Toll-like receptor; VLDL - very low density lipoproteins; WNV - West Nile virus

1 Overview of Nanoparticle Vaccine Delivery Systems

Vaccination is considered the most effective form of medical intervention that humans have ever developed, resulting in unprecedented decrease in healthcare costs, human suffering, and death [1]. While advances in vaccine technology have made significant progress in the past, there are numerous other pathogens, especially for intracellular pathogens such as human immunodeficiency virus (HIV) and Hepatitis C virus (HCV), against which there are no available vaccines. In addition, successful application of current vaccine technologies toward immunotherapy against cancer has been an elusive goal. These major challenges facing the vaccine research field are currently being addressed in many laboratories around the world. Importantly, new exciting discoveries that have advanced our understanding of molecular and cellular interactions between pathogens and our immune system are paving ways for innovative approaches to achieve strong immune activation while avoiding tissue damage and toxicities [1].

This chapter focuses on one class of technology that is revolutionizing the field of vaccine research, namely application of nanotechnology toward vaccine development. Nanotechnology, which has transformed the landscape of drug development, has recently been the subject of intense investigations and has made significant strides for vaccine applications [2-5]. Nanoparticle-based vaccination strategy offers key distinct advantages over conventional vaccines. Specifically, (1) nanoparticles can effectively protect their cargo from degradation in physiological conditions characterized by high salt and enzyme concentrations. (2) Size and shape of nanoparticles can be precisely tuned to mimic characteristics of pathogens, allowing effective draining through the lymphatic system to lymphoid tissues and subsequent internalization in antigen-presenting cells (APCs). (3) Moreover, particle size and charge strongly affect biodistribution and retention of particles in lymph nodes and spleens, thus offering additional means to modulate vaccine delivery to APCs and promote effector and memory immune responses [6-9]. (4) Surface modification of nanoparticles also provides additional opportunities for enhancing targeted delivery to specific cells by conjugation of receptor ligands or antibodies [10-12]. (5) Another major benefit includes nanoparticle-based co-delivery of antigens and adjuvants targeted to the same APCs, which leads to optimal immune activation and antigen presentation [13-17]. (6) Additionally, cytosolic delivery of antigens within APCs can be achieved by "smart" nanoparticle systems designed to promote endosomal escape, thus allowing for effective antigen cross-presentation and induction of cytotoxic CD8+ T lymphocyte (CTL) responses, which are notoriously difficult to induce via traditional vaccination methods [18-22]. (7) Also, vaccine nanoparticles composed of biocompatible materials with low immunogenicity can minimize induction of anti-vector immune responses [23]. This approach permits multiple administrations of the same formulation without loss of potency, which is a major advantage over viral vectors. (8) Furthermore, their potential as off-the-shelf vaccine nano-formulations offers an alternative solution to labor-intensive and expensive dendritic cell (DC)-based vaccine strategies.

2 Activation of Innate and Adaptive Immunity

The immune system, which has developed to protect the host from a variety of pathogens such as bacteria and viruses, can be divided into two branches: innate and adaptive immunity. Innate immunity plays an important role in eradication and/or containment of infectious pathogens at an early stage and relies on common and evolutionarily preserved elements of viruses and bacteria known as pathogen-associated molecular patterns (PAMPs) for recognition using pathogen recognition receptors (PRRs) [24]. This early innate immune response acts as the first line of defense against infection and provides time for the adaptive immunity to form. Adaptive immune responses are orchestrated by APCs that sample the environment for immunogenic antigens and present them to effector T cells. Complementary to this, B cells directly recognize cognate antigens and secrete antibodies with help from CD4+ helper T cells.

APCs bridge innate and adaptive immunity by capturing pathogens and delivering antigens to the effector T and B cells. Among subtypes of APCs, DCs are the major class of professional APCs proficient at capturing wide range of antigens and processing them for presentation to effector T cells. DCs function by preferentially recognizing pathogens through the action of PRRs in a similar way to macrophages [25]. However, rather than upregulating phagocytic behavior and trafficking pathogens to lysosomes for degradation as in macrophages, DCs process and conserve antigens and begin their migration to the local lymph node [26]. Once there, antigens are presented on the DC surface along with an array of stimulatory transmembrane proteins in order to activate antigen-specific naïve T cells. In particular, certain DC subsets possess a unique ability for cross-presentation, a process involving presentation of extracellular antigens on major histocompatibility complex class I (MHC-I), strictly necessary for CTL responses [27].

MHC-I is expressed by every cell in the body and in a traditional antigen-presentation pathway functions as a surveillance mechanism for viral and intracellular infections [28]. Cells continuously turn over cytosolic proteins through the action of proteasome, which cleaves them into short amino acid chains. These peptides are then transported into the endoplasmic reticulum (ER) via TAP1/2, where they bind to MHC-I producing unique peptide-MHC-I complexes, which are then trafficked to the plasma membrane, allowing healthy cells to present self-antigens on their surface [29]. Infected cells contain bacterial or viral proteins, which are processed in the exact same way, leading to foreign antigen display on the plasma membrane. These cells are then readily recognized and killed by antigen-specific CTLs.

Normal cells fail to express immunostimulatory molecules such as CD80/86 to activate and expand the antigen-specific CTL clone. Thus, for the traditional antigen-presentation pathway to work, APCs would have to be infected directly, which may be a rare and unreliable event. Because of this, a process of cross-presentation has evolved, where APCs can take up antigens from the extracellular space and present them on MHC-I [27]. This complex and still not fully understood process involves antigen capture and trafficking into endosomes, which can briefly fuse with the ER and extract the necessary machinery for cross-presentation. Proteins can then be degraded directly by endosomal proteases and loaded onto MHC-I, which is known as the vacuolar pathway, or they can be translocated into the cytoplasm with the assistance of Sec61 protein, degraded by the proteasome, and trafficked back via TAP1/2 complex to be loaded onto MHC-I [27].

Cross-presentation often occurs due to naturally occurring cues, such as Toll-like receptor (TLR) signaling [13]. However, cross-presentation is very difficult to induce for subunit antigens as extracellular antigens phagocytosed into APCs are usually trapped in endosomes, eventually leading to degradation of antigens in lysosomes and presentation in the context of MHC-II [30]. Hence, cross-presentation in APCs has been recognized as the major bottleneck for achieving strong CTL responses in response to protein vaccination [30]. Recent studies have shown that nanoparticle-based vaccine delivery systems can address these challenges by disrupting endosomal membrane and thus allowing antigens to enter the cytosol. One of these approaches involves conjugation of fusogenic or cell-penetrating peptides to the surface of nanoparticles [19]. These peptides are generally composed of positively charged amino acids and can adsorb and disrupt the negatively charged lipid bilayer. When chicken ovalbumin (OVA)-loaded liposomes were modified with octaarganine (R8), this resulted in 5- and 55-fold increase in OVA presentation on MHC-I compared to cationic liposomes and soluble OVA, respectively [19]. Another strategy employed conjugation of pH-dependent fusogenic polymer (linear 3-methylglutarylated poly(glycidol)) to the liposomal surface, which promoted efficient OVA delivery to the cytosol. This strategy led to increased CTL responses, providing protective immunity against challenge with OVA-expressing lymphoma and delaying growth of established tumors in mice [31]. Another complementary approach involves biomimetic incorporation of listeriolysin O, a pathogenic element that induces endosomal lysis in response to acidic endolysosomal pH, leading to increased cytosolic delivery of antigens and efficient induction of antigen cross-presentation [18].

3 Lipid-Based Nanoparticles for Vaccine Delivery

Vaccine nanoparticles can be composed of variety of materials ranging from metals and synthetic polymers to natural polymers and lipid-based systems [5]. Their characteristics can be finely tuned although risks and benefits must be carefully weighed. In particular, lipid-based nanoparticles offer advantages of material biocompatibility, biodegradability, and safety, which are ideal characteristics for a vaccine delivery system [32]. Despite significant progress made over the last three decades with liposomal vaccines [32-37], their clinical development has been challenging. This is partially attributed to the fact that lipid vesicles are prone to degradation and premature release of encapsulated biomacromolecules [38,39]. In this section, we discuss two prime examples of lipid-based nanoparticle systems that are addressing these limitations of conventional lipid-based vehicles and showing early signs of promise in pre-clinical stages of vaccine development.

3.1 Interbilayer-Crosslinked Multilamellar Vesicles (ICMVs)

Toward the goal of producing a stable vaccine delivery vehicle that can promote cross-presentation, a new class of vaccine delivery system, termed ICMV, has been developed [23]. ICMVs are formed by stabilizing multilamellar vesicles with short covalent crosslinks linking lipid headgroups across the opposing faces of adjacent tightly-stacked bilayers within the vesicle walls (Fig. 1a) [23]. As ICMVs have aqueous core surrounded by multilayers of lipid wall, they can serve as a delivery system for co-delivery of both hydrophilic and hydrophobic antigens and adjuvants. In addition, their thick lipid wall renders the particles highly resistant to serum-mediated degradation. Importantly, this feature enhances their in vivo stability and allows stable delivery of vaccine components to lymphoid tissues, thus generating potent CTL and antibody responses as presented below.

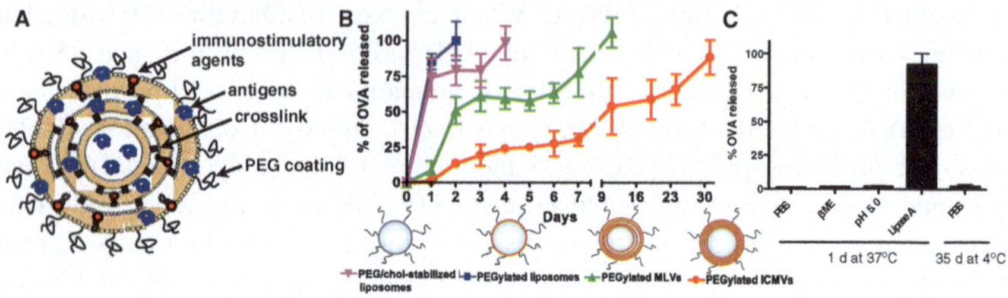

Fig. 1 (A) Schematic illustration of interbilayer-crosslinked multilamellar vesicles (ICMVs) co-loaded with antigens and immunostimulatory agents. (B) ICMVs carrying fluorophore-tagged OVA showed steady release of protein antigen over 30 days at 37°C in 10% FBS RPMI. In contrast, other lipid vesicles showed rapid release of OVA, thus demonstrating serum-stability of ICMVs. (C) ICMVs were incubated with endolysosomal conditions and examined for OVA release. Lipase A, an enzyme found at high concentration in APCs, was able to degrade the particles within 1 day, showing lipase-sensitivity of ICMVs. Figure modified from [23].

3.1.1 ICMVs for Parenteral Vaccination

ICMV synthesis involves preparation of lipid films consisting of maleimide-functionalized lipids, followed by fusion of liposomes into multilamellar structures and crosslinking of lipid multilayers. Briefly, simple liposomes are formed by hydrating lipid films with an aqueous buffer containing water-soluble antigens and adjuvants. Antigen/adjuvant-loaded liposomes are then fused using divalent cations to form multilamellar structures, which are then crosslinked using dithiol linker covalently binding to maleimide-functionalized lipids on apposing lipid layers within multilamellar vesicles. Notably, crosslinked multilayers of lipid wall significantly increased stability of ICMVs, compared with conventional lipid vesicles, leading to stable and prolonged protein release over 30 days in serum-containing media at 37°C (Fig. 1b) [23]. Remarkably, in the presence of lipase A, a common enzyme found in endolysosomes, ICMVs were rapidly degraded and cargo was released within 24 hours (Fig. 1c). These results suggested that ICMVs would undergo rapid degradation upon cellular internalization, leading to delivery of antigens to cytosol [23]. The final step of ICMV synthesis includes conjugation of polyethylene glycol (PEG) to the particle surface in order to decrease aggregation, prevent deposition *in vivo*, and increase *in vivo* trafficking. Aside from superior stability in physiological conditions, ICMVs exhibit a sustained and prolonged draining profile from the site of the injection and accumulated within the local lymph node for over 14 days after vaccine administration [40].

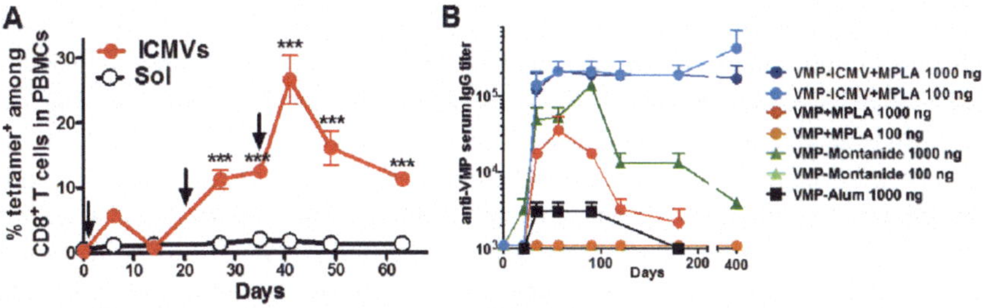

Fig. 2 (A) C57Bl/6 mice were immunized subcutaneously on days 0, 21, and 35 with 10 μg OVA and 0.3 μg MPLA either in ICMVs or soluble form. Assessment of the frequency of OVA-specific T-cells among peripheral blood mononuclear cells indicated strong induction of antigen-specific CTLs in response to ICMV vaccination. (B) C57Bl/6 mice were immunized on days 0 and 21 with vivax malaria sporozoites protein (VMP) in indicated doses formulated in ICMVs or in soluble form admixed with MPLA, Montanide, or Alum. ICMVs elicited long-sustained, high antibody titers against VMP, surpassing those induced by conventional adjuvants. Panel (A) and (B) modified from [23] and [40], respectively.

Subcutaneous vaccination with ICMVs co-loaded with OVA and monophosphoryl lipid A (MPLA, a FDA-approved TLR4 agonist) led to extensive expansion of OVA-specific $CD8^+$ T cells reaching nearly 30% of total CTLs following administration of primary and two booster doses (Fig. 2a) [23]. In addition, CTLs from ICMV-immunized mice expressed significantly higher levels of CD44 and CD62L, indicative of robust memory responses. Finally, ICMV vaccination also promoted high levels of IFNγ secretion by splenocytes after *ex vivo* restimulation [23]. In another study, particles were co-loaded with malaria antigen, VMP001, and MPLA and administered as primary and booster immunizations three weeks apart [40]. This regimen led to high and durable antibody responses lasting over a year with a very low dose of 100 ng of protein and 100 ng of MPLA, with significantly enhanced antibody titers compared with vaccination with currently approved adjuvants, such as Alum and Montanide (Fig. 2b).

3.1.2 ICMVs for Mucosal Vaccination

Intratracheal immunizations with OVA-ICMVs also dramatically expanded OVA-specific $CD8^+$ T cells and reached approximately 12% of systemic CTLs following one booster immunization on day 28 [41]. Formulations were co-loaded with the MPLA adjuvant, but the vaccine was administered along with soluble PolyI:C (TLR3 agonist), which has shown superior responses to MPLA-only particles. These results implicate the possibility of enhancing responses from ICMV-based vaccinations with the use of other adjuvants as well as adjuvant combinations, which need to be explored further in future studies. In response to the pulmonary route of administration, CTLs expressed high levels of integrin $\alpha_4\beta_7$ known as mucosal homing phenotype, and had an effector memory phenotype capable of rapid killing of target cells in the early stages of the infection [41]. In addition to pulmonary route of vaccine delivery, transcutaneous delivery of ICMVs was recently examined via microneedle-mediated vaccination (Fig. 3a) [42]. Microneedle arrays composed of poly(lactide-co-glycolide) acid (PLGA) polymer were coated with alternating layers of negatively charged ICMVs and a biodegradable cationic poly(β-amino ester) (PBAE) via layer-by-layer assembly. After 5 min of transcutaneous administration, PBAE/ICMV films were rapidly transferred from microneedle surfaces into the tissue, targeting antigen-loaded ICMVs to skin-resident DCs. Microneedle-mediated vaccination with ICMVs elicited humoral immune responses that are characterized by a more balanced Th1/Th2 responses with 10-fold higher antibody titers, compared by bolus delivery of soluble or intradermal injection of ICMVs (Fig. 3b). Overall, these studies provide strong evidence that ICMVs designed to co-deliver protein antigens and adjuvant molecules elicit potent CTL and antibody responses when administered via mucosal surfaces, suggesting that non-invasive route of ICMV delivery may provide an alternative method of vaccine administration to facilitate vaccine deployment without the aid of trained healthcare professionals.

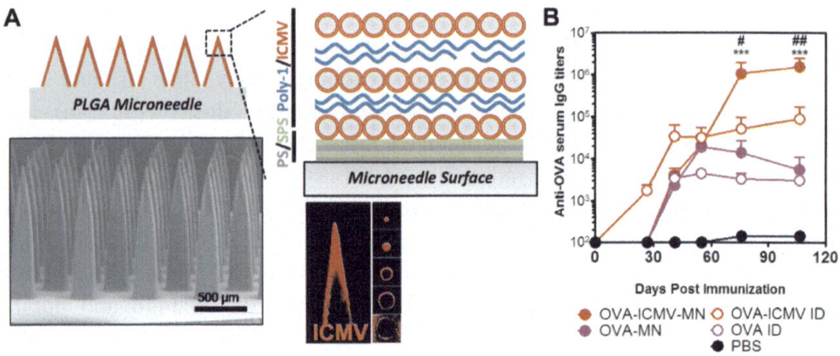

Fig. 3 (A) PLGA microneedle (MN) arrays were coated with ICMVs via layer-by-layer assembly and deposition of ICMVs was confirmed by confocal microscopy. (B) C57Bl/6 mice were immunized on days 0 and 21 with microneedles coated with either OVA-containing ICMVs or soluble OVA. Control groups included OVA-containing ICMVs or soluble OVA given via intradermal injection. Microneedle-mediated transcutaneous vaccination with ICMVs elicited > 10-fold higher anti-OVA serum IgG titers, compared with other groups. Figure reproduced from [42].

In summary, ICMVs are an attractive platform for vaccine delivery due to their potential for efficient co-delivery of hydrophilic and hydrophobic antigens and adjuvants along with their superior stability and persistent draining to local lymph nodes. ICMV vaccination elicits robust effector and effector memory immune responses, characterized by high frequency of CTLs and Th1/Th2 balanced antibody responses, thus highlighting the promise of ICMV-based vaccination strategy against intracellular pathogens and cancers.

3.2 *Lipoproteins*

Lipoproteins are complexes of lipids and proteins [43]. They play important roles in the transport and metabolism of lipids such as triglycerides, free cholesterol, cholesterol esters and phospholipids [44]. There are four major types of lipoproteins, including chylomicrons (CM), very low density lipoproteins (VLDL), low density lipoproteins (LDL) and high density lipoproteins (HDL) [45]. Different types of lipoproteins vary in sizes, lipid, and protein compositions and functions (Table 1) [46].

In addition to the transport of lipids, lipoproteins have also been reported to transport various proteins, hormones, vitamins, and endogenous microRNA and target them to recipient cells, indicating they are multi-functional, endogenous in vivo drug delivery vehicles [47]. Indeed, development of lipoproteins as drug delivery systems aims to utilize their endogenous functions to transport various molecules with the ability to target them to specific cell types. In particular, lipoproteins have major advantages as vaccine delivery systems; (1) lipoproteins are endogenous nanoparticles with minimal safety concerns; (2) lipoproteins have multiple sites for cargo loading, including hydrophobic core and hydrophilic surface that can facilitate delivery of a broad range of molecules; (3) in vivo properties of lipoproteins can be readily tailored by changing their compositions (Table 1), thus offering the capability to effectively modulate their in vivo pharmacokinetic profiles and biodistribution. Recently there has been significant progress in development of lipoproteins as vaccine delivery carriers. In the following section, we will highlight key advances in this field of vaccine research.

Table 1 Compositions of major types of lipoproteins

Lipoprotein	CM	VLDL	LDL	HDL
Density (g/mL)	<0.94	0.94-1.006	1.006-1.063	1.063-1.210
Diameter (nm)	100-1000	30-80	18-28	5-15
Apoprotein	ApoA4, ApoB48	ApoB100, ApoC1, ApoC2, ApoC3, ApoE	ApoB100	ApoA1, ApoA2, ApoAV (5), ApoD, ApoE, ApoM
Total lipids* (wt%)	99	91	80	44
Triacylglycerols	85	55	10	6
Cholesterol esters	3	18	50	40
Cholesterol	2	7	11	7
Phospholpid	8	20	29	46

*The rest are mainly composed of apolipoproteins. Data adapted from [44].

3.2.1 Delivery of Adjuvants by Nanolipoproteins

TLR agonists have been used in pre-clinical settings as immunotherapy for infectious diseases and cancer [48-50]. However, TLR agonists used as monotherapy mediate marginal and short-lived protection, thus requiring frequent administrations of high doses of TLR agonists that often lead to systemic toxicity [51]. To improve the therapeutic efficacy and reduce systemic toxicity, Dina R. et al. used phospholipids and recombinant apolipoproteins (apoE422k) to prepare discoidal nanolipoprotein particles (NLPs) for the delivery of MPLA and CpG (oligonucleotides with unmethylated CpG motifs, a potent TLR9 agonist) [52]. MPLA can naturally be incorporated into NLPs due to its lipophilic nature, while CpG requires modification with lipophilic cholesterol to facilitate particle partitioning. MPLA and CpG were loaded into NLPs at molar ratios of less than 3:1 and 8:1 of TLR agonist to NLP, respectively. By fluorescently labeling apoE422k moiety of NLPs with AF488, the authors have confirmed nanoparticle uptake by mouse primary peritoneal macrophages and J774 cells as well as their localization within lysosomes.

Compared with free CpG and MPLA, CpG:NLPs and MPLA:NLPs increased IL-6, TNF-α, and RANTES cytokines release by both J774 cells and primary peritoneal macrophages in vitro [52]. It is speculated that efficient cellular uptake of NLPs led to trafficking of CpG:NLPs to endosomes where TLR9 is located, thus enhancing TLR9 activation by CpG:NLPs, compared with free CpG. In addition, the authors have reported that empty NLPs without CpG or MPLA also exhibited dose-dependent cytokine release profiles, which might be due to residual endotoxin in NLPs or the intrinsic immunostimulatory properties of structural components of NLPs. The exact mechanisms of immune activation by empty NLPs remain to be investigated.

In vivo administration of CpG:NLPs and MPLA:NLPs also increased levels of serum IL-6, TNF-α, MIP-1α, and RANTES, compared with administration of free CpG and MPLA. At the cellular level, CpG:NLPs and MPLA:NLPs activated macrophages to up-regulate their expression of MHC class II and CD40 as well as DCs to up-regulate co-stimulatory markers, such as CD40, CD80, and CD86.

When CpG molecules in CpG:NLPs were fluorescently labeled by Quasar 670 (CpG-Quasar:NLPs) and intraperitoneally injected into mice, a significant increase in the percent of splenocytes positive for Quasar 670 was observed for CpG-Quasar:NLPs, compared with free CpG-Quasar over 72 h time window post-injection (Fig. 4). These data indicate that complexation of CpG to NLPs can increase their cellular uptake by macrophages and DCs in spleen, leading to up-regulation of activation markers and increased cytokine secretion.

Importantly, CpG:NLPs provided superior protective immunity against infection, compared with free CpG. To demonstrate this, authors treated mice intranasally with 10 μg of CpG:NLPs, CpG, or PBS 24 hours before intranasal infection with a lethal dose of A/PR8/34 strain of influenza. Mice receiving CpG:NLPs survived the challenge until day 14, whereas mice receiving either free CpG or PBS succumbed to viral infection by day 7. These results showed that innate immune responses induced by CpG:NLPs can protect animals against viral infection.

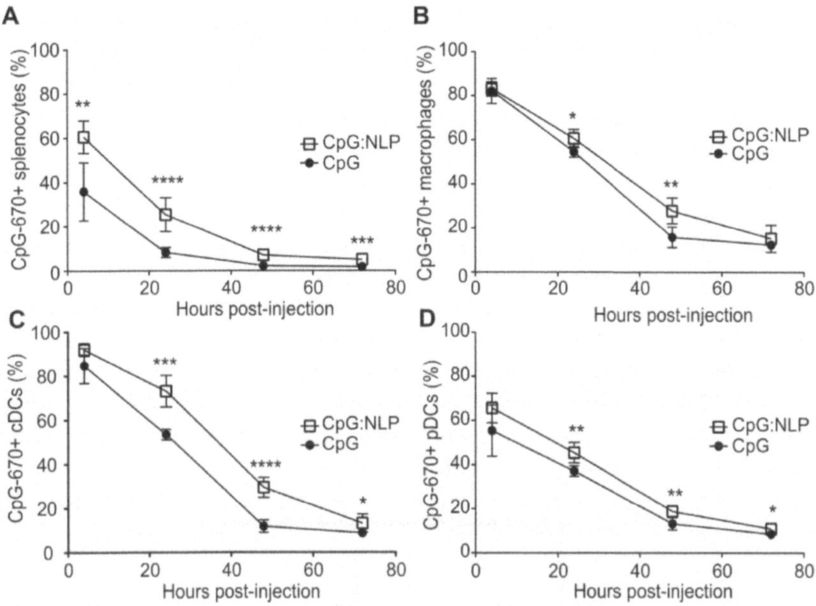

Fig. 4 Uptake of free CpG or CpG:NLP by different cell types at different time points after intraperitoneal injection analyzed by flow cytometry. Conjugation of CpG to NLP promoted the uptake by (A) splenocytes, (B) macrophages (defined as $CD11b^{+}CD11c^{-}$), (C) conventional dendritic cells (cDCs, defined as $CD11c^{+}CD45R^{-}$), and (D) plasmacytoid dendritic cells (pDCs, defined as $CD11c^{+}$ $CD45R^{+}$). CpG:NLP enhanced delivery of CpG to APCs, compared with free CpG. Figure reproduced from [52].

3.2.2 Delivery of Peptide and Protein Antigens by Nanolipoproteins

Peptide and protein antigens are attractive surrogates of live attenuated or inactivated vaccines due to their good safety profiles and relatively simple manufacturing processes. Moreover, each component of peptide and protein antigens can be purified and precisely analyzed, facilitating industrial scale-up and quality control [53] . However, subunit vaccines solely based on peptide and protein antigens typically lack sufficient immunogenicity, thus requiring the use of vaccine adjuvants [54]. Nicholas O. Fischer et al. addressed these issues by applying the same NLP platform technology for delivery of protein antigens [55]. Specifically, phospholipids, recombinant apoE protein, and nickel-chelating lipid 1,2-dioleoyl-sn-glycero-3-(nickel salt) were used to prepare the Nickel-Chelating Nanolipoprotein Particles (NiNLPs). To conjugate West Nile virus (WNV) envelope protein (trE) to NiNLP, trE was tagged by polyhistidine (trE-His) and incubated with NiNLPs for 30 min [56]. Excess, unconjugated antigen was removed, and the presence of surface bound trE-His on NiNLPs was confirmed by SDS-PAGE assay and also by detection of released trE-His upon treatment with EDTA that chelates nickel and disrupts interaction between polyhistidine and nickel containing lipid.

To confirm if conjugation of trE-His to NiNLPs can improve its immunogenicity, 6-week-old mice were immunized once with formulations (free trE-His or NiNLPs-trE-His). Three weeks after vaccination, only NiNLP-trE-His group induced detectable IgG responses to trE whereas free trE-His group failed to achieve sero-conversion. As expected, blank NiNLP did not induce detectable antibodies either, clearly indicating that conjugation of trE-His to NiNLP improves its immunogenicity.

To learn if the vaccination could protect animals from infection, vaccinated animals were challenged with 1000 ffu of live WNV NY99 on week 5 post vaccination. All mice vaccinated with NiNLP-trE-His survived the viral challenge with a minimal weight loss (Fig. 5). In contrast, the majority of animals vaccinated with trE-His, blank NiNLP, or diluent control groups succumbed to the viral challenge. These results provided strong evidence that conjugation of protein antigen to NiNLP could enhance its immunogenicity in vivo and elicit protective immune responses against viral infection.

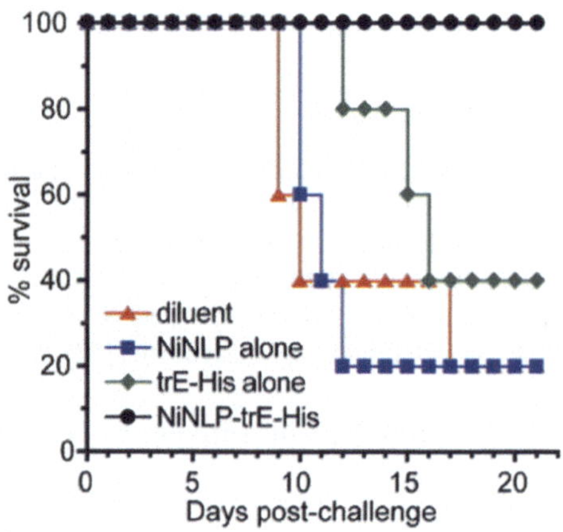

Fig. 5 Mice vaccinated with different formulations (diluent, NiNLP alone, trE-His alone, or NINLP-trE-His) were challenged with 1000 ffu of live WNV NY99 and observed daily for 21 days. Kaplan-Meier survival curves indicate that conjugation of trE-His to NiNLP improved its efficacy and better protects mice from the virus challenge. Figure reproduced from [55].

3.2.3 Co-delivery of Adjuvants and Peptide/Protein Antigens by Lipoproteins

Although conjugation of antigens to nanolipoproteins can improve their immunogenicity, NLPs themselves are not strong adjuvants. Therefore, incorporation of molecular adjuvants together with antigens into nanolipoproteins is a strategy that may further improve immunogenicity of the conjugates. To test this approach, Fischer et al. have used the NPL platform for co-delivery of protein antigens and adjuvants (MPLA or CpG) [58]. Specifically, His-tagged recombinant viral and bacterial antigens, such as influenza hemagglutinin 5 (H5) and Yersinia pestis LcrV, were conjugated to NiNLPs. Furthermore, MPLA or cholesterol-modified CpG were incorporated into antigen-loaded NiNLPs. Quantitative analysis showed that each nanolipoprotein contained ~21 molecules of protein antigens and 3 ± 1 MPLA or 6 ± 1 CpG molecules [58].

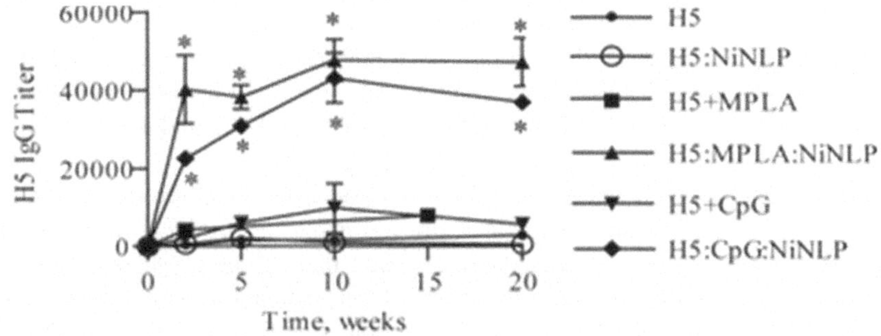

Fig. 6 Mice were vaccinated with different formulations and the kinetics of antibody titers were measured by ELISA at different time points after vaccination. * indicates p<0.001 compared to the physical mixture of free proteins and adjuvants. Figure reproduced from [57].

To learn if co-delivery of antigens and adjuvants by NiNLPs can increase antibody production, mice were immunized with various formulations of antigen and adjuvants, including H5, H5+MPLA (physical mixture), H5+CpG (physical mixture), H5:NiNLPs, H5:MPLA:NiNLPs, and H5:CpG:NiNLPs. As expected, H5:MPLA:NiNLPs and H5:CpG:NiNLPs induced 5-10 times higher H5 IgG titer than H5:NiNLPs, H5+MPLA (physical mixture) or H5+CpG (physical mixture) (Fig. 6). Importantly, antibody responses elicited by H5:MPLA:NiNLPs and H5:CpG:NiNLPs were maintained for at least 20 weeks without additional boost immunization. These results indicate that co-delivery of antigens and adjuvants via NiNLPs can dramatically improve immunogenicity of protein antigens and achieve strong humoral immune responses.

3.2.4 Delivery of Lipid Antigens by Lipoproteins

Unlike peptide antigens that are presented in the context of MHC class I or MHC class II complex, lipid antigens such as α-galactosyl ceramide (α-GC) are presented in the context of CD1 to activate natural killer T (NKT) cells [58,59] that can kill target cells such as infected or cancer cells [60]. Exogenous lipid antigens are trafficked to endocytic compartments, where they can bind to CD1, and presented to the surface of APCs as lipid antigen-CD1 complex. Since lipoproteins are major transporters of lipids, they may be natural, endogenous delivery vehicles for lipid antigens. In order to understand the contribution of lipoproteins in the transport of lipid antigens, Peter van den Elzen et al. used a model CD1d-presented glycolipid, galactosyl(α 1-2)galactosyl ceramide (GGC) to investigate its trafficking patterns in vitro and in vivo [61]. GGC requires uptake and delivery to lysosomes and conversion to the active form α-GC before being presented in the context of CD1d on cell surface.

In order to understand which lipoproteins are involved in GGC trafficking and NKT cell activation, GGC was incubated with human serum in vitro, and various serum fractions, including the VLDL, LDL, and HDL fractions, were separated by fast protein liquid chromatography (FPLC). These fractions were then incubated with DCs to determine if they could activate NKT cells by measuring IFNγ production from NKT cells. Interestingly, only the fraction corresponding to VLDL led to significant NKT cell activation whereas LDL and HDL fractions exhibited little or no NKT cell activation [61] (Fig. 7a,b).

VLDL contains both apolipoprotein B (apoB) and apolipoprotein E (apoE) while LDL and HDL mainly contain apoB and apoA, respectively. Experiments performed with antagonistic antibodies revealed that the presence of antibodies against apoE led to dramatic decrease in NKT cell activation, while the presence of antibodies against apoB and apoA did not significantly affect NKT cell activation (Fig. 7b). Contribution of apoE to NKT cell activation was further confirmed by incubating GGC with apoE-depleted serum, resulting in significantly decreased NKT cell activation (Fig. 7c). NKT activation was restored by adding apoE back to the serum or incubating GGC directly with eluted apoE, further providing the evidence of apoE-mediated delivery of GGC in vitro.

To confirm the importance of apoE in vivo, GGC was intravenously injected into apoE knockout ($apoE^{-/-}$) mice or wild-type (WT) mice and NKT cell activation was analyzed by measuring intracellular IFN-γ production among CD1d-tetramer positive cells. NKT cell activation in $apoE^{-/-}$ mice was dramatically decreased, compared with that in WT mice (Fig. 7d). It is noteworthy that high doses of GGC could still activate NKT cells in $apoE^{-/-}$ mice, suggesting the possibility of alternative pathways other than apoE-mediated delivery of GGC in vivo.

Now that apoE has been confirmed to be the main player in trafficking GGC in vitro and in vivo, the next question is how apoE-bound GGC is recognized by DCs. It is already known that apoE can be taken up into the endocytic system by receptor-mediated endocytosis, which is more efficient than macropinocytosis of free small molecules such as GGC. To confirm this point, DCs were pulsed with either free GGC or apoE-bound GGC and co-cultured with NKT cells. ApoE-bound GGC led to faster and stronger NKT cell activation, compared with free GGC, suggesting that receptor-mediated endocytosis of apoE-bound GGC is more efficient than macropinocytosis of free GGC. ApoE is recognized by several cell-surface receptors, including LDL receptor (LDLR) and LDLR-like protein (LRP). Flow cytometry analysis has confirmed expression of both LDLR and LRP on DCs. Notably, LDLR is significantly up-regulated as monocytes differentiate into DCs, indicating that LDLR may be the main receptor mediating DC uptake of apoE-bound GGC. To confirm this point, apoE3 (the most common human variant with high binding affinity to LDLR) or apoE2 (a variant with low binding affinity to LDLR) was bound to GGC and used to pulse DCs, which were then co-cultured with NKT cells. Compared with apoE3-bound GGC, apoE2-bound GGC only had

40%-47% of the ability to activate NKT cells. In addition, when apoB, the primary component of LDL that can bind to LDLR but not to LRP, was added together with apoE-bound GGC to pulse DCs, NKT cell activation was decreased, compared with control groups. These results clearly showed that LDLR is the main receptor mediating cellular uptake of apoE-bound GGC.

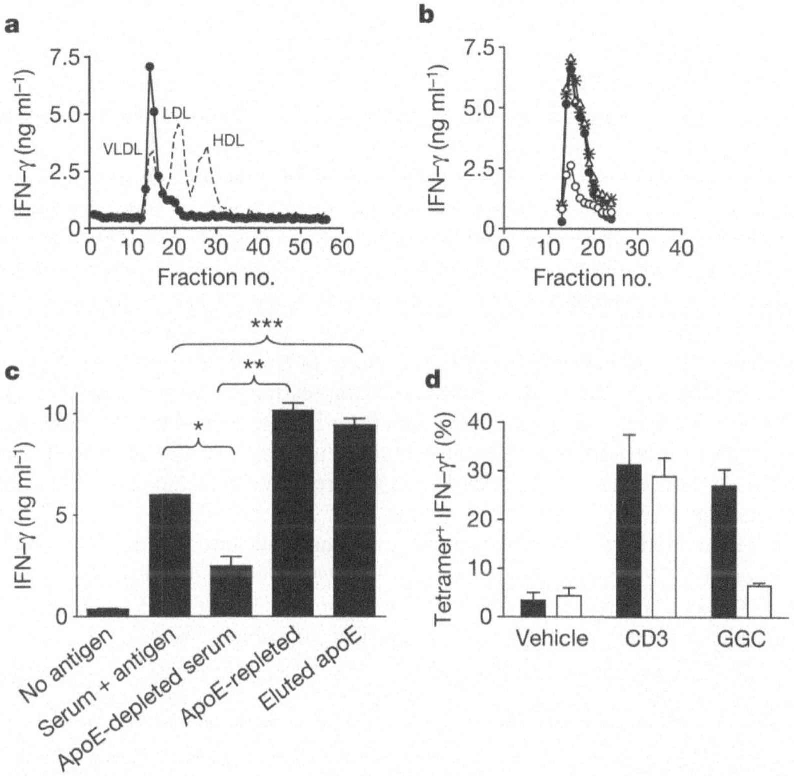

Fig. 7 (A) The lipid antigen GGC was incubated with human serum and different lipoprotein fractions (dashed line) were separated by FPLC. The NKT cell reactivity (solid line) was determined by incubating different lipoprotein fractions with dendritic cells to activate NKT cells. (B) NKT cell reactivity was measured in the absence of anti-ApoE antibodies (filled circles) or presence of anti-apoE (open circles) versus antibodies against apoB (triangles) and apoA-II (asterisks). (C) Incubation of GGC with apoE-depleted serum led to less NKT cell activation, while incubation of GGC with apoE-repleted serum or with eluted apoE directly led to more NKT cell activation. (D) Mice were intravenously injected with GGC, anti-CD3 or vehicle alone. GGC led to less NKT cell activation in apoE$^{-/-}$ mice (open bars) than in wild type mice (filled bars), while CD3 can equally activated NKT cells in apoE$^{-/-}$ mice and wild type mice. Figure reproduced from [61].

To investigate if LDLR is also important for lipid antigen presentation and NKT cell activation in vivo, GGC was intravenously injected into LDLR knockout (LDLR$^{-/-}$) mice or WT mice. NKT cell activation in LDLR$^{-/-}$ mice was significantly decreased, compared with that in WT mice, while non-specific stimulation with anti-CD3 led to equal activation of NKT cells in LDLR$^{-/-}$ mice and WT mice. These results indicate that LDLR plays a major role in trafficking and presentation of lipid antigens in vivo.

3.3 Other Lipoproteins

In addition to being found in vertebrates and insects, lipoproteins are also found in bacteria [62]. Bacterial lipoproteins are a set of membrane proteins with distinct functions, such as adhesion to host cells, modulation of inflammatory processes, and translocation of virulence factors into host cells. These properties of bacterial lipoproteins may be utilized for development of strong vaccine adjuvants.

Chih-Hsiang Leng et al. used patented *E. coli* strains to produce recombinant lipoproteins (rlipo), which were found to be strong stimulators of TLR2 on DCs [63]. In their study, lipoprotein expressed in *E. coli* was conjugated to recombinant tumor protein antigen E7 mutant (rE7m), a viral antigen found in human papilloma virus (HPV)-induced tumors [64]. To test their immunogenicity, mice were prophylactically vaccinated twice with rlipo-E7m or free rE7m with 1 week interval and challenged with 2×10^5 TC-1 tumor cells expressing E7 antigen on day 7 post-vaccination. Mice that were vaccinated with rlipo-E7m remained free of tumor for 60 days, whereas mice administered with either rE7m alone or PBS developed tumors (Fig. 8a). The vaccines were next studied in a more rigorous therapeutic setting. Mice were inoculated with 2×10^5 TC-1 tumor cells and vaccinated on day 7. Mice vaccinated with rlipo-E7m had tumors less than 0.01 cm^3 on day 40, whereas the rE7m or PBS control group had tumors over 2 cm^3 (Fig. 8b). These results indicated fusion of rE7m to recombinant lipoproteins produced in *E. coli* can dramatically improve immunogenicity and efficacy of subunit vaccines.

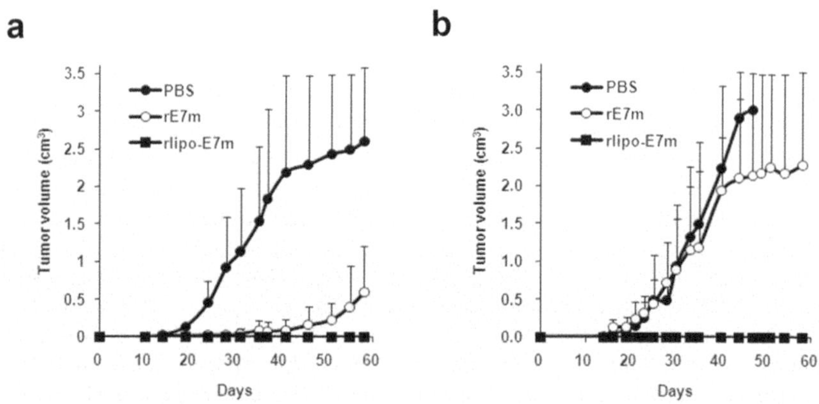

Fig. 8 Prophylactic effect of different formulations: mice were prophylactically vaccinated with different formulations twice with one week interval and inoculated with TC-1 tumor cells 7 days after the last vaccination. (b) Therapeutic effect of different formulations: mice were inoculated with TC-1 tumor cells on day 0 and vaccinated once with different formulations on day 7. Figure reproduced from [64].

4 Conclusions

Lipid-based nanoparticles are promising vaccine delivery systems. As highlighted in this chapter, their inherent biocompatibility, low toxicity and immunogenicity, combined with ease of scale-up and industrial precedence of liposome-based nanomedicines are major attributes that make lipid-based nanoparticles an ideal platform for vaccine applications. With advances in nanomaterial design and our understanding of molecular adjuvants, we can employ a reductionist approach for rational vaccine design and synthesize nanostructures that closely mimic features of pathogens without toxicities associated with them. Overall, nanoparticle-based vaccine strategies that use biocompatible materials to deliver vaccine components, target immune cells, and elicit desired immune responses will have a tremendous impact on the current status of vaccine design and development.

References

1. Pulendran, B., Ahmed, R.: Immunological mechanisms of vaccination. Nat. Immunol. **12**, 509–517 (2011)
2. Swartz, M.A., Hirosue, S., Hubbell, J.A.: Engineering approaches to immunotherapy. Sci. Transl. Med. **4**, 148–149 (2012)
3. Moon, J.J., Huang, B., Irvine, D.J.: Engineering nano- and microparticles to tune immunity. Adv. Mater. **24**, 3724–3746 (2012)
4. Irvine, D.J., Swartz, M.A., Szeto, G.L.: Engineering synthetic vaccines using cues from natural immunity. Nat. Mater. **12**, 978–990 (2013)
5. Sahdev, P., Ochyl, L.J., Moon, J.J.: Biomaterials for nanoparticle vaccine delivery systems. Pharm. Res. **31**, 2563–2582 (2014)
6. Reddy, S.T., van der Vlies, A.J., Simeoni, E., Angeli, V., Randolph, G.J., O'Neil, C.P., Lee, L.K., Swartz, M.A., Hubbell, J.A.: Exploiting lymphatic transport and complement activation in nanoparticle vaccines. Nat. Biotechnol. **25**, 1159–1164 (2007)
7. Zhuang, Y., Ma, Y., Wang, C., Hai, L., Yan, C., Zhang, Y., Liu, F., Cai, L.: PEGylated cationic liposomes robustly augment vaccine-induced immune responses: Role of lymphatic trafficking and biodistribution. Journal of Controlled Release **159**, 135–142 (2012)
8. Badiee, A., Khamesipour, A., Samiei, A., Soroush, D., Shargh, V.H., Kheiri, M.T., Barkhordari, F., McMaster, W.R., Mahboudi, F., Jaafari, M.R.: The role of liposome size on the type of immune response induced in BALB/c mice against leishmaniasis: rgp63 as a model antigen. Experimental Parasitology **132**, 403–409 (2012)
9. Nakamura, T., Yamazaki, D., Yamauchi, J., Harashima, H.: The nanoparticulation by octaarginine-modified liposome improves alpha-galactosylceramide-mediated antitumor therapy via systemic administration. Journal of Controlled Release **171**, 216–224 (2013)
10. Kwon, Y.J., James, E., Shastri, N., Frechet, J.M.: In vivo targeting of dendritic cells for activation of cellular immunity using vaccine carriers based on pH-responsive microparticles. Proc. Natl. Acad. Sci. U. S. A. **102**, 18264–18268 (2005)

11. Cui, L., Cohen, J.A., Broaders, K.E., Beaudette, T.T., Frechet, J.M.: Mannosylated dextran nanoparticles: a pH-sensitive system engineered for immunomodulation through mannose targeting. Bioconjug. Chem. **22**, 949–957 (2011)
12. Trumpfheller, C., Longhi, M.P., Caskey, M., Idoyaga, J., Bozzacco, L., Keler, T., Schlesinger, S.J., Steinman, R.M.: Dendritic cell-targeted protein vaccines: a novel approach to induce T-cell immunity. Journal of Internal Medicine **271**, 183–192 (2012)
13. Burgdorf, S., Scholz, C., Kautz, A., Tampe, R., Kurts, C.: Spatial and mechanistic separation of cross-presentation and endogenous antigen presentation. Nat. Immunol. **9**, 558–566 (2008)
14. Demento, S.L., Cui, W., Criscione, J.M., Stern, E., Tulipan, J., Kaech, S.M., Fahmy, T.M.: Role of sustained antigen release from nanoparticle vaccines in shaping the T cell memory phenotype. Biomaterials **33**, 4957–4964 (2012)
15. Moon, J.J., Suh, H., Polhemus, M.E., Ockenhouse, C.F., Yadava, A., Irvine, D.J.: Antigen-displaying lipid-enveloped PLGA nanoparticles as delivery agents for a Plasmodium vivax malaria vaccine. PLoS One **7**, e31472 (2012)
16. Bershteyn, A., Hanson, M.C., Crespo, M.P., Moon, J.J., Li, A.V., Suh, H., Irvine, D.J.: Robust IgG responses to nanograms of antigen using a biomimetic lipid-coated particle vaccine. J. Control. Release **157**, 354–365 (2012)
17. Kamath, A.T., Mastelic, B., Christensen, D., Rochat, A.F., Agger, E.M., Pinschewer, D.D., Andersen, P., Lambert, P.H., Siegrist, C.A.: Synchronization of dendritic cell activation and antigen exposure is required for the induction of Th1/Th17 responses. Journal of Immunology **188**, 4828–4837 (2012)
18. Andrews, C.D., Huh, M.S., Patton, K., Higgins, D., Van Nest, G., Ott, G., Lee, K.D.: Encapsulating immunostimulatory CpG oligonucleotides in listeriolysin O-liposomes promotes a Th1-type response and CTL activity. Molecular Pharmaceutics **9**, 1118–1125 (2012)
19. Nakamura, T., Moriguchi, R., Kogure, K., Shastri, N., Harashima, H.: Efficient MHC class I presentation by controlled intracellular trafficking of antigens in octaarginine-modified liposomes. Molecular Therapy **16**, 1507–1514 (2008)
20. Hu, Y., Litwin, T., Nagaraja, A.R., Kwong, B., Katz, J., Watson, N., Irvine, D.J.: Cytosolic delivery of membrane-impermeable molecules in dendritic cells using pH-responsive core-shell nanoparticles. Nano Lett. **7**, 3056–3064 (2007)
21. Hu, Y., Atukorale, P.U., Lu, J.J., Moon, J.J., Um, S.H., Cho, E.C., Wang, Y., Chen, J., Irvine, D.J.: Cytosolic delivery mediated via electrostatic surface binding of protein, virus, or siRNA cargos to pH-responsive core-shell gel particles. Biomacromolecules **10**, 756–765 (2009)
22. Wilson, J.T., Keller, S., Manganiello, M.J., Cheng, C., Lee, C.C., Opara, C., Convertine, A., Stayton, P.S.: pH-Responsive nanoparticle vaccines for dual-delivery of antigens and immunostimulatory oligonucleotides. ACS Nano **7**, 3912–3925 (2013)
23. Moon, J.J., Suh, H., Bershteyn, A., Stephan, M.T., Liu, H., Huang, B., Sohail, M., Luo, S., Um, S.H., Khant, H., Goodwin, J.T., Ramos, J., Chiu, W., Irvine, D.J.: Interbilayer-crosslinked multilamellar vesicles as synthetic vaccines for potent humoral and cellular immune responses. Nature Materials **10**, 243–251 (2011)
24. Kumar, H., Kawai, T., Akira, S.: Pathogen Recognition by the Innate Immune System. International Reviews of Immunology **30**, 16–34 (2011)
25. Banchereau, J., Steinman, R.M.: Dendritic cells and the control of immunity. Nature **392**, 245–252 (1998)

26. Palm, N.W., Medzhitov, R.: Pattern recognition receptors and control of adaptive immunity. Immunological Reviews **227**, 221–233 (2009)
27. Joffre, O.P., Segura, E., Savina, A., Amigorena, S.: Cross-presentation by dendritic cells. Nature Reviews Immunology **12**, 557–569 (2012)
28. Barry, M., Bleackley, R.C.: Cytotoxic T lymphocytes: All roads lead to death. Nature Reviews Immunology **2**, 401–409 (2002)
29. Villadangos, J.A., Schnorrer, P.: Intrinsic and cooperative antigen-presenting functions of dendritic-cell subsets in vivo. Nature Reviews Immunology **7**, 543–555 (2007)
30. Heath, W.R., Belz, G.T., Behrens, G.M., Smith, C.M., Forehan, S.P., Parish, I.A., Davey, G.M., Wilson, N.S., Carbone, F.R., Villadangos, J.A.: Cross-presentation, dendritic cell subsets, and the generation of immunity to cellular antigens. Immunol. Rev. **199**, 9–26 (2004)
31. Yuba, E., Harada, A., Sakanishi, Y., Watarai, S., Kono, K.: A liposome-based antigen delivery system using pH-sensitive fusogenic polymers for cancer immunotherapy. Biomaterials **34**, 3042–3052 (2013)
32. Torchilin, V.P.: Recent advances with liposomes as pharmaceutical carriers. Nat. Rev. Drug Discov. **4**, 145–160 (2005)
33. Gregoriadis, G., Gursel, I., Gursel, M., McCormack, B.: Liposomes as immunological adjuvants and vaccine carriers. Journal of Controlled Release **41**, 49–56 (1996)
34. Vangala, A., Bramwell, V.W., McNeil, S., Christensen, D., Agger, E.M., Perrie, Y.: Comparison of vesicle based antigen delivery systems for delivery of hepatitis B surface antigen. J. Control. Release **119**, 102–110 (2007)
35. Steers, N.J., Peachman, K.K., McClain, S., Alving, C.R., Rao, M.: Liposome-encapsulated HIV-1 Gag p24 containing lipid A induces effector CD4 + T-cells, memory CD8 + T-cells, and pro-inflammatory cytokines. Vaccine **27**, 6939–6949 (2009)
36. Bhowmick, S., Mazumdar, T., Sinha, R., Ali, N.: Comparison of liposome based antigen delivery systems for protection against Leishmania donovani. J. Control. Release **141**, 199–207 (2010)
37. Kim, M., Song, L., Moon, J., Sun, Z.Y., Bershteyn, A., Hanson, M., Cain, D., Goka, S., Kelsoe, G., Wagner, G., Irvine, D., Reinherz, E.L.: Immunogenicity of membrane-bound HIV-1 gp41 MPER segments is dominated by residue accessibility and modulated by stereochemistry. J. Biol. Chem. (2013)
38. Allen, T.M., Mumbengegwi, D.R., Charrois, G.J.: Anti-CD19-targeted liposomal doxorubicin improves the therapeutic efficacy in murine B-cell lymphoma and ameliorates the toxicity of liposomes with varying drug release rates. Clin. Cancer Res. **11**, 3567–3573 (2005)
39. Cashion, M.P., Long, T.E.: Biomimetic Design and Performance of Polymerizable Lipids. Accounts of Chemical Research **42**, 1016–1025 (2009)
40. Moon, J.J., Suh, H., Li, A.V., Ockenhouse, C.F., Yadava, A., Irvine, D.J.: Enhancing humoral responses to a malaria antigen with nanoparticle vaccines that expand T-fh cells and promote germinal center induction. Proceedings of the National Academy of Sciences of the United States of America **109**, 1080–1085 (2012)
41. Li, A.V., Moon, J.J., Abraham, W., Suh, H.Y., Elkhader, J., Seidman, M.A., Yen, M.M., Im, E.J., Foley, M.H., Barouch, D.H., Irvine, D.J.: Generation of Effector Memory T Cell-Based Mucosal and Systemic Immunity with Pulmonary Nanoparticle Vaccination. Science Translational Medicine **5** (2013)

42. DeMuth, P.C., Moon, J.J., Suh, H., Hammond, P.T., Irvine, D.J.: Releasable layer-by-layer assembly of stabilized lipid nanocapsules on microneedles for enhanced transcutaneous vaccine delivery. ACS Nano **6**, 8041–8051 (2012)
43. Biochemistry of Lipids, Lipoproteins and Membranes, *5th Edition*, pp. 1–640 (2008)
44. Plasma lipoproteins: composition, structure and biochemistry
45. Rensen, P.C.N., de Vrueh, R.L.A., Kuiper, J., Bijsterbosch, M.K., Biessen, E.A.L., van Berkel, T.J.C.: Recombinant lipoproteins: lipoprotein-like lipid particles for drug targeting. Advanced Drug Delivery Reviews **47**, 251–276 (2001)
46. Ng, K.K., Lovell, J.F., Zheng, G.: Lipoprotein-Inspired Nanoparticles for Cancer Theranostics. Accounts of Chemical Research **44**, 1105–1113 (2011)
47. Vickers, K.C., Palmisano, B.T., Shoucri, B.M., Shamburek, R.D., Remaley, A.T.: MicroRNAs are transported in plasma and delivered to recipient cells by high-density lipoproteins. Nature Cell Biology **13**, 423–433 (2011)
48. Krieg, A.M.: Toll-like receptor 9 (TLR9) agonists in the treatment of cancer. Oncogene **27**, 161–167 (2008)
49. Goldstein, M.J., Varghese, B., Brody, J.D., Rajapaksa, R., Kohrt, H., Czerwinski, D.K., Levy, S., Levy, R.: A CpG-loaded tumor cell vaccine induces antitumor CD4 + T cells that are effective in adoptive therapy for large and established tumors. Blood **117**, 118–127 (2011)
50. Houot, R., Levy, R.: T-cell modulation combined with intratumoral CpG cures lymphoma in a mouse model without the need for chemotherapy. Blood **113**, 3546–3552 (2009)
51. Fischer, N.O., Blanchette, C., Rasley, A.: Enhancing the efficacy of innate immune agonists: could nanolipoprotein particles hold the key? Nanomedicine **9**, 369–372 (2014)
52. Weilhammer, D.R., Blanchette, C.D., Fischer, N.O., Alam, S., Loots, G.G., Corzett, M., Thomas, C., Lychak, C., Dunkle, A.D., Ruitenberg, J.J., Ghanekar, S.A., Sant, A.J., Rasley, A.: The use of nanolipoprotein particles to enhance the immunostimulatory properties of innate immune agonists against lethal influenza challenge. Biomaterials **34**, 10305–10318 (2013)
53. Xu, Z.H., Ramishetti, S., Tseng, Y.C., Guo, S.T., Wang, Y.H., Huang, L.: Multifunctional nanoparticles co-delivering Trp2 peptide and CpG adjuvant induce potent cytotoxic T-lymphocyte response against melanoma and its lung metastasis. Journal of Controlled Release **172**, 259–265 (2013)
54. Aguilar, J.C., Rodriguez, E.G.: Vaccine adjuvants revisited. Vaccine **25**, 3752–3762 (2007)
55. Fischer, N.O., Infante, E., Ishikawa, T., Blanchette, C.A., Bourne, N., Hoeprich, P.D., Mason, P.W.: Conjugation to Nickel-Chelating Nanolipoprotein Particles Increases the Potency and Efficacy of Subunit Vaccines to Prevent West Nile Encephalitis. Bioconjugate Chemistry **21**, 1018–1022 (2010)
56. Blanchette, C.D., Fischer, N.O., Corzett, M., Bench, G., Hoeprich, P.D.: Kinetic Analysis of His-Tagged Protein Binding to Nickel-Chelating Nanolipoprotein Particles. Bioconjugate Chemistry **21**, 1321–1330 (2010)
57. Fischer, N.O., Rasley, A., Corzett, M., Hwang, M.H., Hoeprich, P.D., Blanchette, C.D.: Colocalized Delivery of Adjuvant and Antigen Using Nanolipoprotein Particles Enhances the Immune Response to Recombinant Antigens. Journal of the American Chemical Society **135**, 2044–2047 (2013)
58. Barral, D.C., Brenner, M.B.: CD1 antigen presentation: how it works. Nature Reviews Immunology **7**, 929–941 (2007)

59. Brennan, P.J., Brigl, M., Brenner, M.B.: Invariant natural killer T cells: an innate activation scheme linked to diverse effector functions. Nature Reviews Immunology **13**, 101–117 (2013)
60. Vivier, E., Ugolini, S., Blaise, D., Chabannon, C., Brossay, L.: Targeting natural killer cells and natural killer T cells in cancer. Nature Reviews Immunology **12**, 239–252 (2012)
61. van den Elzen, P., Garg, S., Leon, L., Brigl, M., Leadbetter, E.A., Gumperz, J.E., Dascher, C.C., Cheng, T.Y., Sacks, F.M., Illarionov, P.A., Besra, G.S., Kent, S.C., Moody, D.B., Brenner, M.B.: Apolipoprotein-mediated pathways of lipid antigen presentation. Nature **437**, 906–910 (2005)
62. Kovacs-Simon, A., Titball, R.W., Michell, S.L.: Lipoproteins of Bacterial Pathogens. Infection and Immunity **79**, 548–561 (2011)
63. Chen, H.W., Liu, S.J., Liu, H.H., Kwok, Y., Lin, C.L., Lin, L.H., Chen, M.Y., Tsai, J.P., Chang, L.S., Chiu, F.F., Lai, L.W., Lian, W.C., Yang, C.Y., Hsieh, S.Y., Chong, P., Leng, C.H.: A novel technology for the production of a heterologous lipoprotein immunogen in high yield has implications for the field of vaccine design. Vaccine **27**, 1400–1409 (2009)
64. Huang, C.Y., Chen, J.J.W., Shen, K.Y., Chang, L.S., Yeh, Y.C., Chen, I.H., Chong, P., Liu, S.J., Leng, C.H.: Recombinant Lipidated HPV E7 Induces a Th-1-Biased Immune Response and Protective Immunity against Cervical Cancer in a Mouse Model. PLoS One **7** (2012)

Bioengineered Human Heart and Skeletal Muscles on Chips: Methods and Applications

Ki-Hwan Nam, Mikael Perla, Alec S.T. Smith and Deok-Ho Kim

Abstract This chapter introduces innovative organ-on-chip platforms for chemical assay and toxicity testing that measures the physiological properties of live, engineered muscular tissue samples. The advantages of using such engineered tissues for drug screening compared to more conventional cell-based assays are discussed. Specifically, this chapter will outline recent developments and applications of cardiac and skeletal muscle organ-on-chip systems. Recent advances in micro- and nanofabrication techniques, along with their biological applications with regard to organ-on-chips, are also reviewed in this chapter.

Keywords: Organ-on-chip · Heart-on-chip · Muscle-on-chip · Micro/ nanofabrication

1 Introduction

Current biological analyses used to evaluate the efficacy and/or toxicity of new drug candidates, or to measure human drug responses, are typically 2D cell-based assays, even though such systems do not accurately mimic the complexity of 3D *in vivo*

M. Perla · A.S.T. Smith · D.-H. Kim
Department of Bioengineering, University of Washington, Seattle, WA 98195, USA

K.-H. Nam
Optical Instrumentation Development Team, The Korea Basic Science Institute, Daejeon 34133, Korea

D.-H. Kim
Institute for Stem Cell and Regenerative Medicine, University of Washington, Seattle, WA 98195, USA

D.-H. Kim(✉)
Center for Cardiovascular Biology, University of Washington, Seattle, WA 98195, USA
e-mail: deokho@uw.edu

© Springer International Publishing Switzerland 2016
H. Jo et al. (eds.), *Biomedical Engineering: Frontier Research and Converging Technologies*, Biosystems & Biorobotics 9,
DOI: 10.1007/978-3-319-21813-7_9

tissues. Screening systems that fail to represent human tissue niches effectively can result in false positive or negative data with regard to compound action on overall tissue function. Many established systems screen for drug effects by analyzing single pathways or protein targets, which often over simplifies the model system. Established difficulties in maintaining 2D monolayers for long-term physiological drug screening applications only serve to exacerbate the problems with conventional models, since chronic compound action is difficult to assess. As a potential solution to these issues, next generation tissue culture platforms and their functional analytical devices, called organ-on-chips, are being investigated as potential preclinical models for drug development applications. Such systems could be used to investigate the responses of individual organs or tissues to toxic metabolite or physiologic waste material exposure, thereby enabling a more predictive approach to preclinical assessment of new chemical entities (Table 1). (1-3)

In order to account for tissue-tissue cross-talk and off-target metabolism, next-generation *in vitro* analysis of human responses to external biological, chemical, and electrical stimuli, is likely to rely on the linkage of multiple individual "organs-on-chips," leading to the establishment of more predictive "body-on-a-chip" platforms. (2; 3) Such systems are being designed to be dynamic, functional, and to replicate how nutrients and drugs are exposed to specific organs *in vivo*. (4-8) Given the difficulty in maintaining multiple cell types under constant culture conditions, the use of mathematical and computational modeling to provide multilevel simulations of human pharmacological responses is being investigated. (4) It is believed that physiologically accurate tissue engineering platforms, based on effective computational models of human systems, will improve our ability to investigate the absorption, distribution, metabolism, and excretion of drugs prior to their progression to clinical trials. The introduction of better modeling techniques should in turn improve the accuracy and efficiency of clinical drug development, resulting in reduced cost, and shorter lead times for new compounds between discovery and commercial availability. (5)

Cardiac and skeletal muscle are two organs for which particularly advanced on-chip designs have been developed. The relative simplicity of the cellular population and tissue architecture, coupled with the ability to generate physical and electrophysiological read-outs of function, have led to rapid advancements in the production of such models for *in vitro* analysis. (6-15) Given the importance of these systems for investigating cardiac defects and debilitating muscle-wasting conditions, advances in these on-chip screening systems could yield substantial benefits to patients in the near future. Details of the current state-of-the-art with regard to cardiac and skeletal muscle modeling are discussed in more detail below.

Table 1 Applications of organ-on-chips for biological assays. Reprinted from (16) with permission.

Type	Function	Disease Model	Material	Drug Candidates
Heart	Provides blood circulation to the rest of the body	Cardiovascular disease, Arrhythmia, Mitochondrial cardiomyopathy, cardiotoxicity	Hydrolitic degradable scaffold such as polyglycolic acid (PGA), Engineered iPSC-CM tissue, Self-assembled fibronectin, PIPAAm	Beta Blockers, Protein kinases Isoproterenol and metoprolol (β-adrenergic agonist) E-4031 (hERG blocker) Verapamil (multi-ion channel blocker)
Brain	sensation, perception, action, interaction and cognition	Alzheimer Regeneration of cognitive process Brain tumor	Hybrid nanoparticle-organic transistor, Layered cortical circuits, CMOS neuron components	Prophylactic antiepileptic drugs
Lung	Provides oxygen and expels carbon dioxide	Asthma Pulmonary edema	Decellularized ECM	TRPV 4 blockers
Kidney	Regulates body fluid, pH, excretion of metabolic waste, migration	Drug toxicology Immuno-reactivity Neophoro-toxicity	Biomedical hydrogels Decelluarized porcine kidney Matrigel and collagen I	Amiodarone-toxicity Ibuprofen sodium dodecyl sulfate Cisplatin, Gentamicin, Doxorubicin
Skin	Provides physical barrier and protects body against pathogens	Burn injuries Necrotizing fasciitis Subdermal injury	Decellularized matrices, Alloderm, Apligraf, Dermagraft, Epiccl, Integra Dermal Regeneration Template, OrCd, and TransCyt	Sodium lauryl sulfate (SLS) and dodecyl pyridinium chloride (DPC)
Muscle	Provides form, support, stability, and capability of directed movement to the body	Becker muscular dystrophy Duchenne muscular dystrophy Emery-Dreifuss muscular dystrophy Facioscapulohumeral muscular dystrophy Limb-girdle muscular dystrophy Myotoniacongenita Myotonic dystrophy	Biodegradable polymers such as poly (lactide-co-glycolide), poly (caprolactone), polyphosphazenes, and composites of these polymers with each other and various inorganic compounds	AVI-4658, Eteplirsen
Liver	Performs digestion, metabolism, immunity and the storage of nutrients within the body	Oxygen consumption and transport Hepatic disorders Liver toxicology and metabolism	Polyethylene glycol hydrogels Collagen	Doxorubicin

2 Heart Tissues on Chips

The major impediment to the advancement of the cardiac engineering field is the inability to test parameters in a high-throughput and combinatorial manner. Current methods used to measure changes in fine-scale cytoskeletal and extracellular architectures in response to chemical or physical challenges, as well as the corresponding functional changes to cardiac contractility, are still limited. Centimeter-scale constructs require histological sectioning to visualize intracellular and extracellular architecture, and they are difficult to generate in a high-throughput manner due to their size. (9) Such issues dramatically increase the cost associated with the production of such models, which precludes their adoption in commercial applications. Previous studies have reported a bioengineering approach to fabricate collagen or fibrin-based mini-engineered heart tissue (m-EHTs) and cardiac microtissues (CMTs). (9; 10; 17) The authors demonstrated that these 3D bioengineered cardiac assays are suitable for high-throughput drug screening and tissue morphogenesis studies. The micrometer scale of these CMTs facilitates the rapid penetration of soluble effectors into the constructs, enabling rapid assessment of compound action. In addition, the production of small segments of heart tissue in these models makes the measurement of contractile activity at the tissue level possible, rather than relying on the analysis of individual cells. Legant *et al.* reported the development of micro-scale devices for studying fibroblast-induced collagen reorganization, (17) but later reported the optimization of this tool for reliable CMT formation. The authors demonstrated an ability to generate hundreds of functional CMTs routinely, and monitored their spontaneous contraction frequencies, duration, and force production. They examined the importance of substrate and extracellular matrix (ECM) stiffness on the remodeling capacity and contractility of CMTs. They then modified their initial system to study the effect of broad field electrical stimulation on cell alignment and force generation within CMTs. Finally, they demonstrated the potential of this system for high-throughput drug screening applications by analyzing the dose-response effect of chronotropic and inotropic drugs on CMTs. The obtained results support the use of this model for investigating the effects of various stimuli, such as mechanical preload, matrix stiffness, electrical stimulation, and soluble factors, on the structural and functional properties of engineered CMTs. (8) The authors suggest that their system will provide a valuable platform with which to elucidate how biomechanical, electrical, biochemical, and genetic/epigenetic cues modulate stem cell derived cardiomyocyte differentiation and maturation. (17) Microtissue engineering in combination with stem cell differentiation in this manner could lead to the generation of high-quality functional myocardial tissues from stem cells, which will in turn enable the development of more predictive preclinical assays.

High throughput *in vitro* assays currently play an active role in assessing the safety of cardiac drugs, and have been used to demonstrate that cardiomyocytes cultured in 2D are able to generate significant levels of contractile force. However, many of the experimental models that have been developed so far are not scalable to generate 3D voluminous tissues, which limits their ability to accurately model the native tissue. Boudou *et al.* recently generated 3D cardiac tissue constructs and examined the chips for drug effects (7). Although the described system was 3D, issues arising from a lack of biomimicry in terms of tissue architecture and recapitulation of *in vivo* structures mean the applicability of this system for drug testing are still limited. To mimic the tissue architecture of the heart more effectively, the Healy group have developed a microphysiological system (MPS) that mimics the human myocardium. (10) This system is composed of three major parts to mimic the *in vivo* microenvironment: 1) Aligned 3D human induced pluripotent stem cell-derived cardiomyocytes (hiPSC-CMs) mimicking the structure and geometry of the human myocardium. 2) Effective mimicry of nutrient, drug compound, and metabolic waste transport *in vivo* through integration of a microfluidic circulation system. 3) Protection of the engineered tissue from shear stresses, and controlled diffusive transport of nutrients to the tissue (Fig. 1). Pharmacological studies demonstrated that half-maximal inhibitory/effective concentration values recorded using this system were more consistent with data from tissue models rather than cellular-scale studies. Healy *et al.* (10) suggest that this success is attributable to their unique design, which facilitates the alignment of cells within the microtissue, and recapitulates the mass transport properties of the native myocardium effectively.

The discussed works represent important steps toward the development of a reliable *in vitro* cardiac drug-screening assay. Furthermore, compared to classical 2D culture systems, the results suggest that heart-on-chip models are better predictors of clinical cardiology. As the reliability and reproducibility of these models increases, the utility of these bioinspired systems for generating physiologically relevant human 3D aligned cardiac microtissues will likewise increase. Next-generation preclinical evaluation of new chemical entities will likely rely heavily on such models to predict the efficacy and toxicity of new drugs prior to their inclusion in clinical trials. The current focus on such systems for predicting the arrhythmogenic properties of new drugs, serves to highlight still further the important role these models will assume, as the technology gets closer to accurately recreating the structural and functional properties of the human myocardium *in vitro*.

3 Skeletal Muscles on Chips

The engineering of contractile skeletal muscle tissues from rodent and human cell sources is relatively advanced when compared with other organs. Although several methods have been developed for producing contractile 3D skeletal muscle *in vitro,* (13; 15; 18) the most significant contribution so far to this field in terms of preclinical drug screening was made by Vandenburgh *et al.* (14) The authors reported a new prototype 96-well assay system in which force measurements could be taken from tissue-engineered skeletal muscle using image-based motion detection. The described approach is nondestructive and sensitive enough to measure nN changes in muscle force generation over extended periods of drug exposure. Using this system, the authors demonstrated that the cultured muscle cells align in

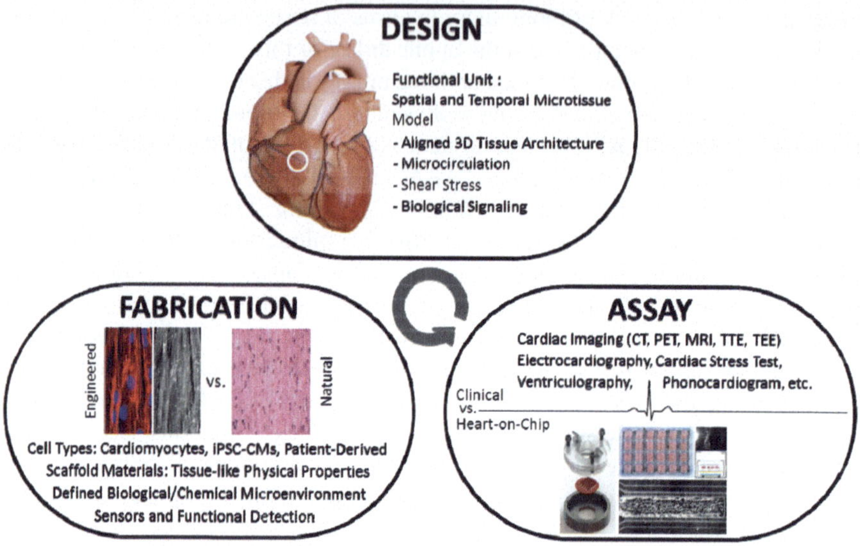

Fig. 1 Heart-on-chip overview: Engineered tissues grown in biomimetic environments better mimic complex tissue physiology and functionalities having potential use as tissue physiology or disease pathology models for drug screening assays

parallel to the passive tensions that develop within the collagen scaffold as it is remodelled. (12; 14; 15)Using this system, the authors demonstrated that these so-called "miniature bioartificial muscles" (mBAMs) undergo myofiber hypertrophy and active force increases in response to treatment with insulin-like growth factor 1. Conversely, exposure of mBAMs to cholesterol-lowering statins was found to promote the deterioration and weakness of the cultured muscle cells. These results suggest that the integration of skeletal muscle tissue engineering techniques with the means to perform real-time biomechanical testing in a single platform is possible, and a suitable method for screening compounds for their capacity to affect muscle performance.

Although there has been substantial progress with regards to generating skeletal muscle assays for screening new compounds, more advanced and integrated models are required to facilitate accurate assessment of compound toxicity/efficacy *in vitro*. The development of fully matured skeletal muscle fibers that can adequately model adult physiological 3D architectures remains the most significant challenge. However, correct interaction with principal supporting tissues; most notably, vascular, myotendinous, and neuronal connections to engineered muscle tissues also must be addressed. The successful generation of skeletal muscle tissues capable of producing force levels comparable to those of native tissue has been demonstrated. (12) However, the published work relied on rodent tissue as the cell source, and no comparable study has been published using human cells. Similarly, vascular infiltration, (12; 19) neuromuscular junction formation, (20; 21) and myotendinous junction connectivity (22) have all been shown to be possible in engineered muscle constructs. The use of nonhuman cell sources or a lack of high-throughput functionality in all of these published models remains an issue, and adaptation of these systems to account for these limitations is crucial.

4 Towards Integrated, Functional Organs-on-Chips for Drug Screening

Flat and highly rigid culture surfaces, with poorly defined surface chemistries, are present in most conventional multi-well plates and culture flasks. It is widely accepted that such models are incapable of recreating biomimetic mechanical or chemical microenvironments that are capable of promoting the advanced functional maturation of seeded cells. As such, a substantial amount of work has been focused on the development of novel nanofabrication techniques for creating *in vitro* cell culture models that are able to provide physiologically representative mechanical and structural cues to cultured cells in order to mimic the *in vivo* cellular microenvironment more closely. In addition to conventional photolithographic methods, techniques employing tunable and biocompatible polymers have been used to fabricate micro- and nanotopographic substrates for advancing cellular maturation. (9; 23-37) Mainstream adoption and application of these systems is required to facilitate their continued development toward more biologically relevant and functionally competent engineered tissue models. In order to become truly representative of human systems, organs-on-chips must eventually be able to reflect the complexity of the human physiology and pathology they are designed to model. As discussed in this review, engineered heart and skeletal muscle organs-on-chips are capable of mimicking the complex tissue physiology and functionalities that these organs possess, and have amazing potential for use in studying tissue physiology, disease pathology, or compound efficacy and toxicity. Critically, however, such single-organ models are incapable of recreating the complexity of living systems, as they do not account for off-target metabolism and systemic activities. This limitation severely inhibits the ability of these models to predict compound action *in vivo* effectively. Consequentially, the key to developing new *in vitro* tools for attaining more predictive preclinical drug discovery data is the development of high-throughput multiorgan models supported by accurate computational modeling. The development of such advanced preclinical platforms has already begun, and will drive the production of cheaper and more effective drug development assays in the near future (38).

Acknowledgements D. H. Kim thanks the Department of Bioengineering at the University of Washington for support from the new faculty startup fund. D. H. Kim is also supported by an American Heart Association Scientist Development Grant (13SDG14560076), a National Institutes of Health R21 Grant (R21AR064395) and a Muscular Dystrophy Association (MDA) Research Grant (MDA 255907) (D.-H.K.). This work was supported by International Collaborative R&D Program through 380 KIAT grant funded by the MOTIE (N0000894).

References

1. Huh, D., Torisawa, Y.S., Hamilton, G.A., Kim, H.J., Ingber, D.E.: Microengineered physiological biomimicry: Organs-on-Chips. Lab on a Chip **12**, 2156–2164 (2012)
2. Esch, M.B., Smith, A.S., Prot, J.M., Oleaga, C., Hickman, J.J., Shuler, M.L.: How multi-organ microdevices can help foster drug development. Advanced Drug Delivery Reviews **69–70**, 158–169 (2014)
3. Sung, J.H., Esch, M.B., Prot, J.M., Long, C.J., Smith, A., et al.: Microfabricated mammalian organ systems and their integration into models of whole animals and humans. Lab on a Chip **13**, 1201–1212 (2013)
4. Abaci, H.E., Shuler, M.L.: Human-on-a-chip design strategies and principles for physiologically based pharmacokinetics/ pharmacodynamics modeling. Integrative Biology **7**, 383–391 (2015)
5. Erson, E.Z., Cavusoglu, M.C.: A software framework for multiscale and multilevel physiological model integration and simulation. IEEE Engineering in Medicine and Biology Society. Annual Conference **2008**, 5449–5453 (2008)
6. Agarwal, A., Goss, J.A., Cho, A., McCain, M.L., Parker, K.K.: Microfluidic heart on a chip for higher throughput pharmacological studies. Lab on a Chip **13**, 3599–3608 (2013)
7. Boudou, T., Legant, W.R., Mu, A.B., Borochin, M.A., Thavandiran, N., et al.: A microfabricated platform to measure and manipulate the mechanics of engineered cardiac microtissues. Tissue Eng. Pt. A **18**, 910–919 (2012)
8. Hansen, A., Eder, A., Bonstrup, M., Flato, M., Mewe, M., et al.: Development of a drug screening platform based on engineered heart tissue. Circulation Research **107**, 35–U70 (2010)
9. Kim, D.H., Kshitiz, Smith R.R., Kim, P., Ahn, E.H., et al.: Nanopatterned cardiac cell patches promote stem cell niche formation and myocardial regeneration. Integrative Biology **4**, 1019–1033 (2012)
10. Mathur, A., Loskill, P., Shao, K.F., Huebsch, N., Hong, S., et al.: Human iPSC-based cardiac microphysiological system for drug screening applications. Scientific Reports **5**, 8883 (2015)
11. Wang, G., McCain, M.L., Yang, L.H., He, A.B., Pasqualini, F.S., et al.: Modeling the mitochondrial cardiomyopathy of Barth syndrome with induced pluripotent stem cell and heart-on-chip technologies. Nature Medicine **20**, 616–623 (2014)
12. Juhas, M., Engelmayr Jr, G.C., Fontanella, A.N., Palmer, G.M., Bursac, N.: Biomimetic engineered muscle with capacity for vascular integration and functional maturation in vivo. Proceedings of the National Academy of Sciences of the United States of America **111**, 5508–5513 (2014)
13. Shansky, J., DelTatto, M., Chromiak, J., Vandenburgh, H.: A simplified method for tissue engineering skeletal muscle organoids in vitro. In Vitro Cellular and Developmental Biology **33**, 659–661 (1997)
14. Vandenburgh, H.: High-content drug screening with engineered musculoskeletal tissues. Tissue Engineering Part B **16**, 55–64 (2010)

15. Vandenburgh, H., Shansky, J., Benesch-Lee, F., Barbata, V., Reid, J., et al.: Drug-screening platform based on the contractility of tissue-engineered muscle. Muscle Nerve **37**, 438–447 (2008)
16. Nam, K.-H., Smith, A.S.T., Lone, S., Kwon, S., Kim, D.-H.: Biomimetic 3D tissue models for advanced high-throughput drug screening. Journal of Laboratory Automation **20**, 201–215 (2015)
17. Legant, W.R., Pathak, A., Yang, M.T., Deshpande, V.S., McMeeking, R.M., Chen, C.S.: Microfabricated tissue gauges to measure and manipulate forces from 3D microtissues. Proceedings of the National Academy of Sciences of the United States of America **106**, 10097–10102 (2009)
18. Bell, E., Ivarsson, B., Merrill, C.: Production of a tissue-like structure by contraction of collagen lattices by human-fibroblasts of different proliferative potential in vitro. Proceedings of the National Academy of Sciences of the United States of America **76**, 1274–1278 (1979)
19. Ye, X.F., Lu, L., Kolewe, M.E., Park, H., Larson, B.L., et al.: A biodegradable microvessel scaffold as a framework to enable vascular support of engineered tissues. Biomaterials **34**, 10007–10015 (2013)
20. Smith, A.S., Long, C.J., Pirozzi, K., Hickman, J.J.: A functional system for high-content screening of neuromuscular junctions. Technology **1**, 37–48 (2013)
21. Umbach, J.A., Adams, K.L., Gundersen, C.B., Novitch, B.G.: Functional neuromuscular junctions formed by embryonic stem cell-derived motor neurons. PLoS One **7**, e36049 (2012)
22. Larkin, L.M., Calve, S., Kostrominova, T.Y., Arruda, E.M.: Structure and functional evaluation of tendon-skeletal muscle constructs engineered in vitro. Tissue Engineering **12**, 3149–3158 (2006)
23. Jiao, A., Trosper, N.E., Yang, H.S., Kim, J., Tsui, J.H., et al.: Thermoresponsive nanofabricated substratum for the engineering of three-dimensional tissues with layer-by-layer architectural control. ACS Nano **8**, 4430–4439 (2014)
24. Kim, D.H., Lee, H., Lee, Y.K., Nam, J.M., Levchenko, A.: Biomimetic nanopatterns as enabling tools for analysis and control of live cells. Advanced Materials **22**, 4551–4566 (2010)
25. Kim, D.H., Lipke, E.A., Kim, P., Cheong, R., Thompson, S., et al.: Nanoscale cues regulate the structure and function of macroscopic cardiac tissue constructs. Proceedings of the National Academy of Sciences of the United States of America **107**, 565–570 (2010)
26. Kim, D.H., Provenzano, P.P., Smith, C.L., Levchenko, A.: Matrix nanotopography as a regulator of cell function. Journal of Cell Biology **197**, 351–360 (2012)
27. Kim, H.N., Kang, D.H., Kim, M.S., Jiao, A., Kim, D.H., Suh, K.Y.: Patterning methods for polymers in cell and tissue engineering. Annals of Biomedical Engineering **40**, 1339–1355 (2012)
28. Kim, P., Yuan, A., Nam, K.H., Jiao, A., Kim, D.H.: Fabrication of poly(ethylene glycol): gelatin methacrylate composite nanostructures with tunable stiffness and degradation for vascular tissue engineering. Biofabrication **6**, 024112 (2014)
29. Ahn, E.H., Kim, Y., Kshitiz, An S.S., Afzal, J., et al.: Spatial control of adult stem cell fate using nanotopographic cues. Biomaterials **35**, 2401–2410 (2014)
30. Chaterji, S., Kim, P., Choe, S.H., Tsui, J.H., Lam, C.H., et al.: Synergistic Effects of Matrix Nanotopography and Stiffness on Vascular Smooth Muscle Cell Function. Tissue Engineering Part A **20**, 2115–2126 (2014)

31. Kim, E.S., Ahn, E.H., Dvir, T., Kim, D.H.: Emerging nanotechnology approaches in tissue engineering and regenerative medicine. International Journal of Nanomedicine **9**, 1–5 (2014)
32. Kim, H.N., Jiao, A., Hwang, N.S., Kim, M.S., Kang, D.H., et al.: Nanotopography-guided tissue engineering and regenerative medicine. Advanced Drug Delivery Review **65**, 536–558 (2013)
33. Kim, J., Kim, H.N., Lim, K.T., Kim, Y., Seonwoo, H., et al.: Designing nanotopographical density of extracellular matrix for controlled morphology and function of human mesenchymal stem cells. Scientific Reports **3**, 3552 (2013)
34. Kshitiz, Park J., Kim, P., Helen, W., Engler, A.J., et al.: Control of stem cell fate and function by engineering physical microenvironments. Integrative Biology **4**, 1008–1018 (2012)
35. Macadangdang, J., Lee, H.J., Carson, D., Jiao, A., Fugate, J. et al.: Capillary force lithography for cardiac tissue engineering. Journal of Visualized Experiments: JoVE **88** (2014) doi:10.3791/50039
36. Nemeth, C.L., Janebodin, K., Yuan, A.E., Dennis, J.E., Reyes, M., Kim, D.H.: Enhanced chondrogenic differentiation of dental pulp stem cells using nanopatterned PEG-GelMA-HA hydrogels. Tissue Engineering Part A **20**, 2817–2829 (2014)
37. Yang, H.S., Ieronimakis, N., Tsui, J.H., Kim, H.N., Suh, K.Y., et al.: Nanopatterned muscle cell patches for enhanced myogenesis and dystrophin expression in a mouse model of muscular dystrophy. Biomaterials **35**, 1478–1486 (2014)
38. Cavero, I., Holzgrefe, H.: Comprehensive in vitro Proarrhythmia Assay, a novel in vitro/in silico paradigm to detect ventricular proarrhythmic liability: a visionary 21st century initiative. Expert Opinion on Drug Safety **13**, 745–758 (2014)

Organ-on-a-Chip Platforms for Drug Screening and Tissue Engineering

Zongjie Wang, Roya Samanipour and Keekyoung Kim

Abstract Developments in micro- and nanofluidic technologies have led to new kinds of cell culture and screening systems that are collectively termed organ-on-a-chip systems. Organ-on-a-chip systems are *in vitro* microfabricated devices that mimic dynamic interactions of *in vivo* microenvironments. In addition to existing two-dimensional and three-dimensional cell tissues, organ-on-a-chip systems can mimic the biomechanical and biochemical microenvironment of *in vivo* tissues as well as the interaction effect of the microenvironment on cell and tissue functions. Due to these features, organ-on-a-chip systems have become excellent platforms for drug screening and delivery test and tissue engineering. In this review, specific examples of various types of organ-on-a-chip devices and their applications in tissue engineering and drug delivery test are discussed. The ready feasibility and performance of current state-of-the-art organ-on-a-chip systems, including lung-on-a-chip, heart-on-a-chip, liver-on-a-chip, vessel-on-a-chip, and tumor-on-a-chip are also covered in this chapter. The limitations of conventional systems, basic fabrication process of organ-on-a-chip devices, and future prospective of organ-on-a-chip are discussed.

Keywords: Organ-on-a-chip · Microfluidics system · Drug screening and delivery · Cellular microenvironment · Tissue engineering

1 Introduction

To date, a typical drug development flow contains pre-clinical, clinical and post-marketing phases [1]. *In vitro* drug screening (bench test) and drug delivery test (currently, mostly animal test) are the two most important pre-clinical phase steps [1, 2]. Those steps aims to select the drugs most likely to work in humans via

Z. Wang · R. Samanipour · K. Kim(✉)
School of Engineering, University of British Columbia, Kelowna, BC V1V 1V7, Canada
e-mail: keekyoung.kim@ubc.ca

© Springer International Publishing Switzerland 2016

H. Jo et al. (eds.), *Biomedical Engineering: Frontier Research and Converging Technologies*, Biosystems & Biorobotics 9,
DOI: 10.1007/978-3-319-21813-7_10

studying human cell behavior with drugs *in vitro*, and the drug effect in live animals, respectively. To better monitor cell behavior with drugs *in vitro*, platforms mimicking *in vivo* environments of the human body have emerged. *In vitro* models can support almost all types of drug screening methods, including free drugs, drug-delivery agents, implanted depots, liposomes, and nanoparticles [3]. At the early stage, most *in vitro* platforms are based on two-dimensional (2D) cell culturing platforms. Those platforms first seed monolayer of cells on the culturing flask. Then, drugs for screening are added into the cell culture media. After certain times, chemical and biological signals produced by cells are recorded and analyzed. Apparently, 2D *in vitro* platforms provide a direct procedure to screen interactions between drugs and cells. However, many studies have shown that 2D cell monolayer cannot adequately mimic three-dimensional (3D) cellular microenvironments in the human body, which results in the oversimplification of the extracellular matrix (ECM). Fig. 1A illustrates the schematic of the typical 2D and 3D cell culture. As shown in Fig. 1B and C, the embryoid body (EB) cell differentiation in 2D and 3D environment are significantly different after 5 days culturing [4]. Similarly, 2D and 3D culturing resulted in totally different cell organization in SKOV-3 cells (Fig. 1D and E) [5].

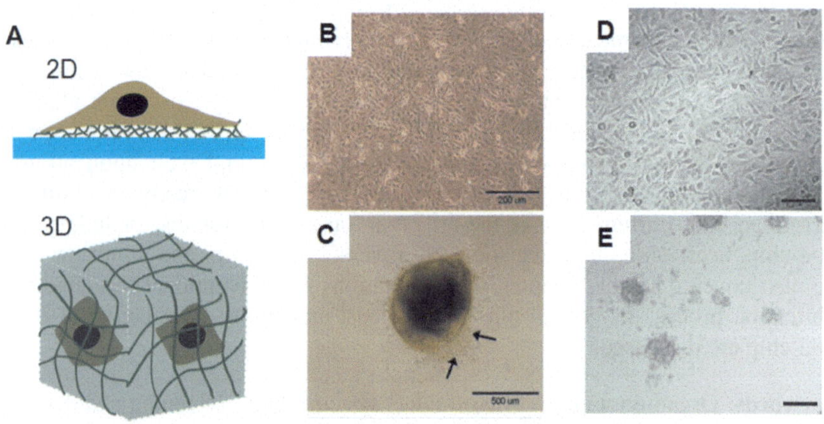

Fig. 1 Examples of 2D and 3D cell culturing. (A) Illustration of 2D flat cell culturing platform and 3D cell culturing. (B) Phase contrast micrographs of 2D monolayer embryoid body cells at day 5 (scale bar: 200 μm) and (C) embryoid body cells in 3D scaffold at day 5 (scale bar: 500 μm). Arrows show cells migrated from an aggregate. The migrated cells became cordlike and multilayer structures, which were not existed in 2D cultures (Adapted from [4]). (D) Phase contrast micrographs of SKOV-3 cells grown in 2D and (E) in 3D environment (Adapted from [5]). Scale bar: 100μm.

In addition to the shape and distribution, it has been found that the tumor cells cultured in in vitro 3D scaffolds were more functionally active after transplanted back to *in vivo* environment [6]. The reason why 2D and 3D culturing are with so many differences is that 2D cell culturing usually results in the oversimplification of the extracellular matrix (ECM). ECM properties of *in vitro* and *in vivo* structures are remarkably different [6]. Thus, how the drugs play a role in the *in vivo* microenvironment is difficult to elucidate by the 2D *in vitro* tests.

Many 3D culturing platforms have been developed in response to this limitation [7–10]. The most common 3D cell culturing platforms are cell aggregations [11], spheroids [7], and cell sheets [12, 13]. Cells accumulate to form specific 3D structures, such as aggregates or multi-cell layers. In such structures, cells contact each other in three directions. These platforms have successfully created the 3D ECM microenvironment *in vitro*. However, the interaction between the cellular microenvironment does not only include chemical responses, but also mechanical and electrical interactions that can affect cell behaviors significantly [14]. One explicit example is the lung. During inhalation, the volume of lung changes markedly, subjecting the lung cells to non-negligible mechanical stress [15]. Also, the changes in a volume play an important role in increasing the surface area to volume ratio, which is a critical parameter for air exchange. To study *in vivo* cell responses to drugs, we have to consider dynamic mechanical and electrical changes in cells, as well as chemical variations. Hence, similar to 2D platforms, passing 3D *in vitro* screening test cannot guarantee that drugs are safe for humans due to the absence of data on mechanical and other essential cell effects [16, 17].

In addition to the drug screening, a drug delivery test determines the metabolism and toxicological effects of drugs [18]. The drug delivery test is considered as the most crucial step before clinical test because it can predict the performance of new drugs inside the human body [19]. Traditional drug delivery testing is based on animal models. During the animal test, the effect of the inputs (drugs) is evident by outputs, such as the concentration changes of specific chemicals in animal body fluids [18]. Although the *in vivo* animal microenvironment is the closest to humans in nature, in practice, animal models have many limitations as a preclinical test platform for drugs. It is difficult to achieve reproducible drug delivery test results using animal models [20]. The characteristics of each animal are not controllable and animal-to-animal variation results in different test outcomes that can appreciably reduce the reliability of animal models. Additionally, the metabolism of drugs is hard to deduce by animal tests, since animal models only guarantee case-by-case results [21]. In most cases, researchers can only investigate the effect of drugs by monitoring the changes in the concentrations of chemical compounds [22]. Animal testing often causes ethical problem and discarding the tested animals is both time- and resource-consuming [20]. Such an indirect monitoring method obstructs the deeper understanding of toxicity and interaction mechanisms as well as long-term stability of drugs. Magnetic resonance imaging (MRI) of the whole body is one of the possible solutions to trace the drug delivery process. However, the expensive cost and limited resolution, reduce its potential usage for drug delivery test [18].

Although many drugs are discovered every year, few can be commercialized because their potential effects on the human body remain unclear [20]. Thus, if we fully understand the problems how *in vivo* microenvironments are constructed and how cells interact with other cells and the surrounding environments, the investigation of the human body at the tissue and cellular levels [23], as well as the development of new drugs for disease therapies will be facilitated [24]. Fortunately, with the development of nanotechnology, scientists can precisely manipulate cellular microenvironments at the micro- and nanoscale. Also, using microfabrication techniques, such as soft lithography, highly accurate devices can be developed at reasonable costs with the marked reduction of the use of chemicals and biological materials [25, 26]. However, rebuilding the cellular microenvironment of particular tissues *in vitro* in a simple and straightforward way remains challenging.

To overcome the challenges, organ-on-a-chip systems have been developed. Organ-on-a-chip devices aim to recapitulate the function and microstructure of multicellular human organs *in vitro* in a microfluidics chip [23, 27]. Such devices include the biological function of organs and the biochemical, bioelectrical, and biomechanical properties of the cellular microenvironments and extracellular matrixes (ECMs). Based on advancements of tissue engineering and micro- and nanofabrication technology, the organ-on-a-chip platforms can facilitate the study of cell-cell [15], cell-ECM [28], and cell-tubular flow [29] interactions. The platforms allow scientists to perform drug screening and delivery testing and analysis *in vitro*. In addition, organ-on-a-chip systems can be employed to characterize cell physical properties in microenvironments [30]. The advantages of organ-on-a-chip systems include low-cost, high-controllability, and the ability to mimic organ functions dynamically [15, 20, 23, 31]. Recent studies have demonstrated the feasibility of organ-on-a-chip systems, which can closely mimic *in vivo* tissues and provide platforms for drug delivery test and biological cell characterization [20, 23, 32].

In this chapter, specific examples of different types of organ-on-a-chip devices for drug screening and tissue engineering, such as lung-on-a-chip, heart-on-a-chip, vessel-on-a-chip, liver-on-a-chip, and tumor-on-a-chip, are discussed. The working principles of those devices as well as their applications are provided. As well, basic fabrication method of microfluidics devices, and future prospects of organ-on-a-chip are discussed.

2 Microfabrication of Organ-on-a-Chip Platforms

To precisely mimic *in vivo* microenvironments, organ-on-a-chip platforms rely on the fabrication processes. Microfabrication techniques have been widely used to fabricate organ-on-a-chip platforms because of the possibility of building physiological environments at the micro-scale [33]. To date, the most commonly used microfabrication technology in organ-on-a-chip platforms is the soft-lithography. Different from the traditional microfabrication for solid state integrated circuits and MEMS devices, in soft-lithography, there is no etching or deposition involved in the majority of fabrication process. Instead, soft elastomeric molds are applied to build desired 3D microstructure. The main advantages of soft-lithography include the low cost and convenient during fabrication, the easy control of deformation and structure in micro-scale, and the high compatibility with a broad range of biomaterials.

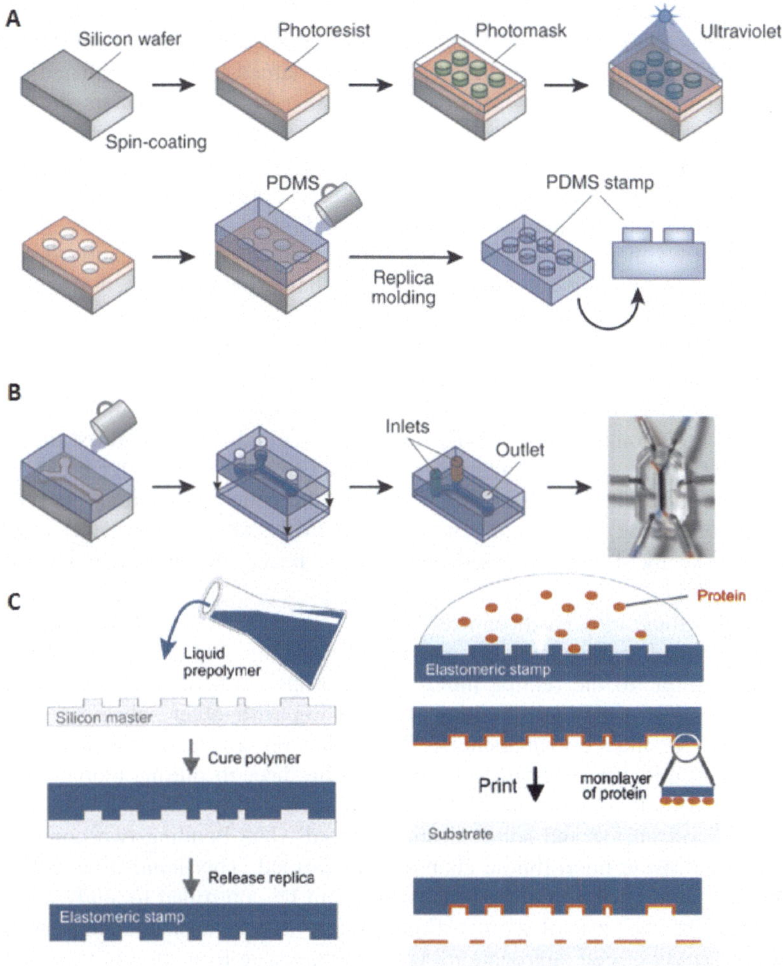

Fig. 2 Microfabrication techniques for organ-on-a-chip platforms. Use of replica molding to create PDMS (A) stamps and (B) channel with two inlets and one outlet (Adapted from [34]). (C) Microcontact printed proteins used to pattern cell (Adapted from [35]).

The most widely used materials in soft-lithography is Poly(Dimethylsiloxane) (PDMS) [25], since the PDMS is transparent, biocompatible, and gas permeable, which are essential properties for creating an organ-on-a-chip. Especially, the gas permeability of PDMS allows oxygen to permeate inside the organ-on-a-chip platforms for maintaining cells or tissues alive [36]. The transparency of PDMS is from 240–1100 nm wavelength. Therefore, the wavelength range from long-wave UV to infrared radiation (IR) can pass through PDMS, which allows a variety of the microscope system to capture high-quality images. Moreover, PDMS is an elastic material with Young's modulus of 750kPa, which makes it suitable for generating the microscale dynamic strain. Another important issue for choosing PDMS as a base material of organ-on-a-chip platforms is the handling-easiness and the cost-effectiveness [36]. In addition to PDMS, other elastomers, including polyurethane (PU) and polyimide, are also compatible with soft-lithography process [37]. Recently, a new class of polymers named highly fluorinated elastomers has been developed for soft-lithography [38, 39]. Such polymers show great release properties and have excellent inertness to swelling by monomers and organic solvents. Highly fluorinated elastomers partly overcome the low bioresistance of PDMS, which is referred as the biggest problem of employing PDMS as the materials for organ-on-a-chip platforms [23, 40].

Fig. 2 shows main techniques of soft-lithography for the fabrication of organ-on-a-chip platforms. Replica molding technique has been widely used in patterning the microstructure of PDMS, such as stamps, channels, and porous membranes. The PDMS stamp shown in Fig. 2A was fabricated using the replica molding technique [41]. A silicon wafer is coated with a uniform layer of photoresist. The photoresist can be either crosslinked or uncrosslinked by the exposure of ultraviolet (UV) light. The use of a photomask permits only the desired parts of the photoresist to be illuminated. After this step, the pattern is transferred to the silicon wafer mold, which becomes a mold. The PDMS is then poured on the patterned mold and is polymerized during baking in an oven. The casted PDMS stamp is peeled off from the substrate. The mold can be reused many times; hence, the process is termed the replica molding.

The soft-lithography has been used to fabricate PDMS channels to mimic *in vivo* fluids through replica molding and PDMS bonding (Fig. 2B). The fabrication steps are similar to the replica molding of stamps, with the difference that the PDMS device is negatively molded and bonded with other substrates (usually glass slides or other complementary PDMS layers). In detail, the hydrophobic surface of PDMS is converted to be hydrophilic through plasma oxidation treatment. After this treatment, the PDMS can be irreversibly bonded to itself, glass, silicon, silicon nitride and some plastic materials. Via bonding two components together, the inside microfluidic channels are formed. The liquid flow in the microfluidic channels is a laminar flow, which can be controlled to study cell-flow interaction. Hence, use of a microfluidic chip with controllable laminar flows allows the study of cell aggregation [42], cardiac tissue formation [43], as well as stem cell differentiation [44].

The soft-lithography technique also allows the customized pattern of proteins on the PDMS porous membranes to study the interactions between both sides of a barrier (e.g. the vascular endothelium and parenchymal tissues) [15]. The pattern of cultured cells can be controlled by microcontact printing technology [35], as shown in Fig. 2C. After fabricating a PDMS membrane with micro-scale pores using replica molding, an elastomeric PDMS stamp is generated and coated with a monolayer of proteins. By stamping the PDMS stamp on the substrate, desired areas of the proteins can be transferred to the membrane substrate. Since cells only grow in the area containing proteins, by controlling the pattern of printed proteins, cells on the membrane can be patterned as well.

Microwells, fluidic channels, and porous membranes are most widely used components for organ-on-a-chip platforms to study *in vivo* cell-cell interactions, cell behaviors, and cell responses to chemical and mechanical stimulation [41].

Microfabrication technology can build and assemble the components of organ-on-a-chip platforms easily, facilitating the design of *in vitro* systems to investigate *in vivo* microenvironment effects.

3 Development of Organ-on-a-Chip Platforms

In this section, various existing state-of-art organ-on-a-chip platforms are introduced to show the feasibility and potential of the system for biomedical applications. In the sub-sections, which include lung-on-a-chip, heart-on-a-chip, liver-on-a-chip, vessel-on-a-chip, and tumor-on-a-chip platforms, working principles and device structures are discussed.

3.1 *Lung-on-a-Chip*

The lung is vital organ where external oxygen gas is exchanged with carbon dioxide in the blood. Investigating cell-cell interactions, cell-blood flow, and cell-gas flow in the respiratory tract is essential for both physiological research and drug delivery testing. Some microfluidics-based platforms are designed to culture lung cancer cells and screen its response to drugs [45] and electrical stimulation [46]. However, these platforms do not mimic the deformation of the lung during breathing, as well as the bi-cell barriers between blood and air. Thus, this type of microfluidic system is limited to cell level characterizations.

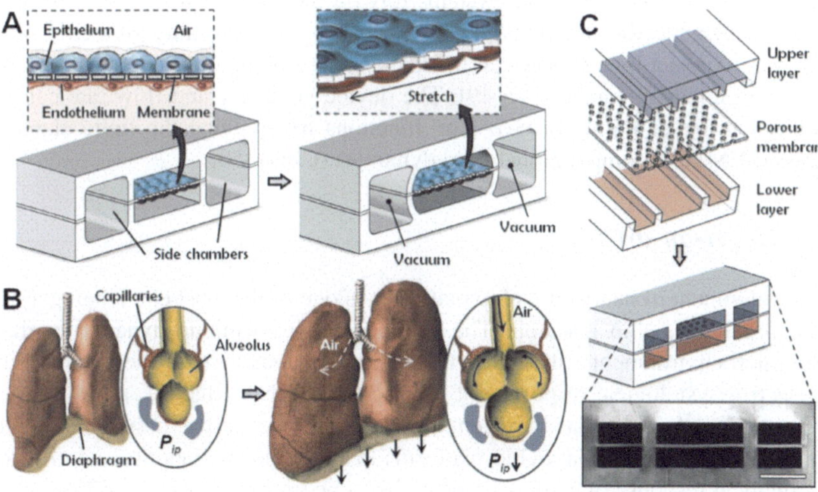

Fig. 3 Lung-on-a-chip platform (Adapted from [15]). (A) PDMS-based membrane mimicking the alveolar-capillary barrier and vacuum based deformation controller mimicking the breathing movements of the lung. (B) Schematic representations of lung size change during the inhalation process. (C) The bonding and alignment of three layer PDMS structures to form the devices.

Huh et al. presented the first lung-on-a-chip system [15]. The structure of the lung-on-a-chip system is shown in Fig. 3. This device consists of three channels: two side channels and one main channel divided by a membrane. The membrane and side channels are the main components of this system. After seeding endothelial and epithelial cells on each side of the porous membrane, the main channel can mimic the alveolar-capillary barrier in the lung, allowing the interaction between two types of cell and two flows in the main channel. For example, immune cells from blood flow can cross through micro holes of the membrane barrier and kill Escherichia coli in the airway flow. The suction pressure on each side of chamber can be controlled to adjust the membrane stretch in the main channel, which mimics the lung breathing movement. Then the relationship between mechanical strain during the breathing movement and cellular response to the nanoparticles were screened and analyzed *in vitro*. Great improvements have been made to verify the *in vitro* microfluidic lung-on-a-chip platform. Compared to traditional platforms, the lung-on-a-chip allows the investigation of complex chemical, mechanical, and biological effects, providing a tangible platform for characterizing dynamic cell behaviors *in vitro* as well as the comprehensive test of drug effects. Also, this research introduced a standard device with the capability of mimicking the interaction barrier between two types of cells. By seeding different types of cells into the device and controlling the input suction pressure, such device can be also suitable for the study of many *in vivo* interactions and transformation, such as the endothelium-epithelium interface in vessels [15] and the capillary-lymphatic interaction in cellular microenvironments [28].

Other types of lung-on-a-chip systems rely on the static characteristics of lung cells to realize specific functions [47–50]. Most of the static lung-on-a-chip platforms aim to detect lung cancer cell markers or screen lung tumor cell response to microenvironmental changes [47–49]. The device geometry and flow rate can be optimized to better mimic *in vivo* lung functions by applying the finite element method (FEM) and computational fluid dynamics (CFD) [50].

3.2 Heart-on-a-Chip

The development of heart-on-a-chip platforms is one of the most challenging areas in the organ-on-a-chip research. The challenges are to mimic dynamic cardiac tissue microenvironments. It is difficult to simultaneously study the contractility and electrophysiological responses of cardiac tissues. Nevertheless, a few studies have been done with cardiac tissue engineering to mimic cardiac tissue microenvironments [51] and heart-on-a-chip platforms to study the contractility and electrophysiological response of cardiac tissue at the same time (Fig. 4 A-C) [52].

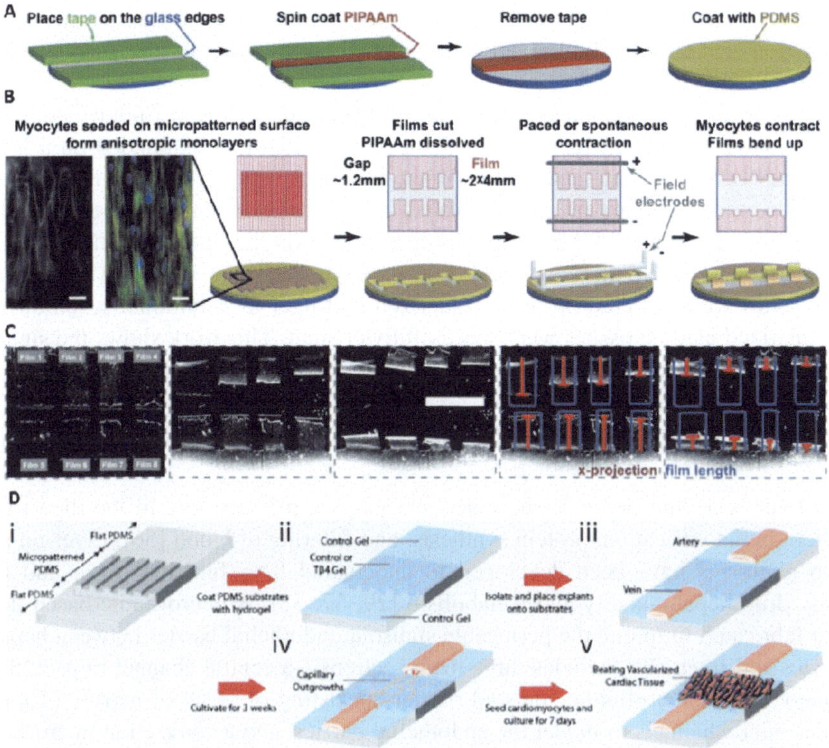

Fig. 4 Heart-on-a-chip platforms. (A) Fabrication of 25 mm round substrates. (C) Contractility assay was operated by anisotropically layered cardiomyocytes. (D) Series of optical micrographs of contractility experiment and analysis (Adapted from [52]). Scale bar: 5 mm. (D) Cardiomyocytes were cultured on vascularized polydimethylsiloxane (PDMS) substrates to model cardiac tissue (Adapted from [53]).

At the early stage of the development of the heart-on-a-chip, a PDMS microfluidic network was combined with planar electrode array to measure the extracellular potential from individual adult cardiomyocytes [54]. Another microfluidic device with an array of electrodes was developed to electrically measure the metabolic profile of cardiomyocytes and optically measure cell contractility [55]. Grosberg et al. first introduced a tissue level heart-on-a-chip to measure the contractility of neonatal cardiac muscle tissue [52]. In the design, eight muscular thin films (MTF) were fabricated on a chip. A layer of poly(N-isopropylacrylamide) (PIPAAm) dissolved at below 35 °C is spin-coated on top of a glass slide (Fig. 5A). Subsequently, a PDMS layer was coated on top of the PIPAAM layer. The PDMS layer was used to seed neonatal rat ventricular cardiomyocytes. The substrate seeded with cells is placed in the bath and the film layers were manually cut to fabricate an array of two opposite rows of four rectangular film layers of MTFs. The MTFs are peeled off after PIPAAm is dissolved as a solution when kept below 35 °C. Finally, electrodes are placed on the top and the bottom of MTFs.

The MTFs start to curl up when electrical stimulation from the field electrode is applied (Fig. 4B). Finally, the deflection of cells responding to electrical stimuli can be measured (Fig. 4C).

Chiu et al. [53] fabricated cardiac tissue by encapsulating cardiomyocytes in microfluidic channels (Fig. 4D). A micropatterned PDMS was bonded on a thin layer of PDMS substrate and coated by the hydrogel with Tβ4, which is an angiogenic peptide. Arteries and veins isolated from mice were placed at the two ends of the micropatterned PDMS. The hydrogel coated on the micropatterned PDMS was vascularized by the artery and vein explants for three weeks. After cardiomyocytes were cultured on the vascularized hydrogel for additional seven days, vascularized cardiac tissues were successfully created. This work shows the significance of the vascularization for tissue engineered organs.

3.3 Liver-on-a-Chip

The liver is a vital organ in the body that plays a primary role in the digestive system in detoxification, protein synthesis, and filtering of blood [56]. Liver-on-a-chip platforms have been developed to understand liver function better and to study drug hepatotoxicity and metabolism [29, 57, 58]. A microfluidic-based device fabricated to mimic the permeable sinusoid endothelial barrier between hepatocytes and liver [29] includes three main sections—a central channel to pack the hepatocytes, a microfluidic sinusoid barrier consisting of a set of narrow (2 mm wide) microchannels to model the endothelial barrier, and a convection microfluidic channel surrounding the barrier (Fig. 5A). The flow of nutrients through the convection microfluidic channel feeds the hepatocytes. This system mimics the transportation between blood flow and hepatocytes, and the shear stress over hepatocytes with a patterned set of microchannels.

3D micro-engineered hepatic cell culturing platforms have been introduced to remodel the complex liver microenvironment [59–61]. Use of a perfused multiwell plate enabled the study of drug toxicity and metabolism (Fig. 5B) [58]. This device consists of a multiwell and an array of 12 bioreactors. The bioreactors including hepatocytes, liver sinusoidal endothelial cells, and stellate and Kupffer cells are placed in a scaffold well. The scaffold wells are continuously perfused by means of micro pumps. Cell-cell signaling (paracrine signaling) of hepatocytes and hepatic stellate cells (HSCs) has been investigated [57]. In the study, the fabricated 3D microfluidic liver on a chip consisted of two different chambers. In one chamber, HSCs were cultured on a flat chamber. In the other chamber, hepatocytes were cultured on a concave chamber to avoid direct cell-cell contact. These two different cells were connected with a tube and one osmotic pump and adjusted to control the flow rate of nutrient solution flow (Fig. 5C). Liver metabolism was also studied numerically and experimentally [62]. Hydrogel preploymer encapsulating liver microsome were injected into microfluidic channel. The liver metabolism was characterized by the diffusion and convection of the microsomal enzyme inside the microfluidic device.

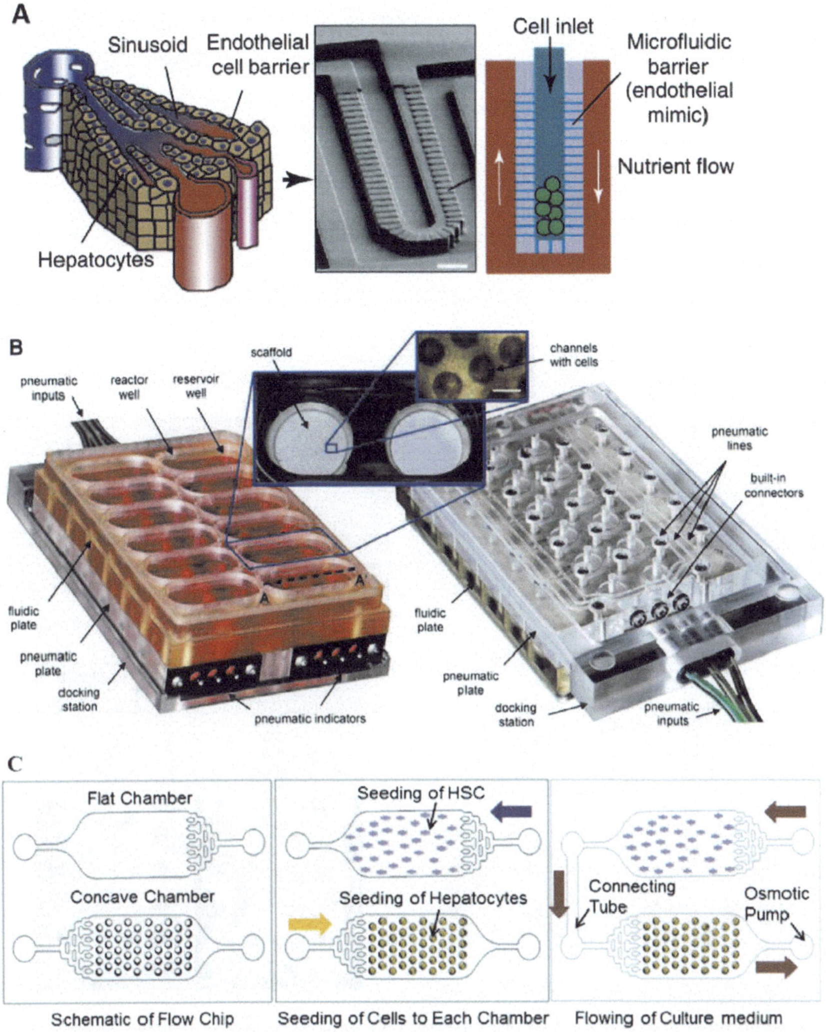

Fig. 5 Liver-on-a-chip platform. (A) A microfluidic liver on a chip consisting of a liver cell chamber in the center surrounded by a nutrient flow chamber to mimic the structure of the liver (Adapted from [29]). (B) Microengineered perfusion of the hepatic culture (Adapted from [58]). (C) 3D microfluidic liver on a chip consists of two different chambers connected with a tube (Adapted from [57]).

3.4 *Vessel-on-a-Chip*

Blood vessels are responsible for transporting blood through the body. Blood is carried away from the heart by arteries. Nanoparticles in the blood vessels diffuse through the capillary endothelial barrier for the exchange of water and chemicals

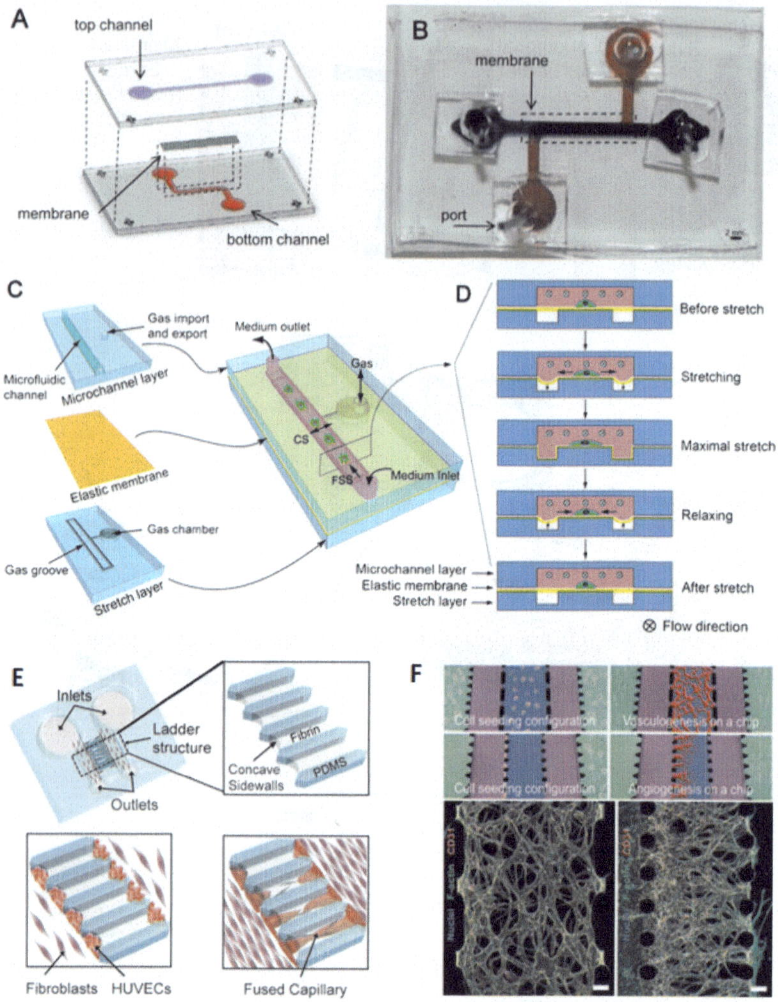

Fig. 6 Examples of vessel-on-a-chip platform. (A) Schematic of two PDMS chamber connected by a membrane and (B) the image of fabricated device (Adapted from [63]). (C) Long microfluidic channel with a consecutive flow and (D) repeatedly stretched and relaxed elastic membrane (Adapted from [64]). (E) Three-dimensional perfusable tubular capillary network in a microfluidic platform (Adapted from [65]). (F) Microvasculare network on a chip filled with fibrin gel inside the main channel (Adapted from [66]).

between tissues and blood. Finally, the blood is carried back from capillaries to the heart by the veins. It is important to study the interaction of these carriers with the vasculature. Studies have investigated the geometric characteristic of vasculature, which include simple channels [67–69], complex parameters [70], and the accumulation of micro- and nanoparticles inside the microchannels [71]. Other important parameters to be considered are the shear stress and cyclic stretch on endothelial cells induced by the blood stream [72–75]. The vessel-on-a-chip is the microfluidic device fabricated to mimic the blood vessels. In most studies, two parallel-flow PDMS chambers separated from a membrane were fabricated to culture vascular endothelial cells on the membrane. Microfluidic chambers were connected to syringe pumps which control flow rates (Fig. 6A and 6B). Shear stresses generated by the flow rate and applied to endothelial cells can be studied [63].

Vascular endothelial cells under cyclic stretching are one of important factors to study [64]. As shown in Fig. 6C and 6D, upper microfluidic channel layer and bottom groove layer were separated by an elastic membrane that applies cyclic stretching strain to cells. A groove layer was attached under the elastic membrane and is connected to a vacuum pump that applies suction pressure to stretch the elastic membrane. Cells were cultured on top of the membrane and fluid flow is passed through the upper channel layer while vacuum pressure is applied to the stretched cells on the elastic membrane. The cells cultured on the membrane were cyclically stretched and relaxed.

Yeon et al. [65] developed a perusable microvessel on a microfluidic chip to form and characterize the capillary network of human umbilical vein endothelial cells (HUVECs) (Fig. 6E). The microfluidic device consisted of two main channels and eight bridge channels. The cells, media and nutrient were perfused through the main channel and eight bridge channels which are filled with fibrin hydrogels. HUVECs were passed through the main channels and attached on the both sides of the bridge channels due to the fibrin gel. Then, fibroblasts were injected into the inlets to co-culture with HUVECs and stimulate angiogenic process. As the angiogenic process, HUVECs migrated into the fibrin gel and the endothelial sprouts of both sides are finally connected after 4 days. Kim et al. [66] also demonstrated the microvessel network structure on a chip (Fig. 6F). The PDMS chip with one main channel and two side channels is fabricated by the conventional replica molding technique. Main channel (blue color) was either filled with a fibrin gel and endothelial cell (EC) mixture for vasculogenic process or filled with the fibrin gel while ECs were attached to the wall of the fibrin gel for angiogenic process. Lung fibroblasts cu-cultured with HUVECs at the side (green color) channels stimulated the formation of vessel networks.

3.5 Tumor-on-a-Chip

The tumor-on-a-chip is one of the major areas of interest in organ-on-a-chip research. The emigration and vascularization mechanisms of tumor cells still remain unclear. Although nanoparticles can be used to deliver drugs to targeted cancer cells, the process and side effects of the nanoparticles are difficult to analyze. Existing tumor-on-a-chip platforms can be divided into three sub-categories. The first monitoring the emigration and the vascularization process of tumor cells, especially antiangiogenesis [76]. The second sub-category aims to screen the effect of nanoparticles encapsulating drugs to the cancer cells [28]. The third subcategory focuses on detecting cancer markers from blood for early stage cancer detection [49, 77–79].

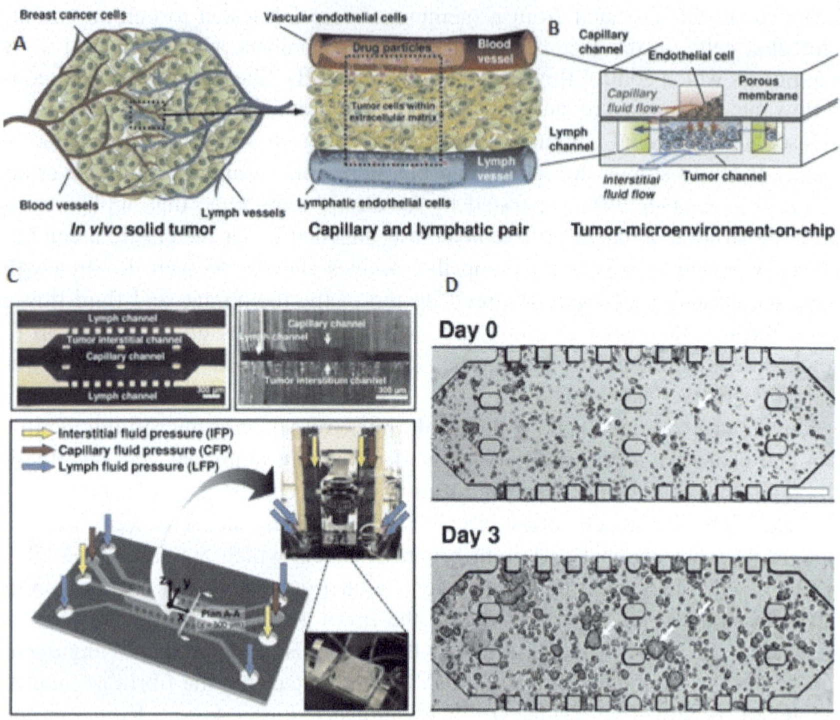

Fig. 7 An example of tumor-on-a-chip devices (Adapted from [28]). (A) *In vivo* breast tumor microenvironment was formed by vascular endothelial cells, lymphatic endothelial cells and tumor cells. (B) Schematic of the tumor-microenvironment-on-chip. (C) Image of fabricated prototype and schematic of the perfusion setup of the tumor-on-a-chip device. (D) Tumor tissue equivalent (1× 107 cells/ml and 6 mg-collagen/ml) was created on the tumor-on-a-chip device and cultured for 3 days. At Day 0, tumor cells, indicated by arrows, loosely aggregated within the collagen matrix. The cells proliferated and bound tightly at Day 3. Scale bar is 300 μm.

In the beginning, microfluidic devices were used to generate 3D tumor spheroids with a controllable density and diameter [80, 81]. Microfluidic devices have also been used to sort out different cancer cells [80] and balance the cell density of multi-inflow [82]. To investigate the interaction between normal cells and tumor cells, several 3D platforms have been developed to co-culture normal cells and tumor cells [48]. Although 3D platforms cannot be considered as organ-on-a-chip devices, they enable further investigate *in vitro* tumor cell behavior.

The most common tumor-on-a-chip platforms are the cancer marker detecting devices. Usually, the microfluidics devices with specially design channel structures can collect circulating tumor cells [49, 77–79]. Nanoparticle screening platforms to create a microenvironment that mimics tumor tissues and nearby vessels were also used [28] in drug screening and delivery systems. A tumor-on-a-chip device mimicking breast cancer is schematized in Fig. 7A and 7B. A group of solid tumor cells and a pair of capillary and lymphatic vessels were mimicked in the 3D structure. The top channel (red channel) mimicked the capillary with endothelium using a monolayer of endothelial cells on a porous membrane. The bottom layer had a center channel mimicking breast tumor cells as well as two side channels mimicking the lymphatic system. Using this platform, the transport of nanoparticles and the dynamic interaction between cancer cells and nanoparticles can be screened lively. During experiments, nanoparticles were carried in the top channel and then pass through the porous membrane to the tumor channel. After the nanoparticles interact with tumors, the byproducts were observed in the side lymphatic channel. This tumor-on-a-chip device provided a visualized way to study the tumor-nanoparticle reaction process. The authors investigated the effects of many parameters, including membrane pore size and fluid pressure to the efficiency of nanoparticles for delivering drugs, providing valuable guidance to design nanoparticles. For example, the effect of nanoparticle size on its transport process was studied. The relationship between the concentration of nanoparticles, the size of nanoparticles, its distance from the boundary, and the infusion time could be observed accurately and clearly. The fluorescent images and visual graphs were impossible to achieve without using organ-on-a-chip technology.

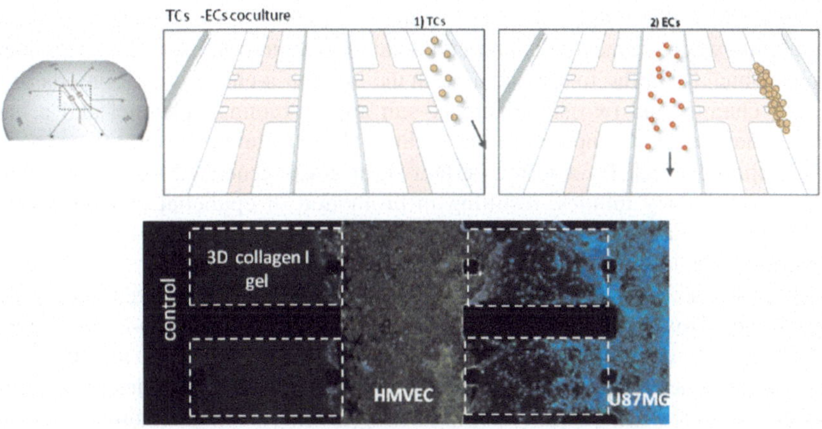

Fig. 8 Schematic of tumor-vessel co-culture devices and fluorescent images showing the invasion process of Brain cancer cells (U87MG). The cancer cells invaded into the collagen type I gel and was approaching the human microvascular endothelial cells (HMVEC).[76]

Compared to the indirect analysis used in traditional method, organ-on-a-chip device offers a faster and much more intuitionistic way to analyze the effects of parameters *in vitro*, helping researcher to understand the reaction process better. Tumor-on-a-chip system will not only accelerate the design of nanoparticles as drug delivery carries, but also improve the reliability and viability of the nanoparticles. In addition to screen the interaction between tumor cells and nanoparticles, the transport phenomena of nanoparticles in vessel tissues has been investigated to help create better nanoparticles [83].

Platforms for screening vascularization are still at the early stage. No long-term study of the formation of blood vessels has been done in microfluidic channels. To investigate short-term relationship between cancer cells and blood vessels, a microfluidic platform shown in Fig. 8 was used [76]. This platform consists of three channels. Human microvascular endothelial cells (HMVEC) were seeded on the central channel to form a vessel; and brain cancer cells, U87MG, were seeded on the side channels. The side channels are connected to the central channel by pathways filled with collagen type I gel. After culturing, the cancer cells accumulate in the nearest pathway to the vessels. Some parts of the collagen have been invaded and the cancer cells were approached to the vessel in the central channel. Aided by organ-on-a-chip devices, one can visualize and record the fluorescent images regarding the invasion process of the cancer cells. The disease-on-a-chip platform has proven to be an excellent *in vitro* tool to examine how cancer cells develop in tissues [84].

3.6 Other Organs-on-a-Chip

In addition to major organs, such as a heart, lung, and liver, kidney- [30], splenon- [85, 86] and breast- [87] on-a-chip devices have been developed. Fig. 9 shows the schematics of these three devices. Kidney-on-a-chip devices include a porous membrane where kidney cells and epithelial are cultured in each side. This membrane, which is similar to the one used in the lung-on-a-chip device, consists of the central channel and two sub-channels—an apical 'luminal' channel and a basal 'interstitial' space. Compared to the traditional microfluidic system, the exposure of the epithelial cell layer to the certain shear stress generated by inflow mimics the *in vivo* kidney tubules, resulting in promotion of epithelial cell polarization and primary cilia formation. This platform is useful for the study of kidney toxicity during drug development.

Breast-on-a-chip and splenon-on-a-chip systems have also utilized microfluidic technology. However, their capacities are limited. The breast-on-a-chip devices only mimic the distribution of anatomical vessels of a breast rather than the whole microenvironment containing glandular, fatty, and vascular tissues. The device can be integrated with other tumor-on-a-chip devices to further mimic tumor development in the breast. The splenon-on-a-chip system just simulates the blood filtering function of *in vivo* splenon using a microfluidic two-phase flow structure. Those chips still lack of accuracy to mimic tissue and cellular level.

Although aforementioned three organ-on-a-chip devices do not precisely mimic corresponding tissues at a microenvironmental level, such devices simulate either anatomical geometries or biological functions using microfluidic techniques. These high-level similarities between organ-on-a-chip devices and genuine organs are helpful for researchers to develop a human-on-a-chip system—a comprehensive *in vitro* system to study drug effects to the whole human body. Currently, most organ-on-a-chip devices mimic only one or two types of organ. Even those

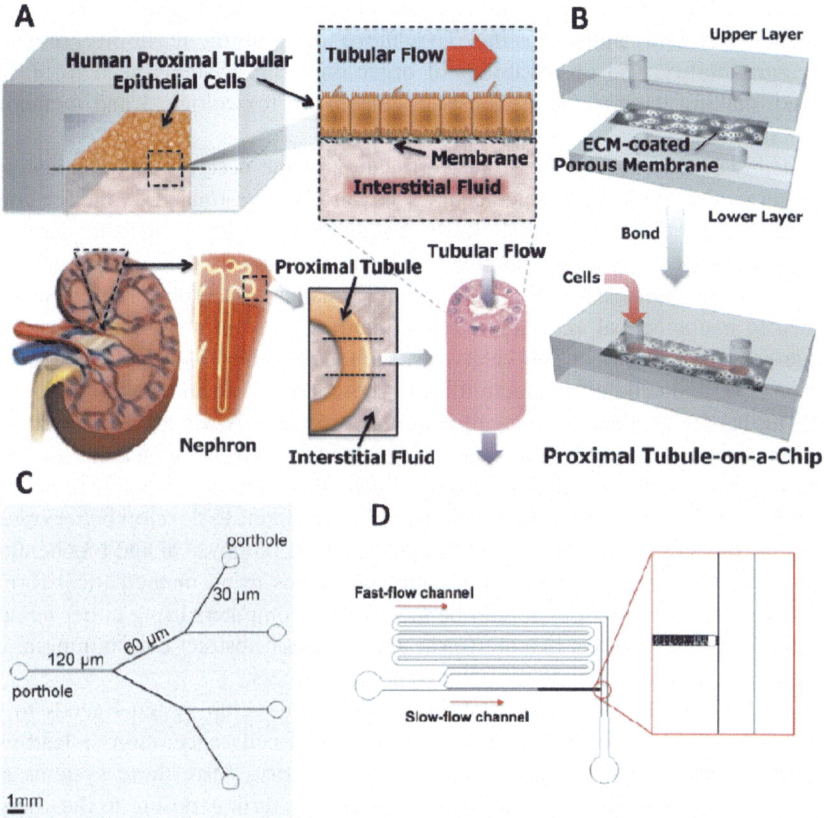

Fig. 9 Kidney-, splenon-, and breast-on-a-chip systems. (A) Conceptual design of kidney tubule-on-a-chip and (B) fabrication steps of kidney-on-a-chip (Adapted from [30]). (C) Schematic of breast-on-a-chip (Adapted from [87]). (D) Layout of splenon-on-a-chip (Adapted from [86]).

particular types of organs are valid for studying the individual organ function in cell and tissue level, the systematic effect of drugs on the human body remains unknown. Human-on-a-chip aims to mimic the whole body functions by the combination of a group of organ-on-a-chip devices. In that case, the interaction between organs can be investigated and the systematic effect of the drug to the human body may be observed more clearly.

4 Current Challenges and Future Direction

Despite the development of many organ-on-a-chip systems to functionally mimic a variety of tissues and organs in recent years, much work remains ahead to manufacture organ-on-a-chip systems. In general, the ultimate goal of organ-on-a-chip system is to build human-on-a-chip systems using human cells and tissues that are capable of replacing animal testing. To achieve this goal, the feasibility, reliability, controllability, and observability of organ-on-a-chip systems need to be improved, making them a comprehensive platform for toxicological and metabolic test of drugs.

For this goal to become feasible, verification of the design, mathematical and physical models with acceptable accuracy have to be investigated. Although computational fluid dynamics (CFD) have developed for many years and there are many CFD-based simulation models [83], the analysis of complex flows with living cells in composite cell media remains challenging. This problem is also related to mathematical analysis of cell properties. Current models just focus on single property of cells or cell behaviors, such as fusion [88–90], ECM-cell interaction [91], and cell-cell interaction [92]. Those simple models need to be developed further to simulate *in vivo* cell activities precisely. Artificial neural network and machine learning technology can be accessed to address the demand of complex models of human-on-a-chip systems. Such algorithms are also able to aid and optimize the design of human-on-a-chips. Taken together, to develop better organ-on-a-chip devices, researchers need to understand the biophysical and biochemical properties of organs and express the nature of organs using mathematical forms. Moreover, these expressions should be modeled by computers using either numerical computation methods (Monte Carlo and FEM) or abstract data mining algorithm to verify and accelerate the detailed design process.

For reliability, the long-term stability of organ-on-a-chip systems needs to be verified. It might not become a potential inducer of cell canceration or lead any undesired change in cell properties and gene integrity. Also, these systems are required to maintain its structural integrity after long-term exposure to the chemical and biological solutions and materials. Widely-used PDMS materials need to be further improved or replaced to achieve better resistance to chemicals, as well as maintain high compatibility with soft lithography techniques and biological cells [40].

For controllability, automation of both organ-on-a-chip fabrication and screening need to be standardized and optimized. Bio-computer-aided design and manufacturing (Bio-CAD and Bio-CAM) should be introduced to accelerate the design flow of organ-on-a-chip system. In addition, to simulate drug metabolism process, novel interconnection methods to integrate several organ-on-a-chip devices as well as the methods to better mimic organ's inherent functions at different levels are highly desired. Future designs of human-on-a-chip systems will require the adoption of modularized methodology—a method to rapidly build highly reliable systems with a variety of desired functions. A practical system will be realized by combining various pre-built modules. This modularized technique requires researchers in the development of organ-on-a-chip to provide well-designed libraries covering all levels and types of organs.

For observability, real-time high-efficient monitoring will remain a challenge for a long time. An ideal human-on-a-chip platform needs to include many basic sensors to record microenvironmental signals, such as pH and ion concentration.

Customized sensors to detect proteins, chemicals, and even circulating tumor cells are also highly demanded. It is difficult to develop sensors at the micro scale with high accuracy, high biocompatibility, and high compatibility with a soft lithography process. Biosensors are potential solutions for this problem [20]. However, both sensitivity and throughput rate of bio-sensors need to be improved.

In the long term, organ-on-a-chip systems in combining with tissue engineering could be more than an alternative solution for drug screening and delivery test. This approach also has the potential to be used as high-speed, highly customized, and non-invasive platform to address patient-to-patient variation of drug response. In addition to clinical applications, organ-on-a-chip systems can also play important roles in biological research. Combined with other nanofabrication technologies, an organ-on-a-chip platform can be comprehensive, highly synthesized platform for cell-, tissue-, and organ-level screening and characterization. This platform will help biologists understand the mystery of microenvironments inside human beings, animals, and plants.

In conclusion, organ-on-a-chip platforms are *in vitro* microfabricated devices mimicking organ functions. These platforms could be the promising tools for discovering drugs and investing tissue functions at the micro-scale for cellular and molecular biology research.

Acknowledgement This work was supported by Natural Sciences and Engineering Research Council of Canada Discovery Grant.

References

1. Tonkens, R.: An overview of the drug development process. Physician Exec. **31**, 48–52 (2005)
2. Djulbegovic, B., Hozo, I., Ioannidis, J.P.A.: Improving the drug development process: more not less randomized trials. JAMA **311**, 355–356 (2014). doi:10.1001/jama.2013.283742
3. Bhise, N.S., Shmueli, R.B., Sunshine, J.C., Tzeng, S.Y., Green, J.J.: Drug delivery strategies for therapeutic angiogenesis and antiangiogenesis. Expert Opin Drug Deliv. **8**, 485–504 (2011). doi:10.1517/17425247.2011.558082
4. Baharvand, H., Hashemi, S.M., Kazemi Ashtiani, S., Farrokhi, A.: Differentiation of human embryonic stem cells into hepatocytes in 2D and 3D culture systems in vitro. Int. J. Dev. Biol. **50**, 645–652 (2006). doi:10.1387/ijdb.052072hb
5. Tibbitt, M.W., Anseth, K.S.: Hydrogels as extracellular matrix mimics for 3D cell culture. Biotechnol. Bioeng. **103**, 655–663 (2009). doi:10.1002/bit.22361
6. Bhise, N.S., Gray, R.S., Sunshine, J.C., Htet, S., Ewald, A.J., Green, J.J.: The relationship between terminal functionalization and molecular weight of a gene delivery polymer and transfection efficacy in mammary epithelial 2-D cultures and 3-D organotypic cultures. Biomaterials **31**, 8088–8096 (2010). doi:10.1016/j.biomaterials.2010.07.023

7. Tung, Y.-C., Hsiao, A.Y., Allen, S.G., Torisawa, Y., Ho, M., Takayama, S.: High-throughput 3D spheroid culture and drug testing using a 384 hanging drop array. Analyst **136**, 473–478 (2011). doi:10.1039/c0an00609b
8. Wilkening, S., Stahl, F., Bader, A.: Comparison of primary human hepatocytes and hepatoma cell line HepG2 with regard to their biotransformation properties. Drug Metab. Dispos. **31**, 1035–1042 (2003). doi:10.1124/dmd.31.8.1035
9. Karp, J.M., Yeh, J., Eng, G., Fukuda, J., Blumling, J., Suh, K.-Y., Cheng, J., Mahdavi, A., Borenstein, J., Langer, R., Khademhosseini, A.: Controlling size, shape and homogeneity of embryoid bodies using poly(ethylene glycol) microwells. Lab. Chip. **7**, 786–794 (2007). doi:10.1039/b705085m
10. Patel, N.G., Zhang, G.: Stacked stem cell sheets enhance cell-matrix interactions. Organogenesis **10**, 170–176 (2014). doi:10.4161/org.28990
11. Karp, J.M., Yeh, J., Eng, G., Fukuda, J., Blumling, J., Suh, K.-Y., Cheng, J., Mahdavi, A., Borenstein, J., Langer, R., Khademhosseini, A.: Controlling size, shape and homogeneity of embryoid bodies using poly(ethylene glycol) microwells. Lab. Chip. **7**, 786–794 (2007). doi:10.1039/b705085m
12. Papenburg, B.J., Liu, J., Higuera, G.A., Barradas, A.M.C., de Boer, J., van Blitterswijk, C.A., Wessling, M., Stamatialis, D.: Development and analysis of multi-layer scaffolds for tissue engineering. Biomaterials **30**, 6228–6239 (2009). doi:10.1016/j.biomaterials.2009.07.057
13. Patel, N.G., Zhang, G.: Stacked stem cell sheets enhance cell-matrix interactions. Organogenesis **10**, 170–176 (2014). doi:10.4161/org.28990
14. Douville, N.J., Zamankhan, P., Tung, Y.-C., Li, R., Vaughan, B.L., Tai, C.-F., White, J., Christensen, P.J., Grotberg, J.B., Takayama, S.: Combination of fluid and solid mechanical stresses contribute to cell death and detachment in a microfluidic alveolar model. Lab. Chip. **11**, 609–619 (2011). doi:10.1039/c0lc00251h
15. Huh, D., Matthews, B.D., Mammoto, A., Montoya-Zavala, M., Hsin, H.Y., Ingber, D.E.: Reconstituting organ-level lung functions on a chip. Science **328**, 1662–1668 (2010). doi:10.1126/science.1188302
16. Mahto, S.K., Yoon, T.H., Rhee, S.W.: A new perspective on in vitro assessment method for evaluating quantum dot toxicity by using microfluidics technology. Biomicrofluidics **4**, 1–9 (2010). doi:10.1063/1.3486610
17. Cho, E.C., Zhang, Q., Xia, Y.: The effect of sedimentation and diffusion on cellular uptake of gold nanoparticles. Nat. Nanotechnol. **6**, 385–391 (2011). doi:10.1038/nnano.2011.58
18. Stirland, D.L., Nichols, J.W., Miura, S., Bae, Y.H.: Mind the gap: a survey of how cancer drug carriers are susceptible to the gap between research and practice. J. Control Release **172**, 1045–1064 (2013). doi:10.1016/j.jconrel.2013.09.026
19. Lammers, T., Kiessling, F., Hennink, W.E., Storm, G.: Drug targeting to tumors: principles, pitfalls and (pre-) clinical progress. J. Control Release **161**, 175–187 (2012). doi:10.1016/j.jconrel.2011.09.063
20. Bhise, N.S., Ribas, J., Manoharan, V., Zhang, Y.S., Polini, A., Massa, S., Dokmeci, M.R., Khademhosseini, A.: Organ-on-a-chip platforms for studying drug delivery systems. J. Control Release **190**, 82–93 (2014). doi:10.1016/j.jconrel.2014.05.004
21. U.S.Food and Drug Administration: Guidance for industry safety testing of drug guidance for industry safety testing of drug metabolites (2008)
22. Fröde, T.S., Medeiros, Y.S.: Animal models to test drugs with potential antidiabetic activity. J. Ethnopharmacol. **115**, 173–183 (2008). doi:10.1016/j.jep.2007.10.038

23. Moraes, C., Mehta, G., Lesher-Perez, S.C., Takayama, S.: Organs-on-a-chip: a focus on compartmentalized microdevices. Ann. Biomed Eng. **40**, 1211–1227 (2012). doi:10.1007/s10439-011-0455-6
24. Zimmerlin, L., Donnenberg, A.D., Rubin, J.P., Basse, P., Landreneau, R.J., Donnenberg, V.S.: Regenerative therapy and cancer: in vitro and in vivo studies of the interaction between adipose-derived stem cells and breast cancer cells from clinical isolates. Tissue Eng. Part A **17**, 93–106 (2011). doi:10.1089/ten.TEA.2010.0248
25. Duffy, D.C., McDonald, J.C., Schueller, O.J., Whitesides, G.M.: Rapid prototyping of microfluidic systems in Poly(dimethylsiloxane). Anal. Chem. **70**, 4974–4984 (1998). doi:10.1021/ac980656z
26. Leclerc, E., Furukawa, K.S., Miyata, F., Sakai, Y., Ushida, T., Fujii, T.: Fabrication of microstructures in photosensitive biodegradable polymers for tissue engineering applications. Biomaterials **25**, 4683–4690 (2004). doi:10.1016/j.biomaterials.2003.10.060
27. Huh, D., Torisawa, Y., Hamilton, G.A., Kim, H.J., Ingber, D.E.: Microengineered physiological biomimicry: organs-on-chips. Lab. Chip. **12**, 2156–2164 (2012). doi:10.1039/c2lc40089h
28. Kwak, B., Ozcelikkale, A., Shin, C.S., Park, K., Han, B.: Simulation of complex transport of nanoparticles around a tumor using tumor-microenvironment-on-chip. J. Control Release **194**, 157–167 (2014). doi:10.1016/j.jconrel.2014.08.027
29. Lee, P.J., Hung, P.J., Lee, L.P.: An artificial liver sinusoid with a microfluidic endothelial-like barrier for primary hepatocyte culture. Biotechnol. Bioeng. **97**, 1340–1346 (2007). doi:10.1002/bit.21360
30. Jang, K.-J., Mehr, A.P., Hamilton, G.A., McPartlin, L.A., Chung, S., Suh, K.-Y., Ingber, D.E.: Human kidney proximal tubule-on-a-chip for drug transport and nephrotoxicity assessment. Integr. Biol. (Camb) **5**, 1119–1129 (2013). doi:10.1039/c3ib40049b
31. Huh, D., Hamilton, G.A., Ingber, D.E.: From 3D cell culture to organs-on-chips. Trends Cell Biol. **21**, 745–754 (2011). doi:10.1016/j.tcb.2011.09.005
32. Ghaemmaghami, A.M., Hancock, M.J., Harrington, H., Kaji, H., Khademhosseini, A.: Biomimetic tissues on a chip for drug discovery. Drug Discov. Today **17**, 173–181 (2012). doi:10.1016/j.drudis.2011.10.029
33. Jiang, B., Zheng, W., Zhang, W., Jiang, X.: Organs on microfluidic chips: A mini review. Sci. China Chem. **57**, 356–364 (2013). doi:10.1007/s11426-013-4971-0
34. Therriault, D., White, S.R., Lewis, J.A.: Chaotic mixing in three-dimensional microvascular networks fabricated by direct-write assembly. Nat. Mater **2**, 265–271 (2003). doi:10.1038/nmat863
35. Bernard, A., Renault, J.P., Michel, B., Bosshard, H.R., Delamarche, E.: Microcontact printing of proteins. Adv. Mater. **12**, 1067–1070 (2000). doi:10.1002/1521-4095(200007)12:14<1067:Aid-Adma1067>3.0.Co;2-M
36. Lee, H., Chung, M., Jeon, N.L.: Microvasculature: An essential component for organ-on-chip systems. MRS Bull **39**, 51–59 (2014). doi:10.1557/mrs.2013.286
37. Gates, B.D., Xu, Q., Stewart, M., Ryan, D., Willson, C.G., Whitesides, G.M.: New approaches to nanofabrication: Molding, printing, and other techniques. Chem. Rev. **105**, 1171–1196 (2005). doi:10.1021/cr030076o
38. Rolland, J.P., Van Dam, R.M., Schorzman, D.A., Quake, S.R., DeSimone, J.M.: Solvent-resistant photocurable "Liquid Teflon" for microfluidic device fabrication. J. Am. Chem. Soc. **126**, 2322–2323 (2004). doi:10.1021/ja031657y

39. Rolland, J.P., Hagberg, E.C., Denison, G.M., Carter, K.R., De Simone, J.M.: High-resolution soft lithography: Enabling materials for nanotechnologies. Angew. Chemie. – Int. Ed. **43**, 5796–5799 (2004). doi:10.1002/anie.200461122
40. Berthier, E., Young, E.W.K., Beebe, D.: Engineers are from PDMS-land, Biologists are from Polystyrenia. Lab. Chip. **12**, 1224–1237 (2012). doi:10.1039/c2lc20982a
41. Bhatia, S.N., Ingber, D.E.: Microfluidic organs-on-chips. Nat. Biotechnol. **32**, 760–772 (2014). doi:10.1038/nbt.2989
42. Prentice-Mott, H.V., Chang, C.-H., Mahadevan, L., Mitchison, T.J., Irimia, D., Shah, J.V.: Biased migration of confined neutrophil-like cells in asymmetric hydraulic environments. Proc. Natl. Acad. Sci. USA **110**, 21006–21011 (2013). doi:10.1073/pnas.1317441110
43. Radisic, M., Deen, W., Langer, R., Vunjak-Novakovic, G.: Mathematical model of oxygen distribution in engineered cardiac tissue with parallel channel array perfused with culture medium containing oxygen carriers. Am. J. Physiol. Heart Circ. Physiol. **288**, H1278–H1289 (2005). doi:10.1152/ajpheart.00787.2004
44. Cimetta, E., Cannizzaro, C., James, R., Biechele, T., Moon, R.T., Elvassore, N., Vunjak-Novakovic, G.: Microfluidic device generating stable concentration gradients for long term cell culture: application to Wnt3a regulation of β-catenin signaling. Lab. Chip. **10**, 3277–3283 (2010). doi:10.1039/c0lc00033g
45. Zhao, L., Wang, Z., Fan, S., Meng, Q., Li, B., Shao, S., Wang, Q.: Chemotherapy resistance research of lung cancer based on micro-fluidic chip system with flow medium. Biomed. Microdevices **12**, 325–332 (2010). doi:10.1007/s10544-009-9388-3
46. Huang, C.-W., Cheng, J.-Y., Yen, M.-H., Young, T.-H.: Electrotaxis of lung cancer cells in a multiple-electric-field chip. Biosens Bioelectron **24**, 3510–3516 (2009). doi:10.1016/j.bios.2009.05.001
47. Xu, Z., Gao, Y., Hao, Y., Li, E., Wang, Y., Zhang, J., Wang, W., Gao, Z., Wang, Q.: Application of a microfluidic chip-based 3D co-culture to test drug sensitivity for individualized treatment of lung cancer. Biomaterials **34**, 4109–4117 (2013). doi:10.1016/j.biomaterials.2013.02.045
48. Hou, H.-S., Tsai, H.-F., Chiu, H.-T., Cheng, J.-Y.: Simultaneous chemical and electrical stimulation on lung cancer cells using a multichannel-dual-electric-field chip. Biomicrofluidics **8**, 052007 (2014). doi:10.1063/1.4896296
49. Huang, T., Jia, C.-P., Jun-Yang, Sun W.-J., Wang, W.-T., Zhang, H.-L., Cong, H., Jing, F.-X., Mao, H.-J., Jin, Q.-H., Zhang, Z., Chen, Y.-J., Li, G., Mao, G.-X., Zhao, J.-L.: Highly sensitive enumeration of circulating tumor cells in lung cancer patients using a size-based filtration microfluidic chip. Biosens Bioelectron **51**, 213–218 (2014). doi:10.1016/j.bios.2013.07.044
50. Long, C., Finch, C., Esch, M., Anderson, W., Shuler, M., Hickman, J.: Design optimization of liquid-phase flow patterns for microfabricated lung on a chip. Ann. Biomed. Eng. **40**, 1255–1267 (2012). doi:10.1007/s10439-012-0513-8
51. Shin, S.R., Jung, S.M., Zalabany, M., Kim, K., Zorlutuna, P., Kim, S.B., Nikkhah, M., Khabiry, M., Azize, M., Kong, J., Wan, K.T., Palacios, T., Dokmeci, M.R., Bae, H., Tang, X., Khademhosseini, A.: Carbon-nanotube-embedded hydrogel sheets for engineering cardiac constructs and bioactuators. ACS Nano **7**, 2369–2380 (2013). doi:10.1021/nn305559j
52. Grosberg, A., Alford, P.W., McCain, M.L., Parker, K.K.: Ensembles of engineered cardiac tissues for physiological and pharmacological study: heart on a chip. Lab. Chip. **11**, 4165–4173 (2011). doi:10.1039/c1lc20557a

53. Chiu, L.L.Y., Montgomery, M., Liang, Y., Liu, H., Radisic, M.: Perfusable branching microvessel bed for vascularization of engineered tissues. Proc. Natl. Acad. Sci. USA **109**, E3414–E3423 (2012). doi:10.1073/pnas.1210580109
54. Werdich, A.A., Lima, E.A., Ivanov, B., Ges, I., Anderson, M.E., Wikswo, J.P., Baudenbacher, F.J.: A microfluidic device to confine a single cardiac myocyte in a sub-nanoliter volume on planar microelectrodes for extracellular potential recordings. Lab. Chip. **4**, 357–362 (2004). doi:10.1039/b315648f
55. Cheng, W., Klauke, N., Sedgwick, H., Smith, G.L., Cooper, J.M.: Metabolic monitoring of the electrically stimulated single heart cell within a microfluidic platform. Lab. Chip. **6**, 1424–1431 (2006). doi:10.1039/b608202e
56. Ohashi, K., Yokoyama, T., Yamato, M., Kuge, H., Kanehiro, H., Tsutsumi, M., Amanuma, T., Iwata, H., Yang, J., Okano, T., Nakajima, Y.: Engineering functional two- and three-dimensional liver systems in vivo using hepatic tissue sheets. Nat. Med. **13**, 880–885 (2007). doi:10.1038/nm1576
57. Lee, S.-A., No, D.Y., Kang, E., Ju, J., Kim, D.-S., Lee, S.-H.: Spheroid-based three-dimensional liver-on-a-chip to investigate hepatocyte-hepatic stellate cell interactions and flow effects. Lab. Chip. **13**, 3529–3537 (2013). doi:10.1039/c3lc50197c
58. Domansky, K., Inman, W., Serdy, J., Dash, A., Lim, M.H.M., Griffith, L.G.: Perfused multiwell plate for 3D liver tissue engineering. Lab. Chip. **10**, 51–58 (2010). doi:10.1039/b913221j
59. Toh, Y.-C., Lim, T.C., Tai, D., Xiao, G., van Noort, D., Yu, H.: A microfluidic 3D hepatocyte chip for drug toxicity testing. Lab. Chip. **9**, 2026–2035 (2009). doi:10.1039/b900912d
60. Toh, Y.-C., Zhang, C., Zhang, J., Khong, Y.M., Chang, S., Samper, V.D., van Noort, D., Hutmacher, D.W., Yu, H.: A novel 3D mammalian cell perfusion-culture system in microfluidic channels. Lab. Chip. **7**, 302–309 (2007). doi:10.1039/b614872g
61. Goral, V.N., Hsieh, Y.-C., Petzold, O.N., Clark, J.S., Yuen, P.K., Faris, R.A.: Perfusion-based microfluidic device for three-dimensional dynamic primary human hepatocyte cell culture in the absence of biological or synthetic matrices or coagulants. Lab. Chip. **10**, 3380–3386 (2010). doi:10.1039/c0lc00135j
62. Lee, J., Kim, S.H., Kim, Y.C., Choi, I., Sung, J.H.: Fabrication and characterization of microfluidic liver-on-a-chip using microsomal enzymes. Enzyme. Microb. Technol. **53**, 159–164 (2013). doi:10.1016/j.enzmictec.2013.02.015
63. Srigunapalan, S., Lam, C., Wheeler, A.R., Simmons, C.A.: A microfluidic membrane device to mimic critical components of the vascular microenvironment. Biomicrofluidics **5**, 13409 (2011). doi:10.1063/1.3530598
64. Zheng, W., Jiang, B., Wang, D., Zhang, W., Wang, Z., Jiang, X.: A microfluidic flow-stretch chip for investigating blood vessel biomechanics. Lab. Chip. **12**, 3441–3450 (2012). doi:10.1039/c2lc40173h
65. Yeon, J.H., Ryu, H.R., Chung, M., Hu, Q.P., Jeon, N.L.: In vitro formation and characterization of a perfusable three-dimensional tubular capillary network in microfluidic devices. Lab. Chip. **12**, 2815 (2012). doi:10.1039/c2lc40131b
66. Kim, S., Lee, H., Chung, M., Jeon, N.L.: Engineering of functional, perfusable 3D microvascular networks on a chip. Lab. Chip. **13**, 1489–1500 (2013). doi:10.1039/c3lc41320a
67. Kusunose, J., Zhang, H., Gagnon, M.K.J., Pan, T., Simon, S.I., Ferrara, K.W.: Microfluidic system for facilitated quantification of nanoparticle accumulation to cells under laminar flow. Ann. Biomed. Eng. **41**, 89–99 (2013). doi:10.1007/s10439-012-0634-0

68. Korin, N., Kanapathipillai, M., Matthews, B.D., Crescente, M., Brill, A., Mammoto, T., Ghosh, K., Jurek, S., Bencherif, S.A., Bhatta, D., Coskun, A.U., Feldman, C.L., Wagner, D.D., Ingber, D.E.: Shear-activated nanotherapeutics for drug targeting to obstructed blood vessels. Science **337**, 738–742 (2012). doi:10.1126/science.1217815
69. Kim, D., Finkenstaedt-Quinn, S., Hurley, K.R., Buchman, J.T., Haynes, C.L.: On-chip evaluation of platelet adhesion and aggregation upon exposure to mesoporous silica nanoparticles. Analyst **139**, 906–913 (2014). doi:10.1039/c3an01679j
70. Rosano, J.M., Tousi, N., Scott, R.C., Krynska, B., Rizzo, V., Prabhakarpandian, B., Pant, K., Sundaram, S., Kiani, M.F.: A physiologically realistic in vitro model of microvascular networks. Biomed. Microdevices **11**, 1051–1057 (2009). doi:10.1007/s10544-009-9322-8
71. Namdee, K., Thompson, A.J., Charoenphol, P., Eniola-Adefeso, O.: Margination propensity of vascular-targeted spheres from blood flow in a microfluidic model of human microvessels. Langmuir **29**, 2530–2535 (2013). doi:10.1021/la304746p
72. Samuel, S.P., Jain, N., O'Dowd, F., Paul, T., Kashanin, D., Gerard, V.A., Gun'ko, Y.K., Prina-Mello, A., Volkov, Y.: Multifactorial determinants that govern nanoparticle uptake by human endothelial cells under flow. Int. J. Nanomedicine **7**, 2943–2956 (2012). doi:10.2147/IJN.S30624
73. Lamberti, G., Tang, Y., Prabhakarpandian, B., Wang, Y., Pant, K., Kiani, M.F., Wang, B.: Adhesive interaction of functionalized particles and endothelium in idealized microvascular networks. Microvasc. Res. **89**, 107–114 (2013). doi:10.1016/j.mvr.2013.03.007
74. Tesfamariam, B., DeFelice, A.F.: Endothelial injury in the initiation and progression of vascular disorders. Vascul. Pharmacol. **46**, 229–237 (2007). doi:10.1016/j.vph.2006.11.005
75. Li, Y.-S.J., Haga, J.H., Chien, S.: Molecular basis of the effects of shear stress on vascular endothelial cells. J. Biomech. **38**, 1949–1971 (2005). doi:10.1016/j.jbiomech.2004.09.030
76. Zervantonakis, I.K., Kothapalli, C.R., Chung, S., Sudo, R., Kamm, R.D.: Microfluidic devices for studying heterotypic cell-cell interactions and tissue specimen cultures under controlled microenvironments. Biomicrofluidics **5**, 13406 (2011). doi:10.1063/1.3553237
77. Hyun, K.-A., Lee, T.Y., Jung, H.-I.: Negative enrichment of circulating tumor cells using a geometrically activated surface interaction chip. Anal. Chem. **85**, 4439–4445 (2013). doi:10.1021/ac3037766
78. Li, P., Stratton, Z.S., Dao, M., Ritz, J., Huang, T.J.: Probing circulating tumor cells in microfluidics. Lab. Chip. **13**, 602–609 (2013). doi:10.1039/c2lc90148j
79. Yamamura, S., Yatsushiro, S., Abe, K., Baba, Y., Kataoka, M.: Development of a cell microarray chip for detection of circulating tumor cells. J. Phys. Conf. Ser. **352**, 012041 (2012). doi:10.1088/1742-6596/352/1/012041
80. Alshareef, M., Metrakos, N., Juarez Perez, E., Azer, F., Yang, F., Yang, X., Wang, G.: Separation of tumor cells with dielectrophoresis-based microfluidic chip. Biomicrofluidics **7**, 11803 (2013). doi:10.1063/1.4774312
81. Sun, D., Lu, J., Chen, Z., Yu, Y., Li, Y.: A novel three-dimensional microfluidic platform for on chip multicellular tumor spheroid formation and culture. Microfluid Nanofluidics **17**, 831–842 (2014). doi:10.1007/s10404-014-1373-3
82. Elliott, N.T., Yuan, F.: A microfluidic system for investigation of extravascular transport and cellular uptake of drugs in tumors. Biotechnol. Bioeng. **109**, 1326–1335 (2012). doi:10.1002/bit.24397

83. Tan, J., Shah, S., Thomas, A., Ou-Yang, H.D., Liu, Y.: The influence of size, shape and vessel geometry on nanoparticle distribution. Microfluid Nanofluidics **14**, 77–87 (2013). doi:10.1007/s10404-012-1024-5
84. Vidi, P.-A., Maleki, T., Ochoa, M., Wang, L., Clark, S.M., Leary, J.F., Lelièvre, S.A.: Disease-on-a-chip: mimicry of tumor growth in mammary ducts. Lab. Chip. **14**, 172–177 (2014). doi:10.1039/c3lc50819f
85. Duque-Muñoz, L., Aguirre-Echeverry, C.A., Castellanos-Domínguez, G.: EEG rhythm analysis using stochastic relevance. IFMBE Proc. **41**, 658–661 (2014). doi:10.1007/978-3-319-00846-2
86. Rigat-Brugarolas, L.G., Elizalde-Torrent, A., Bernabeu, M., De Niz, M., Martin-Jaular, L., Fernandez-Becerra, C., Homs-Corbera, A., Samitier, J., del Portillo, H.A.: A functional microengineered model of the human splenon-on-a-chip. Lab. Chip. **14**, 1715–1724 (2014). doi:10.1039/c3lc51449h
87. Grafton, M.M.G., Wang, L., Vidi, P.-A., Leary, J., Lelièvre, S.A.: Breast on-a-chip: mimicry of the channeling system of the breast for development of theranostics. Integr. Biol. (Camb) **3**, 451–459 (2011). doi:10.1039/c0ib00132e
88. Yang, X., Mironov, V., Wang, Q.: Modeling fusion of cellular aggregates in biofabrication using phase field theories. J. Theor. Biol. **303**, 110–118 (2012). doi:10.1016/j.jtbi.2012.03.003
89. Yang, X., Sun, Y., Wang, Q.: A phase field approach for multicellular aggregate fusion in biofabrication. J. Biomech. Eng. **135**, 71005 (2013). doi:10.1115/1.4024139
90. Thomas, G.L., Mironov, V., Nagy-Mehez, A., Mombach, J.C.M.: Dynamics of cell aggregates fusion: Experiments and simulations. Phys. A Stat. Mech. its Appl. **395**, 247–254 (2014). doi:10.1016/j.physa.2013.10.037
91. Ki, C.S., Lin, T.-Y., Korc, M., Lin, C.-C.: Thiol-ene hydrogels as desmoplasia-mimetic matrices for modeling pancreatic cancer cell growth, invasion, and drug resistance. Biomaterials **35**, 9668–9677 (2014). doi:10.1016/j.biomaterials.2014.08.014
92. Abdulla, T, Imms, RA, Schleich, J-M, Summers, R: Towards multiscale systems modeling of endocardial to mesenchymal transition. In: Conf Proc. Annu Int Conf IEEE Eng Med Biol Soc IEEE Eng Med Biol Soc Annu Conf 2011, pp. 449–52 (2011). doi:10.1109/IEMBS.2011.6090062

Part IV
Biomechanics, Mechanobiology and Disease

Hemodynamics and Mechanobiology of Aortic Valve Calcification

Joan Fernandez Esmerats, Jack Heath, Amir Rezvan and Hanjoong Jo

Abstract Calcific aortic valve disease (CAVD) is a significant cause of mortality among the aging population. Because disease mechanisms are still unclear, the only current treatment option for this disease is aortic valve replacement. In order to understand the disease and develop treatment targets, we must explore the hemodynamic and mechanical forces affecting the valve, as well as the biological pathways altered by those mechanics ultimately leading to tissue calcification. The unique structure of the valve allows it to adapt to the constantly changing mechanical environment and the wide array of forces exerted upon the tissue during the cardiac cycle. CAVD results in an impairment in the structure and function of the aortic valve, resulting in pathological systemic changes. A variety of mechanosensors at the valvular endothelial cell surface are responsible for transmission of mechanical signals to alter cell phenotype and ultimate tissue function. Recently, valvular microRNAs have been discovered to be mechanosensitive, possibly regulating key signaling in the pathology of CAVD. Our understanding of the aortic valve and CAVD is evolving rapidly, and future therapies rely on further exploration of this complex disease.

The aortic valve regulates the flow between the left ventricle and the aorta (Figure 1A) supplying oxygenated blood to the body. It opens during systole as the ventricle contracts, forwarding blood to the circulatory system, and closes as the ventricle relaxes—due to the beating of the heart known as the cardiac cycle. Calcific aortic valve disease encompasses a number of clinical conditions ranging from hardening and thickening of the aortic valve (aortic valve sclerosis) to severe reduction of the aortic valve area (aortic valve stenosis), resulting in obstruction of

---

J.F. Esmerats · J. Heath · H. Jo(✉)
Department of Biomedical Engineering, Emory University and Georgia Institute of Technology, Atlanta, USA
e-mail: hjo@emory.edu

A. Rezvan
Emory School of Medicine, Atlanta, USA

© Springer International Publishing Switzerland 2016
H. Jo et al. (eds.), *Biomedical Engineering: Frontier Research and Converging Technologies*,
Biosystems & Biorobotics 9,
DOI: 10.1007/978-3-319-21813-7_11

the blood flowing from the left ventricle [1]. This disease causes more than 28,000 deaths a year and 48,000 hospitalizations in the United States. CAVD is present in 25-29% of the population over 65 and is associated with a 50% increase in myocardial infarction and cardiovascular death [2-6]. By the time it is clinically recognized the only available treatment is valve replacement or repair surgery [7]. Traditionally, open-heart surgery was required for valve replacement; however, new technologies have allowed for less invasive methods, such as transcatheter aortic valve implantation (TAVI) [8, 9]. Of the 300,000 prosthetic valves implanted worldwide in 2010, a third (100,000) were in North America [10]; projections estimate that by 2050, 850,000 implants will occur per year [11].

CAVD is associated with subclinical inflammation, which leads to thickening of the valve and fibrosis and ends with calcium deposition within the valve [12]. The region of the valve that is preferentially calcified is the fibrosa, the side facing the aorta. This side of the valve experiences higher mechanical stimuli combined with lower magnitude oscillatory shear stress. Regions of the cardiovascular system exposed to these mechanical conditions are considered especially susceptible to inflammation leading to cardiovascular disease, including during atherosclerotic plaque development [13-15]. In fact, the role of shear stress in development of valvular dysfunction is clear when considering the individual valve leaflets and the flow experienced by each. The noncoronary cusp (NCC, Figure 1B) is subjected to a lower magnitude of shear stress than the other two leaflets due to the absence of diastolic coronary flow, and this leaflet is commonly the first to exhibit calcification [16].

Early inflammation in the aortic valve endothelium is believed to occur due to a wide range of events, usually as a consequence of valvular endothelial dysfunction due to pathological flow, chronic inflammatory disease, or other inflammatory conditions [17, 18]. The thickening of the aortic valve follows the inflammatory phase and is due to lipoprotein accumulation, cellular infiltration and extracellular matrix (ECM) formation [2]. As a result of extensive endothelial inflammation in early stages of aortic valve calcification, T lymphocytes and monocytes infiltrate the tissue and are activated, releasing cytokines such at TGF-β1 or IL-1β [19-21]. These cytokines contribute in later stages of the disease by promoting ECM formation and remodeling, and at the later stages local calcification is observed as calcium phosphate crystals are deposited in the ECM (Figure 1C) [22, 23]. Calcification of the aortic valve is the result of valvular interstitial cell (VIC) differentiation into osteoblasts, characterized by an upregulation of bone-related intermediates and transcription factors, including bone morphogenetic proteins, transforming growth factor β1, osteopontin, bone sialoprotein, osteocalcin, alkaline phosphatase and Runx2 [24].

In order to understand the mechanisms leading to CAVD, this chapter explores the mechanical and biological aspects of the valvular environment. First, a general overview of the valve structure, including the crucial extracellular components, gives a picture of the complex construction allowing for the tissue's physiological function. The following discussion explores the biomechanical forces exerted on the valve, giving insight into new tissue engineering approaches useful for studying CAVD. Finally, a thorough examination of the biological pathways involved in initiation of the pathological disease, with a focus on the mechanosensory mechanisms within the endothelium, links the mechanical forces in the tissue environment to the resulting pathological inflammation and ultimate valve calcification.

1 Basic Aortic Valve Structure

The aortic valve is a complex heterogeneous tissue with three layers that define its function and behavior. In humans, the leaflets have a thickness smaller than one millimeter and are described as three separate interacting layers (as shown in Figure 1D):

1. *Fibrosa:* This layer faces the aortic side of the valve and is comprised of valvular endothelial cells (VECs, in direct contact with the blood), valvular interstitial cells (VICs, residing below the endothelial cells), a small amount of elastin fibers and circumferential type I and type III fibrillar collagen, which serves as the load-bearing layer. The collagen fibers are aligned in a circumferential manner to allow for high strain under low loads and for a higher elastic modulus when more load is applied during the cardiac cycle [25]. In porcine aortic valves, the tissue is six times stiffer in the circumferential direction than the radial direction (Figure 2A) [26, 27]; the aortic valve stretches considerably in the radial direction to ensure good coaptation in order to avoid regurgitation during the diastolic phase.
2. *Spongiosa:* This layer is placed between fibrosa and ventricularis and is comprised of VICs, a small amount of elastin fibers, glycosaminoglycans and proteoglycans. These two last components serve as lubrication for the fibrosa and ventricularis layers as they deform during the cardiac cycle. The main component of the extracellular matrix in this layer is hyaluronan, which binds large amounts of water to form a foam-like structure. During compression of the aortic valve this layer absorbs energy and helps arrange elastin fibers on the ventricularis and collagen fibers on the fibrosa [27, 28].
3. *Ventricularis:* This layer faces the ventricular side of the aortic valve and is comprised of valvular endothelial cells, VICs, collagen and radially oriented elastin fibers [5, 27, 29]. Collagen fibers are stretched during the diastolic phase and relax during the systolic phase as the valve compresses. Elastin fibers in this layer aid in fast compression during the opening phase to ensure that the valve opens and closes in a fast and consistent manner [28].

The unique structure of the valve described above and shown in Figure 1D allows for consistent long-term physiological movement due to the presence of elastin fibers throughout the tissue structure. The amount of collagen and elastin in the valve is 55% and 13% (dry weight), respectively [30], and although collagen fibrils are stiff and have a maximum strain of 2% before yielding [31], they are in a crimped and wavy configuration to allow for strain of up to 40% without yielding [26]. Elastin serves as a scaffold between the different layers and helps intertwine collagen fibers between the spongiosa and fibrosa; a "honeycomb" arrangement of elastin surrounds collagen fibers and links them within the tissue [32]. Without this interconnection, the aortic valve would be at risk of delamination and separation of its different layers as a consequence of the dynamic mechanical environment experienced throughout a cardiac cycle [33].

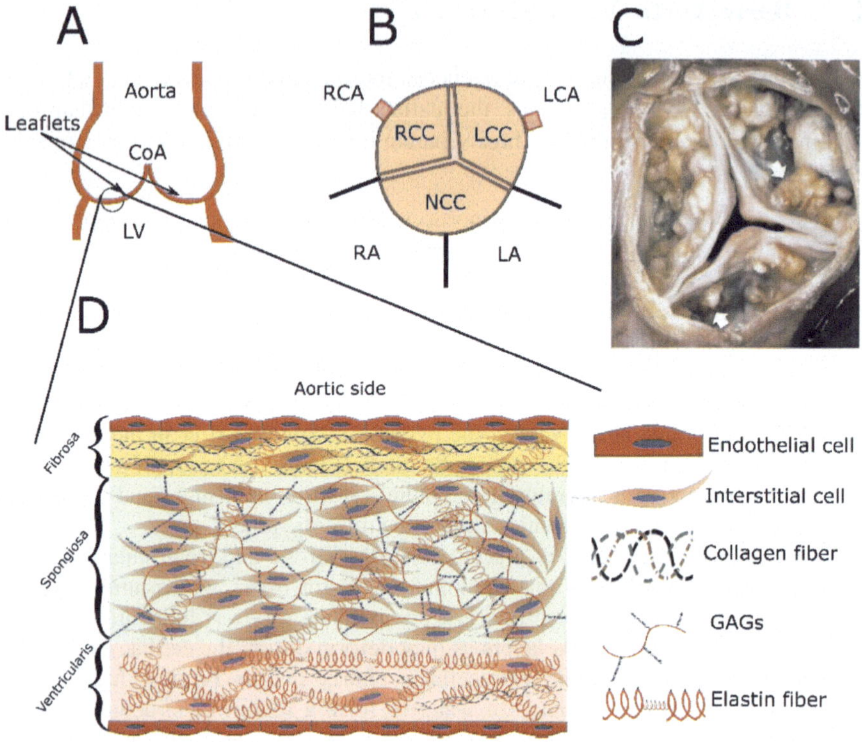

Fig. 1 Aortic valve anatomy and structure. A: The aortic valve is located between the left ventricle (LV) and the aorta, regulating systemic blood delivery from the heart. The coaptation area (CoA) denotes the joining of the three leaflets to result in a tight seal. B: Valve leaflet nomenclature. The right coronary cusp (RCC) is the leaflet nearest to the right main coronary artery (RCA). The left coronary cusp (LCC) is proximal to the left main coronary artery (LCA). The noncoronary cusp (NCC) is the third leaflet, adjacent to the right (RA) and left atria (LA). C: Calcified nodules (white arrows) deposited on the fibrosa of the aortic valve (image borrowed from [34]). This widespread calcification is usually not observed until later stages of CAVD. D: Layers of the aortic valve. The fibrosa layer faces the aorta and contains less elastin and more collagen fibers. Interstitial cells are dispersed throughout the layers of the valve excluding the endothelium. The middle layer, the spongiosa, contains most of the interstitial cells, as well as glycosaminoglycans (GAGs) and a small amount of elastin. That elastin spans the spongiosa and the ventricularis, which is the side of the valve facing the left ventricle of the heart. This layer contains a larger amount of elastin than the others and a small amount of collagen.

2 Aortic Valve Hemodynamics

As discussed previously, the aortic valve is in constant motion, resulting in an ever-changing mechanical environment. Understanding of the dynamic mechanics of the valve is crucial for the study of the biological pathways that regulate CAVD. There are four types of forces, summarized in Figure 2, exerted upon the aortic valve:

1. *Pressure*: The two faces of the valve are subjected to contrasting pressures as the valve opens and closes, which affects the tissue shape and geometry [35]. During diastole, pressure in the ventricle rapidly drops and pressure in the aorta remains fairly stable, pushing the valve closed. Under physiological conditions this transvalvular pressure, defined as the difference between the aortic and left ventricular pressure during diastole, ranges between 80 and 120 mmHg. In hypertensive patients, the systemic blood pressure is increased, and the transvalvular pressure can reach values greater than 180 mmHg [36]. The second manifestations of arterial disease study (SMART) found that systolic blood pressure has a 9% correlation with total cardiovascular calcifications, including aortic valve calcification, indicating how systemic changes in the cardiovascular system can affect the aortic valve [37]. Another study conducted on elderly patients found that increased ambulatory diastolic blood pressure was the most significant independent factor contributing to advanced aortic valve calcification [38]; this increase in pressure enhances the effect of low shear stress and cyclic stretch and accelerates the progression of aortic valve calcification. These two studies and others [39-41] reflect the important contribution of systemic pressure to the onset and progression of aortic valve calcification. In the future, specific animal studies, using models described later in this chapter, should be performed in order to truly understand the effect increased pressure in the molecular mechanisms of calcification in the valve.
2. *Bending stress*: Bending stress is defined as the stress generated on a solid when it is deformed. In the valve, bending stress causes the fibrosa layer, which is concave, to be compressed whereas the convex ventricularis layer is under tension [42]. The most precisely developed model of bending stress has been shown by Corden et al [43], who tested trileaflet polyurethane valves used in ventricular assist devices and artificial hearts under pulsatile flow. With this system, the group found that the bending stress for the polymeric valves ranges between 0.4 and 0.7 MPa, with a bending strain ranging from 4.7 to 8.5% at peak systole compared to the diastolic position.

   In the search for clearer models of aortic valve function, the examination of the bicuspid aortic valve has proven invaluable. Bicuspid aortic valve (BAV) has a prevalence of 0.5% to 2%, with a male predominance of 3 to 1, making it the most common congenital heart valve defect [44]. BAV patients develop symptomatic aortic stenosis at the age of 50-60 years old

[45] while patients with tricuspid valves normally present symptomatic AS at the age of 70-80 years old [46, 47]. Therefore, this condition is useful for us to study aortic valve calcification, as we know that patients with BAV develop faster and more severe aortic valve stenosis. Katayama et al [48] studied the bending strain, defined as deformation of the tissue caused by bending stress, in stenotic and non-stenotic tricuspid and bicuspid aortic valves using a computational model and concluded two findings: the bending strain increased in stenotic tricuspid and bicuspid valves compared to nonstenotic valves; and a greater increase was observed in bicuspid valves, with strain concentrations at the midline of the leaflets. The group hypothesized that excessive bending strain on the bicuspid leaflets could be responsible for the rapid progression of calcification in bicuspid versus tricuspid valves. Another study by Conti et al [49] found that bicuspid aortic valves experience an eight-fold increase in strain compared to tricuspid valves. Bending stress is clearly a critical factor in the valve environment, and it may be important to control the bending stress in order to prevent valve dysfunction.

3. *Axial stress*: Valves change their shape and length throughout the cardiac cycle to adapt to the mechanical environment. Axial stress, defined as tension that the valve experiences perpendicular to the valve face, is paramount in creating a tight seal to prevent regurgitation of blood into the ventricle [42]. Axial stress is separated into two categories in the valve: circumferential and radial, as visualized in Figure 2A. In the aortic valve of many different species—ovine, porcine and human—the circumferential direction presents a higher elastic modulus compared to the radial direction [50]. In canine aortic valves, leaflets elongate by 11% in the circumferential direction and by 31% in the radial direction from diastole to systole [51]. This behavior is expected as the leaflets need to stretch in the radial direction to form a tight seal and prevent regurgitation. Maximum axial stress in the aging valve is located in the area at which the valve leaflet attaches to the aortic root, which is also the region where calcification is first observed [52]. This increase in stress is associated with development and progression of valve calcification; in turn, disease progression leads to increased axial stress, generating a "vicious cycle" accelerating valve dysfunction.
4. *Shear stress*: Shear stress is exerted on the valve by blood flowing across the face of the valve tissue. In basic terms, shear stress is defined as the component of the stress parallel to the surface of interest. The aortic valve experiences completely different shear profiles on each side of the leaflet, tightly correlating with the preferential calcification of the fibrosa [16, 53]. Shear stress at the leaflet surface is experienced in a cyclical manner—due to the blood flow during the typical cardiac cycle. The ventricularis is subjected to unidirectional shear stress as blood is ejected from the ventricle to the aorta whereas the fibrosa experiences oscillatory shear stress. Oscillatory shear stress at the fibrosa has been directly associated with valve dysfunction and CAVD [16, 53, 54]. Aboelkassem et al [55] derived a ma-

thematical model describing shear stress during systole, which combines aortic valve dynamics, systemic vascular resistance and sinus vortex local pressure. The model uses a Hill's classical semi-spherical vortex model and an aortic pressure–area compliance constitutive relationship to effectively mimic the oscillatory flow experienced at the fibrosa, with a change in flow direction between the opening and closing sequence of the valve. This model nicely shows the flow forces on the valve during systole and the relationship between the blood flow and the deformation of the valve during the cardiac cycle. In agreement with the mathematical model, the peak shear stresses have been measured by another group using laser Doppler velocimetry: 20 dyn/cm^2 on the aortic side and 64-71 dyn/cm^2 on the ventricular side [56]. The results from this study correlate with the theory that the fibrosa layer preferentially calcifies due to the low and oscillatory shear stress the same way as atherosclerosis, which presents similar mechanisms of disease onset such as endothelial dysfunction and inflammation [57], developing in regions of low and oscillatory flow [58, 59].

Because endothelial cells reside at the face of the valve in direct contact with blood flow, shear stress is a key regulator of endothelial cell function, communicating systemic changes in the body to the valvular tissue. Laminar shear stress, as observed at the ventricularis, regulates cell shape and microfilament structure as cells align in the direction of flow [60], correlated with synthesis of endothelial nitric-oxide synthase and production of nitric oxide, a potent vasodilator. Further discussion of the biological pathways altered by differential shear stresses is discussed later in the chapter.

A study conducted by Saikrishnan et al [61] showed reduced wall shear stress and increased turbulence in bicuspid valves utilizing bicuspid aortic valves constructed from porcine tricuspid valves. The study concluded that these differences in hemodynamics may play a key role in amplifying the biological responses of the valvular leaflets, which the authors elucidate as being one of the key components of the accelerated progression of calcification in bicuspid aortic valves. Another elegant valve computer model examined bicuspid aortic valves in a fluid-structure computational model, showing that the fused leaflets of the bicuspid valve exhibited strong abnormalities in shear stress compared to the tricuspid controls [62]. The same group studied these abnormalities in ex vivo porcine aortic valve leaflets and found that abnormal shear stress increased endothelial inflammation [63]. These functional studies, coupled with the mechanobiological findings discussed later in the chapter, emphasize the importance of endothelial shear stress in tissue function, and future studies may finally show how abnormal shear stress directly contributes to the onset of calcific aortic valve disease.

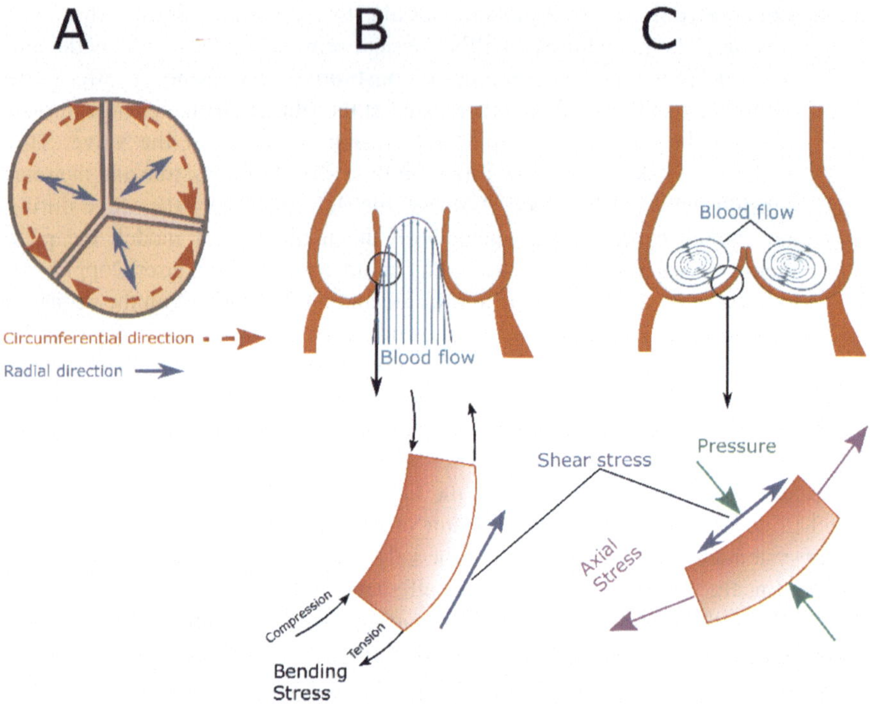

Fig. 2 Forces affecting the aortic valve in systole and diastole. A: A cross-section of the aortic valve is shown indicating the circumferential (red dashed arrows) and radial direction (blue solid arrows) of the valve. B: The main forces acting on the valve during systole (when the valve is open) are shown. The fibrosa is compressed while the ventricularis is stretched and is subjected to tensile bending stress. The ejection of blood from the ventricle exerts unidirectional laminar shear stress on the ventricularis (represented by the parallel blue arrows in the top panel and the single blue arrow in the lower panel). C: The main forces acting during diastole (valve is closed) are shown. Pressure (green arrows) exerted from cycling blood flow pulls the valve shut, and axial stress (purple arrows) seals the valve to prevent regurgitation. The fibrosa, the aortic side of the valve, experiences oscillatory shear stress (represented by the blue circling arrows in the top panel and the double-headed blue arrow in the lower panel).

3 CAVD: Changes in Hemodynamic and Mechanical Environment

Calcific aortic valve disease is characterized by the formation of calcification nodules on the leaflets of the valve. The calcification of the valve alters its mechanical properties and narrows the area of valve opening, resulting in changes in the mechanical forces experienced by the valve. Thus, the change in valve shape and hemodynamics can be used to diagnose the severity of aortic stenosis (AS). There are four parameters that are widely used by clinicians in the diagnosis of AS: aortic valve peak velocity; aortic valve area (AVA); mean aortic valve gradient (MAVG), corresponding to the pressure loss of the blood flow as it crosses the valve; and left ventricular ejection fraction (LVEF), corresponding to the percentage of blood that leaves the left ventricle. These parameters are used to place the degree of AS in one of four stages, as shown in Table 1 [64]:

Table 1 Degrees of aortic valve stenosis. Shown are the hemodynamic parameters used to evaluate AS, the most common pathology observed in each stage and the symptoms that the patient may exhibit. AVA: aortic valve area, MAVG: mean aortic valve gradient, LVEF: left ventricular ejection fraction. Table adapted from [65]

Stage	AV peak velocity (m/s)	AVA (cm^2)	MAVG (mmHG)	LVEF (%)	Valve pathology	Patient symptoms
A: At risk of AS	<2m/s	>1cm^2	<20 mmHG	>50%	Aortic valve sclerosis	Asymptomatic
B: Progress-ive AS	>2m/s and <4m/s	>1cm^2	>20 mmHG and <40mmHg	>50%	Mild-to-moderate calcification	Asymptomatic
C: Asymp-tomatic severe AS	>4m/s	<1 cm^2	>40 mmHg	>50%	Severe calcification and severely reduced leaflet opening	Asymptomatic
D: Symp-tomatic severe AS	>4m/s	<1 cm^2	>40 mmHg	<50%	Severe calcification and severely reduced leaflet opening	Exertional dyspnea, exertional angina and syncope of angina

4 Animal Models of CAVD

To study CAVD in a controlled system, animal models of valvular dysfunction are utilized in the laboratory setting. The same principles used to diagnose AS in humans may be applied in the study of the progression of this disease in a wide array of animal models. Valvular cells extracted from bovine sources have been widely used due to the ease of extraction and because the structure of the bovine aortic valve is similar to the human aortic valve [65-67]. Porcine models have been utilized to a large extent for studies of the aortic valve, since the heart and cardiovascular system are nearer in size to the human equivalents [68-70]. A relevant porcine model for aortic valve calcification has been developed by Balachandran et al [71] in which the effects of cyclic stretch were studied using porcine aortic valve leaflets in a unidirectional stretch bioreactor. Calcification was significantly increased under higher stretch conditions, along with increases in bone marker expression, cellular apoptosis, total calcium content and alkaline phosphatase activity. These phenomena correlate with those observed in human valve calcification, and prove the utility of this model for the study of disease mechanisms and treatments. In another useful animal model, hypercholesteremia is induced in rabbits by feeding a high cholesterol (0.25%) diet for prolonged periods of time, which induces the development of atherosclerosis and aortic valve calcification [72-74]. In each of the cases listed above, the cost of maintaining, breeding and genetically modifying such large animals is often prohibitive. Thus, the mouse has been explored as a tool for further studies in the valve by our group and many others.

The mouse has been widely used to study aortic valve disease progression. Various techniques may be used to measure murine aortic valve function, including ultrasound, magnetic resonance imaging (MRI), catheterization and ex vivo tissue staining. Overall, the best technique to assess the degree of aortic stenosis is MRI, because it causes no injury to the animal and gives information about valve morphology, severity of stenosis and anatomy and function of the heart. However, MRI has a high cost and requires a high level of sedation. Catheterization is a powerful technique but requires the mice to be fully sedated and may also cause internal injuries [75]. Confirmation of the extent of calcification of the valve leaflets requires post-mortem sectioning of the aortic valve and staining with dyes such as Alizarin red [76, 77] or von Kossa stain [71, 78]. However, this technique requires sacrifice of the animal to harvest the heart and vasculature, which is does not allow for monitoring of disease progression. Thus, ultrasound echocardiographic imaging, discussed at length below, is most widely used. This method conveys the same information as MRI with less precision. However, the advantages of echocardiography are low levels of sedation required for mice, decreasing the likelihood that the method may interfere with biological study, and the relatively low total cost of the imaging.

Our group and others have proven the utility of echocardiography to assess the status of aortic stenosis in mice. Two images obtained with echo are shown below. The top panel of Figure 3 shows the peak aortic valve velocity for a young (2 months of age) ApoE$^{-/-}$ mice where can observe a peak velocity of 900 mm/s. The bottom panel shows the peak aortic valve velocity for aged ApoE$^{-/-}$ (15 months) mice with three months of high fat diet, which is much higher, in the range of 2200 mm/s, indicating that the aortic valve is stenosed. Post-mortem calcification staining on these leaflets showed that valves with severe stenosis also have calcification nodules on the leaflets and the aortic valve root, whereas mice with no stenosis have non-calcified valves and root.

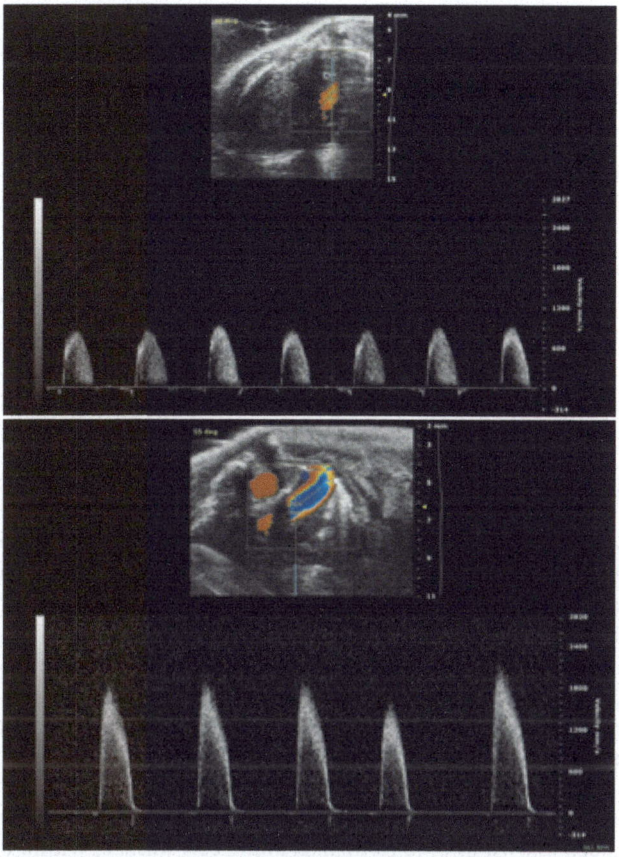

Fig. 3 Changes in aortic valve peak velocity caused by valve stenosis recorded by echocardiography. Top panel: Doppler image of the aortic valve of a young (2 months) $ApoE^{-/-}$ mouse, which presents a peak velocity of 900 mm/s. Bottom panel: the aortic stenosis of an aged $ApoE^{-/-}$ mouse fed for three months with high fat diet results in a peak velocity on the order of 2200 m/s.

5 Computational Models of CAVD

Increasingly, alternative models are used to visualize the dynamics of valve function and dysfunction, including several powerful computer imaging techniques. A recent study by Weinberg et al [79, 80] proposes a computational model for the aging aortic valve by varying the leaflet thickness and the extensibility from data obtained experimentally. Progression of CAVD is modeled by adding calcified zones to the valve where the sites of calcification initiation are the regions of high flexure (these regions show preferential calcification formation in explanted aortic valves). This model is able to predict how the peak velocity in the aortic valve and the aortic valve area evolve over time. This study can be translated clinically into a more accurate prediction of the degradation of the valve function and a method to determine the best tissue parameters to target in order to diminish the rate of calcification. The data from this paper could be used to create in vitro and in vivo models that emulate specific phases of the disease so that the different pathologies experienced by the valve can be better studied. A comparison between the model by Weinberg et al and experimental data [81] shows a good correlation between the two, proving the utility of computational models for the study of the disease progression.

Engineered and computer models of valve function and dysfunction have been expansive and comprehensive, giving us new insights into the hemodynamics of CAVD and offering novel tools for future research. These ideas, integrated with the biology of the valve endothelium described below, will propel us forward to a clearer picture of the entire disease environment.

6 Mechanobiology of Aortic Valve Endothelium

Cells sense the physical forces transmitted by neighboring cells or the surrounding extracellular matrix through a series of mediators at the cell membrane that are called mechanosensors. These may be transmembrane proteins, cytoskeletal proteins, glycans and other sugars, ion channels, or a combination of these. The complex array of cellular mechanical sensors has yet to be fully realized, but their importance for cell function and dysfunction has been made clear through multiple studies. The majority of these studies have been carried out in vascular endothelium, but there is a lack of information in valvular endothelial cells. The following is an exploration of the key mechanosensory mechanisms utilized by the endothelium, including a brief description of our current understanding of their role in aortic valve disease.

Integrins, focal adhesion complexes and the cytoskeleton

Integrins are a superfamily of transmembrane cell adhesion receptors that have traditionally been known to bind to ligands at the cell surface, within the extracellular matrix, and within the cytoplasm. Integrins transduce signals from the extracellular environment to the cell interior; in this way, changes in the concentrations of key signaling molecules in the plasma are communicated to the cell. In addition, integrins have more recently been recognized as sensors of the mechanical environment surrounding the cell, which also results in changes in intracellular signal transduction pathways [82, 83]. In addition, integrins receive intracellular signals that regulate their ligand-binding affinity, fine-tuning the communication between the cell membrane and other cellular compartments [84]. In coordinating cues from the extracellular environment with intracellular signaling, integrins play an important role in cell adhesion, migration, proliferation and survival of the cell.

The mechanism by which integrins transmit signals to the cell is dependent on their binding to the cytoskeleton [85]. Along with cytoskeletal proteins such as actin, a large number of binding partners colocalize with integrins near the cell surface, forming complexes known as focal adhesions [86, 87]. Focal adhesion proteins in these complexes mediate the bidirectional mechanical transduction among the integrins, external stimuli and the cell [88]. Extensive research has found that the integrin-focal adhesion complexes have the ability to sense forces important to the aortic valve, including cyclic stretch [89], shear stress [90] and hypotonic stress [91]. The complexes at the focal adhesions are also critical in signaling changes in the cell phenotype in response to mechanical and other cues, including signals for cell and tissue remodeling [92, 93], cell proliferation and apoptosis [93, 94] and cell migration and angiogenesis [90, 95].

Integrins encompass but one of the families of mechanosensors, but the signals they transmit through focal adhesion complexes are critical for cell phenotype. Other classes transmit signals specific to cell-cell communication and inflammation, and all of the adhesion molecules work in tandem in the valve endothelium to determine the fate of the cell and the entire tissue.

G protein-coupled receptors (GPCR)

Similar to the other transmembrane proteins discussed above, GPCR bind to a variety of extracellular ligands to activate downstream signaling, namely through G-protein activated signal transduction pathways [96]. These receptors have also been shown to be strongly sensitive to changes in flow across the surface of the endothelium [97]. While some work has shown specific methods of subunit recruitment utilized by the GPCR to become activated under mechanical stimuli [98], the role of GPCR in the endothelium has not been fully realized, either in the vasculature or in the valve. These receptors have known functions in many disease models [99], and thus their potential importance in the valve is clear. More work must be focused here to integrate these mechanical receptors into the lexicon of valvular mechanobiology.

Caveolae

Caveolae are specialized lipid-rich invaginations on the surface of mammalian cells that sense hemodynamic and mechanical changes in the environment and translate it into biochemical signals that regulate cell function [100, 101]. The protein responsible for the maintenance of the caveolar structure, caveolin-1, interacts with proteins in the cell to transmit mechanical signals and alter cellular phenotype [102]. While little work has been performed studying these in the valve, the role of caveolae has been previously studied in the pathology of cardiovascular disease. In a carotid ligation mouse model of atherosclerosis, the caveloae-specific protein caveolin-1 was found to be critical for remodeling of the vessel wall [103]. In a longer-term atherosclerosis model, the $ApoE^{-/-}$ mouse, with a genetic deletion of caveolin-1, mice experienced a 70% reduction in the formation of aortic lesions compared to controls [104], possibly linking the mechanical forces experienced by the atherosclerotic endothelium to the resulting pathologies within the vasculature. In some of the only studies in the aortic valve, caveolin-1 was shown to be highly expressed in the valve endothelium, and the protein is translocated to the cell membrane after aortic valve endothelial cells are exposed to oxidized LDL [101]. With the strong evidence of the mechanosensory role of caveloae and their structural protein in other diseases involving differential flow, the caveolae are a strong candidate for further study in CAVD. Lipid metabolism in the valve is poorly understood, and it may be the caveolae which link mechanosensing to this degenerative phenomenon.

Glycocalyx

Endothelial cells are coated with a sugar-rich layer formed of proteoglycans and glycoproteins, bound to long chains of carbohydrates, called glycosaminoglycans (GAGs) which provides a physical barrier between the blood and the endothelial cells [105, 106]. This network of sugars provides a luminal mesh where ions and other molecules in the blood are "trapped" to allow for interaction with membrane channels and receptors [107]. The role of glycocalyx in the endothelium is multi-faceted: apart from aiding in ligand-receptor interactions, it may also regulate vascular permeability via deformation in response to mechanical force [108]. When the plasma constituents bind they reveal an important mechanical component of the glycocalyx: their binding may alter the stiffness and/or movement of the GAGs, transmitting signals to the membrane proteins and acting as a mechanical stimulus to the cell [109-111]. Under disturbed flow or oscillatory shear stress, the endothelial glycocalyx collapses, which may result in an impairment of its function as a scaffold for signaling molecules [112] In addition, the inflammatory conditions associated with disturbed flow have been shown to cause changes in the glycocalyx structure which promote pro-inflammatory processes in the endothelial cells [113]. The role of glycocalyx in the development of aortic valve calcification has not been well studied; however, the studies of Sarphie et al [114, 115] found that the aortic valves of rabbits fed a hypercholestermic diet exhibit macromolecular changes in the endothelial glycocalyx, increasing the binding of LDL molecules and IgGs, known mediators in inflammation and progression of valve disease. The LDL and antibody binding was more acute on the fibrosa, the aortic side of the valve more susceptible to calcification.

The paucity of studies of the glycocalyx in the valve leaves much to be desired. However, from the extensive characterization of this network in the vascular endothelium, we can imagine the opportunity for advances in valve biology through future studies of the glycocalyx. The mechanosensitive properties of the proteoglycans and glycoproteins at the cell surface deserve more thorough investigation in order to garner a clear picture of how mechanical forces affect valve endothelial biology.

Other mechanosensory molecules

Endothelial cells express a variety of other mechanosensors, such as PECAM-1 (related with adherence and migration function), P-selectin and E-selectin (regulating the rolling of endothelial cells), and cadherins (located at the intercellular junction) [116-121]. Cadherins have been well studied as mechanosensors [122]. While general cadherin expression in the valve is increased in CAVD [123], no studies have shown the mechanical properties of these proteins in aortic valve endothelial function and valve stenosis.

PECAM-1 is a well-known mechanically regulated protein in the vascular endothelium; it is critical for vascular morphogenesis, blood pressure and atherosclerosis in response to flow [124, 125]. Tensional forces in PECAM-1 cause a plethora of signaling changes in endothelial cells by cell-wide activation of integrins and the small GTPase RhoA, facilitating changes in cytoskeletal architecture and stiffening [126]. Pulmonary endothelial PECAM-1 expression is responsible for neutrophil infiltration and inflammation [127], which may reflect its role in the inflammation of the valve endothelial cells. Under unidirectional shear stress, PECAM-1 in valvular endothelial cells recruits and binds to vimentin in the cytoplasm leading to cytoskeletal remodeling [128]. Given its mechanosensitivity and characterization in other endothelial models, PECAM-1 represents an opportune therapeutic target in the valve. Further exploration should lead to novel insights into the role of this and other CAMs in valvular mechanical signaling.

Another mechanosensor well-studied in the vascular endothelium is the primary cilia. Primary cilia, resembling rods covered by membranes, extend from the surface of the cells [129]. Nauli et al [130] have studied the role of vascular endothelial cilia by targeting polycystin-1, required for cilia function, and Polaris, required for their structure. Cells without functional cilia lack the ability to translate extracellular shear stress into intracellular calcium signaling and nitric oxide synthesis. The primary cilia are critical for the normal function of the endothelium, which certainly warrants further analysis into their role in valve dysfunction and CAVD.

As summarized in Figure 4, the role of mechanosensors in the aortic valve is a potentially exciting direction for the field of valve biology and engineering. Future thorough exploration of these proteins and their interaction with the mechanical environment will clearly link hemodynamics and signaling.

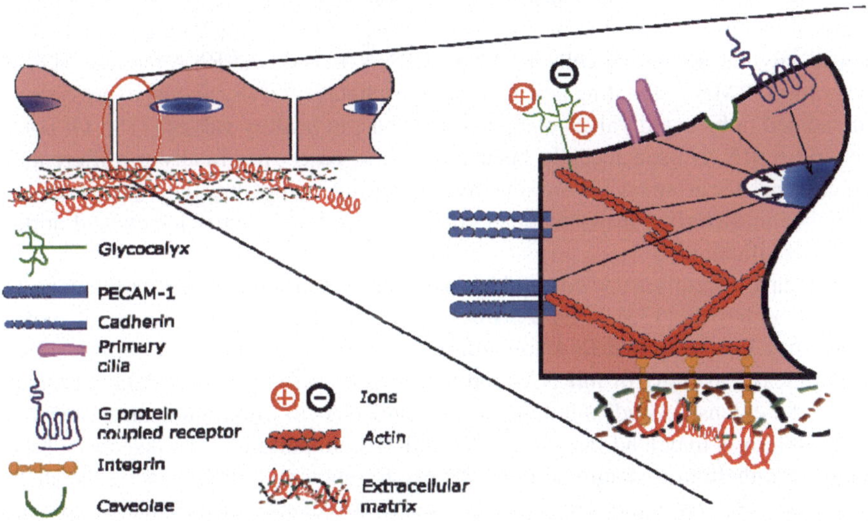

Fig. 4 Known and predicted mechanosensors in valvular endothelial cells. Integrins bind to the extracellular matrix to garner information about the valve environment. The glycocalyx traps ions and other molecules in the bloodstream to aid in signal transduction. Lipid-rich caveolar structures can sense mechanical changes in the endothelial environment, leading to downstream phenotypic changes in the cell. Cell adhesion molecules (CAMs), such as PECAM-1 and cadherins, may sense mechanical changes between cells and at the cell surface upon changes in blood flow, as they are known to do in the vascular endothelium. These signals are transmitted within the cell through the actin cytoskeleton and other signaling cascades (indicated by blue arrows), leading to transcriptional changes in the nucleus and an ultimate alteration in the cell phenotype. G protein coupled receptors and cilia, well-studied mechanosensors in the vasculature, may similarly communicate mechanical forces in the cell environment in this way.

7 microRNAs in CAVD

microRNAs (miRNAs) are short (18-22) nucleotide sequences which bind to the 3'-untranslated region (UTR) of mRNA, regulating protein expression mainly by degradation of mRNA targets or suppression of translation [131]. In a wide range of diseases, miRNAs are key regulators of signaling pathways: some miRNAs prevent disease progression and some induce further pathological signaling. The role of several miRNAs has been characterized in many cardiovascular diseases, including atherosclerosis [132, 133], heart failure [134], diabetes [135], and hypertension [136].

The role of these small molecules in the aortic valve is a relatively new frontier. miRNAs have been studied in the interstitial cells and two miRNAs have been found to play a role in the valve calcification process. MiRNA-30b has been found to be an inhibitor of osteoblast differentiation and its expression is reduced in stenotic valves compared to healthy valves [137]. Another miRNA also studied in aortic valvular interstitial cells has been miRNA-141, this miRNA reduces TGF-β activation, BMP-2 signaling and alkaline phosphatase activity and its expression is attenuated in bicuspid valves compared to tricuspid healthy valves [138]. Of note, several miRNAs have recently been reported to react specifically to differential mechanical shear stress (shear-sensitive miRNAs), indicating their potential importance in disturbed flow-mediated conditions such as atherosclerosis and aortic valve calcification.

We have recently reviewed endothelial shear-sensitive miRNAs in the context of atherosclerosis [133] in an expansive analysis of the ways these small molecules may link disturbed flow to cardiovascular disease. This review emphasizes the clinical feasibility of miRNAs and points to their future in translational medicine. The extensive studies linking specific miRNAs to differential shear stress in atherosclerosis may lend insight into the miRNAs important in the response of the valve endothelium to disturbed flow. An exciting potential target among the shear sensitive miRNAs is the miR-712/205 family. This family of miRNAs is upregulated by disturbed flow and contributes to endothelial inflammation and vascular hyperpermeability by targeting tissue inhibitor of metalloproteinase-3 (TIMP3). Expression and activity of TIMP3 regulates the breakdown of the extracellular matrix [139]. In addition, using the partial carotid ligation mice model with ApoE$^{-/-}$ mice [140], our group has shown that silencing of miR-712 results in rescue of TIMP3 and prevention of atherosclerosis. This model is highly translatable, and the profound results from these studies point to the therapeutic potential of this miRNA in a clinical setting in the future. Extracellular matrix remodeling is a critical step in the progression of aortic valve stenosis as it will lead to an increase in leaflet stiffness and to the activation of cell signaling pathways responsible for disease progression. Of paramount importance in the extracellular matrix degradation are MMPs. In particular, MMP-9 is upregulated in calcified regions of stenotic valves and is involved in the fibrotic remodeling of the ECM [141].

Our group has also recently shown a large number of endothelial valve miRNAs whose expression is dependent on changes in shear stress. A miRNA microarray was performed using human aortic valve endothelial cells (HAVECs) from either the fibrosa or ventricularis. Among 30 miRNAs significantly altered under oscillatory shear stress, 4 were validated using qPCR. One molecule in particular, miRNA-486-5p, was notable in its robust response in both of these settings. It was further studied by a series of experiments, including overexpression in HAVECs and exposure of these cells to differential shear stress. These studies firmly proved that miRNA-486-5p expression in HAVECs increases cell migration and decreases apoptosis, and provoking our interest in further studies examining its potential therapeutic effects in valvular dysfunction and disease. The in vitro results were validated and confirmed using fresh porcine aortic valve tissue in an ex vivo shear bioreactor [142], in which the fibrosa or the ventricularis side of the leaflet were exposed to different shear conditions [143].

Endothelial miRNAs are critical regulatory mechanosensitive molecules in cardiovascular diseases. Therefore, they are a fascinating new avenue of research in the field of valvular biology and CAVD. As our group and others delve into the role of miRNAs in mechanosensory mechanisms, the possibilities are vast for future miRNA therapeutics to prevent and/or attenuate disease onset and progression.

8 Aortic Valve Mechanobiology: Future Opportunities in Research

This chapter discusses the current state of our understanding of aortic valve mechanics, endothelial mechanosensing, and their effects on CAVD. As reflected here, we have a somewhat basic understanding of the ways the aortic valve transmits signals from extracellular mechanical forces to intracellular signaling. However, there are no strong treatments used to reverse or prevent the endothelial dysfunction and tissue calcification resulting from these signaling mechanisms. In the future, targeting the mechanosensors summarized above or viable mechanosensitive candidates downstream, such as the microRNAs, should reveal solid therapeutic options to maintain valvular health and possibly reverse the onset of CAVD.

Acknowledgments Research in the Jo lab is supported by funding from NIH grants HL095070, HL114772, and HL113451 to H. Jo. The work is also supported by the National Heart, Lung, and Blood Institute of the NIH through a Program of Excellence in Nanotechnology award (HHSN268201000043C to H. Jo).

References

1. Rajamannan, N.M., et al.: Calcific Aortic Valve Disease: Not Simply a Degenerative Process: A Review and Agenda for Research From the National Heart and Lung and Blood Institute Aortic Stenosis Working GroupExecutive Summary: Calcific Aortic Valve Disease – 2011 Update. Circulation **124**(16), 1783–1791 (2011)
2. Otto, C.M., et al.: Association of aortic-valve sclerosis with cardiovascular mortality and morbidity in the elderly. N. Engl. J. Med. **341**(3), 142–147 (1999)
3. Hsu, S.Y., et al.: Aortic valve sclerosis is an echocardiographic indicator of significant coronary disease in patients undergoing diagnostic coronary angiography. International Journal of Clinical Practice **59**(1), 72–77 (2005)
4. Mohler, E.R., et al.: Development and progression of aortic valve stenosis: atherosclerosis risk factors–a causal relationship? A clinical morphologic study. Clin. Cardiol. **14**(12), 995–999 (1991)
5. Leopold, J.A.: Cellular mechanisms of aortic valve calcification. Circ. Cardiovasc. Interv. **5**(4), 605–614 (2012)
6. Writing Group, M., et al.: Heart Disease and Stroke Statistics—2012 Update: A Report From the American Heart Association. Circulation **125**(1), e2-e220 (2012)
7. Baxley, W.A.: Aortic valve disease. Curr. Opin. Cardiol. **9**(2), 152–157 (1994)
8. Moat, N.E., et al.: Long-term outcomes after transcatheter aortic valve implantation in high-risk patients with severe aortic stenosis: the U.K. TAVI (United Kingdom Transcatheter Aortic Valve Implantation) Registry. J. Am. Coll. Cardiol. **58**(20), 2130–2138 (2011)
9. Gilard, M., et al.: Registry of transcatheter aortic-valve implantation in high-risk patients. N. Engl. J. Med. **366**(18), 1705–1715 (2012)
10. Mohammadi, H., Mequanint, K.: Prosthetic aortic heart valves: modeling and design. Med. Eng. Phys. **33**(2), 131–147 (2011)
11. Yacoub, M.H., Takkenberg, J.J.: Will heart valve tissue engineering change the world? Nat. Clin. Pract. Cardiovasc Med. **2**(2), 60–61 (2005)
12. Czarny, M.J., Resar, J.R.: Diagnosis and Management of Valvular Aortic Stenosis. Clinical Medicine Insights: Cardiology, 15–24 (2014). (4465-CMC-Diagnosis-and-Management-of-Valvular-Aortic-Stenosis.pdf)
13. Cunningham, K.S., Gotlieb, A.I.: The role of shear stress in the pathogenesis of atherosclerosis. Lab. Invest. **85**(1), 9–23 (2004)
14. Ku, D.N., et al.: Pulsatile flow and atherosclerosis in the human carotid bifurcation. Positive correlation between plaque location and low oscillating shear stress. Arteriosclerosis, Thrombosis, and Vascular Biology **5**(3), 293–302 (1985)
15. Peiffer, V., Sherwin, S.J., Weinberg, P.D.: Does low and oscillatory wall shear stress correlate spatially with early atherosclerosis? A systematic review **99,** 242-250 (2013)
16. Freeman, R.V., Otto, C.M.: Spectrum of Calcific Aortic Valve Disease: Pathogenesis, Disease Progression, and Treatment Strategies. Circulation **111**(24), 3316–3326 (2005)
17. Gould, S.T., et al.: Hemodynamic and cellular response feedback in calcific aortic valve disease. Circ. Res. **113**(2), 186–197 (2013)
18. Carabello, B.A.: Introduction to aortic stenosis. Circ. Res. **113**(2), 179–185 (2013)
19. Otto, C.M., et al.: Characterization of the early lesion of 'degenerative' valvular aortic stenosis. Histological and immunohistochemical studies. Circulation **90**(2), 844–853 (1994)

20. Olsson, M., et al.: Accumulation of T lymphocytes and expression of interleukin-2 receptors in nonrheumatic stenotic aortic valves. J. Am. Coll. Cardiol. **23**(5), 1162–1170 (1994)
21. Wallby, L., et al.: T lymphocyte infiltration in non-rheumatic aortic stenosis: a comparative descriptive study between tricuspid and bicuspid aortic valves. Heart **88**(4), 348–351 (2002)
22. Kaden, J.J., et al.: Interleukin-1 beta promotes matrix metalloproteinase expression and cell proliferation in calcific aortic valve stenosis. Atherosclerosis **170**(2), 205–211 (2003)
23. Jian, B., et al.: Progression of aortic valve stenosis: TGF-beta1 is present in calcified aortic valve cusps and promotes aortic valve interstitial cell calcification via apoptosis. Ann. Thorac. Surg. **75**(2), 457–465 (2003). discussion 465-6
24. Rajamannan, N.M., et al.: Human Aortic Valve Calcification Is Associated With an Osteoblast Phenotype. Circulation **107**(17), 2181–2184 (2003)
25. Sacks, M.S., Schoen, F.J., Mayer, J.E.: Bioengineering challenges for heart valve tissue engineering. Annu. Rev. Biomed. Eng. **11**, 289–313 (2009)
26. Vesely, I., Lozon, A.: Natural preload of aortic valve leaflet components during glutaraldehyde fixation: effects on tissue mechanics. J. Biomech. **26**(2), 121–131 (1993)
27. Wiltz, D., et al.: Extracellular Matrix Organization, Structure, and Function. Calcific Aortic Valve Disease (2013)
28. Schoen, F.J.: Aortic valve structure-function correlations: role of elastic fibers no longer a stretch of the imagination. J. Heart Valve Dis. **6**(1), 1–6 (1997)
29. Chen, J.H., Simmons, C.A.: Cell-matrix interactions in the pathobiology of calcific aortic valve disease: critical roles for matricellular, matricrine, and matrix mechanics cues. Circ. Res. **108**(12), 1510–1524 (2011)
30. Bashey, R.I., Torii, S., Angrist, A.: Age-related collagen and elastin content of human heart valves. J. Gerontol. **22**(2), 203–208 (1967)
31. Kastelic, J., Baer, E.: Deformation in tendon collagen. Symp. Soc. Exp. Biol. **34**, 397–435 (1980)
32. Scott, M., Vesely, I.: Aortic valve cusp microstructure: the role of elastin. Ann. Thorac. Surg. **60**(2 Suppl), S391–S394 (1995)
33. Tseng, H., Grande-Allen, K.J.: Elastic fibers in the aortic valve spongiosa: a fresh perspective on its structure and role in overall tissue function. Acta Biomater. **7**(5), 2101–2108 (2011)
34. Li, R., et al.: A dynamic model of calcific nodule destabilization in response to monocyte- and oxidized lipid-induced matrix metalloproteinases. Am. J. Physiol. Cell Physiol. **302**(4), C658–C665 (2012)
35. Bäck, M., et al.: Biomechanical factors in the biology of aortic wall and aortic valve diseases **99,** 232–241 (2013)
36. Then, K.L., Rankin, J.A.: Hypertension: a review for clinicians. Nursing Clinics of North America **39**(4), 793–814 (2004)
37. Takx, R.A.P., et al.: The interdependence between cardiovascular calcifications in different arterial beds and vascular risk factors in patients at high cardiovascular risk. Atherosclerosis **238**(1), 140–146 (2015)
38. Iwata, S., et al.: Higher Ambulatory Blood Pressure Is Associated With Aortic Valve Calcification in the Elderly: A Population-Based Study. Hypertension **61**(1), 55–60 (2013)
39. Bermejo, J.: The effects of hypertension on aortic valve stenosis. Heart **91**(3), 280–282 (2005)

40. Kaden, J.J., Haghi, D.: Hypertension in aortic valve stenosis–a Trojan horse. Eur. Heart J. **29**(16), 1934–1935 (2008)
41. Ivanovic, B., Tadic, M., Dincic, D.: The effects of arterial hypertension on aortic valve stenosis. Vojnosanit. Pregl. **67**(7), 588–592 (2010)
42. Chester, A.H., et al.: The living aortic valve: From molecules to function. Global Cardiology Science & Practice **2014**(1), 52–77 (2014)
43. Corden, J., David, T., Fisher, J.: Determination of the curvatures and bending strains in open trileaflet heart valves. Proc. Inst. Mech. Eng. H. **209**(2), 121–128 (1995)
44. Siu, S.C., Silversides, C.K.: Bicuspid aortic valve disease. J. Am. Coll. Cardiol. **55**(25), 2789–2800 (2010)
45. Beppu, S., et al.: Rapidity of progression of aortic stenosis in patients with congenital bicuspid aortic valves. Am. J. Cardiol. **71**(4), 322–327 (1993)
46. Stewart, B.F., et al.: Clinical factors associated with calcific aortic valve disease. Cardiovascular Health Study. J. Am. Coll. Cardiol. **29**(3), 630–634 (1997)
47. Boon, A., et al.: Cardiac valve calcification: characteristics of patients with calcification of the mitral annulus or aortic valve. Heart **78**(5), 472–474 (1997)
48. Katayama, S., et al.: Bicuspid aortic valves undergo excessive strain during opening: A simulation study. The Journal of Thoracic and Cardiovascular Surgery **145**(6), 1570–1576 (2013)
49. Conti, C.A., et al.: Biomechanical implications of the congenital bicuspid aortic valve: A finite element study of aortic root function from in vivo data. The Journal of Thoracic and Cardiovascular Surgery **140**(4), 890–896.e2 (2010)
50. Martin, C., Sun, W.: Biomechanical characterization of aortic valve tissue in humans and common animal models. Journal of biomedical materials research. Part A **100**(6) (2012). doi:10.1002/jbm.a.34099
51. Abrams, J.: The aortic valve by Mano Thubrikar Crc Press, Inc., Boca Raton (1990) 221 pages, illustrated, $97.50 ISBN: 0–8493–4771–8. Clinical Cardiology **14**(4), 364a-365 (1991)
52. Singh, R., et al.: Age-related changes in the aortic valve affect leaflet stress distributions: implications for aortic valve degeneration. J. Heart Valve Dis. **17**(3), 290–298 (2008). discussion 299
53. Richards, J., et al.: Side-specific endothelial-dependent regulation of aortic valve calcification: interplay of hemodynamics and nitric oxide signaling. Am. J. Pathol. **182**(5), 1922–1931 (2013)
54. Ankeny, R.F., et al.: Preferential Activation of SMAD1/5/8 on the Fibrosa Endothelium in Calcified Human Aortic Valves - Association with Low BMP Antagonists and SMAD6. PLoS ONE **6**(6), e20969 (2011)
55. Aboelkassem, Y., Savic, D., Campbell, S.G.: Mathematical modeling of aortic valve dynamics during systole. Journal of Theoretical Biology **365**, 280–288 (2015)
56. Weston, M.W., LaBorde, D.V., Yoganathan, A.P.: Estimation of the shear stress on the surface of an aortic valve leaflet. Ann. Biomed. Eng. **27**(4), 572–579 (1999)
57. Agmon, Y., et al.: Aortic valve sclerosis and aortic atherosclerosis: different manifestations of the same disease? Insights from a population-based study. Journal of the American College of Cardiology **38**(3), 827–834 (2001)
58. Ku, D.N., et al.: Pulsatile flow and atherosclerosis in the human carotid bifurcation. Positive correlation between plaque location and low oscillating shear stress. Arteriosclerosis **5**(3), 293–302 (1985)
59. Cunningham, K.S., Gotlieb, A.I.: The role of shear stress in the pathogenesis of atherosclerosis. Lab. Invest. **85**(1), 9–23 (2005)

60. Malek, A.M., Izumo, S.: Mechanism of endothelial cell shape change and cytoskeletal remodeling in response to fluid shear stress. J. Cell Sci. **109**(Pt 4), 713–726 (1996)
61. Saikrishnan, N., Mirabella, L., Yoganathan, A.P.: Bicuspid aortic valves are associated with increased wall and turbulence shear stress levels compared to trileaflet aortic valves. Biomech. Model. Mechanobiol. (2014)
62. Chandra, S., Rajamannan, N.M., Sucosky, P.: Computational assessment of bicuspid aortic valve wall-shear stress: implications for calcific aortic valve disease. Biomech. Model. Mechanobiol. **11**(7), 1085–1096 (2012)
63. Sucosky, P., et al.: Altered shear stress stimulates upregulation of endothelial VCAM-1 and ICAM-1 in a BMP-4- and TGF-beta1-dependent pathway. Arterioscler. Thromb. Vasc. Biol. **29**, 254–260 (2009)
64. Nishimura, R.A., et al.: 2014 AHA/ACC Guideline for the Management of Patients With Valvular Heart Disease: a report of the American College of Cardiology/American Heart Association Task Force on Practice Guidelines. Circulation **129**(23), e521–e643 (2014)
65. Ortolani, F., et al.: Pro-calcific responses by aortic valve interstitial cells in a novel in vitro model simulating dystrophic calcification. Ital. J. Anat. Embryol. **115**(1–2), 135–139 (2010)
66. Bertacco, E., et al.: Proteomic analysis of clonal interstitial aortic valve cells acquiring a pro-calcific profile. J. Proteome Res. **9**(11), 5913–5921 (2010)
67. Renato, M., et al.: Proteomic analysis of interstitial aortic valve cells acquiring a pro-calcific profile. Methods Mol. Biol. **1005**, 95–107 (2013)
68. Li, F., et al.: Oxidized Low-Density Lipoprotein Promotes Osteoblastic Differentiation of Valvular Interstitial Cells through RAGE/MAPK. Cardiology **130**(1), 55–61 (2015)
69. Rattazzi, M., et al.: Extracellular pyrophosphate is reduced in aortic interstitial valve cells acquiring a calcifying profile: Implications for aortic valve calcification. Atherosclerosis **237**(2), 568–576 (2014)
70. Sider, K.L., et al.: Evaluation of a porcine model of early aortic valve sclerosis. Cardiovascular Pathology **23**(5), 289–297 (2014)
71. Balachandran, K., et al.: Elevated Cyclic Stretch Induces Aortic Valve Calcification in a Bone Morphogenic Protein-Dependent Manner. The American Journal of Pathology **177**(1), 49–57 (2010)
72. Cai, H., Harrison, D.: Endothelial dysfunction in cardiovascular diseases: the role of oxidant stress. Circ. Res. **87**, 840–844 (2000)
73. Armstrong, Z.B., et al.: Effects of an Angiotensin II Type 1 Receptor Blocker on Aortic Valve Sclerosis in a Preclinical Model. Canadian Journal of Cardiology **30**(9), 1096–1103 (2014)
74. Hamilton, A.M., et al.: Statin treatment of hypercholesterolemic-induced aortic valve sclerosis. Cardiovascular Pathology **20**(2), 84–92 (2011)
75. Weiss, R.M., Miller, J.D., Heistad, D.D.: Fibrocalcific Aortic Valve Disease: Opportunity to Understand Disease Mechanisms Using Mouse Models. Circulation Research **113**(2), 209–222 (2013)
76. Honda, S., et al.: A Novel Mouse Model of Aortic Valve Stenosis Induced by Direct Wire Injury. Arteriosclerosis, Thrombosis, and Vascular Biology **34**(2), 270–278 (2014)
77. Le Quang, K., et al.: Early Development of Calcific Aortic Valve Disease and Left Ventricular Hypertrophy in a Mouse Model of Combined Dyslipidemia and Type 2 Diabetes Mellitus. Arteriosclerosis, Thrombosis, and Vascular Biology **34**(10), 2283–2291 (2014)

78. Galeone, A., et al.: Aortic valvular interstitial cells apoptosis and calcification are mediated by TNF-related apoptosis-inducing ligand. International Journal of Cardiology **169**(4), 296–304 (2013)
79. Weinberg, E.J., Schoen, F.J., Mofrad, M.R.: A computational model of aging and calcification in the aortic heart valve. PLoS One **4**(6), e5960 (2009)
80. Weinberg, E.J., Kaazempur, M.R.: Mofrad, Transient, three-dimensional, multiscale simulations of the human aortic valve. Cardiovasc. Eng. **7**(4), 140–155 (2007)
81. Piper, C., et al.: Can progression of valvar aortic stenosis be predicted accurately? The Annals of Thoracic Surgery **76**(3), 676–680 (2003)
82. Yang, B., Rizzo, V.: Shear Stress Activates eNOS at the Endothelial Apical Surface Through 1 Containing Integrins and Caveolae. Cell. Mol. Bioeng. **6**(3), 346–354 (2013)
83. Zhao, F., et al.: Roles for GP IIb/IIIa and alphavbeta3 integrins in MDA-MB-231 cell invasion and shear flow-induced cancer cell mechanotransduction. Cancer Lett. **344**(1), 62–73 (2014)
84. Takada, Y., Ye, X., Simon, S.: The integrins. Genome Biol. **8**(5), 215 (2007)
85. Ziegler, W.H., et al.: Integrin connections to the cytoskeleton through talin and vinculin. Biochem. Soc. Trans. **36**(Pt 2), 235–239 (2008)
86. Parsons, J.T., et al.: Focal adhesion kinase: a regulator of focal adhesion dynamics and cell movement. Oncogene **19**(49), 5606–5613 (2000)
87. Ciobanasu, C., Faivre, B., Le Clainche, C.: Integrating actin dynamics, mechanotransduction and integrin activation: the multiple functions of actin binding proteins in focal adhesions. Eur. J. Cell Biol. **92**(10–11), 339–348 (2013)
88. Zebda, N., Dubrovskyi, O., Birukov, K.G.: Focal adhesion kinase regulation of mechanotransduction and its impact on endothelial cell functions. Microvasc. Res. **83**(1), 71–81 (2012)
89. Lehoux, S., et al.: Differential regulation of vascular focal adhesion kinase by steady stretch and pulsatility. Circulation **111**(5), 643–649 (2005)
90. Li, S., et al.: The role of the dynamics of focal adhesion kinase in the mechanotaxis of endothelial cells. Proc. Natl. Acad. Sci. USA **99**(6), 3546–3551 (2002)
91. Hirakawa, M., et al.: Sequential activation of RhoA and FAK/paxillin leads to ATP release and actin reorganization in human endothelium. J. Physiol. **558**(Pt 2), 479–488 (2004)
92. Hsu, H.J., et al.: Stretch-induced stress fiber remodeling and the activations of JNK and ERK depend on mechanical strain rate, but not FAK. PLoS One **5**(8), e12470 (2010)
93. Sokabe, M., et al.: Mechanotransduction and intracellular signaling mechanisms of stretch-induced remodeling in endothelial cells. Heart Vessels (Suppl. 12), 191-3 (1997)
94. Wu, C.C., et al.: Directional shear flow and Rho activation prevent the endothelial cell apoptosis induced by micropatterned anisotropic geometry. Proc. Natl. Acad. Sci. USA **104**(4), 1254–1259 (2007)
95. Wang, J.G., et al.: Uniaxial cyclic stretch induces focal adhesion kinase (FAK) tyrosine phosphorylation followed by mitogen-activated protein kinase (MAPK) activation. Biochem. Biophys. Res. Commun. **288**(2), 356–361 (2001)
96. Katritch, V., Cherezov, V., Stevens, R.C.: Structure-function of the G protein-coupled receptor superfamily. Annu. Rev. Pharmacol. Toxicol. **53**, 531–556 (2013)

97. Chachisvilis, M., Zhang, Y.L., Frangos, J.A.: G protein-coupled receptors sense fluid shear stress in endothelial cells. Proc. Natl. Acad. Sci. USA **103**(42), 15463–15468 (2006)
98. Cuerrier, C.M., et al.: Effect of thrombin and bradykinin on endothelial cell mechanical properties monitored through membrane deformation. J. Mol. Recognit. **22**(5), 389–396 (2009)
99. Christopoulos, A.: Advances in G protein-coupled receptor allostery: from function to structure. Mol. Pharmacol. **86**(5), 463–478 (2014)
100. Chatterjee, S., Fisher, A.B.: Mechanotransduction in the endothelium: role of membrane proteins and reactive oxygen species in sensing, transduction, and transmission of the signal with altered blood flow. Antioxid. Redox Signal. **20**(6), 899–913 (2014)
101. Rajamannan, N.M., et al.: Localization of caveolin 1 in aortic valve endothelial cells using antigen retrieval. J. Histochem. Cytochem. **50**(5), 617–628 (2002)
102. Radel, C., Rizzo, V.: Integrin mechanotransduction stimulates caveolin-1 phosphorylation and recruitment of Csk to mediate actin reorganization. Am. J. Physiol. Heart Circ. Physiol. **288**(2), H936–H945 (2005)
103. Yu, J., et al.: Direct evidence for the role of caveolin-1 and caveolae in mechanotransduction and remodeling of blood vessels. J. Clin. Invest. **116**(5), 1284–1291 (2006)
104. Frank, P.G., et al.: Genetic ablation of caveolin-1 confers protection against atherosclerosis. Arterioscler. Thromb. Vasc. Biol. **24**(1), 98–105 (2004)
105. Becker, B.F., Chappell, D., Jacob, M.: Endothelial glycocalyx and coronary vascular permeability: the fringe benefit. Basic Res. Cardiol. **105**(6), 687–701 (2010)
106. Alphonsus, C.S., Rodseth, R.N.: The endothelial glycocalyx: a review of the vascular barrier. Anaesthesia **69**(7), 777–784 (2014)
107. Afratis, N., et al.: Glycosaminoglycans: key players in cancer cell biology and treatment. FEBS J. **279**(7), 1177–1197 (2012)
108. Mehta, D., Ravindran, K., Kuebler, W.M.: Novel regulators of endothelial barrier function. Am. J. Physiol. Lung Cell. Mol. Physiol. **307**(12), L924–L935 (2014)
109. Jacob, M., et al.: The endothelial glycocalyx affords compatibility of Starling's principle and high cardiac interstitial albumin levels. Cardiovasc. Res. **73**(3), 575–586 (2007)
110. Gouverneur, M., et al.: Fluid shear stress stimulates incorporation of hyaluronan into endothelial cell glycocalyx. Am. J. Physiol. Heart Circ. Physiol. **290**(1), H458–H452 (2006)
111. Peters, W., et al.: Nanomechanics and sodium permeability of endothelial surface layer modulated by hawthorn extract WS 1442. PLoS One **7**(1), e29972 (2012)
112. Kolářová, H., et al.: Modulation of Endothelial Glycocalyx Structure under Inflammatory Conditions. Mediators of Inflammation **2014**, 694312 (2014)
113. Chien, S.: Molecular and mechanical bases of focal lipid accumulation in arterial wall. Prog. Biophys. Mol. Biol. **83**(2), 131–151 (2003)
114. Sarphie, T.G.: Interactions of IgG and beta-VLDL with aortic valve endothelium from hypercholesterolemic rabbits. Atherosclerosis **68**(3), 199–212 (1987)
115. Sarphie, T.G.: A cytochemical study of the surface properties of aortic and mitral valve endothelium from hypercholesterolemic rabbits. Exp. Mol. Pathol. **44**(3), 281–296 (1986)
116. Kobayashi, H., Boelte, K.C., Lin, P.C.: Endothelial cell adhesion molecules and cancer progression. Curr. Med. Chem. **14**(4), 377–386 (2007)

117. Szmitko, P.E., et al.: New markers of inflammation and endothelial cell activation: Part I. Circulation **108**(16), 1917–1923 (2003)
118. Privratsky, J.R., Newman, P.J.: PECAM-1: regulator of endothelial junctional integrity. Cell Tissue Res. **355**(3), 607–619 (2014)
119. Tarbell, J.M., et al.: Fluid Mechanics, Arterial Disease, and Gene Expression. Annu. Rev. Fluid Mech. **46**, 591–614 (2014)
120. Kutlar, A., Embury, S.H.: Cellular adhesion and the endothelium: P-selectin. Hematol. Oncol. Clin. North Am. **28**(2), 323–339 (2014)
121. Gavard, J.: Endothelial permeability and VE-cadherin: a wacky comradeship. Cell. Adh. Migr. **8**(2), 158–164 (2014)
122. Wang, N., Tytell, J.D., Ingber, D.E.: Mechanotransduction at a distance: mechanically coupling the extracellular matrix with the nucleus. Nat. Rev. Mol. Cell Biol. **10**(1), 75–82 (2009)
123. Zhou, J., et al.: Cadherin-11 expression patterns in heart valves associate with key functions during embryonic cushion formation, valve maturation and calcification. Cells Tissues Organs **198**(4), 300–310 (2013)
124. Hahn, C., Schwartz, M.A.: Mechanotransduction in vascular physiology and atherogenesis. Nat. Rev. Mol. Cell Biol. **10**(1), 53–62 (2009)
125. Chiu, Y.J., McBeath, E., Fujiwara, K.: Mechanotransduction in an extracted cell model: Fyn drives stretch- and flow-elicited PECAM-1 phosphorylation. J. Cell Biol. **182**(4), 753–763 (2008)
126. Collins, C., et al.: Localized tensional forces on PECAM-1 elicit a global mechanotransduction response via the integrin-RhoA pathway. Curr. Biol. **22**(22), 2087–2094 (2012)
127. Noel, J., et al.: PECAM-1 and caveolae form the mechanosensing complex necessary for NOX2 activation and angiogenic signaling with stopped flow in pulmonary endothelium. Am. J. Physiol. Lung Cell. Mol. Physiol. **305**(11), L805–L818 (2013)
128. Conway, D.E., et al.: Fluid shear stress on endothelial cells modulates mechanical tension across VE-cadherin and PECAM-1. Curr. Biol. **23**(11), 1024–1030 (2013)
129. Ando, J., Yamamoto, K.: Flow detection and calcium signalling in vascular endothelial cells. Cardiovasc. Res. **99**(2), 260–268 (2013)
130. Nauli, S.M., et al.: Endothelial Cilia Are Fluid Shear Sensors That Regulate Calcium Signaling and Nitric Oxide Production Through Polycystin-1. Circulation **117**(9), 1161–1171 (2008)
131. Guo, H., et al.: Mammalian microRNAs predominantly act to decrease target mRNA levels. Nature **466**(7308), 835–840 (2010)
132. Nazari-Jahantigh, M., et al.: MicroRNA-specific regulatory mechanisms in atherosclerosis. Journal of Molecular and Cellular Cardiology (0)
133. Kumar, S., et al.: Role of Flow-Sensitive microRNAs in Endothelial Dysfunction and Atherosclerosis: Mechanosensitive Athero-miRs. Arteriosclerosis, Thrombosis, and Vascular Biology **34**(10), 2206–2216 (2014)
134. Kalozoumi, G., Yacoub, M., Sanoudou, D.: MicroRNAs in heart failure: Small molecules with major impact. Glob. Cardiol. Sci. Pract. **2014**(2), 79–102 (2014)
135. Moura, J., Borsheim, E., Carvalho, E.: The Role of MicroRNAs in Diabetic Complications-Special Emphasis on Wound Healing. Genes (Basel) **5**(4), 926–956 (2014)
136. Neves, V.J., et al.: Exercise training in hypertension: Role of microRNAs. World. J. Cardiol. **6**(8), 713–727 (2014)

137. Zhang, M., et al.: MicroRNA-30b is a multifunctional regulator of aortic valve interstitial cells. The Journal of Thoracic and Cardiovascular Surgery **147**(3), 1073–1080.e2 (2014)
138. Yanagawa, B., et al.: miRNA-141 is a novel regulator of BMP-2–mediated calcification in aortic stenosis. The Journal of Thoracic and Cardiovascular Surgery **144**(1), 256–262.e2 (2012)
139. Nagase, H., Visse, R., Murphy, G.: Structure and function of matrix metalloproteinases and TIMPs. Cardiovasc. Res. **69**(3), 562–573 (2006)
140. Nam, D., et al.: A model of disturbed flow-induced atherosclerosis in mouse carotid artery by partial ligation and a simple method of RNA isolation from carotid endothelium. J. Vis. Exp.(40) (2010)
141. Perrotta, I., et al.: Matrix Metalloproteinase-9 Expression in Calcified Human Aortic Valves: A Histopathologic, Immunohistochemical, and Ultrastructural Study. Appl. Immunohistochem. Mol. Morphol. (2014)
142. Platt, M.O., et al.: Cyclic pressure and shear stress regulate matrix metalloproteinases and cathepsin activity in porcine aortic valves. J. Heart Valve Dis. **15**(5), 622–629 (2006)
143. Holliday-Ankeny, C.J., et al.: The function of shear-responsive and side-dependent microRNA-486-5p in aortic valve endothelium. Cardiovascular Pathology **22**(3), e50

Functional Stem Cell Biomechanics: Application of Biophysical Techniques and Multi-content 3D Image Analysis

Shan Sun, Amit Paul, John Kanagaraj and Michael Cho

Abstract The mechanical properties of the cell - cytoskeleton elasticity, membrane tension, and adhesion strength - play an important role in the regulation of stem cell differentiation. While the cellular mechanical properties are significantly altered during stem cell specification to a particular phenotype, the complexity of events associated with transformation of these precursor cells leaves many questions unanswered about morphological, structural, proteomic, and functional changes in differentiating stem cells. However, control of cell behaviors might be feasible through manipulation of the cellular mechanical properties using external physical stimuli and manipulation of mechanically sensitive signaling molecules. Biomechanical regulation of stem cell differentiation can minimize the number of chemicals and growth factors that would otherwise be required for tissue engineering. Coupled with a thorough understanding of stem cell behavior, both experimentally and computationally, development of more effective approaches is a feasible way to expand stem cells and to regulate their phenotypic commitment. We recently developed a high-content/high-throughput screening algorithm that offers significant improvements in 3D quantitative analysis at the single cell level. A consistent pattern observed in all types of stem cell differentiation indicates the cytoskeleton remodeled significantly before lineage-specific cellular changes occurred. This demonstrates that cellular mechanical transformations are a precursor to stem cell differentiation and to phenotypic functionality.

Keywords: Stem cell · Mechanobiology · Cellular mechanics · 3D image reconstruction · Multi content analytics

S. Sun · A. Paul · J. Kanagaraj · M. Cho(✉)
Department of Bioengineering, University of Illinois at Chicago, 851 S. Morgan St., MC 063 Chicago, IL 60607, USA
e-mail: mcho@uic.edu

© Springer International Publishing Switzerland 2016

H. Jo et al. (eds.), *Biomedical Engineering: Frontier Research and Converging Technologies*, Biosystems & Biorobotics 9,
DOI: 10.1007/978-3-319-21813-7_12

1 Cellular BIomechanics

1.1 Theoretical Models

Living cells and tissue are subjected to mechanical loading and deformations that are caused either by external stimuli or by changing internal physiological conditions. Not only do the cells readjust themselves to respond and withstand multiple deformations, they also modulate many cellular functions and processes that are designed to bear mechanical loading [Wang, 2006]. The mechanisms by which the cells mechanically respond to physical loads and maintain their structural stability (e.g., mechanotransduction) are complex and yet to be fully explored. A thorough understanding of these mechanisms is critical because they convert mechanical signals to useful biological and biochemical signals in order to regulate cell homeostasis [Ingber, 2006]. Many theoretical models have been developed over the last few decades to describe the experimentally observed mechanical behaviors of adherent or suspended cells. Either continuum mechanics or microstructure-based approaches have been utilized to provide some detailed explanation of potential signal conversion pathways [Lim, 2006; Stamenović, 2002]. Typically, the continuum approach treats the cell as an effective single or multi-phase material. This model simplifies the structural organization inside the cell and therefore has the limitation of providing an understanding at the whole cell level. It nonetheless allows the researcher to predict the overall mechanical deformation of living cells [Stamenović, 2002], including the rheological and mechanical properties. A continuum mechanical model can offer a detailed explanation on the distribution of stresses and strains impacting the cell which, in turn, could be used to determine the potential transmission of these forces to the cytoskeletal and subcellular components.

As a specific example of the continuum mechanics model, the liquid drop models were originally developed to account for the rheology of suspended cells in micropipette aspiration experiments [Evans, 1989]. Briefly, in the Newtonian liquid drop model, the cell interior is assumed to be a homogeneous Newtonian viscous liquid and the cell cortex is treated as an anisotropic viscous fluid layer. This model is straightforward and easy to conceptualize and does serve as a good approximation for a wide range of experimental conditions such as large deformation in the white blood cell [Tsai, 1993]. Disagreements between the Newtonian liquid drop model and experiment results highlight its limitations [Hochmuth, 1993; Thoumine, 1999]. As an alternative, a competitive model, known as the Maxwell liquid drop model, was developed [Dong, 1988] that substituted the viscous Newtonian liquid with viscoelastic cytoplasm. While such an improvement can account for small or initial deformation during a micropipette aspiration of the cell, it is still considered inadequate to explain the experimental results [Dong, 1991].

If the cell is assumed to be a homogeneous incompressible linear elastic or viscoelastic material [Theret, 1988], the solid models may be applied to determine the mechanical parameters at equilibrium. For example, the elasticity of a viscoelastic material is known to depend on both the loading rate and hysteresis.

Within a range of acceptable loading rate, the linear elastic approximation can accurately provide a framework for interpretation of data obtained from a variety of biophysical techniques – micropipet aspiration [Kuznetsova, 2007], atomic force microscope (AFM) indentation [Titushkin, 2007], and magnetic twisting cytometry [Vliet, 2003] – developed and refined to measure and characterize the cellular mechanical properties.

The linear approximation is also able to describe creep and stress relaxation experiments in suspended cells (e.g., leukocytes) and several types of anchorage-dependent cells [Koay, 2003; Sato, 1990]. Taking the continuum theory to another level of complexity, the cytoplasm may be considered to consist of both solid polymeric contents and interstitial fluid. Then it will be appropriate to treat the two phases separately as in the biphasic model [Humphrey, 2001]. This model has been widely used to study musculoskeletal cell mechanics, especially single chondrocytes and their interaction with the extracellular cartilage matrix [Shieh, 2002; Shieh, 2003]. However, the biphasic theory and irregular geometry often render analytical solution too challenging [Shieh, 2003].

Unlike the continuum mechanical models, the microstructural approaches rely on the cellular cytoskeleton as a discrete architectural basis of the cell shape and structural integrity, potentially providing more insights into detailed subcellular and molecular mechanics events. Abundant evidence can be found demonstrating that the cellular cytoskeleton plays a major role in transmitting mechanical stresses from the cell surface, across the cytoplasm and into the nucleus [Chen, 1999]. Despite its complex and dynamic architecture, the cytoskeleton is usually modeled as idealized structural geometry, material isotropy, homogeneity, and elasticity. One particular microstructural model that has been developed and extensively used is the tensegrity. It presume there exists in the cell pre-existing tensile stress (e.g., borne by microfilaments) in order to maintain the structural stability [Galli, 2005; Ingber, 1997; Stamenović, 2002]. The pre-stress is postulated to be balanced either by the external structure attachments (e.g., focal adhesion) or by its internal compression-supporting elements (e.g., microtubules) – more likely by a combination of the two. A mechanically distinct feature of all stress-supported structures is evident that their elasticity increases approximately linearly with increasing pre-stress, i.e., they exhibit pre-stress induced hardening very much like live adherent cells [Wang, 2002]. It should be noted here that the existence of compression-bearing elements (e.g., microtubules) within the cytoskeleton has long been a point of controversy. Recent studies indicate that, depending on the extent of cell spreading, cell shape, and cell type, microtubules may or may not play a critical role in balancing cytoskeleton pre-stress [Stamenović, 2002; Titushkin, 2007]. Nevertheless, the tensegrity model provides a plausible and more realistic description of the cytoskeleton mechanics of adherent cells. It can account for a number of experimental observations during mechanical testing including pre-stress induced stiffening, strain hardening, and the effect of cell spreading on cell deformability. As of the date, the tensegrity is considered to be the most relevant model of the cell in its natural environment.

In summary, while the relatively simple continuum models may provide a good explanation of the overall mechanical characteristics of living cells, a fuller description of subcellular deformations and interactions within the cell requires more complex microstructure-based models. In spite of many foreseen challenges, integration of continuum and micro-structural approaches into a hybrid method would allow the development of comprehensive mechanical models not only for the whole cell but also for the subcellular regions and intracellular components. Finally, it is important to point out that proper choice of theoretical models to interpret the experimental data is crucial as it can strongly affect and influence the derived cell mechanical properties. This of course requires the researcher to be fully aware of the limitations of theoretical models.

1.2 Experimental Techniques for Single Cell Manipulation

Living cells are mechanically active. They respond to mechanical cues in their physiological environments and generate forces by subcellular processes. These activities appear to mediate numerous cellular functions such as, to name just a few, locomotion, cell adhesion, differentiation, and proliferation [Hoffman, 2009]. The magnitude of physiological forces experienced by the cells *in vivo* has been estimated to span > 15 orders of magnitude from pN-range forces in cochlea hair cells [Liao, 2007] to kN-range tensile forces in tendons [Maffulli, 2003]. The mechanical behavior of a living cell cannot be described simply in terms of fixed mechanical parameters, however. The cell adapts actively and dynamically its structure and mechanical status to its local microenvironment. In order to understand this complex interrelationship between intracellular and extracellular biomechanical processes, an array of experimental tools has been developed that can measure forces with high temporal and special resolution [Addae-Mensah, 2008; Van Vliet; 2003]. Among these tools, our laboratory has successfully integrated and applied the combination of atomic force microscopy and laser optical tweezers to probe the cellular biomechanics.

1.2.1 Atomic Force Microscopy

The atomic force microscopy (AFM) relies on a small probe (tip) attached to a flexible cantilever [Radmacher, 2002]. The tip can be used either to image the live cell topography or to locally indent the cell with ~ 0.5 μm indentation depth. Relative deformation of the cell, as detected by diffraction pattern of a laser beam reflected off from the cantilever, can then be used to estimate the indentation force and calculate the local elastic modulus of the cell. The cantilever spring constant can be calibrated independently by the thermal fluctuation method to compute the indentation forces in the 0.1-100 nN range [Costa, 2006]. AFM has been widely used for cellular mechanical testing at the single cell level because it provides the capability of nanometer-scale positioning accuracy, topographical imaging and mechanical manipulation of a single cell with pN to nN-scale forces. When combined with a conventional imaging modality (e.g., fluorescence microscopy),

AFM can track biological processes in near physiological environments over time, and visualize and correlate the structural deformation of the indented cell with subcellular processes.

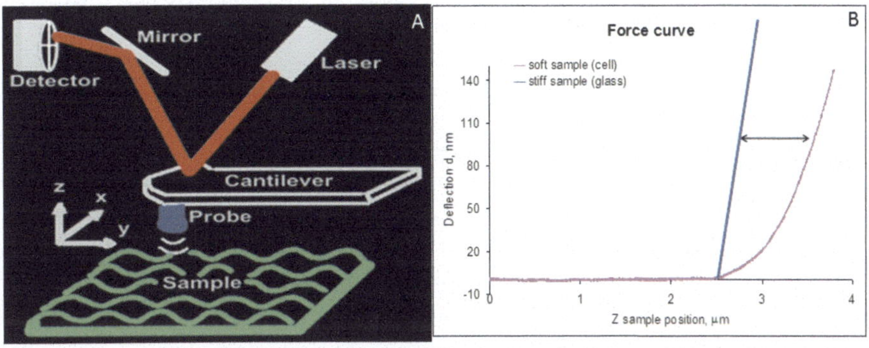

Fig. 1 Single cell elasticity measurements with microindentation technique. (A) AFM allows both live cell imaging and mechanical testing (e.g., microindentation) in near physiological conditions. The AFM tip attached to a cantilever descents slowly to a surface and causes an indentation. The depth of indentation is detected by laser light diffraction pattern. (B) Typical force-distance curves are obtained for hard and soft surfaces and usually analyzed with the classical Hertz model that relates the applied force to the indentation depth.

To measure the local cell elasticity (i.e., Young's modulus), a force-distance curve is acquired by indenting the cell with a tip probe (Fig. 1). Briefly, the cantilever deflection remains unchanged until the probe comes in contact with a surface. As the probe indents the surface, the deflection is measured optically and determined. A hard surface should demonstrate a sharp rise in the deflection measurement, whereas a gradual increase is observed on indenting a softer surface (e.g., living cell). The rate at which the cantilever deflection rises (e.g., the slope) is critical for the calculation of elastic modulus. The apparent local Young's modulus is then obtained by fitting the experimental force-curves to the Hertz model [Costa, 1999; Radmacher, 2002], which relates the loading force with the depth of indentation by probes of different geometries (e.g., spherical, conic, pyramidal, etc). Although the Hertz model has been validated repeatedly for the cell microindentation analysis, care must be taken in interpretation of the AFM indentation data [Costa, 2006]. Specifically, assumptions implicitly built into the Hertz model treat the sample as a homogeneous, semi-infinite, isotropic, and linear elastic material. Intrinsic fibrillar cellular cytoskeleton structure can and often do result in a spatial heterogeneity of cell elasticity. Our laboratory identified the potential pitfall of using nm-size sharp AFM tips and has been using 10 µm spherical tips. This technical correction distributes the indenting load over several μm^2 area that tends to average the contribution of multiple cytoskeletal fibers and therefore makes the Hertz analysis more accurate [Titushkin, 2007]. To treat the cell as semi-infinite material, indentations of the cell should be small enough (< 500 nm) such that the measurements are not influenced by the cell boundary and underlying substrate. The Hertz model can adequately characterize the simple elastic cell behavior in quasi-static loading conditions with an indenter actuated at a low speed (~1 μm/s) to exclude viscous effects of buffering solution and the cell. Table 1 provides a range of Young's moduli that have been measured and reported using AFM and different cell types.

Table 1 I AFM Measurement of Young's Moduli

Cell Type	Young's Modulus (kPa)	
3T3 Fibroblasts	1.6 ± 0.2	Mahaffy, 2004
Human Fibroblasts	1.8 ± 1.1	Titushkin, 2010*
Human Osteoblast	1 to 2	Takai, 2005
Human Osteoblast	1.7 ± 1.0	Titushkin, 2009b*
Chondrocytes (superficial zone)	0.55 ± 0.23	Darling, 2006
Endothelial Cells	0.9 ± 0.2	Ohashi, 2002
Endothelial Cells	1.4 to 6.8	Mather, 2001
Cardiac Muscles	100.3 ± 10.7	Mather, 2001
Skeletal Muscles	24.7 ± 3.5	Mather, 2001
Human MSCs	3.2 ± 1.4	Titushkin, 2007*
Mouse Myoblasts	2.9 ± 0.8	Sun, 2014*
Human Neuroblastoma Cells	0.4 ± 0.1	Titushkin, 2010*
Human Hepatocytes	0.9 ± 0.2	Sun, 2013*

Atomic force microscopy was used in the indentation mode to measure the cantilever deflection. With several assumptions described in the text (see section I.B), the local elasticity can be inferred. The range of Young's moduli reported by several research groups (* indicates measurements from our laboratory) should be noted and may be attributed to different cell types as well as not identical experimental conditions.

1.2.2 Laser Optical Tweezers

Laser optical tweezers (LOT) are a technique developed based on non-contact optical manipulation of dielectric objects that are either attached to the cell membrane or placed inside the cell [Sheetz, 1998]. An optical tweezer can be formed by focusing typically a laser beam to a diffraction limited spot with a high numerical aperture lens. A tightly focused laser beam exerts two forces on a micron-size dielectric particle of high refractive index: a gradient force and a scattering force. While the gradient force attracts the particle to the beam's focal point (the region of the highest intensity), the scattering force deflects the particle along the direction of the light propagation. With a sufficient laser intensity, the gradient force dominates and overcome the scattering force, and the particle is captured in a 3D trap (Fig. 2). The resultant force applied to a bead may be

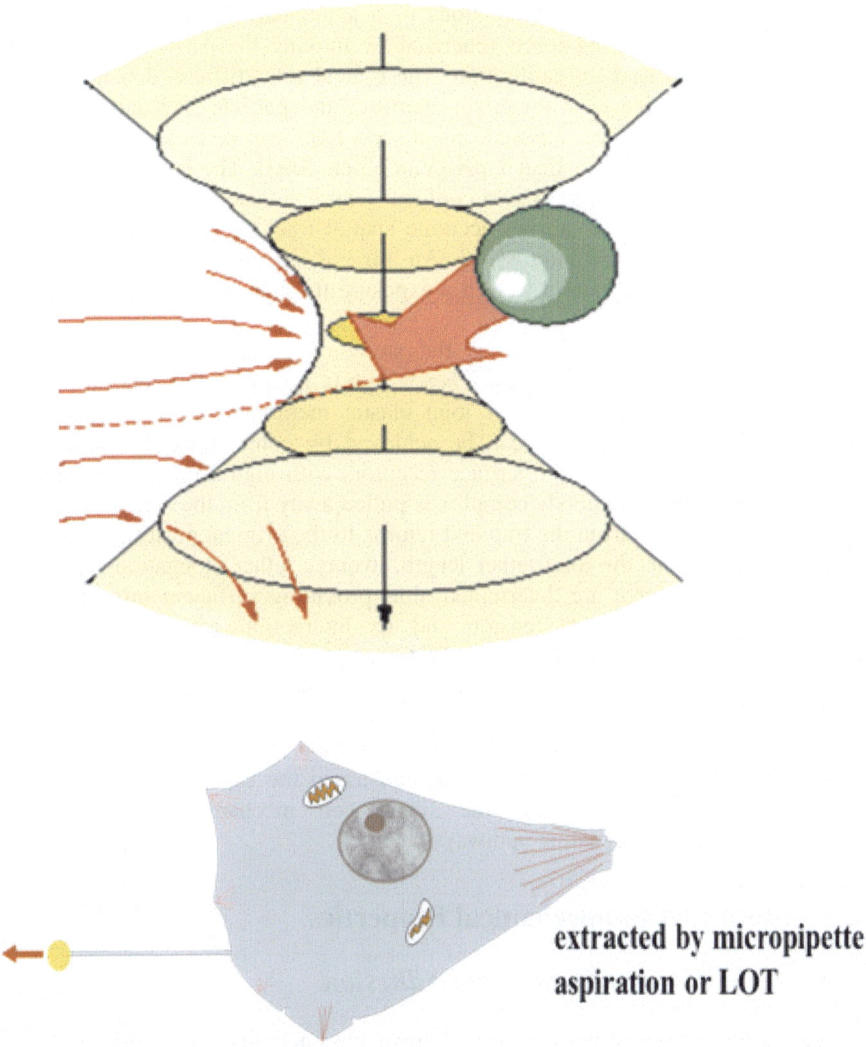

Fig. 2 Schematic diagram of laser optical tweezers (LOT) system. (Top panel) An infrared laser beam was steered to trap and move μm-size particles in the focal plane. At a sufficiently large laser intensity, the gradient force dominates over the scattering force. (Lower panel) Membrane tethers are extracted from live cells. Dielectric beads (0.5 μm in diameter) were conjugated with antibodies against integrins and attached to the cell membrane. LOT traps one bead and pulls it away from the cell, as indicated by the arrow. Thin membrane tethers extending from the beads to the cell body appear. Using conventional brightfield microscopy, the tether length and its diameter are measured which, in turn, provide an estimate of the plasma membrane tension.

estimated using elastic spring approximation, valid for small displacements of the trapped particle in a harmonic potential [Neuman, 2004]. The force calibration to determine the LOT stiffness (e.g., spring constant) is performed by analyzing the power spectrum or thermal fluctuations of a small particle caught in the trap. Alternately, a Stokes drag force generated by moving the particle in a viscous medium has been used for calibration. The optical trap stiffness depends on the laser power, objective lens numerical aperture, and particle size and properties [Wright, 1994]. The force imposed on cells via LOT can be as large as 500 pN, with force resolution better than 1 pN [Van Vliet, 2003]. The optical trap can be manipulated using a system of two confocal lenses actuated by a high precision motorized translator. This allows accurate spatial tracking of a trapped particle position by typical optical microscopy. An infrared laser is preferred to minimize potential photodamage caused by cell's exposure to high intensity laser radiation [Peterman, 2003].

One notable LOT application is the study of the cell plasma membrane mechanics. In this experiment, latex beads tightly bound to the cell surface are used as handles for extraction of long elastic membrane tethers [Dai, 1998; Raucher, 1999] (Fig. 2). This can be achieved by conjugating the beads with antibodies that bind to the cell surface receptors with high affinity. An optically trapped bead/receptor-antibody complex is pulled away from the cell at a constant speed until it escapes from the trap and returns to the original position. From this tracking experiment, the total tether length, average tether elongation force, and maximum escape force are determined, thus providing sufficient information to characterize the membrane tension and its interaction with the underlying cytoskeleton [Sheetz, 2006; Titushkin, 2006].

It should be noted that in many earlier macroscopic methods, mechanical stress is typically applied to the whole cell population (e.g., substrate stretching, flow chambers), and the heterogeneity among cell responses was largely ignored. By combining AFM with LOT, we have demonstrated the capability to manipulate *individual* cells and to study the single cell mechanical responses and mechanotransduction signaling pathways.

2 Stem Cell Biomechanical Properties

2.1 Actin Structure and Focal Adhesion

Among different types of stem cells, human mesenchymal stem cells (hMSCs) provide a robust model system to probe, determine, and delineate differences in the cellular mechanical properties. The actin cytoskeleton organization in hMSCs, for example, is quite different from that of their default descendants – osteoblasts [Titushkin, 2007; Titushkin, 2009a]. Using confocal optical sectioning of adhered cells, hMSCs have been shown to demonstrate numerous thick actin bundles, or stress fibers, extending throughout the cytoplasm and terminating at focal contacts on the cell membrane (Fig. 3). In contrast, osteoblasts have fewer stress fibers and show predominantly a thin dense meshwork of microfilaments. However, the average actin concentration, estimated from fluorescent images, appears to be similar in both cell types [Titushkin, 2009b]. Implication is that while the total actin content does not vary appreciably, there is an intricate dynamics between the

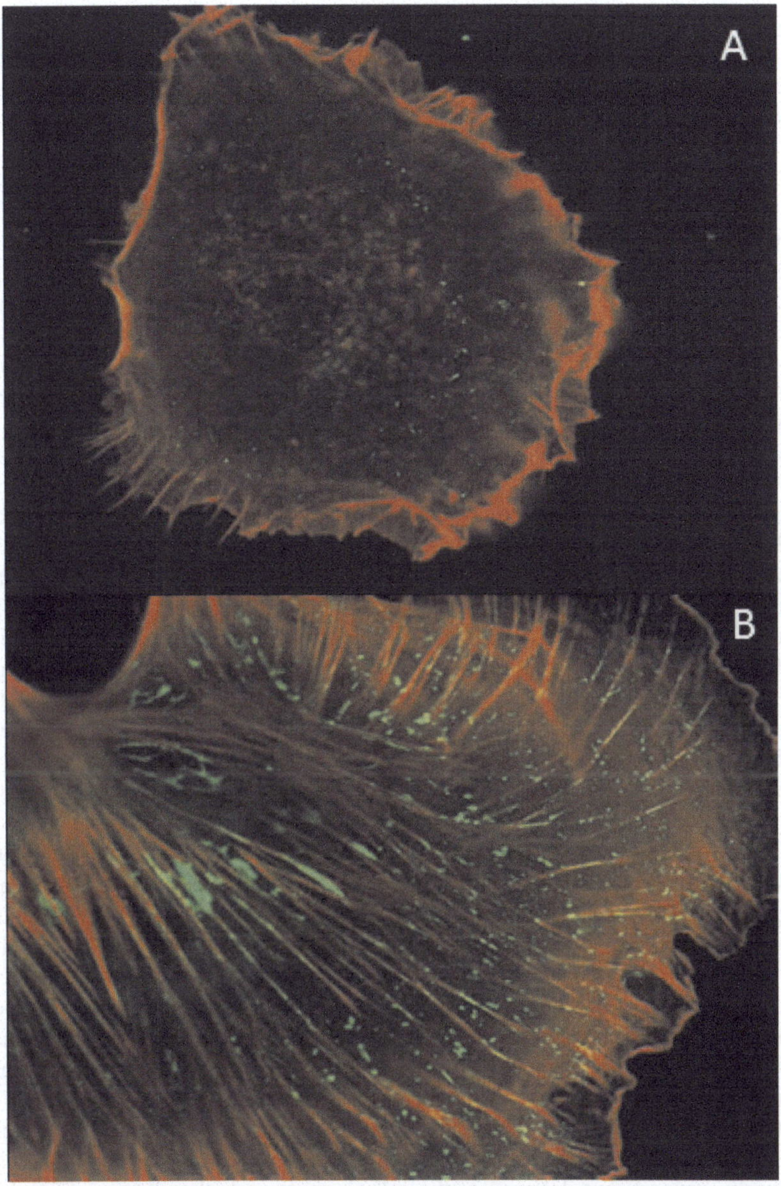

Fig. 3 Actin cytoskeleton and focal adhesion. F-actins were visualized using rhodamine phalloidin (red), and vinculins found in focal adhesions were stained using fluorescently conjugated antibodies. (A) Immunofluorescent images of a thin filamentous actin meshwork and vinculins showed that osteoblasts contain fewer and smaller focal adhesions. (B) In contrast, hMSCs showed thick actin stress fibers, and multiple and large adhesion contacts.

F- and G-actins. Moreover, focal adhesion organization and distribution appear also cell type-dependent. Because vinculin has been identified as one of key proteins in found in the focal adhesion, it can fluorescently labeled to visualize and characterize the focal adhesion. Large focal adhesions are usually associated with actin stress fibers in adherent stem cells. On the contrary, in mature cells like osteoblasts, the surface density of focal adhesions is lower and they are physically smaller in size than in stem cells. It is consistent with our observation that MSC subculture protocols require a longer incubation time to trypsinize than other cell types including osteoblasts and fibroblasts. In addition, we also observed that MSCs cultured on 2D substrate do not easily migrate, thus lending further support for the postulate of stronger stem cell adhesion to the substrate. Differential adhesion of hMSCs and osteoblasts can then be attributed to multiple focal contacts and stress fibers in stem cell. As a result, hMSCs have a much higher surface area on a 2D substrate than osteoblasts as we reported earlier and demonstrate a jagged edgy cell morphology than mature osteoblasts as indicated by a lower form-factor [Titushkin, 2009b]. Our biophysical studies using single protein tracking techniques indicate that some important cell membrane proteins involved in the regulation of focal adhesion (e.g., integrins) are laterally immobile in undifferentiated stem cells and do not show diffusion [Chen, 2007]. They either form clusters or are bound to the cytoskeleton with high affinity. However, upon induction of stem cell differentiation, the integrins overcome the lateral mobility constraints and exhibit diffusion in the cell membrane. It is indeed interesting to note that the integrin diffusion coefficient is correlated with the differentiation stage and, in the case of osteodifferentiation, becomes similar to the different coefficient measured in mature osteoblasts [Chen, 2007].

Different features in actin and focal adhesion organization are likely responsible for the distinct mechanical properties of hMSCs as compared to fully differentiated cells. To characterize the elasticity of the cellular cytoskeleton, the AFM microindentation technique with a 10 μm-diameter spherical indenter was applied. The average elastic modulus of hMSCs, 3.2±1.4 kPa, is almost 2-fold higher than 1.7±1.0 kPa for mature osteoblasts [Titushkin, 2007; Titushkin, 2009a]. The stem cells were found to be stiffer than many other cell types including fibroblasts, keratinocytes, and neuronal-like cells [Kuznetsova, 2007]. This distinctive stem cell cytoskeletal mechanics is likely due to multiple thick stress fibers present in the hMSC cytoplasm. When treated with an actin polymerization inhibitor (cytochalasin D), the elastic modulus decreases to 0.7±0.3 kPa in stem cells and to 0.9±0.5 kPa in osteoblasts with no statistical difference. The structural actin organization is indeed important and dictates the cellular mechanical state. Moreover, microtubule disruption with nocodazole results in only a minor statistically insignificant decrease in the cell elasticity in both cell types [Titushkin, 2007]. The residual elasticity after treatment with both drugs is statistically indistinguishable in both cell types and could have been provided by intermediate filaments and other intracellular components. Therefore, the cell elasticity appears to be predominantly regulated by the actin organization with microtubules and intermediate filaments providing only a minor contribution in hMSCs. These findings support the previous results that actin is the major determinant of the cytoskeleton elasticity rather than microtubules in many cell types including chondrocytes, myocytes, and fibroblasts [Takai, 2005; Trickey, 2004].

2.2 *Plasma Membrane Tension*

To explore the plasma membrane mechanical properties, 0.5 μm diameter fluorescent polystyrene beads were attached to the cell surface and pulled with a laser optical trap to extract thin lipid membrane tethers (see Fig. 2). The typical force versus length profile shows that an average 3-4 pN optical force exerted on the bead was constant during tether elongation, suggesting that an additional membrane is drawn from the buffered reservoir [Titushkin, 2006; Titushkin, 2007]. When the tether reached the maximum length, the force abruptly rose up to ~10 pN until the bead escaped from the trap, indicating that the membrane reservoir could have been depleted. Assuming the tether elongation was associated with membrane stretching, one would expect the force would increase with the tether length. Based on our observation, we favor the postulate of the involvement of membrane reservoir over stretching. Conventional optical microscopy was used to determine the average tether length in undifferentiated hMSCs (10.6±1.1μm), which is about 2.5 fold higher than that in osteoblasts (4.0±1.1μm) or in fibroblasts (3.0±0.5 μm).

At least two explanations can be offered to account for the differences observed from the membrane tether experiments. First, the apparent membrane tension may be regulated by cholesterol which is known to play an important role in determining the lipid bilayer membrane rigidity [Morris, 2001]. Cholesterol depletion from the cell plasma membrane with methyl-β-cyclodextrin increases the membrane tether lengths in many types of cells [Titushkin, 2006]. Second, membrane interaction with the cytoskeleton may affect the extracted tether length. When actin microfilament polymerization was inhibited by cytochalasin D, the length of tethers extracted from fibroblasts and osteoblasts increased more than a 4-fold as compared to untreated fibroblasts. However, disruption of the actin microfilaments in hMSCs did not significantly affect the tether length. This result is likely due to different actin cytoskeleton arrangements in hMSCs and fibroblasts leading to a stronger coupling between the membrane and actins in fibroblasts than in stem cells [Titushkin, 2007]. The dense thin actin meshwork in load-bearing cells should yield low cell elastic modulus, but there might be multiple binding sites for coupling between the membrane and cytoskeleton. We identified a family of linkers proteins, including the well- characterized ezrin, radixin and moesin (ERM proteins) that can serve as the mechano-sensitive molecules regulating the strength of membrane-cytoskeletal coupling [Sun, 2013; Sun, 2014; Titushkin, 2011; Titushkin, 2013]. In load- bearing cells, the sum of many weak interactions forms a stronger membrane-cytoskeleton linkage. On the other hand, in hMSCs thick actin stress fibers provide high Young's modulus but decrease the number of linker protein binding sites (Fig. 4). Thus, the overall strength of membrane-cytoskeleton coupling in stem cells is less than that found in osteoblasts or fibroblasts as assessed by the membrane tethers. This suggests a weaker physical coupling between the cytoskeleton and the plasma membrane in hMSCs than in osteoblasts or fibroblasts. The role of ERM linkers in the membrane-cytoskeleton association is confirmed by inhibiting phosphorylation of the ERM proteins through ATP depletion, which results in a significant membrane-cytoskeleton separation in osteoblasts, but not in mesenchymal stem cells [Titushkin, 2009b].

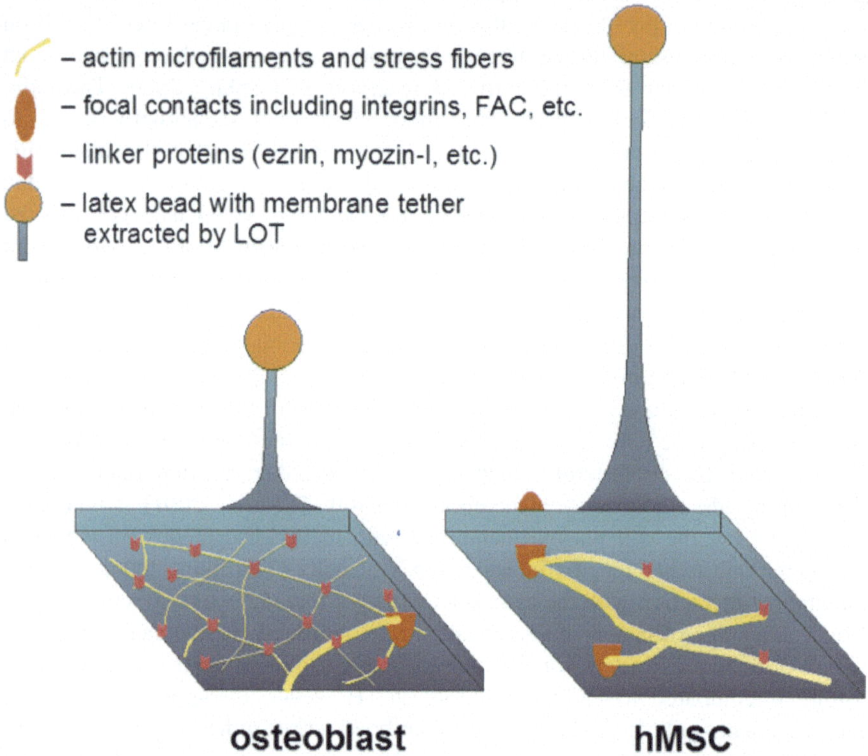

Fig. 4 Working model of stem cell biomechanics. Thin and dense actin filaments in osteoblasts are presumed tightly bound to the plasma membrane through multiple linker proteins (e.g., ERM proteins) and focal adhesion sites. In contrast, thick stress fibers in hMSCs are associated with the membrane mostly at focal contacts due to the lower contact area with the membrane, and thus availability of protein linker binding sites are substantially decreased. Such an arrangement should lead to the overall weaker membrane-cytoskeleton adhesion. This postulate is supported by the extracted membrane tethers that are longer than those in osteoblasts. It is also consistent with lower membrane tension in hMSCs. However, thick bundled actin structure in hMSCs provides a higher Young's modulus of cytoskeleton.

3 Multi-content Image Analysis

Based on collective findings from our laboratory and others, changes in the mechanical properties of stem cells are not merely a consequence of differentiation but may actually initiate and specify the stem cell fate. However, heterogeneity among a given population of stem cells poses a daunting challenge to validate such postulate. To overcome such a challenge, our laboratory has begun implementing the single cell analysis that has recently emerged as a highly interdisciplinary field. The goal is to be able to provide insights into the complexity of morphological, structural, proteomic, and functional changes in differentiating stem cells. By profiling individual cells it may be plausible to delineate the characteristics of each stem cell from ensemble measurements. To this end, we have successfully developed custom-designed multi (or typically referred to as high) content analysis algorithms to detect, analyze and predict the fate of heterogeneous stem cells. These results, briefly outlined in the following section, should be of particular interest for control of the fate of stem cells in tissue engineering, regenerative and personalized medicine, and other cell-based therapeutic applications.

To demonstrate the heterogeneity of hMSCs during differentiation into at least two morphologically distinctive phenotypes (e.g., round vs. spread), a group of cells was captured at day 7 of adipogenesis (Fig. 5A). The morphologies of these cells were not completely consistent with the mature adipocytes but three of the four cells expressed a significant amount of lipid droplets. In addition, the microfilaments appear to have been reorganized to the cortical network formation, leaving the intracellular space virtually free of actins and filling it with lipid droplets in those 3 cells. The fourth cell did not express lipid droplets, while still undergoing actin reorganization. To illustrate the capability of our algorithms to segment and extract features from each cell, we chose the one cell that most expressed lipid droplets and performed segmentation (Fig. 5B). After successful segmentation, the morphological and molecular signatures of each cell were determined using our recently developed a 58-parameter metric (referred to as multi-content analysis). Since we cannot list all of the parameters, only 18 of them are shown for illustration purposes in the table. To further demonstrate our approach of characterizing the stem cell response and heterogeneity that is lineage-dependent, we performed similar experiments while directing hMSCs to the myogenic lineage (Fig. 5C and 5D). As expected, the myoblast-like cells maintained the spindle shape morphology and expressed an abundance of MyoD1 proteins by day 7. Again, following the same segmentation process, the 18 parameters for myogenesis were determined and provided in the corresponding table. Since primary cells should provide the "gold standards", the differentiation probability can be quantified based on the 58-parameter analysis performed using human primary cells and statistically comparing the two sets of the extracted parameters. However, there is no a priori validity of each of 58 extracted features. Thus it remains as a goal to minimize the number of parameters that is required to accurately predict the stem cell fate.

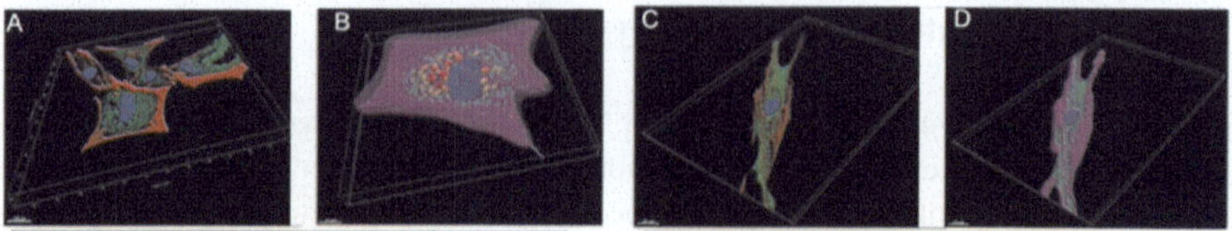

Whole Cell Total Volume (µm³)	34861.70
Nuclei Total Volume (µm³)	1093.57
Lipid Total Volume (µm³)	1199.66
Actin Total Volume (µm³)	868.99
Whole Cell Surface Area (µm²)	11261.90
Nuclei 3D Surface Area (µm²)	972.61
Lipid Total 3D Surface Area (µm²)	2687.24
Actin 3D Surface Area (µm²)	7271.78
Whole Cell Ellipticity (Oblate)	0.95
Whole Cell Ellipticity (Prolate)	0.01
Whole Cell Sphericity	0.76
Nuclei Sphericity	0.53
Nuclei Mean Intensity (AFU)	129.59
Lipids Mean Intensity (AFU)	27.12
Actin Mean Intensity (AFU)	14.07
Nuclei Sum Intensity (AFU)	44901400.00
Lipids Sum Intensity (AFU)	11464100.00
Actin Sum Intensity (AFU)	15798000.00

Whole Cell Total Volume (µm³)	33203.90
Nuclei Total Volume (µm³)	843.41
MyoD1 Total Volume (µm³)	4761.77
Actin Total Volume (µm³)	2394.24
Whole Cell Surface Area (µm²)	13818.00
Nuclei 3D Surface Area (µm²)	643.23
MyoD1 Total 3D Surface Area (µm²)	6656.11
Actin 3D Surface Area (µm²)	12209.80
Whole Cell Ellipticity (Oblate)	0.06
Whole Cell Ellipticity (Prolate)	0.06
Whole Cell Sphericity	0.26
Nuclei Sphericity	0.67
Nuclei Mean Intensity (AFU)	106.15
MyoD1 Mean Intensity (AFU)	78.90
Actin Mean Intensity (AFU)	19.97
Nuclei Sum Intensity (AFU)	28522800.00
MyoD1 Sum Intensity (AFU)	119688000.00
Actin Sum Intensity (AFU)	27297900.00

Fig. 5 Example of 3D multi-content analyses. The two specific lineages of adipocytes and myoblasts were examined. (A) Heterogeneity of adipogenic cells (bar = 150 µm) and (B) segmented and data mined by our unique algorithms (bar = 10 µm). Not only can each cell be segmented by the actin organization (red), but the individual lipid droplets (green) are identified and visualized in different colors. (C) and (D) Myoblasts were found to reorganize actins in a very different manner, and the myogenic biomarker (MyoD1, green) identifies these cells as the "mostly likely" sub-population that should differentiate to the myogenic lineage (bar = 20 µm).

Finally, it is not surprising that the cytoskeleton of hMSCs alters dramatically during differentiation. The cytoskeletal remodeling is both time- and lineage-dependent. A recent observation provided a new way of characterizing and perhaps predicting the fate of individual stem cells at an early stage of differentiation. Stem cells do exhibit unique cytoskeletal patterns along the various stages of differentiation. Similar to the identification method based on fingerprints, our laboratory is working on the "cytoskeleton fingerprint" that predicts the probability of achieving the intended differentiation based on detailed image analysis. To develop new methodologies, we utilized the well-known biometric fingerprinting technique to delineate the lineage-specific actin organization. For example, human fingerprints are identified by characterizing their pattern features (Fig. 6A and 6B, [Chakraborty, 2013; Mosaliganti, 2012]) and comparing them to an established database. The same identification technique was applied to stem cells by first skeletonizing the F-actin organization (i.e., actin fingerprinting; Fig. 6C to 6F) and assessing the likelihood of intended differentiation by matching the F-actin fingerprinting to that found in mature primary cells. While full validation is still required, this unique methodology should be capable of predicting the differentiation probability of individual live stem cells using only one parameter. Therefore, we observe individual cell behavior which can be incorporated into new and more complete biomechanical model of stem cell differentiation.

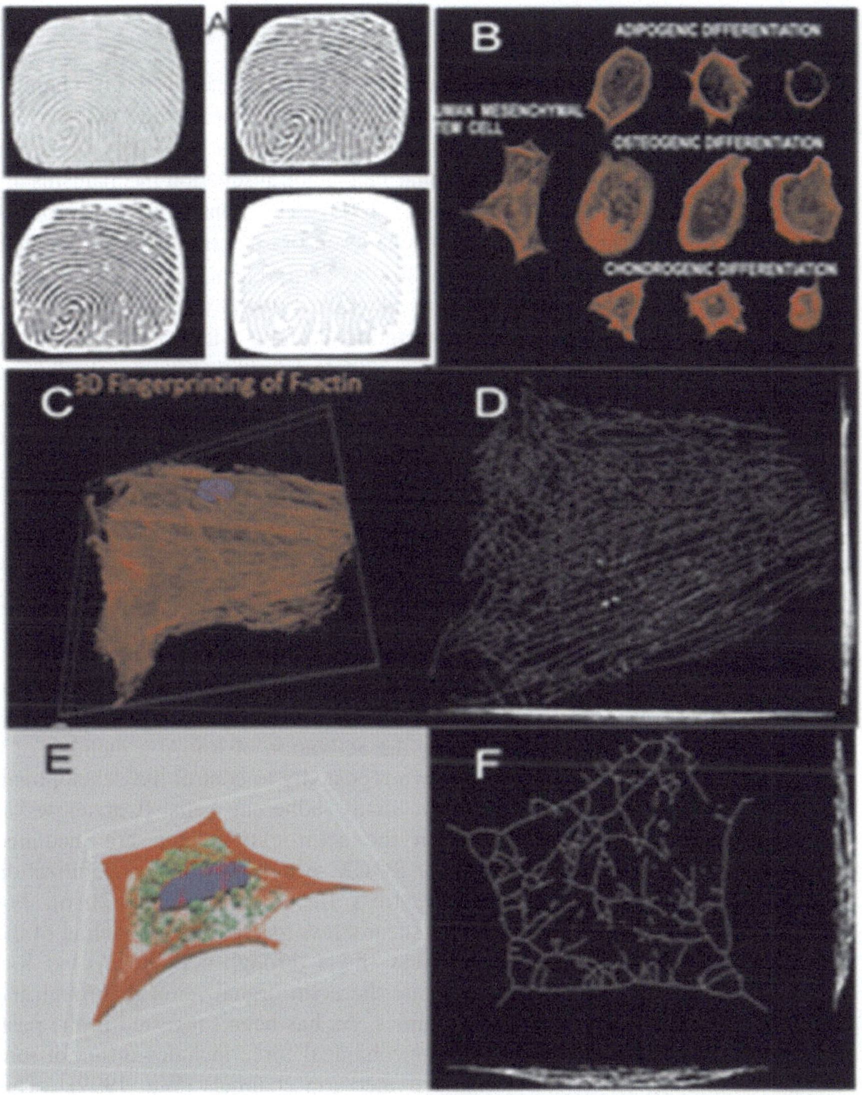

Fig. 6 Skeletonization of F-actin structures. (A) Example of pattern analysis of fingerprint identification. (B) The F-actin organization is lineage-specific and a function of time points (4, 10, and 15-21 day). (C) and (D); images of F-actin in an undifferentiated hMSC and the result of our fingerprinting algorithm. (E) and (F); images of hMSC directed to adipogenesis (lipid droplets in green) and the result of the F-actin fingerprinting analysis. The F-actin fingerprints of undifferentiated hMSC (D) and adipocyte-like cell (F, rotated 180°) are readily distinguishable.

4 Mechanotransduction Pathways

Cells are presumed to detect, transmit and respond to changes in the mechanical environment through the mechano-sensitive molecules that include cell surface receptors, cytoskeletal proteins, ion channels, and various secondary signaling molecules. Excellent reviews on the cellular mechanotransduction mechanisms are available elsewhere [Ingber, 2006; Silver, 2003]. The transmembrane proteins integrins are an essential family of the mechanosensors that transduce mechanical signals by themselves or in concert with the cytoskeletal proteins and focal adhesions [Katsumi, 2004]. Integrins in adherent cells physically connect the extracellular matrix proteins to the cytoskeleton and focal adhesion, and therefore serve as universal signaling receptors [Chen, 2004]. In addition, the focal adhesion molecules (FAK, src, vinculin, paxillin, etc.), typically associated with the actin stress fibers, may also sense mechanical forces either by altering their relative positions or by conformational changes of specific focal adhesion molecules [Wozniak, 2004]. Mechanical forces are postulated to promote the assembly of focal adhesion and trigger integrin-dependent activation of mitogen-activated protein kinases (MAPK). In addition to regulation of adhesion-dependent gene expression, the role of MAPK in cell proliferation and differentiation has long been established [Chen, 2001].

The actin cytoskeleton is responsible for transmission of external physical force to the cell nucleus. Continuous myosin-generated cytoskeletal tension allows rapid and coordinated response to a physical stimulus and a long range force transmission [Wang, 2009]. In particular, the activation of RhoA, a member of a family of GTPase proteins, has been shown repeatedly to control the development of focal adhesion and stress fibers in many adherent cells [Chrzanowska-Wodnicka, 1996]. Effects of RhoA on the actin cytoskeleton are mediated primarily through its downstream effector ROCK, which may induce stabilization of filamentous actin and formation of stress fibers [Bershadsky, 2006]. For example, mechanical stretching and fluid shear stress may cause remodeling of the actin cytoskeleton with formation of stress fibers through both RhoA and Rac GTPase signaling. Another way to regulate the actin organization is through the modulation of intracellular calcium dynamics. As has been suggested in the past, an early cellular response to the applied mechanical force includes influx of ions through not voltage-gated but stretch-activated ion channels [Wu, 1999]. This family of stretch-activated channels is very interesting in that they are mechanically operated and selectively allow influx of cations, thus rapidly altering the membrane potential. Calcium, a ubiquitous second messenger that regulates important cellular and molecular processes such as proliferation, differentiation, cell biomechanics, cytoskeletal reorganization, gene expression and metabolism [Berridge, 2004; Titushkin, 2009c], is then spatially and temporally modulated through influx of ions from the extracellular matrix and/or emptying intracellular calcium store [Cho, 2002a; Cho, 2002b, Khatib, 2004; Titushkin, 2004]. Among many cell types, hMSCs demonstrate robust calcium activities and exhibit regular intracellular calcium oscillations. [Foreman, 2006; Sun, 2007]. The Ca^{2+} dynamics in hMSCs strongly depends on the cell differentiation stage, substrate stiffness, and actomyosin contractile tension [Kim, 2009]. It has been shown that intracellular Ca^{2+} dynamics may also be regulated by specific voltage-gated calcium channels, integrins, as well as autocrine and paracrine ATP signaling through purinergic receptors [Berridge, 2003; Kawano, 2006].

Another plausible mechanotransduction mechanism may be activated at the cell membrane involving G-protein-coupled receptors (GPCR) [Sarasa-Renedo, 2005]. Indeed, GPCR-mediated activation of phospholipase C (PLC) is a key event that modulates altered calcium dynamics and subsequent production of yet another secondary messenger, diacylglycerol (DAG). DAG can lead to activation of protein kinase C (PKC) that, in turn, activates MAP kinase cascades [López-Toledano, 2004]. Finally, recent data indicate an important role of the ezrin/radixin/moesin (ERM) proteins mediating cellular mechanotransduction. The ERM proteins appear to function not only as the mechanical linkage between actin microfilaments and the plasma membrane, but also as regulators of multiple signaling pathways including Rho GTPase activities [Hughes, 2007; Mangeat, 1999]. As a result, the ERM linkers can control many mechanical properties of the cell including the membrane-cytoskeleton interaction, actin organization, cytoskeleton elasticity, and focal adhesion distributions [Louvet-Vallée, 2000]. In summary, while this brief survey of mechanotransduction mechanisms cannot fully explain how the cell decides to integrate multiple mechanical signals and modify its cellular, subcellular and genetic responses, intensive efforts are underway to elucidate detailed biomolecular mechanotransduction mechanisms that might be either cell type-dependent or possibly stem cell-specific. The anticipated advances in our understanding of mechanotransduction pathways particularly pertinent to stem cells would undoubtedly elevate the current state of regenerative stem cell tissue engineering.

5 Stem Cell Mechanics and Differentiation

Stem cell differentiation is usually achieved by using molecular mediators and soluble factors (e.g., dexamethasone, insulin, etc.) that can induce the stem cell lineage commitment [Deng, 2008]. Over the past decade the important role of other environmental factors in the regulation of stem cell fate has been established, including mechanical signals transmitted from the extracellular matrix [Guilak, 2009]. In particular, the importance of substrate stiffness as a means for modulating cell shape, phenotype and lineage commitment has been increasingly appreciated [Estes, 2004]. In groundbreaking research, Engler et al. demonstrated that the hMSC lineage specification can be determined by the mechanical properties of the substrate alone without use of any soluble factors [Engler, 2006; Rehfeldt, 2007]. The soft substrates mimicking the elasticity of brain tissue (0.1-1 kPa) appeared to drive hMSCs to the neurogenic lineage, while the same hMSCs assumed the myogenic phenotype on stiffer matrices (8-17 kPa). On even stiffer substrates resembling osteoid matrix (25-40 kPa), osteogenic commitment was observed. During the initial week in culture, reprogramming of these lineages was possible with addition of soluble induction factors, but after few weeks in culture, the cells committed irreversibly to the lineage specified by the matrix elasticity. As would be expected, combining inputs of matrix stiffness and soluble factors that favor the same phenotype resulted in synergistic expression of markers of differentiated stem cells [Engler, 2006; Even-Ram, 2006]. Cell contractility mediated by non-muscle myosin II was demonstrated essential for the cell fate determination in response to specific matrix elasticity. Interestingly, inhibition of myosin II blocked the contribution of matrix compliance only when hMSCs received a combination of matrix and soluble factor regulatory inputs. Consistent with this study, the lineage commitment of neural stem cell has been shown to depend on the effective stiffness of underlying substrate [Saha, 2006]. While softer substrates promoted neuronal differentiation, stiffer substrates led to glial differentiation of adult neural stem cells. The potential complexity of mechanically-driven stem cell lineage specification is emphasized by the fact that different tissues exhibit similar mechanical properties [Wang, 2006]. It should not be surprising that the contributions of specific ligand-receptor interactions, soluble growth factors, and physical properties of local cell microenvironment together determine the stem cell behaviors and fate. A goal yet to be achieved is to be able to prescribe a specific combination of these various differentiation cues to optimally drive stem cells to the intended lineage.

5.1 Lineage-Dependent Changes in the Stem Cell Biomechanics

The lineage-dependent changes in the stem cell biomechanics is perhaps best illustrated by examining two cell types that are mechanically and functionally different – osteoblasts and adipocytes. Many laboratories including ours have demonstrated repeatedly that hMSCs can differentiate into the osteogenic or adipogenic lineages [Sun, 2013; Titushkin, 2009, Titushkin, 2011, Titushkin, 2013]. While osteoblasts are expected to assume mechanical and functional characteristics to bear force, the major role of adipocytes is to participate in the metabolic regulation. In addition, osteoblasts are presumed to acquire irregular cell shape, adipocytes are more spherical, suggesting the cytoskeletal reorganization is also lineage-dependent. Such key differences in the cell morphology and functionality of the two cell types should be evident in and reflected by the mechanical properties that are phenotypically specific. It is worth repeating here that undifferentiated stem cells must integrate various physicochemical cues to respond to and initiate the lineage commitment at an early stage of differentiation that requires morphological and biomechanical changes. This raises a rather important question – are the cytoskeletal reorganization, altered cellular morphology and mechanical properties consequences of stem cell differentiation, or are they pre-requisite for the lineage commitment?

In order to address the question without ambiguity, our laboratory applied the RNA interference technique to modulate and measure the mechanical properties of hMSCs undergoing either the osteogenic or adipogenic commitment. One specific goal was to alter the stem cell mechanical properties before or during the induction of biochemically directed stem cell differentiation, and to examine the role of potential mechanotransduction pathways. Since RNAi provides temporal down-regulation of the targeted genes [Zhou, 2006], it is a useful tool to test the role of key molecules presumed to be involved in the mechanotransduction signaling. As described earlier, the ERM linker proteins represent a target molecule. As expected, during induction by soluble factors, the stem cell mechanical properties are adjusted to those of fully differentiated cells. We found and reported that transient suppression of ERM expression by RNAi results in a temporary decrease of the cell elasticity and the membrane-cytoskeleton adhesion, mediated by the disassembly of the actin stress fibers and the focal adhesions in hMSCs and progenitor cells [Titushkin, 2011; Titushki, 2013, Sun, 2014]. The decrease in ERM protein expression in hMSCs on day 3 after ERM siRNA transfection correlates with the partial disassembly of the actin stress fibers and the focal adhesions. However, the effect of ERM knockdown disappears by day 6 due to the transient nature of the RNAi-mediated gene silencing. Since the introduced siRNA cannot replicate itself, it disappears quickly due to degradation and dilution by cell division [Zhou, 2006]. The maximal gene knockdown by the siRNA transient transfection is typically achieved within 48–72 hrs and lasts for just 24–96 hrs. On day 2 after ERM siRNA transfection, the residual ERM linkers can provide a weaker mechanical link between the membrane and the cytoskeleton than in normal stem cells. The ERM knockdown experiments show that intact cytoskeleton and/or focal adhesions appear to be a pre-requisite for early osteogenic commitment of hMSCs. The disturbance of actin and focal adhesion complexes by ERM siRNA treatment prior to the cell induction to the osteogenic lineage interferes with the expression of tissue specific molecular markers. Apparently, the stem cell stress fibers and focal adhesion-mediated signaling are critical for biochemical induction of the early osteogenic pathways. Indeed, it has been shown by different investigators that actin cytoskeleton disruption prevents mesenchymal stem cells from osteogenic differentiation [Rodríguez, 2004; Yourek, 2007). Moreover, it has been suggested that the cytoskeleton tension regulates osteogenic versus adipogenic specification of stem cells [McBeath, 2004]. It is possible that ERM knockdown affects actin dismantling by releasing the cell tension through down-regulation of Rho activity, but the exact mechanisms remain to be identified. In osteogenically induced cells, ERM knockdown on either day 0 or day 3 seems to interfere with osteogenic commitment as confirmed by lower levels of the osteogenic markers' expressions on day 18 compared to the RNAi-untreated hMSCs. Although ERM knockdown in the growth medium decreases the stem cell elasticity to the level typical for mature osteoblasts (~2 kPa), the significant membrane-cytoskeleton separation induced by ERM siRNA treatment (tether length ~13 μm) is inconsistent with a strong membrane-cytoskeleton adhesion in osteoblasts (tether length 4.0±1.1 μm; [Titushkin, 2007]. The mechanical status of ERM-deficient stem cells is not relevant to that of fully differentiated osteoblasts, leading to the suppression of osteogenic commitment by transient ERM down-regulation. No difference between the effects of ERM siRNA treatment on day 0 and day 3 is likely attributed to substantially different time-scales of osteodifferentiation and RNAi protocols. Indeed, while the osteogenic differentiation and up-regulation of the osteogenic markers typically occur within 3-4 weeks [Pittenger, 1999], the transient siRNA transfection effect usually lasts for a few days only [Zhou, 2006]. To observe ERM-mediated modulation of the cell mechanics and osteogenic commitment, a longer-lasting gene silencing protocol may be required, e.g., repeated ERM RNAi treatment with 4-6 day intervals. Any changes in the osteodifferentiating cell mechanics are not observed until after day 6 of osteogenesis. Therefore, ERM knockdown might need to start later (about day 6) during osteodifferentiation, for the most effective modulation of cellular mechanics. These findings emphasize the importance of the dynamical changes in cellular mechanics for the cell commitment to a particular lineage. Manipulation of stem cell mechanics for control of cell differentiation should be preceded by the initial induction of biochemical signaling pathways by soluble factors. The temporal dynamics of cell morphological, mechanical and functional changes should be carefully considered to design and optimize RNAi-based and other biophysical approaches for regulation of the stem cell commitment.

Similar ideas can be tested using a cell type that is morphologically, mechanically and functionally different than osteoblasts. We applied RNAi technique to suppress the ERM linker protein expressions in hMSCs while they are undergoing biochemically-directed adipogenesis. We have recently shown that the adipogenically differentiating stem demonstrate essentially the same mechanical properties by day 12 of adipogenesis, independent of the siRNA treatment [Titushkin, 2013]. However, interference with the stem cell biomechanics significantly prevented *or* facilitated adipogenesis. For example, early intervention with siRNA to suppress the ERM protein expressions significantly reduced the adipogenesis. In contrast the same siRNA treatment applied at day 3 of adipogenic differentiation increased adipogenesis. An interesting finding is that preservation of the initial cellular mechanical homeostasis (e.g., day 1) is required for proper stem cell differentiation. On the other hand, one can intervene at a later day (e.g., day 3) to facilitate the required remodeling of the morphology and cytoskeletal organization that appear to facilitate the intracellular and genetic processes consistent with adipogenesis. An implication may be the ERM knockdown-mediated mechanical changes are more relevant to adipogenesis (e.g., low elastic modulus, weak membrane-cytoskeleton interaction) than osteogenesis (higher Young's modulus). The dynamics of siRNA mediated ERM down-regulation (~ 1 week) is more comparable to the time course of adipogenic differentiation (~ 2 weeks) rather osteogenic differentiation (> 4 weeks). Comparison of the final cell morphology of adipocytes and osteoblabsts suggests that rapid cytoskeletal rearrangement required for adipocytes should be better facilitated by modulation of the ERM linker proteins. These observations may explain, in part, more pronounced effects of RNAi-mediated ERM knockdown on adipogenesis than osteogenesis [Titushkin, 2011].

As reported earlier, polymeric actin is the major determinant of the cytoskeleton elasticity rather than microtubules or intermediate filaments in many cell types

[Takai, 2005; Trickey, 2004; Titushkin, 2007; Titushkin, 2011, Titushkin, 2013]. The actin cytoskeleton participates in the stem cell fate regulation by controlling multiple intracellular signaling pathways [Guilak, 2009]. First, it has been shown recently that ROCK-dependent cytoskeletal tension and the contractile forces play important roles in the stem cell fate regulation [McBeath, 2004]. The ability of Rho-ROCK signaling to direct osteogenic versus adipogenic hMSCs differentiation relies on non-muscle myosin II activity, which emphasizes the role of cytoskeleton in mechanotransduction [Engler, 2006]. Second, polymeric actin in the cell is closely associated with the focal adhesions providing a mechanical link to the extracellular environment, and therefore playing a key role in mediating cellular responses to external physical cues [Docheva, 2007]. Third, ERM proteins-mediated actin interaction with the plasma membrane determines the membrane mechanical parameters (e.g., surface tension), and regulates the transport of active agents by endocytosis and exocytosis [Morris, 2001]. The ERM proteins in this strategic position might regulate the whole cell mechanics not only by providing physical membrane-cytoskeleton coupling, but also by participating in signaling mechanisms and affecting the cytoskeleton structure and cell adhesion [Mangeat, 1999; Poullet, 2001]. Indeed, the ERM involvement in Rho transduction pathway was demonstrated in several studies [Mackay, 1997; Mangeat, 1999]. The Rho-dependent phosphorylation and conformational activation of the ERM proteins is required for association of the ERM family and vinculin with the membrane [(Kotani, 1997]. However, the role of ERM as a downstream Rho effector in the reorganization of actin cytoskeleton has been established [Louvet-Vallée, 2000]. The Rho enzyme, in turn, has been shown to regulate formation of the stress fibers and the focal adhesions in many types of cultured cells [Albiges-Rizo, 2009]. Moreover, the ERM family proteins have been implicated in the assembly of focal adhesions [Poullet, 2001]. For example, ezrin has been shown to induce FAK (focal adhesion kinase) activation independently of the cell-matrix adhesion. The dual role of ERM proteins as both mechanical linkers and regulators of the intracellular signaling cascades make them potent candidates for modulation of the stem cell biomechanics and cell commitment to a particular phenotype. For adipogenesis, time-dependent suppression of the ERM proteins is critical for facilitating or inhibiting the intended stem cell differentiation. A possible differential role of the individual ERM proteins involved in the regulation of stem cell biomechanics is currently under investigation. The development of biomechanical therapeutic approaches for soft tissue regeneration appears feasible in the near future.

5.2 Orthogonal Cues

It is evident that stem cells respond both to biological cues (e.g., growth factors) and to physical stimulus (e.g., shear stress, electric field and etc.). However, biological and physical cues are perceived as orthogonal in the sense that one is derived from physical forces or stresses while the other is proportional to a chemical concentration or gradient – two phenomena that should have limited, if any at all, interactions. While it is possible to induce conformation changes of growth factors by physical stimulus and therefore this might alter the biological function of these potent molecules, an unphysiologically large magnitude of physical stimulus would be required. In contrast, one may contemplate exploiting the cell's ability to detect both types of cues, and introduce a combination of cues for additive or synergistic effects. Additive stem differentiation implies that the effects (e.g., quantifiable tissue-specific markers) produced by multiple cues should be the same as the sum of those produced by each cue used alone. In comparison, synergistic differentiation implies that a combination of the physicochemical cues causes a greater effect than the sum of effects produced by each cue alone. A significant advantage would be to direct hMSCs to differentiate with the highest possible efficiency into the prescribed phenotype by recreating the native cell environment, with biology and engineering interacting at multiple levels [Burdick, 2009]. Since many physical, chemical, and genetic cues seem to promote stem cell differentiation, strategies to optimize these cues to facilitate or interfere with the intended fate of stem cells is indeed crucial for regenerative tissue engineering. Specific differentiation cues must be identified and precisely controlled to define the accurate cellular microenvironment [Metallo, 2007]. As previous studies in our laboratory and other groups have repeatedly demonstrated, stem cell differentiation may be facilitated in response to a combination of soluble factors and externally applied physical stimuli, which may additively or synergistically enhance the intended stem cell commitment [Mangeat, 1999; Titushkin, 2010]. For example, a combination of soluble factors and an electrical stimulation induces synergistic osteodifferentiation, while the use of an electrical stimulation alone fails to osteogenically differentiate hMSCs [Sun, 2007]. Both mechanical and electrical stimuli have been applied separately or combined with soluble factors to facilitate stem cell differentiation [Titushkin, 2010]. A diverse variety of currently available electromagnetic or mechanical stimulation modalities suggests multiple, plausible and effective tools for selective manipulation of the stem cell biophysical properties into a particular pre-selected cell phenotype by precisely controlling the physical stimulus characteristics. Synergistic interaction may also accelerate the rate at which the intended stem cell differentiation proceeds. In addition to demonstrating synergistic stem cell differentiation in response to a combination of soluble factors and externally applied physical stimuli, elucidation of the potential coupling mechanisms would have to be studied and elucidated. It is fascinating to observe that while these biophysical and biochemical cues are often thought of as orthogonal, when applied simultaneously to stem cells, the orthogonal cues may target separate mechanotransduction pathways or amplify the same signaling mechanisms. Much more research would be required to delineate the intricate detection/amplification systems built into the living cells and tissue.

Finally, it should be emphasized that physical or biomechanical regulation of stem cell differentiation can greatly minimize the number of growth factors that would otherwise be required for successful tissue engineering. Targeting specific combinations of physicochemical cues should allow exploiting optimally the unique biological and biophysical properties of stem cells for the prescribed differentiation. Establishment of a new paradigms in which stem cells can be manipulated physicochemically will undoubtedly expedite the current effort towards designing and engineering functional tissue constructs for stem cell-based regenerative medicine.

6 Conclusions

Stem cells are likely to play an important role in the rapidly developing field of regenerative medicine and functional tissue engineering. There is urgency to be able to regulate, control, and manipulate the potent stem cells. While biologists have long appreciated the important role for soluble factors in the regulation of stem cell differentiation, physical cues may be just as important in influencing the fate of stem cells. Distinctive biological and biophysical properties of stem cells and how they are impacted by the external differentiation signals are yet to be fully explored. Stem cells also exhibit a unique set of biomechanical features that are quite different from other mature cells. Since cellular biomechanics is known to be intimately involved in many cellular functions and activities, elucidation of the coupling mechanisms between biological and mechanical transformations in differentiating stem cells is of great interest among cell biologists, biomedical engineers and clinicians. A number of theoretical models, useful tools (e.g., AFM and LOT) and techniques (e.g., multi-content image analysis) have been developed to mechanically probe and manipulate stem cells and interpret the results properly. These technological advances allow the researcher to explore the spatial organization and elasticity of cytoskeleton structures in connection with membrane mechanics and overall cellular mechanical properties and responses in various cell types in near physiological conditions. This could open exciting possibilities to control stem cell responses and fate for various tissue engineering applications. Understanding the detailed mechanisms of stem cell biomechanical and molecular responses to external cues represents will undoubtedly lead to general models of stem cell mechanobiology as well as to stem cell-based regenerative medicine. Future work should include execution and full analysis of multifactorial design experiments in which the presumed orthogonal cues (e.g., soluble biochemical and physical force are non-interactive) might interact to produce higher order effects. However, these cues would have to be optimized for a particular phenotype that is distinctively different from others morphologically and functionally (e.g., excitable vs. non-excitable tissues). In spite of such difficulty, advances in the molecular, biophysical, and imaging tools should allow the establishment of new paradigms in which stem cells can be manipulated physicochemically for the intended differentiation. With the abundance of quantitative experimental data that are now available from sophisticated imaging modalities, powerful computation techniques are being developed that accurately analyze and predict the cellular and nuclear transformations which are potentially required to initiate the onset of stem cell differentiation into multiple lineages. The combinatorial approach of 3D high-resolution imaging and computational modeling should provide a robust system to better understand hMSC differentiation and ultimately reduce heterogeneity in stem cell populations.

Finally, it is important to raise the question whether our current understanding of manipulation and regulation of hMSCs could be applicable to other types of stem cells, particularly inducible pluripotent stem cells (iPSCs). By genetically reprogramming mature adults cells to dedifferentiate, iPSCs are believed to resemble and have the capability similar to embryonic stem cells. For example, different phenotypic cells of the three germ layers can be generated by iPSCs [Armstrong, 2010; Tonge, 2014]. The potential for iPSCs to be utilized for drug testing/development and regenerative tissue engineering is only beginning to emerge. One major obstacle that must be overcome before iPSCs can be fully exploited is to substantially increase the transfection efficiency and to develop alternative delivery mechanisms (e.g., non-viral delivery) that are more compatible with in vivo and clinical studies. With respect to biomechanics and mechanotransduction mechanisms, it remains virtually unknown how iPSCs may behave and respond to the external biophysical cues. Our laboratory is currently launching a series of studies to characterize and compare the mechanical properties of hMSCs and human iPSCs and elucidate possible different mechanosignaling pathways that might be stem cell type-dependent.

Acknowledgments The authors acknowledge funding sources: National Institutes of Health grant (HL083298) and a grant from the Office of Navy Research (N00014-13-1-0404), and a grant from the Hong Kong Research Grants Council.

References

Addae-Mensah, K.A., Wikswo, J.P.: Measurement techniques for cellular biomechanics in vitro. Exp. Biol. Med. **233**, 792–809 (2008)

Albiges-Rizo, C., Destaing, O., Fourcade, B., Planus, E., Block, M.R.: Actin machinery and mechanosensitivity in invadopodia, podosomes and focal adhesions. J. Cell Sci. **122**, 3037–3049 (2009)

Armstrong, L., Tilgner, K., Saretzki, G., Atkinson, S.P., Stojkovic, M., Moreno, R., Przyborski, S., Lako, M.: Human induced pluripotent stem cell lines show stress defense mechanisms and mitochondrial regulation similar to those of human embryonic stem cells. Stem Cells **28**, 661–673 (2010)

Berridge, M.J., Bootman, M.D., Roderick, H.L.: Calcium signaling: dynamics, homeostasis and remodelling. Nat. Rev. Mol. Cell Biol. **4**, 517–529 (2003)

Berridge, M.J.: Calcium signal transduction and cellular control mechanisms. Biochim. Biophys. Acta **1742**, 3–7 (2004)

Bershadsky, A.D., Ballestrem, C., Carramusa, L., Zilberman, Y., Gilquin, B., Khochbin, S., et al.: Assembly and mechanosensory function of focal adhesions: experiments and models. Eur. J. Cell Biol. **85**, 165–173 (2006)

Burdick, J.A., Vunjak-Novakovic, G.: Review: Engineered microenvironments for controlled stem cell differentiation. Tissue Eng. Part A **2**, 205–219 (2009)

Chakraborty, A., Perales, M.M., Reddy, G.V., Roy-Chowdhury, A.K.: Adaptive geometric tessellation for 3D reconstruction of anisotropically developing cells in multilayer tissues from sparse volumetric microscopy images. PLoS ONE (2013). doi:10.1371/journal.pone.0067202

Chen, C.S., Ingber, D.E.: Tensegrity and mechanoregulation: from skeleton to cytoskeleton. Osteoarthr. Cartil. **7**, 81–94 (1999)

Chen, C.S., Tan, J., Tien, J.: Mechanotransduction at cell-matrix and cell-cell contacts. Ann. Rev. Biomed. Eng. **6**, 275–302 (2004)

Chen, H., Titushkin, I., Stroscio, M., Cho, M.: Altered membrane dynamics of quantum dot-conjugated integrins during osteogenic differentiation of human bone marrow derived progenitor cells. Biophys. J. **92**, 1399–1408 (2007)

Chen, Z., Gibson, T.B., Robinson, F., Silvestro, L., Pearson, G., Xu, B., et al.: MAP kinases. Chem. Rev. **101**, 2449–2476 (2001)

Cho, M.R., Marler, J.P., Thatte, H.S., Golan, D.E.: Control of calcium entry in human fibroblasts by frequency-dependent electrical stimulation. Frontiers Biosci. **7**, 1–8 (2002)

Cho, M.R.: A review of electrocoupling mechanisms mediating facilitated wound healing. IEEE Trans. Plasma Sci. **30**, 1504–1515 (2002). Invited Review

Chrzanowska-Wodnicka, M., Burridge, K.: Rho-stimulated contractility drives the formation of stress fibers and focal adhesions. J. Cell Biol. **133**, 1403–1415 (1996)

Collinworth, A.M., Zhang, S., Kraus, W.E., Truskey, G.A.: Apparent elastic modulus and hysteresis of skeletal muscle cells throughout differentiation. Am. J. Physiol. Cell Physiol. **283**, C1219–C1227 (2002)

Costa, K.D., Yin, F.C.: Analysis of indentation: implications for measuring mechanical properties with atomic force microscopy. J. Biomech. Eng. **121**, 462–471 (1999)

Costa, K.D.: Imaging and probing cell mechanical properties with the atomic force microscope. Meth. Mol. Biol. **319**, 331–361 (2006)

Dai, J., Sheetz, M.P., Wan, X., Morris, C.E.: Membrane tension in swelling and shrinking molluscan neurons. J. Neurosci. **18**, 6681–6692 (1998)

Darling, E.M., Zauscher, S., Guilak, F.: Viscoelastic properties of zonal articular chondrocytes measured by atomic force microscopy. Osteoarth. Cartil. **14**, 571–579 (2006)

Deng, Z., Sharff, K.A., Tang, N., Song, W., Luo, J., Luo, X., et al.: Regulation of osteogenic differentiation during skeletal development. Front. Biosci. **13**, 2001–2021 (2008)

Dong, C., Skalak, R., Sung, K.L., Schmid-Schönbein, G.W., Chien, S.: Passive deformation analysis of human leukocytes. J. Biomech. Eng. **110**, 27–36 (1998)

Dong, C., Skalak, R., Sung, K.L.: Cytoplasmic rheology of passive neutrophils. Biorheol. **28**, 557–567 (1991)

Docheva, D., Popov, C., Mutschler, W., Schieker, M.: Human mesenchymal stem cells in contact with their environment: surface characteristics and the integrin system. J. Cell. Molec. Med. **11**, 21–38 (2007)

Estes, B.T., Gimble, J.M., Guilak, F.: Mechanical signals as regulators of stem cell fate. Curr. Top. Dev. Biol. **60**, 91–126 (2004)

Engler, A.J., Sen, S., Sweeney, H.L., Discher, D.E.: Matrix elasticity directs stem cell lineage specification. Cell **126**, 677–689 (2006)

Evans, E., Yeung, A.: Apparent viscosity and cortical tension of blood granulocytes determined by micropipet aspiration. Biophys. J. **56**, 151–160 (1989)

Even-Ram, S., Artym, V., Yamada, K.M.: Matrix control of stem cell fate. Cell **126**, 645–647 (2006)

Foreman, M.A., Smith, J., Publicover, S.J.: Characterization of serum-induced intracellular Ca2 + oscillations in primary bone marrow stromal cells. J. Cell. Physiol. **206**, 664–671 (2006)

Galli, C., Guizzardi, S., Passeri, G., Macaluso, G.M., Scandroglio, R.: Life on the wire: on tensegrity and force balance in cells. Acta Biomed. **76**, 5–12 (2005)

Guilak, F., Jones, W.R., Ting-Beall, H.P., Lee, G.M.: The deformation behavior and mechanical properties of chondrocytes in articular cartilage. Osteoarth. Cartil. **7**, 59–70 (1999)

Guilak, F., Cohen, D.M., Estes, B.T., Gimble, J.M., Liedtke, W., Chen, C.S.: Control of stem cell fate by physical interactions with the extracellular matrix. Cell Stem Cell **5**, 17–26 (2009)

Hochmuth, R.M., Ting-Beall, H.P., Beaty, B.B., Needham, D., Tran-Son-Tay, R.: Viscosity of passive human neutrophils undergoing small deformations. Biophys. J. **64**, 1596–1601 (1993)

Hoffman, B.D., Crocker, J.C.: Cell mechanics: dissecting the physical responses of cells to force. Ann. Rev. Biomed. Eng. **11**, 259–288 (2009)

Hughes, S.C., Fehon, R.G.: Understanding ERM proteins–the awesome power of genetics finally brought to bear. Curr. Opin. Cell Biol. **19**, 51–56 (2007)

Humphrey, J.D.: Stress, strain, and mechanotransduction in cells. J. Biomech. Eng. **123**, 638–641 (2001)

Ingber, D.E.: Tensegrity: the architectural basis of cellular mechanotransduction. Ann. Rev. Physiol. **59**, 575–599 (1997)

Ingber, D.E.: Cellular mechanotransduction: putting all the pieces together again. FASEB J. **20**, 811–827 (2006)

Katsumi, A., Orr, A.W., Tzima, E., Schwartz, M.A.: Integrins in mechanotransduction. J. Biol. Chem. **279**, 12001–12004 (2004)

Kawano, S., Otsu, K., Kuruma, A., Shoji, S., Yanagida, E., Muto, Y., et al.: ATP autocrine/paracrine signaling induces calcium oscillations and NFAT activation in human mesenchymal stem cells. Cell Calcium **39**, 313–324 (2006)

Khatib, L., Golan, D.E., Cho, M.R.: Physiologic electrical stimulation provokes calcium increase mediated by phospholipase C activation in human osteoblasts. FASEB J. (2004). doi:10.1096/fj.04-1814fje

Kim, T., Seong, J., Ouyang, M., Sun, J., Lu, S., Hong, J.P., et al.: Substrate rigidity regulates Ca2 + oscillation via RhoA pathway in stem cells. J. Cell. Physiol. **218**, 285–293 (2009)

Koay, E.J., Shieh, A.C., Athanasiou, K.A.: Creep indentation of single cells. J. Biomech. Eng. **125**, 334–341 (2003)

Kotani, H., Takaishi, K., Sasaki, T., Takai, Y.: Rho regulates association of both the ERM family and vinculin with the plasma membrane in MDCK cells. Oncogene **14**, 1705–1713 (1997)

Kuznetsova, T.G., Starodubtseva, M.N., Yegorenkov, N.I., Chizhik, S.A., Zhdanov, R.I.: Atomic force microscopy probing of cell elasticity. Micron **38**, 824–833 (2007)

Liao, Z., Feng, S., Popel, A.S., Brownell, W.E., Spector, A.A.: Outer hair cell active force generation in the cochlear environment. J. Acoust. Soc. Am. **122**, 2215–2225 (2007)

Lim, C.T., Zhou, E.H., Quek, S.T.: Mechanical models for living cells – A review. J. Biomech. **39**, 195–216 (2006)

López-Toledano, M.A., Redondo, C., Lobo, M.V.T., Reimers, D., Herranz, A.S., Paíno, C.L., et al.: Tyrosine hydroxylase induction by basic fibroblast growth factor and cyclic AMP analogs in striatal neural stem cells: role of ERK1/ERK2 mitogen-activated protein kinase and protein kinase C. J. Histochem. Cytochem. **52**, 1177–1189 (2004)

Louvet-Vallée, S.: ERM proteins: from cellular architecture to cell signaling. Biol. Cell **92**, 305–316 (2000)

Maffulli, N., Wong, J.: Rupture of the Achilles and patellar tendons. Clin. Sports Med. **22**, 761–776 (2003)

Mahaffy, R.E., Park, S., Gerde, E., Kas, J., Shih, C.K.: Quantitative analysis of the viscoelastic properties of thin regions of fibroblasts using atomic force microscopy. Biophys. J. **86**, 1777–1793 (2004)

Mangeat, P., Roy, C., Martin, M.: ERM proteins in cell adhesion and membrane dynamics. Trends Cell Biol. **9**, 187–192 (1999)

Mather, A.B., Collinworth, A.M., Reichert, W.M., Kraus, W.E., Truskey, G.A.: Endothelial, cardiac muscle and skeletal muscle exhibit different viscous and elastic properties as determined by atomic force microscopy. J. Biomech. **34**, 1545–1553 (2001)

McBeath, R., Pirone, D.M., Nelson, C., Bhadriraju, K., Chen, C.S.: Cell shape, cytoskeletal tension, and RhoA regulate stem cell lineage commitment. Develop. Cell **6**, 483–495 (2004)

Mackay, D.J., Esch, F., Furthmayr, H., Hall, A.: Rho- and rac-dependent assembly of focal adhesion complexes and actin filaments in permeabilized fibroblasts: an essential role for ezrin/radixin/moesin proteins. J. Cell Biol. **138**, 927–938 (1997)

Metallo, C.M., Mohr, J.C., Detzel, C.J., de Pablo, J.J., Van Wie, B.J., Palecek, S.P.: Engineering the stem cell microenvironment. Biotechnol. Prog. **23**, 18–23 (2007)

Morris, C.E., Homann, U.: Cell surface area regulation and membrane tension. J. Membr. Biol. **179**, 79–102 (2001)

Mosaliganti, K.R., Noche, R.R., Xiong, F., Swinburne, I.A., Megason, S.G.: ACME: automated cell morphology extractor for comprehensive reconstruction of cell membranes. Biol, PLoS Compu. (2012). doi:10.1371/journal.pcbi.1002780

Neuman, K.C., Block, S.M.: Optical trapping. Rev. Sci. Instrumen. **75**, 2787–2809 (2004)

Ohashi, T., Ishii, Y., Ishikawa, Y., Matsumoto, T., Sato, M.: Experimental and numerical analyses of local mechanical properties measured by atomic force microscopy for sheared endothelial cells. Biomed. Mater. Eng. **12**, 319–327 (2002)

Peterman, E.J.G., Gittes, F., Schmidt, C.F.: Laser-induced heating in optical traps. Biophys. J. **84**, 1308–1316 (2003)

Pittenger, M.F., Mackay, A.M., Beck, S.C., Jaiswal, R.K., Douglas, R., Mosca, J.D., et al.: Ezrin interacts with focal adhesion kinase and induces its activation independently of cell-matrix adhesion. J. Biol. Chem. **276**, 37686–37691 (2001)

Radmacher, M.: Measuring the elastic properties of living cells by the atomic force microscope. Meth. Cell Biol. **68**, 67–90 (2002)

Raucher, D., Sheetz, M.P.: Characteristics of a membrane reservoir buffering membrane tension. Biophys. J. **77**, 1992–2002 (1999)

Rehfeldt, F., Engler, A.J., Eckhardt, A., Ahmed, F., Discher, D.E.: Cell responses to the mechanochemical microenvironment- implications for regenerative medicine and drug delivery. Adv. Drug Deliv. Rev. **59**, 1329–1339 (2007)

Rodríguez, J.P., González, M., Ríos, S., Cambiazo, V.: Cytoskeletal organization of human mesenchymal stem cells (MSC) changes during their osteogenic differentiation. J. Cell. Biochem. **93**, 721–731 (2004)

Saha, S., Ji, L., de Pablo, J.J., Palecek, S.P.: Inhibition of human embryonic stem cell differentiation by mechanical strain. J. Cell. Physiol. **206**, 126–137 (2006)

Sarasa-Renedo, A., Chiquet, M.: Mechanical signals regulating extracellular matrix gene expression in fibroblasts. Scand. J. Med. Sci. Sports **15**, 223–230 (2005)

Sato, M., Theret, D.P., Wheeler, L.T., Ohshima, N., Nerem, R.M.: Application of the micropipette technique to the measurement of cultured porcine aortic endothelial cell viscoelastic properties. J. Biomech. Eng. **112**, 263–268 (1990)

Sheetz, M.P.: Laser Tweezers in Cell Biology. Academic Press (1998)

Sheetz, M.P., Sable, J.E., Döbereiner, H.: Continuous membrane-cytoskeleton adhesion requires continuous accommodation to lipid and cytoskeleton dynamics. Ann. Rev. Biophys. Biomol. Struct. **35**, 417–434 (2006)

Shieh, A.C., Athanasiou, K.A.: Biomechanics of single chondrocytes and osteoarthritis. Crit. Rev. Biomed. Eng. **30**, 307–343 (2002)

Shieh, A.C., Athanasiou, K.A.: Principles of cell mechanics for cartilage tissue engineering. Ann. Biomed. Eng. **31**, 1–11 (2003)

Silver, F.H., Siperko, L.M.: Mechanosensing and mechanochemical transduction: how is mechanical energy sensed and converted into chemical energy in an extracellular matrix? Crit. Rev. Biomed. Eng. **31**, 255–331 (2003)

Simonetti, D.W., Craig, S., Marshak, D.R.: Multilineage potential of adult human mesenchymal stem cells. Science **284**, 143–147 (1999)

Stamenović, D., Ingber, D.E.: Models of cytoskeletal mechanics of adherent cells. Biomech. Model. Mechanobiol. **1**, 95–108 (2002)

Sun, S., Liu, Y., Lipsky, S., Cho, M.: Physical manipulation of calcium oscillations facilitates osteodifferentiation of human mesenchymal stem cells. FASEB J. **21**, 1472–1480 (2007)

Sun, S., Song, Z.Y., Cotler, S.J., Cho, M.: Biomechanics and functionality of hepatocytes in liver cirrhosis. J. Biomech. **47**, 2205–2210 (2013)

Sun, S., Wong, S.W., Mak, A.F.T., Cho, M.: Impact of oxidative stress on cellular biomechanics and Rho signaling in C2C12 myoblasts. J. Biomech. **47**, 3650–3656 (2014)

Takai, E., Costa, K.D., Shaheen, A., Hung, C.T., Guo, X.E.: Osteoblast elastic modulus measured by atomic force microscopy is substrate dependent. Ann. Biomed. Eng. **33**, 963–971 (2005)

Theret, D.P., Levesque, M.J., Sato, M., Nerem, R.M., Wheeler, L.T.: The application of a homogeneous half-space model in the analysis of endothelial cell micropipette measurements. J. Biomech. Eng. **110**, 190–199 (1988)

Thoumine, O., Cardoso, O., Meister, J.J.: Changes in the mechanical properties of fibroblasts during spreading: a micromanipulation study. Euro. Biophys. J. Biophys. Lett. **28**, 222–234 (1999)

Titushkin, I., Rao, V., Cho, M.R.: Mode- and cell-type dependent calcium responses induced by electrical stimulus. IEEE Trans. Plasma Sci. **32**, 1614–1619 (2004). Invited Review

Titushkin, I., Cho, M.: Distinct membrane mechanical properties of human mesenchymal stem cells determined using laser optical tweezers. Biophys. J. **90**, 2582–2591 (2006)

Titushkin, I., Cho, M.: Modulation of cellular mechanics during osteogenic differentiation of human mesenchymal stem cells. Biophys. J. **93**, 3693–3702 (2007)

Titushkin, I., Cho, M.R.: Controlling cellular biomechanics of human mesenchymal stem cells. Conf. Proc. IEEE Eng. Med. Biol Soc. **1**, 2090–2093 (2009a)

Titushkin, I., Cho, M.: Regulation of cell cytoskeleton and membrane mechanics by electric field: Role of linker proteins. Biophys J. **96**, 717–728 (2009b)

Titushkin, I., Rao, V.S., Pickard, W.F., Moros, E.G., Shafirstein, G., Cho, M.R.: Altered calcium dynamics mediates P19-derived neuron-like cell responses to millimeter-wave radiation. 2011. Radiat. Res. **172**, 725–736 (2009c)

Titushkin, I., Shin, J., Sun, S., Cho, M.: Physicochemical control of adult stem cell differentiation: shedding light on potential molecular mechanisms. J. Biomed. Biotechnol. (2010). doi:10.1155/2010/743476

Titushkin, I., Cho, M.: Altered osteogenic commitment of human mesenchymal stem cells by ERM protein-dependent modulation of cellular biomechanics. J. Biomech. **44**, 2692–2698 (2011)

Titushkin, I., Sun, S., Paul, A., Cho, M.: Control of adipogenesis by Ezrin, Radixin and Moesin-dependent biomechanics remodeling. J. Biomech. **46**, 521–526 (2013)

Tonge, P.D., Corso, A.J., Monetti, C., Hussein, S.M., Puri, M.C., Michael, I.P., et al.: Divergent reprogramming routes lead to alternative stem-cell states. Nature **516**, 192–197 (2014)

Trickey, W.R., Vail, T.P., Guilak, F.: The role of the cytoskeleton in the viscoelastic properties of human articular chondrocytes. J. Orthop. Res. **22**, 131–139 (2004)

Tsai, M.A., Frank, R.S., Waugh, R.E.: Passive mechanical behavior of human neutrophils: power-law fluid. Biophys. J. **65**, 2078–2088 (1993)

Van Vliet, K.: The biomechanics toolbox: experimental approaches for living cells and biomolecules. Acta Mater. **51**, 5881–5905 (2003)

Wang, J.H., Thampatty, B.P.: An introductory review of cell mechanobiology. Biomech. Model. Mechanobiol. **5**, 1–16 (2006)

Wang, N., Tolić-Nørrelykke, I.M., Chen, J., Mijailovich, S.M., Butler, J.P., Fredberg, J.J., et al.: Cell prestress. I. Stiffness and prestress are closely associated in adherent contractile cells. Am. J. Physiol. Cell Physiol. **282**, C606–C616 (2002)

Wang, N., Tytell, J.D., Ingber, D.E.: Mechanotransduction at a distance: mechanically coupling the extracellular matrix with the nucleus. Nat. Rev. Mol. Cell Biol. **10**, 75–82 (2009)

Wozniak, M.A., Modzelewska, K., Kwong, L., Keely, P.J.: Focal adhesion regulation of cell behavior. Biochim. Biophys. Acta **1692**, 103–119 (2004)

Wright, W.H., Sonek, G.J., Berns, M.W.: Parametric study of the forces on microspheres held by optical tweezers. Appl. Opt. **33**, 1735–1748 (1994)

Wu, Z., Wong, K., Glogauer, M., Ellen, R.P., McCulloch, C.A.: Regulation of stretch-activated intracellular calcium transients by actin filaments. Biochem. Biophys. Res. Commun. **261**, 419–425 (1999)

Yourek, G., Hussain, M.A., Mao, J.J.: Cytoskeletal changes of mesenchymal stem cells during differentiation. Amer. Soc. Artificial Internal Organs J. **53**, 219–228 (2007)

Zhou, D., He, Q.S., Wang, C., Zhang, J., Wong-Staal, F.: RNA interference and potential applications. Curr. Topics Med. Chem. **6**, 901–911 (2006)

Neuroregeneration: Disease Modeling and Therapeutic Strategies for Alzheimer's and Parkinson's Diseases

Angel J. Santiago-Lopez, Candice M. Hovell, Hyunjung Lee and YongTae Kim

Abstract Alzheimer's disease (AD) and Parkinson's disease (PD) are the two most common neurodegenerative diseases accounting for approximately more than 40 million patients worldwide. Progress in the fundamental understanding and effective therapy of these diseases has been hindered by the failure of currently FDA-approved pharmacological agents and the lack of research models that properly recapitulate disease pathology. One promising approach for the treatment of these diseases is regenerative medicine and its associated technologies. The rise of stem cells technology, cellular replacement, gene therapy, and immunomodulation together present an opportunity for novel combination approaches for AD and PD treatment. Additionally, fundamental studies of

A.J. Santiago-Lopez
School of Chemical and Biomolecular Engineering, Georgia Institute of Technology, Atlanta, GA 30332, USA

A.J. Santiago-Lopez · Y. Kim
Institute for Bioengineering and Bioscience, Georgia Institute of Technology, 345 Ferst Drive (Rm 3134), Atlanta, GA 30332, USA

C.M. Hovell · H. Lee · Y. Kim
Department of Biomedical Engineering, Georgia Institute of Technology, Atlanta, GA 30332, USA

Y. Kim
School of Mechanical Engineering, Georgia Institute of Technology, Atlanta, GA 30332, USA

Y. Kim (✉)
Institute for Electronics and Nanotechnology, Georgia Institute of Technology, Atlanta, GA 30332, USA
e-mail: ytkim@gatech.edu

© Springer International Publishing Switzerland 2016

H. Jo et al. (eds.), *Biomedical Engineering: Frontier Research and Converging Technologies*, Biosystems & Biorobotics 9,
DOI: 10.1007/978-3-319-21813-7_13

human-derived cellular environments may enable truly personalized medicine. In this Chapter, we review these technologies for disease modeling and therapeutic intervention in AD and PD and discuss current challenges associated with their clinical translations.

1 Introduction

Neurodegenerative diseases can be described as neurological disorders, usually progressive in nature, in which severe neural cell loss causes cognitive dysfunction or motor deficits. These diseases include Alzheimer's Disease (AD), Parkinson's Disease (PD), Huntington's Disease (HD), amyotrophic lateral sclerosis, among others. AD and PD are the two most common neurodegenerative diseases and account for approximately 6 million patients in the US and more than 40 million worldwide [1]. At the pace pharmacological therapies are emerging, the future for AD and PD patients is by no means encouraging. The cost of caring for those with neurodegenerative diseases is estimated to total $214 billion in 2014, increasing to $1.2 trillion (in today's dollars) by 2050 for the treatment of 16 million AD patients in the US alone [1]. Currently, there is no perfect cure for these neurodegenerative diseases. This challenge underscores a large unmet need to develop new treatments to alter the course of the diseases and improve the quality of life for AD and PD patients.

Strategies in regenerative medicine may provide the potential for breakthroughs in the treatment of neurodegenerative diseases. In principle, regenerative medicine in neuroscience focuses on the rescue or restoration of neuronal function upon degeneration. Traditionally, the regenerative potential of neural cells was viewed as basically limited, as evidenced by the beliefs of famous neuroanatomist Santiago Ramón y Cajal: "Everything may die. Nothing may be regenerated." Nonetheless, therapeutic strategies can prevent progressive neuronal loss and degeneration in a process called neuroprotection. However, neuroprotective strategies, as well as conventional pharmacological agents, often encounter hurdles in their translation into the clinic. Mainly, lack of physiological relevance in disease model systems prevent the adequate assessment of proposed therapies.

This chapter discusses emerging strategies to more efficiently model neurodegenerative diseases in vitro. Effective and rapid clinical translation is highly dependent on the ability to properly model disease for comprehensive evaluation of potential treatments and in-depth understanding of fundamental mechanisms. This section is primarily focused on current efforts to accurately model disease pathophysiology. Furthermore, therapeutic strategies for neurorestorative and neuroprotective activity are reviewed. Thus, the intention is to provide an overview of potential uses of stem cell technology and other bioengineered strategies for the assessment and treatment of neurodegenerative diseases in the context of AD and PD.

2 Pathophysiology of Alzheimer's and Parkinson's Diseases

2.1 Alzheimer's Disease

AD is considered the most common form of dementia and characterized by the selective degeneration of basal forebrain cholinergic neurons [2]. It is clinically described as a gradually progressive disorder characterized by cognitive symptoms including impairment of episodic memory, impaired judgment, decision-making, and orientation [3]. AD is named after German psychiatrist and neuropathologist Alois Alzheimer. In 1906, Dr. Alzheimer presented the postmortem analysis of a 56-year-old patient who was originally admitted with progressive memory loss, focal symptoms, delusions and hallucinations [4]. Results from the brain histopathological analysis showed substantial formation of neurofibrillary tangles (NFTs) and neuritic plaques – the two defining pathological hallmarks currently recognized for AD. The plaques are accumulated insoluble fibrils composed of a small peptide known as β-amyloid (Aβ).

2.1.1 Amyloid Hypothesis

Masters et al. first identified Aβ as the main component of AD-like senile plaques and opened the door for a new line of Aβ-directed investigation [5]. Subsequent studies published in the early 1990s identified specific point mutations in the sequence of the amyloid precursor protein (APP) leading to a genetic framework for the description of Aβ as an APP breakdown product [6–8]. By virtue of these developments, Hardy and Giggins properly stipulated the –still debated- 'amyloid hypothesis' in an effort to describe AD pathophysiology [9]. The amyloid hypothesis states that Aβ deposition is the causative pathogenic factor of AD and stimulates a cascade of events leading to the formation of NFTs, neuronal loss, and dementia (Fig. 1).

Aβ formation mechanisms have been widely studied for AD treatment. Aβ is produced by a general physiological mechanism known as regulated intramembrane proteolysis (RIP) [10]. In particular for APP, large peptide domains are released to the extracellular environment by proteolytic cleavage of specific membrane-anchored proteases, namely α-secretase and β-secretase [11]. The latter class of protease is often referred to as the β-site APP-cleaving enzyme (BACE), which has received detailed attention as a therapeutic route as discussed later in this chapter. Following BACE cleavage, the still membrane-bounded APP residue is subjected to an intramembrane incision mediated by the γ-secretase complex. The γ-secretase complex is responsible for cleavage at three different sites (γ, ε and ζ) within the transmembrane domain, with the incision at the γ-site being the one responsible for releasing the Aβ peptide. Specifically, the incision after the Aβ 42 amino acid is responsible for the production of $A\beta_{42}$: the Aβ peptide with the most prominent ability to oligomerize and form the characteristic toxic amyloid species in AD cortex [11].

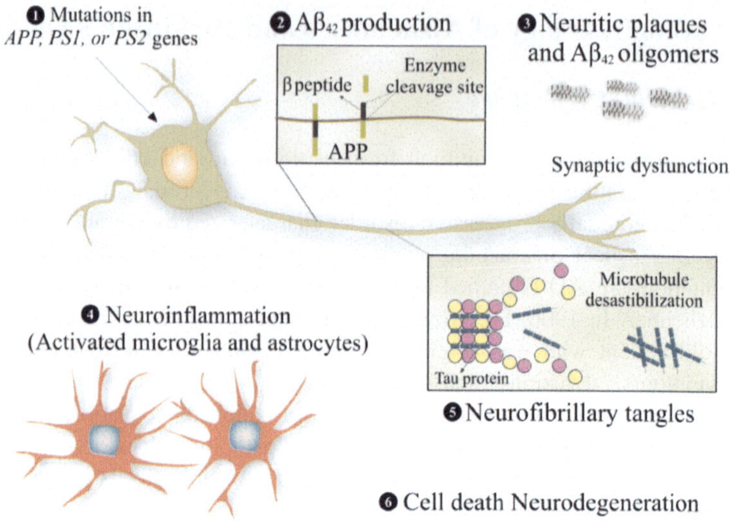

Fig. 1 The amyloid hypothesis. The generation of toxic Aβ peptides is the causative pathogenic factor of AD, and stimulates a cascade of events leading to the formation of NFTs, neuroinflammation, neuronal loss, and dementia.

Several mechanisms have been proposed to explain how Aβ affects synapse activity and cognitive function. One view is that synaptic depression is mediated by glutamate-induced excitotoxicity. The biochemical cascade initiates when Aβ oligomers prevent glutamate uptake, leading to an increased concentration of excitatory signals at the synaptic cleft. This increase in glutamate induces the activation of *N*-methyl D-aspartate receptors (NMDARs) and extrasynaptic NMDARs. Overstimulation of the post-synaptic membrane initiates desensitization of the receptors and, ultimately, synaptic depression [12, 13]. This view is supported by the evidence that the presence of Aβ oligomers disrupts glutamate clearance mechanisms, perturbs synaptic plasticity and promotes synaptic depression [14]. Working concurrently with NMDARs, other types of receptors have been linked to Aβ-induced synaptic dysfunction. These other receptors include metabotropic glutamate receptors, cellular prion proteins (PrP^c), and α7-nicotinic acetylcholine receptors [15]. Despite these findings, the amyloid hypothesis is still highly contested (Fig. 1). The main argument against the hypothesis is that substantial accumulation of senile plaques have been found in brains of healthy aged individuals and that neurofibrillary formation and neurodegeneration may predate any detectable Aβ plaques [16]. Additionally, recent failed clinical trials of Aβ-targeted therapies [17] pose current critical challenges for a mechanistic study on the role of Aβ in AD pathophysiology.

2.1.2 Tau Axis Hypothesis

According to the initial observations by Dr. Alzheimer, one key factor in the definition of AD pathophysiology is the presence of NFTs, which correspond to a self-assembled aggregation of hyperphosphorylated tau protein. Tau is one of the most predominant microtubule-associated proteins (MAPs) in cells of neural origin and is primarily regulated by phosphatases and kinsases. Other post-translational modifications, such as tau acetylation [18], are currently under study to determine their relevance in AD [19]. In addition to its main role in the microtubule (MT) stabilization machinery and in axonal transport, tau may exhibit multifunctional traits of relevance to AD pathophysiology [20, 21]. Tau aggregation is not a pathological factor exclusive of AD; in fact, there are several neurological disorders – termed *tauopathies*- characterized by tau NFTs [22]. These include frontotemporal dementia (FTD), amyotrophic lateral sclerosis, Pick's disease, and amyotrophic lateral sclerosis/parkinsonism–dementia complex.

The mechanisms by which tau imparts neurotoxicity leading to neurodegeneration are not well understood in AD. Equally important, when exactly the formation of NFTs occurs in the cascade of events in AD pathogenesis remains unanswered. The primary notion of the relationship between Aβ-induced toxicity and the formation of NFTs is that tau acts as a downstream accomplice of Aβ-induced neurodegeneration. One proposed mechanism is that Aβ promotes pathological tau filament assembly by triggering the cleavage of tau through caspases that result in a conformational state prone to polymerization [23]. It has also been proposed that tau, considered an axonal protein, could have a dendritic role relevant to Aβ-induced neurotoxicity [20]. Particularly, the identification of tau interactions with tyrosine kinase Fyn was instrumental in elucidating tau-dependent Aβ neurotoxicity, defining a new Aβ-Fyn-tau pathogenic triad [24]. Fyn phosphorylates subunit 2 of the NMDA receptor, resulting in stabilization of this receptor's interaction with PSD95, a scaffolding protein in the dendritic spines of neurons. This stabilization, in turn, strengthens signaling by the excitotoxic neurotransmitter glutamate, which enhances Aβ toxicity [25, 26]. Although this view suggests a tau-dependent Aβ toxicity, several reports point to Aβ and tau working synergistically to promote neurodegeneration, for instance, by disrupting mitochondrial function [27, 28]. On the grounds of the multiple roles that tau can play in AD pathophysiology, Ittner and Götz have formulated a new hypothesis to describe AD pathophysiology termed the 'tau axis hypothesis' (Fig. 2) [27]. Specifically, the tau axis hypothesis states that sustained increasing levels of dendritic tau make neurons vulnerable to Aβ (Fig. 2). While the tau axis hypothesis might explain AD pathophysiology in a more comprehensive manner compared to the amyloid hypothesis, it still remains to be determined whether therapeutics based on these mechanisms will have any profound effect in AD treatment.

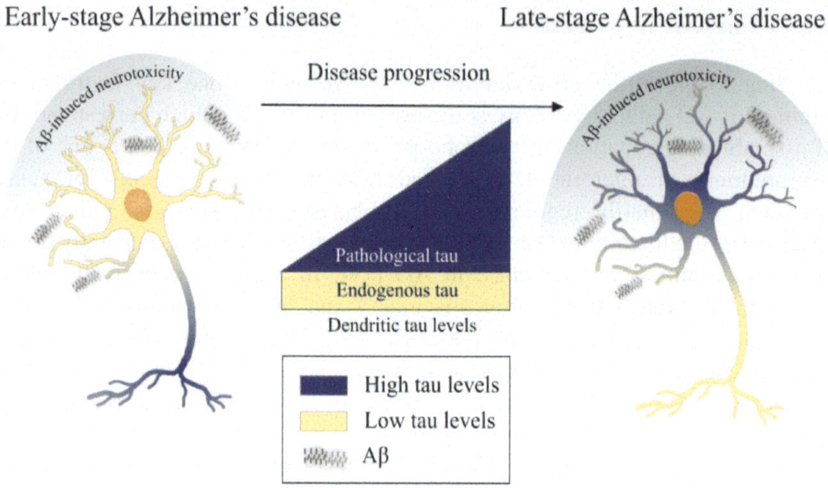

Fig. 2 The **tau axis hypothesis.** In the presence of Aβ, tau becomes increasingly phosphorylated (hyperphosphorylated or pathological tau). During disease progression, the accumulation of pathological tau in subdendritic domains makes neurons vulnerable to Aβ-induced toxicity. Adapted from [27].

2.1.3 Neuroinflammation in AD

Often seen as a secondary event, neuroinflammation is believed to have considerable implications in AD pathophysiology. This notion is supported by the increased number of studies relating AD and inflammation [29]. Some evidence indicates that neuroinflammation is an upstream event that leads to the activation of tau kinases and eventual formation of NFTs [30]. Evidence of an inflammatory response in AD includes the presence of activated microglia and astrogliosis surrounding the senile plaques, which is manifested by an increase in the number, size, and motility of astrocytes [31, 32]. In particular, Aβ-induced glial activation is likely to elicit an inflammatory response and the subsequent release of neurotoxic cytokines [33]. Aβ has been proposed to activate microglia and astrocytes through toll-like receptor 4 (TLR4) signaling that results in a downstream cascade leading to the expression of inflammatory response genes [31]. Perhaps due to their polarized nature, glial cells possess complex biology that has introduced a conflicted perception of their role in AD. For instance, although APP mice with deficient tumor growth factor β (TGF-β) signaling in neurons show more neurodegeneration and Aβ accumulation, other APP mice expressing TGF-β receptor in myeloid cells have reduced disease [29]. At the same time, the interleukin-1 (IL-1) cytokine family has been shown to involve inflammatory response in the AD paradigm. Microglia, astrocytes as well as oligodendrocytes can release neurotoxic mediators such as pro-inflammatory cytokines and reactive oxygen species (ROS) upon binding to IL-1. Moreover, the direct effect of IL-1 on neurons includes the enhancement of seizure activity and of Ca^{2+} entry through the

NMDAR ion channel, which contributes to increased susceptibility of neurons in inflammatory conditions [34].

2.1.4 Genetic Etiology

AD is considered a "polyproteinopathy" in which protein accumulation and structural conformation provide the pathological basis for neurodegeneration [35]. However, the driving forces in protein dynamics are not well understood. Hereditary cases of early-onset AD (<60 years) are identified by mutations in either APP, presenilin 1 (PSEN1), or PSEN2 genes. Whereas, predisposition for late-onset (>60 years) sporadic AD, the most common form of the disease, has been firmly linked to mutations in the apolipoprotein-E ε4 (apoE4) allele[35, 36]. In essence, apoE, a predominant form of apolipoprotein in the brain, serves as a ligand in receptor-mediated endocytosis of lipoprotein particles. Distinct apoE isoforms have been implicated in regulating mechanisms such as protein degradation [37, 38]. For this, we must put in perspective that although neurological disorders are characterized by protein deposits, it could be an impairment in clearance mechanisms rather than abnormal production. Investigation of the final stage of Aβ metabolism represents an important piece of information in AD. In this regard, lipoprotein particles reconstituted with apoE may interact with Aβ and promote amyloid clearance by receptor-mediated endocytosis of the apoE-Aβ complex. Alternatively, apoE may contribute to Aβ removal from the brain to the systemic circulation by acting as a transport protein within the blood-brain barrier (BBB) milieu [37]. Moreover, the role of apoE in cholesterol homeostasis, neuroinflammation and oxidative stress suggest an Aβ-independent function in AD pathogenesis [39].

2.2 *Parkinson's Disease*

Parkinson's disease (PD) is a progressive neurological movement disorder characterized by the degeneration of dopaminergic neurons from the pars compacta of the substantia nigra (SN) [40]. Most notably, PD patients suffer from rigidity, slow movement, impaired motor functions, resting tremor, and postural instability. A cohort of PD patients also suffers from depression-like symptoms as well as impairment at the autonomic and cognitive level. PD is pathologically defined by the presence of intracellular protein aggregates or cellular inclusions (termed Lewy bodies) and dystrophic neurites (Lewy neurites) [41]. Cell loss in the SN is also evidenced by a 'pale midbrain' or diffuse staining patterns because of the loss of melanin-containing dopaminergic neurons [42]. In 1919, Konstantin Nikolaevich Tretiakoff named these cellular inclusions Lewy bodies or 'corps de Lewy' after their initial discovery by Fritz Jakob Heinrich Lewy seven years earlier [43]. This observation served as molecular description to the neurological disorder first described by James Parkinson in 1817.

2.2.1 Mechanism of Neurodegeneration in PD

Lewy bodies in PD are mainly composed of aggregated α-synuclein, an intrinsically disordered protein with an unclear physiological function [44]. The causative effect for these protein-rich cytoplasmic lesions has been the primary focus for most years in PD research. Although familial PD is not the predominant form of the disease (<10%), research on its etiology has provided a better understanding of the neurobiological mechanisms underlying protein misfolding in sporadic cases of PD. Approximately 15 genes from the PARK family have been linked to the familial form of the disease, while sporadic cases exhibit a close correlation to environmental factors as 1-methyl-4-phenyl-1,2,3,6-tetrahydropyridine (MPTP), paraquat and rotenone. Genetic profiling in humans has led to the idea that α-synuclein drives PD pathophysiology either through SNCA dominantly inherited mutations or via gene-multiplication or gene-promoter polymorphisms that increase the susceptibility of α-synuclein expression [42]. Additionally, mutations in parkin-encoding gene PARK2 have been identified to be responsible of disruption of protein degradation mechanisms and maintenance of mitochondrial function [45]. Thus, genetic mutations and dysfunctional protein metabolism seem to be the key aspects of neurodegenerative diseases including PD.

These conclusions must draw our attention to the pathological similarities between these two neurodegenerative diseases, AD and PD. For instance, during self-assembly α-synuclein undergoes a conformational change to a β-sheet structure that is also characteristic of amyloid deposits in AD [46]. Moreover, mutations in MAP tau have been associated with parkinsonism [42]. Conversely, the mechanisms by which cellular inclusions exert neurotoxicity in PD are a topic of intensive research. As opposed to the synaptic dysfunction caused by extracellular Aβ oligomers and neuritic plaques in AD, cytoplasmic α-synuclein is believed to disrupt protein trafficking or deform the cell, causing cell death [47]. One proposed mechanism believed to be instrumental in PD pathophysiology is that mitochondrial dysfunction, followed by oxidative stress, induces a considerable amount of toxic species leading to neurodegeneration (Fig. 3). In this regard, mutations in proteins involved in the maintenance of mitochondrial function (DJ-1, PINK1, parkin) have led to the development of an oxidative stress hypothesis [45, 48]. Recent literature suggest that oxidative stress may lead to a disruption in protein metabolism via the ubiquitin-proteosomal system, increasing the concentration of misfolded α-synuclein aggregates [48, 49].

Concurrent lines of research indicate that kinase hyperactivity mediated by mutations in the leucine-rich repeat kinase 2 (LRRK2 or PARK8) plays a pivotal role in PD, particularly in late-onset cases [48, 50]. A postmortem analysis of six LRRK2 carriers revealed profound brainstem dopaminergic degeneration accompanied by Lewy bodies, Lewy neurites, and other pathological characteristics of parkinsonism [51]. Additionally, mutations in LRRK2 have been implicated in tremor-dominant form of the disease, further highlighting its role in disease progression [45, 52]. Although the downstream physiological effects by mutation are not clear, alterations in mutant LRRK2 showed the reduction of neuronal toxicity in primary cell cultures [53]. In association with the oxidative stress hypothesis, dopaminergic neurons derived from p.G2019S-LRKK2 mutant induced pluripotent stem cells are susceptible to caspase-3 activation and cell death after exposure to stress agents such as hydrogen peroxide, MG-132 and 6-hydroxydopamine [54]. The physiological role of LRKK2 in vivo has been demonstrated in the impairment of protein degradation mechanisms (e.g. autophagy), stimulation of inflammatory products, and oxidative damage resulting in apoptotic cell death [55].

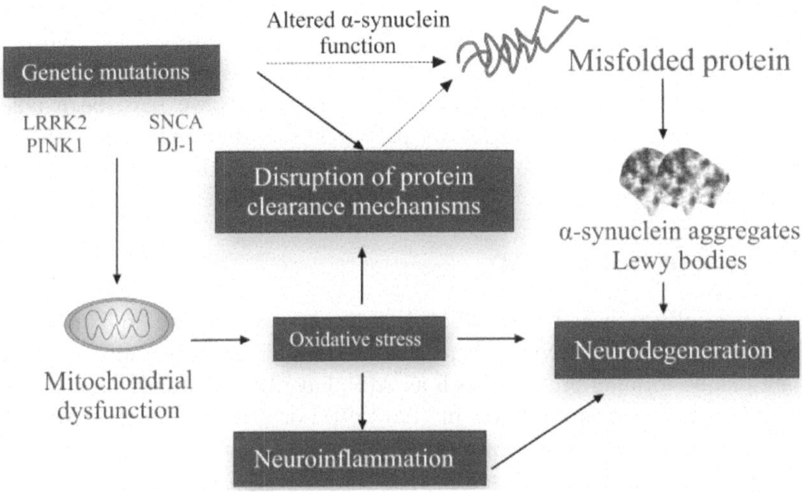

Fig. 3 Mechanisms of neurodegeneration in PD. Pathogenic mutations may disrupt protein clearance mechanisms (e.g. autophagy) and alter protein expression, causing accumulation of misfolded proteins, such as α-synuclein. Protein aggregates may induce neurodegeneration. Oxidative stress caused by mitochondrial dysfunction may also initiate a cascade of events, including neuroinflammation, leading to neurodegeneration.

2.2.2 Neuroinflammation in PD

Neuroinflammation is characterized by activated microglia, increased BBB permeability, pro-inflammatory cytokine release, and leukocyte invasion. Evidence of neuroinflammation in PD pathophysiology was first revealed by postmortem autopsies of PD brains, in which activated microglia, inflammatory markers, and T lymphocytes were identified [56]. SN degeneration in PD is susceptible to oxidative stress that results in production of substantial proinflammatory cytokines. Physiological events based on the oxidative stress hypothesis trigger deleterious inflammatory responses and promote neurodegeneration [57]. Patients with juvenile parkinsonism and PD have an increased concentration in cerebrospinal fluid of proinflammatory cytokines such as TNF and interleukins IL-1β, IL-2, IL-4 and IL-6 [58]. Glial activation and inflammatory response is often seen as a 'neuroprotective' strategy; however, the progression of this response in an uncontrolled manner may elicit higher stress, potentially leading to more severe degeneration.

3 Models for Alzheimer's and Parkinson's Diseases

3.1 *Animal Models*

While many key features of disease pathology have been identified via postmortem analysis of human diseased brains, the underlying molecular mechanisms of AD and PD have largely been studied via animal models mimicking disease relevant pathologies. The complex neurobiology involved in AD and PD pathophysiology has made these diseases challenging to model. However, by trying to model these diseases in animals we have been able to obtain valuable information of disease pathogenesis. More importantly, in-depth understanding gives rise to improved models for drug testing and enhances their translational potentials.

3.1.1 Animal Models for AD

Substantial effort has been directed towards reproducing the two main pathological hallmarks, senile plaques and NFTs. Transgenic mice expressing mutant plaque-forming proteins, such as APP, have been a widely used model to study AD pathogenic mechanisms in vivo [59]. For example, a successful model of Aβ-related pathology was achieved with the expression of high levels of mutant APP (V717F) in transgenic heterozygous mice using a platelet-derived growth factor (PDGF)-β promoter APP. This model allowed for the identification of characteristic features of AD pathology, including the development of neuritic plaques, Aβ deposits, synaptic loss, and activated microglia [60]. It was demonstrated with this model that antibodies against Aβ could cross the BBB and were effective in reducing the amyloid burden [61]. This finding has contributed to the work on active immunization for AD therapy as discussed later in this chapter (section 4.3).

Aβ plaque formation and cognitive impairment have also been established through the incorporation of human tau in animal models [59]. One of the earliest models expressing human tau containing the FTDP-17 mutation (P301L) resulted in motor and behavioral deficits in transgenic mice, with age- and gene-dose-dependent development of NFTs [62]. This model has helped unveil important aspects relevant to FTD and AD including the relationship between the formation of NFTs and the presence of $A\beta_{1\text{-}42}$ fibrils [63], a demonstration of age-dependent NFTs formation, and memory rescue upon suppression of transgenic tau [64]. More comprehensive models of AD pathology have been developed, including a triple-transgenic mouse model that exhibits both Aβ plaques and tau NFTs. This model uses pronuclear microinjection of two independent transgene constructs encoding human APPSwe and tauP301L(4R/0N) into single-cell embryos harvested from mutant homozygous presinilin-1 (PS1M146V) knockin mice[65]. These studies have been useful in elucidating mechanisms of synaptic dysfunction and the Aβ-tau relationship, providing a better understanding of AD features [65–67]. Meanwhile, nontransgenic animal models with AD-like pathology have been used to show that administration of insoluble Aβ in the aged rhesus monkey cerebral cortex resulted in neuronal loss, tau phosphorylation, and microglial activation [68]. Additionally, tau reduction has been shown to ameliorate excitotoxicity in both transgenic and nontransgenic mice [69]. Furthermore, preservation of conservatory mechanisms in AD and PD allows for use of small model organisms such as the nematode *Caenorhabditis elegans* (*C. elegans*), *Saccharomyces cerevisiae* (yeast), the common fruitfly *Drosophila melanogaster*, and oocytes from frog *Xenopus* as reliable models for some AD pathological markers [70–74].

3.1.2 Animal Models for PD

Recapitulating key clinical and neuropathological features of PD is highly demanding [75]. One common approach to modeling PD is to induce considerable loss of dopaminergic neurons from the SN. There are two widely used toxin-induced rodent models of PD. One model is based on the intracerebral administration of the hydroxylated analogue of dopamine, oxidopamine or 6-OHDA. Upon internalization by dopaminergic neurons via the dopamine transporter, 6-OHDA is believed to initiate degeneration through a combination of oxidative stress and mitochondrial dysfunction [76]. Oxidative stress can be generated by enzymatic reaction of 6-OHDA or by susceptibility for auto-oxidation [77]. Despite 6-OHDA's unclear mechanism for neurotoxicity, animal models seem to recapitulate some of PD features well enough to serve as a predictive model. For example, in behavioral tests, 6-OHDA lesioned rats exhibited significant motor deficits akin to parkisonian akinesia or dyskinesia [78, 79]. Moreover, with the exception of antimuscarinic drugs, all drugs in clinical use today have shown efficacy in this model [76].

The other toxin-induced model involves the neurotoxin MPTP. At 3-5 days post-administration of MPTP, cardinal symptoms of parkinsonism are more than evident. These include akinesia, bradykinesia, rigidity of the limb and trunk, and postural abnormalities [76]. Its mechanism of action is proposed to impair mitochondrial respiration via inhibition of complex I of the electron transport chain. Impairment in mitochondrial function is followed by reduced ATP, ROS generation, and subsequent cell death and neurodegeneration [76, 77]. Although widely used in a number of species including mice and primates, MPTP is not effective in rats, and mouse strains vary widely in their sensitivity to the toxin [80]. For these reasons, we will focus our discussion solely on the success of the MPTP primate model. The MPTP primate model has responded positively to most of the dopamine-targeted therapy for PD. This includes responsiveness to L-DOPA (current gold standard in PD treatment), as well as to bromocriptine, pergolide, cabergoline, apomorphine, ropinirole, pramipexole, and piribedil [76]. The success of this model is not limited to pharmacological testing. In fact, recent studies are emerging in which neuromodulatory strategies (e.g. deep brain stimulation) are being tested in MPTP-lesioned primates with expectations of a better translational potential [81, 82].

Genetic manipulation in genes associated to the familial form of the disease (α-synuclein, DJ-1, PINK1, LRRK2) is also employed in the efforts to reproduce the myriad of PD pathological features. One thing to note is that no transgenic model seems to replicate dopaminergic neuronal loss. Mainly, over-expression of α-synuclein in mice using PDGF-β or Thy-1 promoters have shown deficits that, at most, may represent early alterations in motor functions that remain "subclinical" in PD patients [83]. While truncated human α-synuclein shows selective nigral dopaminergic neuron loss in a model, it fails to induce degeneration in a progressive manner. While often seen as a limitation, the lack of dopamine neuron loss in PD animal models can allow scientists to focus on other mechanisms of neurodegeneration. For instance, parkin-deficient animal models of PD provided evidence of the essential role of mitochondrial dysfunction and oxidative damage in PD pathophysiology in the absence of nigral degeneration [84, 85].

Certainly, the challenge in recapitulating the pathological identity of AD and PD is self-evident. Despite much of the understanding we have gained from animal studies, their prediction for clinical translation is still controversial [86]. In fact, failure of AD therapeutics in clinical trials have questioned hypotheses that are well supported in animal studies [17]. To date, no neuroregenerative or neuroprotective agent identified from studies in the PD animal models has been shown to be efficacious in PD patients [75].

3.2 Human Cellular Models of Alzheimer's and Parkinson's Disease

Cell culture is a very distant representation of human physiology. However, due to the emergence of three-dimensional support matrices (e.g. hydrogels), biomimetic microsystems (e.g. organs-on-chips) and stem cell technology, reliable in vitro models of human diseases have become more conceivable. This section in particular is devoted to the discussion of current efforts based on stem cell and reprogramming technology approaches to properly model AD and PD [87, 88]. In this approach, the ultimate goal is to recapitulate most of the disease pathological hallmarks in a physiologically relevant model derived from the patient's own cells or tissue. This technology involves the reprogramming of somatic cells to become a specific cell type known as induced pluripotent stem cells (iPSCs) (Fig. 4) [88, 89]. The generation of disease-specific iPSCs has been described for a variety of diseases including PD, HD, Down syndrome, and type-1 diabetes mellitus, among others [90].

3.2.1 Cellular Models for AD

In contrast to animal models, which are mostly generated through genetic manipulation of known mutations from the familial form of the disease, the generation of iPSCs from patients with sporadic forms of disease (majority of cases) would provide an insight into the formation of neurodegeneration[91]. For example, iPSC-derived neurons reprogrammed from AD patient-specific fibroblasts have been used to show that pathological factors Aβ and phosphorylated tau increase in both familial- and sporadic-derived models [92]. In addition, intracellular Aβ oligomers are produced in patients carrying the pathogenic APP mutation (APP-E693Δ) and patients with sporadic AD [93]. More recently, both Aβ and tau pathology have been reproduced in a human neural progenitor cell (hNPC) three-dimensional culture system, which responded favorably to β and γ–secretase inhibitors [94]. AD pathophysiology has also been modeled with iPSCs derived from patients with Down syndrome [95]. Down syndrome is caused by a trisomy of the chromosome that encodes APP, which might be one explanation of the early-onset AD developed by adults with this syndrome [5]. In addition to the identification of $A\beta_{1\text{-}42}$ oligomers, the presence of dendritic tau in this model further support the tau axis hypothesis as a description of AD [95].

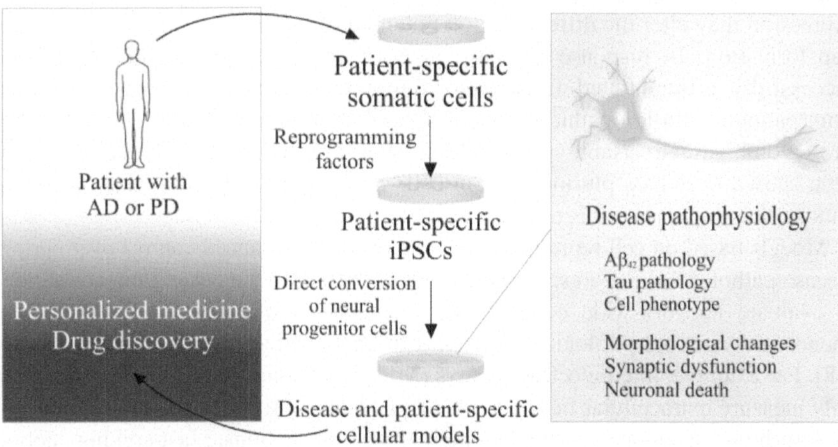

Fig. 4 Using Human Pluripotent Stem Cells for Disease Modeling. Patient-specific somatic cells (e.g. skin fibroblasts) can be reprogrammed into induced pluripotent stem cells (iPSCs) using specific genetic factors. Neuronal cells with specific disease phenotypes can be differentiated to neural stem cells. The development of in vitro cellular models based on patient-specific cells may lead to personalized medicine approaches with minimal side effects and improved pharmacological efficacy.

3.2.2 Cellular Models for PD

Human-derived cellular models hold promises in PD modeling and drug discovery. Development of culture systems based on dopaminergic neurons derived from patients with the familial or sporadic PD may enable the discovery of new neuroprotective candidates to work synergistically with currently approved pharmacological therapies. Initial work in this field includes the differentiation of human embryonic stem cells (hESCs) into neurons of dopaminergic identity [96].

Dopaminergic differentiation of iPSCs derived from human origin has proved to be achievable under media conditions with neurotrophic factors [97]. Isogenic iPSCs that differ at either of two susceptibility variants for PD were developed through the combination of stem cell technology and the zinc finger nuclease (ZFN)-mediated gene editing approach [98]. The ability to accurately model the genetics of a specific patient population represents a pivotal step towards a personalized approach in modeling and treating heterogeneous neurodegenerative diseases. In particular for PD, specific iPSCs carrying the commonly observed LRRK2 mutation have been successfully generated [54, 99]. Notably, studies using these culture systems have revealed autophagic compromise as evidenced by the accumulation of autophagic vesicles. These studies further support the notion that neurodegeneration due to protein deposits is a problem of erratic clearance mechanisms instead of increased protein production.

Reprogramming iPSCs involves heavily-oriented systems biology techniques and biochemical profiling[88]. One limitation in this technology is the use of viral vectors for the delivery of the reprogramming factors since even low vector expression may alter the differentiation potential of the iPSCs or induce malignant transformation. In response to this concern, dopaminergic neurons have been successfully differentiated from factor-free reprogrammed fibroblasts obtained from patients with idiopathic PD using Cre-recombinase excisable viruses [100]. Cre-recombinase excisable viruses (doxycline-induced lentiviral vectors) have been shown to induce pluripotency in cells derived either from mouse or human donors, thus supporting this technology[101, 102].

Models based on cell reprogramming clearly offer enhanced control over certain disease pathological features. However, we have to advance current methodologies to evaluate network-wide connectivity, mechanisms of synaptic plasticity, and phenotypic electrophysiological outcomes to characterize disease more efficiently [88]. For example, microelectrode arrays (MEAs) fall short in this quest as they can only measure extracellular field potentials, providing results that do not account for subthreshold synaptic potentials, membrane oscillations, fast-spiking action potentials, and other measures that could be assessed by intracellular recordings [103]. Certainly, the translation of these efforts would bring impact to the improvement of models for drug design and screening and, more importantly, to the development of a healthcare paradigm based on personalized medicine.

4 Strategies for Neuronal Regeneration

4.1 Pharmacological Therapy

Currently approved treatments for both AD and PD rely mostly on the use of pharmacological drugs. In general, these chemical formulations focus more on ameliorating symptoms rather than curing disease, as their neuroprotective or restorative action is incomplete. The only therapies with the potential of being neuroprotective are those targeting secondary events upon neurotoxicity (e.g. inflammation) as described at the end of this section.

Current drugs for AD target the cholinergic and glutamatergic neurotransmission systems. Given the characteristic loss of cholinergic neurons during AD progression, initial therapeutic strategies focused on counterweighting this depletion. Specifically, acetylcholinesterase inhibitors (e.g. Donepezil, Rivastigmine, Galantamine) or muscarinic and nicotinic acetylcholine receptors could improve cholinergic transmission in AD patients[104]. Logically, investigators have also pursued the quest for drugs targeting Aβ production. In particular, drug development based on the disruption of the Aβ synthesis pathway comprises mostly β- and γ-secretase inhibitors. The molecular basis of the drugs such as thiazolidinediones, rosiglitazone, and pioglitazone is the stimulation of the nuclear peroxisome proliferator-activated receptor γ (PPARγ), leading to a suppression in the expression of β-secretase or BACE1[105]. Similarly, γ-secretase inhibitors in clinical studies seem to impair Aβ production in the CNS and reduce Aβ concentrations in plasma. Interestingly, semagacestat (γ-secretase inhibitor developed by Eli Lilly) advanced up to phase-III clinical trials but failed to have significantly more efficacy than the placebo (NCT00762411). Other Aβ-targeted therapies include bexarotene, a RXR agonist, and have been shown to effectively stimulate Aβ clearance mechanisms and reverse Aβ-induced toxicity through an increased expression of the apolipoprotein apoE[106].

The gold standard treatment to compensate for SN dopamine loss in PD is the dopamine precursor levodopa (dihydroxyphenylalanine or L-DOPA)[80]. Clinical studies have shown that when co-administered with DOPA decarboxylase inhibitors (e.g. carbidopa), L-DOPA slows disease progression[107]. Dopamine agonists have also shown comparable improvement in motor control in PD patients, and monoamine oxidase B inhibitors, such as selegiline and rasagiline, have shown long-lasting benefits when given at early stages of the disease[108]. For severe motor deficits in patients with advanced PD, the NMDAR antagonist amantadine (1-aminoadamantane) is the preferred course of action. It has been shown that amantadine can reduce the severity of dyskinesias by 24% to 56%, although potential side effects include confusion and livedo reticularis[108]. Besides L-DOPA, active pharmacological treatment in PD has found limited success due to variability in results from clinical studies.

Secondary disease events such as oxidative damage and inflammation represent equally important targets for neuroregeneration. One prominent pharmacological agent is the naturally occurring turmeric (*Curcuma longa*) medicinal plant commonly referred to as curcumin. Research on curcumin as an anti-inflammatory and anti-oxidant drug dates back to the beginnings of traditional Indian medicine [109]. The role of curcumin as a neuroprotectant in different models of neuroinflammation and AD has been extensively investigated [110–112]. Although, its effective therapeutic use has been hindered due to poor bioavailability [113]. Non-steroidal anti-inflammatory drugs (NSAIDs) also demonstrate anti-amyloidogenic activity. As an example, AD transgenic mice on a six month treatment with Ibuprofen showed reduced Aβ aggregates and inflammation amelioration, but no change in oxidative damage[114]. NSAIDs mechanisms of action in Aβ pathology focuses on the impairment of Aβ aggregation either via the production of amyloid-binding proteins or through changes in protein conformation[115]. Conversely, coenzyme 10 (Co10) has received attention as a modulator of oxidative stress in AD and PD. Co10, known as ubiquinone, exhibits a plethora of neuroprotective effects ranging from mitochondrial protein uncoupling to antioxidant properties in both mitochondria and lipid membranes[116–118]. Its efficacy is highly dependent on targeting mitochondrial dysfunction, portraying a more suitable candidate for PD treatment rather than for AD. In fact, a phase-II clinical trial with 80 patients diagnosed with early-stage PD showed improved functionality in 46% as a response to different doses of Co10[116].

4.2 Cellular and Gene Therapy

In principle, neuroregeneration aims to restore neuronal function by cell reprogramming, cell replacement, or gene transfer. The theme remains to generate functional neurons in a controlled manner that could potentially replace those impaired by neurodegenerative diseases. Furthermore, the biological strategies to be discussed in this section seek to prevent the characteristic progressive neuronal loss by providing trophic support and neuroprotection.

4.2.1 AD Therapy

Neuroregenerative treatments such as cell transplantation are still in infancy and require further investigation. Recent studies in direct reprogramming technology have shown that the combinatorial expression of neural-lineage-specific transcription factors (*Ascl1*, *Brn2*, *Myt1l*) allows for the direct conversion of mouse fibroblasts into fully functional neurons. Although the molecular mechanisms involved in this direct conversion remain a topic of research, incorporation of this transcriptional program yielded cells capable of expressing neuron-specific proteins, generating action potentials, and forming functional synapses[119]. In addition, reactive glial cells were reprogrammed into functional neurons in vivo using retroviral expression of a single neural transcription factor (NeuroD1). Specifically, astrocytes were reprogrammed into glutamatergic neurons while NG2 cells were reprogrammed into glutamatergic and GABAergic neurons[120]. All reprogrammed neurons conserved electrophysiological properties; for example, cells derived from the intrinsic inflammatory response to the disease are used for regeneration.

One effective regenerative therapy for AD is to replace the substantial loss of cholinergic neurons commonly observed in Alzheimer's diseased brains. One promising strategy is to directly differentiate stem cells into cholinergic-like neurons (Fig. 5). Human umbilical cord mesenchymal stem cells derived from Wharton's jelly could be differentiated into cells of "bulbous" morphology, positively expressing cholinergic markers [121]. Another strategy is to deliver nerve growth factors (NGFs), which are known to counteract cholinergic neuronal death, stimulate cell function, and improve memory [122, 123]. While promising, NGF-based therapies face challenges in the delivery to the brain; that is, NGF excluded from the selective transport across the BBB induces deleterious side effects when delivered directly into the intracerebral region. Hence, alternate delivery schemes have been explored to accomplish neuroregeneration through NGF support. One approach is to employ ex vivo gene therapy in which patient's fibroblasts are genetically modified to produce human NGF and then injected into the lesioned brain (surgical delivery). Results from a phase 1 clinical trial using this technology revealed improved cognitive function in all eight subjects (initially diagnosed with early-stage AD)[124]. Similarly, a phase 1 clinical trial was conducted employing CERE-110, an adeno-associated, virus-based gene delivery vector that encodes for human NGF. Initial studies on rodent AD model showed dose-dependent NGF distribution after delivery and neurorestorative effect on basal forebrain cholinergic neurons[125]. Another NGF delivery scheme that has progressed to clinical trials is NsG0202, an encapsulated cell biodelivery device. The device consists of a cell immobilization matrix (scaffold) composed of polyvinyl alcohol (PVA) and a hollow fiber membrane which allow for the controlled release of the bioactive agent within the implanted region. Studies performed in Göttingen minipigs provided evidence of increased NGF levels in tissue surrounding the device and the capacity of long-term delivery up to 1 year [126]. In a recently finished clinical trial, retrieved implants showed low but persistent NGF release and no inflammatory response associated with the device implantation [127]. Another promising growth factor to AD regenerative strategies is the brain-derived growth factor (BDNF). This growth factor is considered as a potential therapeutic in neurodegenerative diseases due to its role in sustaining physiological processes of the normal adult brain [128]. Particularly, BDNF has been implicated in rescue of spatial learning and memory deficits of the transgenic mouse model of AD, exhibiting both Aβ deposits and tau pathology [129]. Additionally, direct BDNF delivery has proved to exert neuroprotective effects in multiple animal models of AD (Aβ-mice, aged rats and aged primates), supporting further studies on its clinical efficacy [130]. Despite these studies, BDNF therapy does not seem to reduce the amyloidogenic burden or NFTs pathology while improving cognitive function in an Aβ-independent manner[130, 131].

Alternate bioengineered strategies for AD therapy include siRNA delivery, genetic modification (gene therapy), and Aβ degrading enzymes [133–135]. These strategies are more directed towards preventing additional neuronal loss by targeting pathological markers instead of replacing or restoring damaged neurons. For instance, lentiviral vector delivery of BACE1-directed siRNA inactivated Aβ-producing machinery and improved behavioral deficits in APP transgenic mice [136]. Additionally, *Becn1*-expressing lentiviral vector delivery reduced both intracellular and extracellular Aβ pathology in APP transgenic mice [137]. Notably, there is a clear involvement of autophagy in neurodegenerative diseases as it can promote the degradation of misfolded proteins. Other lines of evidence, however, suggest an upstream involvement in disease progression warranting further study [138, 139].

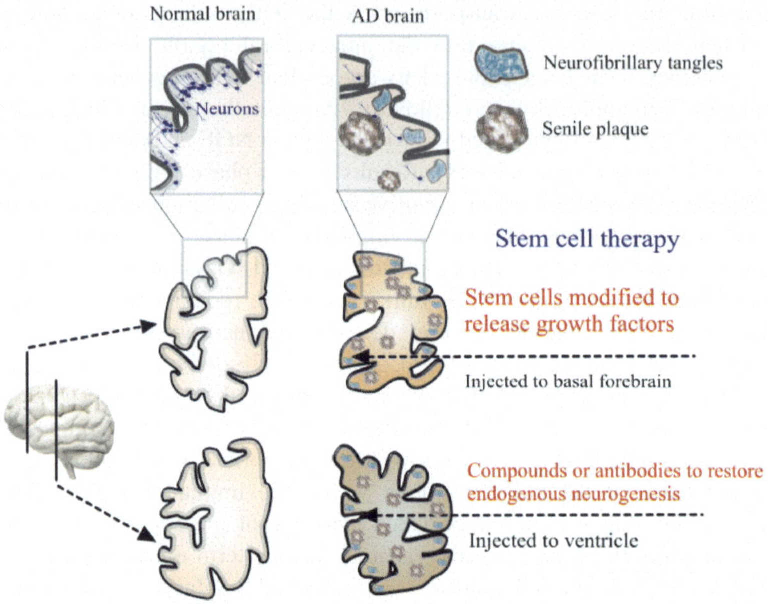

Fig. 5 Schematic of stem cell therapy in AD. Neurodegeneration in AD is characterized by loss of cholinergic neurons in the presence of senile plaques and neurofibrillary tangles. Regenerative strategies for AD treatment include the injection of stem cell-derived cholinergic neurons to replace degenerated cells. Stem cells can also be delivered to the basal forebrain to provide trophic support via the release of growth factors. Alternatives include the administration of antibodies to sequester pathological hallmarks and promote endogenous neurogenesis. Adapted from [132].

4.2.2 PD Therapy

Cell replacement strategies for PD have been contemplated for over a decade. Initial work in this area was focused on fetal nigra graft transplantation; whereas more recent study has focused on the use of reprogramming technology enabled cell transplantation from stem cells. Patients with advanced PD have shown sustained improvement in motor functions upon transplantation of fetal ventral mesencephalic tissue. Furthermore, dense regions of dopaminergic neurons have been revealed during post-mortem neuropathological assessment, and dopamine release seems to remain stable even in the long-term [140, 141]. Despite these promising results, nigra graft transplantation has not become the gold standard for PD treatment due to variable results in clinical trials. The poor outcome after transplantation may be associated with progressive dopaminergic degeneration in areas outside the grafts [142, 143]. These outcomes raise the challenge for a potential cell replacement strategy that could possibly cope with disease progression and migrate to adjacent lesioned regions in response to neurodegenerative cues. Sources of functional neurons for cell transplantation with active proliferation capabilities now include different types of stem cells or reprogrammed somatic cells (Fig. 6). Embryonic stem cells (ESCs) can spontaneously differentiate into dopaminergic neurons and improve motor deficits when transplanted in low doses into a Parkinson rat model [144]. In particular, the feasibility of cell therapy using dopaminergic neurons derived from ESCs is supported by their success in non-human primates [145] as well as their high yield under specific culture conditions [146]. Functional dopaminergic neurons derived from iPSCs (either converted directly or generated after induction of fibroblast-like cell type) have also demonstrated therapeutic potential in PD animal models [147–149]. Furthermore, dopaminergic neurons derived from either ESCs or iPSCs are capable of improving motor function in comparable levels when administered to a rat model of PD [150].

In addition to the ethical considerations with respect to ESCs research, major concerns with this strategy are the possibility of tumor formation and cell viability post-transplantation [147, 151]. One alternative to overcome this limitation is the use of direct conversion strategies to derive dopaminergic neurons from somatic sources. Most notably, human fibroblasts from PD patients can be reprogrammed into functional dopaminergic neurons by distinct combinations of key transcription factors (*Brn2, Myt1l, Ascl1, FoxA2, Nurr1* and *Lmx1a*) involved in neural differentiation and phenotype expression[152, 153]. Clearly, these advancements, although in their infancy, augur hope for the next generation of cell replacement strategies in PD treatment.

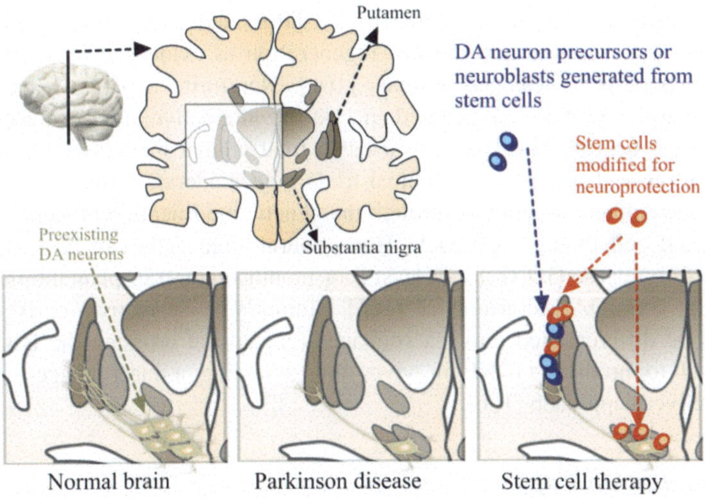

Fig. 6 Schematic of stem cell therapy in PD. Detrimental symptoms in PD result from the substantial loss of dopaminergic (DA) neurons in the substantia nigra (SN). DA neuron precursors derived from stem cells can be delivered to the putamen and the SN for degenerated cell replacement. Adapted from [132].

Alternatively, gene therapy to restore dopaminergic neurons directly or via trophic support is a subject of intensive investigation in PD. Mainly, direct gene therapy seeks to improve dopamine production in lesioned Parkinson brains or ameliorate side effects induced by conventional pharmacological drugs (e.g. L-DOPA). Overall functional improvement is observed in different animal models after the administration of enzymatic dopamine production-genes such as tyrosine hydroxylase, guanosine 5'-triphosphate cyclohydrolase 1, and L-amino acid decarboxylase or AADC [154, 155]. The use of L-DOPA, although beneficial at early stages of the disease, loses its efficacy in the long-term possibly due to a reduction in the activity of dopamine-synthesizing enzymes. In this regard, increased expression of AADC through gene delivery in a monkey PD animal model can improve the conversion of unmetabolized L-DOPA to dopamine, hence improving the clinical efficacy of the currently approved pharmacological agent to treat parkisonian symptoms [156, 157]. In support of this strategy, results from a six month baseline phase-1 clinical trial evaluating the use of adeno-associated virus vector-mediated gene delivery of AADC revealed a 46% improvement of motor functions in all six cases [158]. Non-dopaminergic strategies based on the modulation of GABA have been also introduced [159]. In this regard, gene delivery of glutamic acid decarboxylase (GAD), the enzyme responsible for GABA synthesis, significantly improved the symptoms of patients with late-stage PD [160].

Akin to AD, trophic support via growth factor administration is a desirable therapeutic route for PD that is hindered by the lack of effective administration strategies. Delivery through genetic vectors has enabled their use in PD animal models and human clinical trials. For instance, lentiviral delivery of glial cell line-derived neurotrophic factor (GDNF) to MPTP-treated aged monkeys prevents nigrostritial degeneration and induces regeneration as evidenced by recovery of functional deficits in lesioned animals [161]. Neurturin, a naturally occurring GDNF analog, is also considered a viable alternative for restoration of dopaminergic neurons and has been used to treat animal models of PD and PD patients in phase 1 and phase 2 clinical trials with moderate efficacy [162–164]. An alternate delivery strategy is through the genetic modification of stem cells for an enhanced cell therapy approach. Here, neural stem cells can be genetically engineered to release GDNF for at least four months upon transplantation into the striatum of 6-OHDA-lesioned mice [165]. Similarly, bone marrow-derived stem cells can be differentiated into macrophages (capable of crossing the BBB) and engineered to produce GDNF in vivo, thus enabling a macrophage-mediated GDNF delivery approach [166].

4.3 Immunotherapy

Immune strategies in neurodegenerative diseases seek to stimulate protein degradation and metabolism as means to prevent disease progression. The first in vivo studies of direct immunization against Aβ showed remarkable results in promoting Aβ-plaques clearance, but subsequent human clinical trials failed due to deleterious side effects, such as meningoencephalitis [167]. The need for a more profound understanding of the biological mechanisms underlying these results motivated further investigations. Aβ-plaques in AD may be formed from physiological Aβ in plasma in addition to the peptide production in the brain. The central domain anti-Aβ antibody, monoclonal antibody 266 (mab266), has a direct effect in Aβ removal from plasma and does not bind to Aβ deposits in the brain [168]. Moreover, mab266 acts effectively against memory impairment and AD symptoms without altering brain Aβ levels [169]. In a contradictory view, the intracranial administration of anti-Aβ antibodies results in diminished Aβ-plaques levels [170]. In fact, a placebo controlled phase 1 trial failed to provide significant evidence of neuronal restoration in the long term upon active immunization with anti-Aβ [171]. Although Aβ immunotherapy neutralizes Aβ burden, it has no effect in late-stage tau pathology [172, 173]. In fact, immunization strategies against tau pathology comprise a different approach for AD treatment and are currently under evaluation [174–176]. Tau immunization seeks to restore neuronal function by the removal of pathological tau and has only proven effective in treating tangles in animal models [174].

Therapeutic immunization for PD remains as a currently studied strategy [167]. Efforts in this regard include α-synuclein immunization and neuroprotection induced by the modulation of regulatory T cells. Active immunization in mice with human α-synuclein resulted in the decreased accumulation of aggregated α-synuclein in neuronal cell bodies and synapses [177]. Thus, it follows that raised antibodies from the administered therapy were able to target and promote clearance of protein aggregates in the regions of great relevance for disease progression. Adoptive transfer of copolymer-1 immune cells (T cell delivery) to MPTP-treated mice alone or in combination with α-synuclein immunization has been shown to attenuate microglial inflammatory response, increase expression of GDNF, and promote SN protection [178–180]. Although not a regenerative approach *per se*, combined immunotherapies may prevent progressive neuronal loss through enhanced neuroprotection.

5 Concluding Remarks

Neurodegenerative diseases may represent the most devastating set of diseases of the century with prevalence statistics that could reach epidemiologic levels in the near future. Failure of the currently approved therapeutic strategies and lack of research models that properly recapitulate disease pathology contribute to a scenario of little hope for current patients. In view of the lack of progress, stem cell biology and diverse bioengineered approaches may offer alternatives in disease treatment and management. Patient-specific disease models clearly offer leverage over conventional culture systems or animal models that do not mimic human physiology accurately. Consequently, the process of drug development can become more efficient and hopefully prevent the failure of costly clinical trials.

Overwhelming evidence supports the continued evaluation of regenerative strategies for the treatment of AD and PD. Certainly, it is essential to circumvent difficulties associated with the clinical translation of these therapies [132]. Basic research on safety, clinical efficacy, and potential delivery routes must be extensively conducted to observe the implementation of modern regenerative medicine in neurodegenerative diseases.

References

1. Thies, W., Bleiler, L.: 2013 Alzheimer's disease facts and figures. Alzheimers Dement **9**, 208–245 (2013). doi:10.1016/j.jalz.2013.02.003
2. Whitehouse, P.J., Price, D.L., Struble, R.G., et al.: Alzheimer's disease and senile dementia: loss of neurons in the basal forebrain. Science **215**, 1237–1239 (1982). doi:10.1126/science.7058341
3. Blennow, K., de Leon, M.J., Zetterberg, H.: Alzheimer's disease. Lancet **368**, 387–403 (2006). doi:10.1016/S0140-6736(06)69113-7
4. Goedert, M., Spillantini, M.G.: A Century of Alzheimer's Disease. Science **314**(80), 777–781 (2006)
5. Masters, C.L., Simms, G., Weinman, N.A., et al.: Amyloid plaque core protein in Alzheimer disease and Down syndrome. Proc. Natl. Acad. Sci. USA **82**, 4245–4249 (1985). doi:10.1073/pnas.82.12.4245
6. Goate, A., Chartier-Harlin, M.C., Mullan, M., et al.: Segregation of a missense mutation in the amyloid precursor protein gene with familial Alzheimer's disease. Nature **349**, 704–706 (1991). doi:10.1038/349704a0
7. Mullan, M., Crawford, F., Axelman, K., et al.: A pathogenic mutation for probable Alzheimer's disease in the APP gene at the N-terminus of beta-amyloid. Nat. Genet. **1**, 345–347 (1992). doi:10.1038/ng0892-345
8. Hardy, J.: Framing beta-amyloid. Nat. Genet. **1**, 233–234 (1992). doi:10.1038/ng0792-233
9. Hardy, J.A., Higgins, G.A.: Alzheimer's disease: the amyloid cascade hypothesis. Science **256**, 184–185 (1992). doi:10.1126/science.1566067
10. Brown, M.S., Ye, J., Rawson, R.B., Goldstein, J.L.: Regulated intramembrane proteolysis: a control mechanism conserved from bacteria to humans. Cell **100**, 391–398 (2000). doi:10.1016/S0092-8674(00)80675-3
11. Haass, C., Selkoe, D.J.: Soluble protein oligomers in neurodegeneration: lessons from the Alzheimer's amyloid beta-peptide. Nat. Rev. Mol. Cell Biol. **8**, 101–112 (2007). doi:10.1038/nrm2101
12. Palop, J.J., Mucke, L.: Amyloid-beta-induced neuronal dysfunction in Alzheimer's disease: from synapses toward neural networks. Nat. Neurosci. **13**, 812–818 (2010). doi:10.1038/nn.2583
13. Paoletti, P., Bellone, C., Zhou, Q.: NMDA receptor subunit diversity: impact on receptor properties, synaptic plasticity and disease. Nat. Rev. Neurosci. **14**, 383–400 (2013). doi:10.1038/nrn3504
14. Li, S., Hong, S., Shepardson, N.E., et al.: Soluble Oligomers of Amyloid β Protein Facilitate Hippocampal Long-Term Depression by Disrupting Neuronal Glutamate Uptake. Neuron **62**, 788–801 (2009). doi:10.1016/j.neuron.2009.05.012

15. Shankar, G.M., Li, S., Mehta, T.H., et al.: Amyloid-beta protein dimers isolated directly from Alzheimer's brains impair synaptic plasticity and memory. Nat. Med. **14**, 837–842 (2008). doi:10.1038/nm1782
16. Hardy, J., Selkoe, D.J.: The Amyloid Hypothesis of Alzheimer's Disease: Progress and Problems on the Road to Therapeutics. Science **297**(80) (2002)
17. Karran, E., Mercken, M., De Strooper, B.: The amyloid cascade hypothesis for Alzheimer's disease: an appraisal for the development of therapeutics. Nat. Rev. Drug. Discov. **10**, 698–712 (2011). doi:10.1038/nrd3505
18. Cohen, T.J., Guo, J.L., Hurtado, D.E., et al.: The acetylation of tau inhibits its function and promotes pathological tau aggregation. Nat. Commun. **2**, 252 (2011). doi:10.1038/ncomms1255
19. Martin, L., Latypova, X., Terro, F.: Post-translational modifications of tau protein: Implications for Alzheimer's disease. Neurochem. Int. **58**, 458–471 (2011). doi:10.1016/j.neuint.2010.12.023
20. Ittner, L.M., Ke, Y.D., Delerue, F., et al.: Dendritic function of tau mediates amyloid-beta toxicity in Alzheimer's disease mouse models. Cell **142**, 387–397 (2010). doi:10.1016/j.cell.2010.06.036
21. Morris, M., Maeda, S., Vossel, K., Mucke, L.: The Many Faces of Tau. Neuron **70**, 410–426 (2011). doi:10.1016/j.neuron.2011.04.009
22. Lee, V.M., Goedert, M., Trojanowski, J.Q.: Neurodegenerative Tauopathies. Annu. Rev. Neurosci. **24**, 1121–1159 (2001). doi:10.1146/annurev.neuro.24.1.1121
23. Gamblin, T.C., Chen, F., Zambrano, A., et al.: Caspase cleavage in tau: Linking amyloid and neurofibrillary tangles in Alzheimer's disease. Proc. Natl. Acad. Sci. **100** (2003)
24. Roberson, E.D., Halabisky, B., Yoo, J.W., et al.: Amyloid-β/Fyn-induced synaptic, network, and cognitive impairments depend on tau levels in multiple mouse models of Alzheimer's disease. J. Neurosci. **31**, 700–711 (2011). doi:10.1523/JNEUROSCI.4152-10.2011
25. Lee, G., Thangavel, R., Sharma, V.M., et al.: Phosphorylation of tau by fyn: implications for Alzheimer's disease. J. Neurosci. **24**, 2304–2312 (2004). doi:10.1523/JNEUROSCI.4162-03.2004
26. Haass, C., Mandelkow, E.: Fyn-tau-amyloid: A toxic triad. Cell **142**, 356–358 (2010). doi:10.1016/j.cell.2010.07.032
27. Ittner, L.M., Götz, J.: Amyloid-β and tau–a toxic pas de deux in Alzheimer's disease. Nat. Rev. Neurosci. **12**, 65–72 (2011). doi:10.1038/nrn2967
28. Rhein, V., Song, X., Wiesner, A., et al.: Amyloid-beta and tau synergistically impair the oxidative phosphorylation system in triple transgenic Alzheimer's disease mice. Proc. Natl. Acad. Sci. USA **106**, 20057–20062 (2009). doi:10.1073/pnas.0905529106
29. Wyss-Coray, T., Rogers, J.: Inflammation in Alzheimer disease-A brief review of the basic science and clinical literature. Cold Spring Harb. Perspect. Med. **2**, 1–23 (2012). doi:10.1101/cshperspect.a006346
30. Ballatore, C., Lee, V.M.-Y., Trojanowski, J.Q.: Tau-mediated neurodegeneration in Alzheimer's disease and related disorders. Nat. Rev. Neurosci. **8**, 663–672 (2007). doi:10.1038/nrn2194
31. Glass, C.K., Saijo, K., Winner, B., et al.: Mechanisms underlying inflammation in neurodegeneration. Cell **140**, 918–934 (2010). doi:10.1016/j.cell.2010.02.016

32. Akiyama, H., Barger, S., Barnum, S., et al.: Inflammation and Alzheimer's disease. Neurobiol. Aging (2000). doi:10.1016/S0197-4580(00)00124-X
33. Morales, I., Guzmán-Martínez, L., Cerda-Troncoso, C., et al.: Neuroinflammation in the pathogenesis of Alzheimer's disease. A rational framework for the search of novel therapeutic approaches. Front. Cell. Neurosci. **8**, 112 (2014). doi:10.3389/fncel.2014.00112
34. Zipp, F., Aktas, O.: The brain as a target of inflammation: common pathways link inflammatory and neurodegenerative diseases. Trends Neurosci. **29**, 518–527 (2006). doi:10.1016/j.tins.2006.07.006
35. Huang, Y., Mucke, L.: Alzheimer mechanisms and therapeutic strategies. Cell **148**, 1204–1222 (2012). doi:10.1016/j.cell.2012.02.040
36. Bertram, L., McQueen, M.B., Mullin, K., et al.: Systematic meta-analyses of Alzheimer disease genetic association studies: the AlzGene database. Nat. Genet. **39**, 17–23 (2007). doi:10.1038/ng1934
37. Kim, J., Basak, J.M., Holtzman, D.M.: The role of apolipoprotein E in Alzheimer's disease. Neuron **63**, 287–303 (2009). doi:10.1016/j.neuron.2009.06.026
38. Liu, C.-C., Liu, C.-C., Kanekiyo, T., et al.: Apolipoprotein E and Alzheimer disease: risk, mechanisms and therapy. Nat. Rev. Neurol. **9**, 106–118 (2013). doi:10.1038/nrneurol.2012.263
39. Jofre-Monseny, L., Minihane, A.-M., Rimbach, G.: Impact of apoE genotype on oxidative stress, inflammation and disease risk. Mol. Nutr. Food Res. **52**, 131–145 (2008). doi:10.1002/mnfr.200700322
40. Obeso, J.A., Rodriguez-Oroz, M.C., Goetz, C.G., et al.: Missing pieces in the Parkinson's disease puzzle. Nat. Med. **16**, 653–661 (2010). doi:10.1038/nm.2165
41. Moore, D.J., West, A.B., Dawson, V.L., Dawson, T.M.: Molecular pathophysiology of Parkinson's disease. Annu. Rev. Neurosci. **28**, 57–87 (2005). doi:10.1146/annurev.neuro.28.061604.135718
42. Shulman, J.M., De Jager, P.L., Feany, M.B.: Parkinson's disease: genetics and pathogenesis. Annu. Rev. Pathol. **6**, 193–222 (2011). doi:10.1146/annurev-pathol-011110-130242
43. Goedert, M., Spillantini, M.G., Del Tredici, K., Braak, H.: 100 years of Lewy pathology. Nat. Rev. Neurol. **9**, 13–24 (2013). doi:10.1038/nrneurol.2012.242
44. Lees, A.J., Hardy, J., Revesz, T., Lila, R.: Parkinson's disease. Lancet **373**, 2055–2066 (2009)
45. Nuytemans, K., Theuns, J., Cruts, M., Van Broeckhoven, C.: Genetic etiology of Parkinson disease associated with mutations in the SNCA, PARK2, PINK1, PARK7, and LRRK2 genes: a mutation update. Hum. Mutat. **31**, 763–780 (2010). doi:10.1002/humu.21277
46. Serpell, L.C., Berriman, J., Jakes, R., et al.: Fiber diffraction of synthetic alpha-synuclein filaments shows amyloid-like cross-beta conformation. Proc. Natl. Acad. Sci. USA **97**, 4897–4902 (2000). doi:10.1073/pnas.97.9.4897
47. Dauer, W., Przedborski, S.: Parkinson's Disease: Mechanisms and Models. Neuron **39**, 889–909 (2003)
48. Henchcliffe, C., Beal, M.F.: Mitochondrial biology and oxidative stress in Parkinson disease pathogenesis. Nat. Clin. Pract. Neurol. **4**, 600–609 (2008). doi:10.1038/ncpneuro0924

49. Schapira, A.H.: Mitochondria in the aetiology and pathogenesis of Parkinson's disease. Lancet Neurol. **7**, 97–109 (2008). doi:10.1016/S1474-4422(07)70327-7
50. West, A.B., Moore, D.J., Biskup, S., et al.: Parkinson's disease-associated mutations in leucine-rich repeat kinase 2 augment kinase activity. Proc. Natl. Acad. Sci. USA **102**, 16842–16847 (2005). doi:10.1073/pnas.0507360102
51. Zimprich, A., Biskup, S., Leitner, P., et al.: Mutations in LRRK2 cause autosomal-dominant parkinsonism with pleomorphic pathology. Neuron **44**, 601–607 (2004). doi:10.1016/j.neuron.2004.11.005
52. Clarimón, J., Pagonabarraga, J., Paisán-Ruíz, C., et al.: Tremor dominant Parkinsonism: Clinical description and LRRK2 mutation screening. Mov. Disord. **23**, 518–523 (2008). doi:10.1002/mds.21771
53. Smith, W.W., Pei, Z., Jiang, H., et al.: Kinase activity of mutant LRRK2 mediates neuronal toxicity. Nat. Neurosci. **9**, 1231–1233 (2006). doi:10.1038/nn1776
54. Nguyen, H.N., Byers, B., Cord, B., et al.: LRRK2 mutant iPSC-derived da neurons demonstrate increased susceptibility to oxidative stress. Cell Stem Cell **8**, 267–280 (2011). doi:10.1016/j.stem.2011.01.013
55. Tong, Y., Yamaguchi, H., Giaime, E., et al.: Loss of leucine-rich repeat kinase 2 causes impairment of protein degradation pathways, accumulation of alpha-synuclein, and apoptotic cell death in aged mice. Proc. Natl. Acad. Sci. USA **107**, 9879–9884 (2010). doi:10.1073/pnas.1004676107
56. Hirsch, E.C., Hunot, S.: Neuroinflammation in Parkinson's disease: a target for neuroprotection? Lancet Neurol. **8**, 382–397 (2009). doi:10.1016/S1474-4422(09)70062-6
57. Tansey, M.G., Goldberg, M.S.: Neuroinflammation in Parkinson's disease: its role in neuronal death and implications for therapeutic intervention. Neurobiol. Dis. **37**, 510–518 (2010). doi:10.1016/j.nbd.2009.11.004
58. Mogi, M., Harada, M., Narabayashi, H., et al.: Interleukin (IL)-1??, IL-2, IL-4, IL-6 and transforming growth factor-?? levels are elevated in ventricular cerebrospinal fluid in juvenile parkinsonism and Parkinson's disease. Neurosci. Lett. **211**, 13–16 (1996). doi:10.1016/0304-3940(96)12706-3
59. Götz, J., Ittner, L.M.: Animal models of Alzheimer's disease and frontotemporal dementia. Nat. Rev. Neurosci. **9**, 532–544 (2008). doi:10.1038/nrn2420
60. Games, D., Adams, D., Alessandrini, R., et al.: Alzheimer-type neuropathology in transgenic mice overexpressing V717F beta-amyloid precursor protein. Nature **373**, 523–527 (1995). doi:10.1038/373523a0
61. Bard, F., Cannon, C., Barbour, R., et al.: Peripherally administered antibodies against amyloid beta-peptide enter the central nervous system and reduce pathology in a mouse model of Alzheimer disease. Nat. Med. **6**, 916–919 (2000). doi:10.1038/78682
62. Lewis, J., McGowan, E., Rockwood, J., et al.: Neurofibrillary tangles, amyotrophy and progressive motor disturbance in mice expressing mutant (P301L) tau protein. Nat. Genet. **25**, 402–405 (2000). doi:10.1038/79109
63. Götz, J., Chen, F., van Dorpe, J., Nitsch, R.M.: Formation of neurofibrillary tangles in P3011 tau transgenic mice induced by Abeta 42 fibrils. Science **293**, 1491–1495 (2001). doi:10.1126/science.1062097
64. Santacruz, K., Lewis, J., Spires, T., et al.: Tau suppression in a neurodegenerative mouse model improves memory function. Science **309**, 476–481 (2005). doi:10.1126/science.1113694

65. Oddo, S., Caccamo, A., Shepherd, J.D., et al.: Triple-transgenic model of Alzheimer's Disease with plaques and tangles: Intracellular Aβ and synaptic dysfunction. Neuron **39**, 409–421 (2003). doi:10.1016/S0896-6273(03)00434-3
66. Oddo, S., Caccamo, A., Kitazawa, M., et al.: Amyloid deposition precedes tangle formation in a triple transgenic model of Alzheimer's disease. Neurobiol. Aging **24**, 1063–1070 (2003). doi:10.1016/j.neurobiolaging.2003.08.012
67. Winton, M.J., Lee, E.B., Sun, E., et al.: Intraneuronal APP, not free Aβ peptides in 3xTg-AD mice: implications for tau versus Aβ-mediated Alzheimer neurodegeneration. J. Neurosci. **31**, 7691–7699 (2011). doi:10.1523/JNEUROSCI.6637-10.2011
68. Yamada, K., Nabeshima, T.: Animal models of Alzheimer's disease and evaluation of anti-dementia drugs. Pharmacol. Ther. **88**, 93–113 (2000). doi:10.1016/S0163-7258(00)00081-4
69. Roberson, E.D., Scearce-Levie, K., Palop, J.J., et al.: Reducing endogenous tau ameliorates amyloid beta-induced deficits in an Alzheimer's disease mouse model. Science **316**, 750–754 (2007). doi:10.1126/science.1141736
70. McColl, G., Roberts, B.R., Gunn, A.P., et al.: The Caenorhabditis elegans Aβ-42 model of Alzheimer disease predominantly Expresses Aβ 3-42. J. Biol. Chem. **284**, 22697–22702 (2009). doi:10.1074/jbc.C109.028514
71. Treusch, S., Hamamichi, S., Goodman, J.L., et al.: Functional Links Between A Toxicity, Endocytic Trafficking, and Alzheimer's Disease Risk Factors in Yeast. Science **334**(80), 1241–1245 (2011). doi:10.1126/science.1213210
72. Muqit, M.M.K., Feany, M.B.: Modelling neurodegenerative diseases in Drosophila: a fruitful approach? Nat. Rev. Neurosci. **3**, 237–243 (2002). doi:10.1038/nrn751
73. Limon, A., Reyes-Ruiz, J.M., Miledi, R.: Loss of functional GABAA receptors in the Alzheimer diseased brain. Proc. Natl. Acad. Sci. **109**, 10071–10076 (2012). doi:10.1073/pnas.1204606109
74. Bishop, N.A., Lu, T., Yankner, B.A.: Neural mechanisms of ageing and cognitive decline. Nature **464**, 529–535 (2010). doi:10.1038/nature08983
75. Dawson, T.M., Ko, H.S., Dawson, V.L.: Genetic Animal Models of Parkinson's Disease. Neuron **66**, 646–661 (2010). doi:10.1016/j.neuron.2010.04.034
76. Duty, S., Jenner, P.: Animal models of Parkinson's disease: a source of novel treatments and clues to the cause of the disease. Br. J. Pharmacol. **164**, 1357–1391 (2011). doi:10.1111/j.1476-5381.2011.01426.x
77. Blum, D., Torch, S., Lambeng, N., et al.: Molecular pathways involved in the neurotoxicity of 6-OHDA, dopamine and MPTP: Contribution to the apoptotic theory in Parkinson's disease. Prog. Neurobiol. **65**, 135–172 (2001). doi:10.1016/S0301-0082(01)00003-X
78. Lundblad, M., Andersson, M., Winkler, C., et al.: Pharmacological validation of behavioural measures of akinesia and dyskinesia in a rat model of Parkinson's disease. Eur. J. Neurosci. **15**, 120–132 (2002). doi:10.1046/j.0953-816x.2001.01843.x
79. Deumens, R., Blokland, A., Prickaerts, J.: Modeling Parkinson's disease in rats: an evaluation of 6-OHDA lesions of the nigrostriatal pathway. Exp. Neurol. **175**, 303–317 (2002). doi:10.1006/exnr.2002.7891
80. Betarbet, R., Sherer, T.B., Greenamyre, J.T.: Animal models of Parkinson's disease. Bioessays **24**, 308–318 (2002). doi:10.1002/bies.10067

81. Rosin, B., Slovik, M., Mitelman, R., et al.: Closed-loop deep brain stimulation is superior in ameliorating parkinsonism. Neuron **72**, 370–384 (2011). doi:10.1016/j.neuron.2011.08.023
82. Tass, P.A., Qin, L., Hauptmann, C., et al.: Coordinated reset has sustained aftereffects in Parkinsonian monkeys. Ann. Neurol. **72**, 816–820 (2012). doi:10.1002/ana.23663
83. Meredith, G.E., Sonsalla, P.K., Chesselet, M.-F.: Animal models of Parkinson's disease progression. Acta Neuropathol. **115**, 385–398 (2008). doi:10.1007/s00401-008-0350-x
84. Palacino, J.J., Sagi, D., Goldberg, M.S., et al.: Mitochondrial dysfunction and oxidative damage in parkin-deficient mice. J. Biol. Chem. **279**, 18614–18622 (2004). doi:10.1074/jbc.M401135200
85. Clark, I.E., Dodson, M.W., Jiang, C., et al.: Drosophila pink1 is required for mitochondrial function and interacts genetically with parkin. Nature **441**, 1162–1166 (2006). doi:10.1038/nature04779
86. Van der Worp, H.B., Howells, D.W., Sena, E.S., et al.: Can animal models of disease reliably inform human studies? PLoS Med. **7**, e1000245 (2010). doi:10.1371/journal.pmed.1000245
87. Young, J.E., Goldstein, L.S.B.: Alzheimer's disease in a dish: promises and challenges of human stem cell models. Hum. Mol. Genet. **21**, R82–R89 (2012). doi:10.1093/hmg/dds319
88. Bellin, M., Marchetto, M.C., Gage, F.H., Mummery, C.L.: Induced pluripotent stem cells: the new patient? Nat. Rev. Mol. Cell Biol. **13**, 713–726 (2012). doi:10.1038/nrm3448
89. Takahashi, K., Tanabe, K., Ohnuki, M., et al.: Induction of Pluripotent Stem Cells from Adult Human Fibroblasts by Defined Factors. Cell **131**, 861–872 (2007). doi:10.1016/j.cell.2007.11.019
90. Park, I.-H., Arora, N., Huo, H., et al.: Disease-specific induced pluripotent stem cells. Cell **134**, 877–886 (2008). doi:10.1016/j.cell.2008.07.041
91. Marchetto, M.C., Brennand, K.J., Boyer, L.F., Gage, F.H.: Induced pluripotent stem cells (iPSCs) and neurological disease modeling: progress and promises. Hum. Mol. Genet. **20**, R109–R115 (2011). doi:10.1093/hmg/ddr336
92. Israel, M.A., Yuan, S.H., Bardy, C., et al.: Probing sporadic and familial Alzheimer's disease using induced pluripotent stem cells. Nature **482**, 216–220 (2012). doi:10.1038/nature10821
93. Kondo, T., Asai, M., Tsukita, K., et al.: Modeling Alzheimer's disease with iPSCs reveals stress phenotypes associated with intracellular Aβ and differential drug responsiveness. Cell Stem Cell **12**, 487–496 (2013). doi:10.1016/j.stem.2013.01.009
94. Choi, S.H., Kim, Y.H., Hebisch, M., et al.: A three-dimensional human neural cell culture model of Alzheimer's disease. Nature **515**, 274–278 (2014). doi:10.1038/nature13800
95. Shi, Y., Kirwan, P., Smith, J., et al.: A Human Stem Cell Model of Early Alzheimer's Disease Pathology in Down Syndrome. Sci. Transl. Med. **4**, 1–9 (2012)
96. Perrier, A.L., Tabar, V., Barberi, T., et al.: Derivation of midbrain dopamine neurons from human embryonic stem cells. Proc. Natl. Acad. Sci. USA **101**, 12543–12548 (2004). doi:10.1073/pnas.0404700101
97. Swistowski, A., Peng, J., Liu, Q., et al.: Efficient generation of functional dopaminergic neurons from human induced pluripotent stem cells under defined conditions. Stem Cells **28**, 1893–1904 (2010). doi:10.1002/stem.499

98. Soldner, F., Laganière, J., Cheng, A.W., et al.: Generation of isogenic pluripotent stem cells differing exclusively at two early onset parkinson point mutations. Cell **146**, 318–331 (2011). doi:10.1016/j.cell.2011.06.019
99. Sánchez-Danés, A., Richaud-Patin, Y., Carballo-Carbajal, I., et al.: Disease-specific phenotypes in dopamine neurons from human iPS-based models of genetic and sporadic Parkinson's disease. EMBO Mol. Med. **4**, 380–395 (2012). doi:10.1002/emmm.201200215
100. Soldner, F., Hockemeyer, D., Beard, C., et al.: Parkinson's disease patient-derived induced pluripotent stem cells free of viral reprogramming factors. Cell **136**, 964–977 (2009). doi:10.1016/j.cell.2009.02.013
101. Zhou, B.Y., Ye, Z., Chen, G., et al.: Inducible and reversible transgene expression in human stem cells after efficient and stable gene transfer. Stem Cells **25**, 779–789 (2007). doi:10.1634/stemcells.2006-0128
102. Brambrink, T., Foreman, R., Welstead, G.G., et al.: Sequential Expression of Pluripotency Markers during Direct Reprogramming of Mouse Somatic Cells. Cell Stem Cell **2**, 151–159 (2008). doi:10.1016/j.stem.2008.01.004
103. Yu, D.X., Marchetto, M.C., Gage, F.H.: Therapeutic translation of iPSCs for treating neurological disease. Cell Stem Cell **12**, 678–688 (2013). doi:10.1016/j.stem.2013.05.018
104. Scarpini, E., Schelterns, P., Feldman, H.: Treatment of Alzheimer's disease; current status and new perspectives. Lancet Neurol. **2**, 539–547 (2003). doi:10.1016/S1474-4422(03)00502-7
105. Mangialasche, F., Solomon, A., Winblad, B., et al.: Alzheimer's disease: clinical trials and drug development. Lancet Neurol. **9**, 702–716 (2010). doi:10.1016/S1474-4422(10)70119-8
106. Cramer, P.E., Cirrito, J.R., Wesson, D.W., et al.: ApoE-directed therapeutics rapidly clear β-amyloid and reverse deficits in AD mouse models. Science **335**, 1503–1506 (2012). doi:10.1126/science.1217697
107. Fahn, S., Oakes, D., Shoulson, I., et al.: Levodopa and the progression of Parkinson's disease. N. Engl. J. Med. **351**, 2498–2508 (2004). doi:10.1007/s11910-005-0069-1
108. Schapira, A.H.V.: Treatment options in the modern management of Parkinson disease. Arch. Neurol. **64**, 1083–1088 (2007). doi:10.1001/archneur.64.8.1083
109. Shishodia, S., Sethi, G., Aggarwal, B.B.: Curcumin: getting back to the roots. Ann. N. Y. Acad. Sci. **1056**, 206–217 (2005). doi:10.1196/annals.1352.010
110. Begum, A.N., Jones, M.R., Lim, G.P., et al.: Curcumin structure-function, bioavailability, and efficacy in models of neuroinflammation and Alzheimer's disease. J. Pharmacol. Exp. Ther. **326**, 196–208 (2008). doi:10.1124/jpet.108.137455
111. Cole, G.M., Teter, B., Frautschy, S.A.: Neuroprotective effects of curcumin. In: Aggarwal, B.B., Surh, Y.-J., Shishodia, S. (eds) Mol. Targets Ther. Uses Curcumin Heal. Dis., pp 197–212. Springer US (2007)
112. Lim, G.P., Chu, T., Yang, F., et al.: The Curry Spice Curcumin Reduces Oxidative Damage and Amyloid Pathology in an Alzheimer Transgenic Mouse. J. Neurosci. **21**, 8370–8377 (2001)
113. Anand, P., Kunnumakkara, A.B., Newman, R.A., Aggarwal, B.B.: Bioavailability of curcumin: problems and promises. Mol. Pharm. **4**, 807–818 (2007). doi:10.1021/mp700113r

114. Lim, G.P., Yang, F., Chu, T., et al.: Ibuprofen effects on Alzheimer pathology and open field activity in APPsw transgenic mice. J. Neurosci. **22**, 983–991 (2001)
115. Gasparini, L., Ongini, E., Wenk, G.: Non-steroidal anti-inflammatory drugs (NSAIDs) in Alzheimer's disease: old and new mechanisms of action. J. Neurochem. **91**, 521–536 (2004). doi:10.1111/j.1471-4159.2004.02743.x
116. Beal, M.F.: Mitochondrial Dysfunction and Oxidative Damage in Alzheimer's and Parkinson's Diseases and Coenzyme Q 10 as a Potential Treatment. J. Bioenerg. Biomembr. **36**, 381–386 (2004)
117. Cleren, C., Yang, L., Lorenzo, B., et al.: Therapeutic effects of coenzyme Q10 (CoQ10) and reduced CoQ10 in the MPTP model of Parkinsonism. J. Neurochem. **104**, 1613–1621 (2008). doi:10.1111/j.1471-4159.2007.05097.x
118. Beal, M.F.: Mitochondria, Oxidative Damage, and Inflammation in Parkinson's Disease. Ann. N. Y. Acad. Sci. **991**, 120–131 (2003)
119. Vierbuchen, T., Ostermeier, A., Pang, Z.P., et al.: Direct conversion of fibroblasts to functional neurons by defined factors. Nature **463**, 1035–1041 (2010). doi:10.1038/nature08797
120. Guo, Z., Zhang, L., Wu, Z., et al.: In Vivo direct reprogramming of reactive glial cells into functional neurons after brain injury and in an Alzheimer's disease model. Cell Stem Cell **14**, 188–202 (2014). doi:10.1016/j.stem.2013.12.001
121. Liang, J., Wu, S., Zhao, H., et al.: Human umbilical cord mesenchymal stem cells derived from Wharton's jelly differentiate into cholinergic-like neurons in vitro. Neurosci. Lett. **532**, 59–63 (2013). doi:10.1016/j.neulet.2012.11.014
122. Fischer, W., Wictorin, K., Björklund, A., et al.: Amelioration of cholinergic neuron atrophy and patial memory impariment in aged rats by nerve growth factor. Nature **329**, 65–68 (1987)
123. Lindvall, O., Kokaia, Z.: Stem cells for the treatment of neurological disorders. Nature **441**, 1094–1096 (2006). doi:10.1038/nature04960
124. Tuszynski, M.H., Thal, L., Pay, M., et al.: A phase 1 clinical trial of nerve growth factor gene therapy for Alzheimer disease. Nat. Med. **11**, 551–555 (2005). doi:10.1038/nm1239
125. Bishop, K.M., Hofer, E.K., Mehta, A., et al.: Therapeutic potential of CERE-110 (AAV2-NGF): targeted, stable, and sustained NGF delivery and trophic activity on rodent basal forebrain cholinergic neurons. Exp. Neurol. **211**, 574–584 (2008). doi:10.1016/j.expneurol.2008.03.004
126. Fjord-Larsen, L., Kusk, P., Tornøe, J., et al.: Long-term delivery of nerve growth factor by encapsulated cell biodelivery in the Göttingen minipig basal forebrain. Mol. Ther. **18**, 2164–2172 (2010). doi:10.1038/mt.2010.154
127. Wahlberg, L.U., Lind, G., Almqvist, P.M., et al.: Targeted delivery of nerve growth factor via encapsulated cell biodelivery in Alzheimer disease: a technology platform for restorative neurosurgery. J. Neurosurg. **117**, 340–347 (2012). doi:10.3171/2012.2.JNS11714
128. Nagahara, A.H., Tuszynski, M.H.: Potential therapeutic uses of BDNF in neurological and psychiatric disorders. Nat. Rev. Drug Discov. **10**, 209–219 (2011). doi:10.1038/nrd3366

129. Blurton-jones, M., Kitazawa, M., Martinez-coria, H., et al.: Neural stem cells improve cognition via BDNF in a transgenic model of Alzheimer disease. Proc. Natl. Acad. Sci. **106**, 1–6 (2009)
130. Nagahara, A.H., Merrill, D.A., Coppola, G., et al.: Neuroprotective effects of brain-derived neurotrophic factor in rodent and primate models of Alzheimer's disease. Nat. Med. **15**, 331–337 (2009). doi:10.1038/nm.1912
131. Shihabuddin, L.S., Aubert, I.: Stem cell transplantation for neurometabolic and neurodegenerative diseases. Neuropharmacology **58**, 845–854 (2010). doi:10.1016/j.neuropharm.2009.12.015
132. Lindvall, O., Kokaia, Z.: Stem cells in human neurodegenerative disorders — time for clinical translation? J. Clin. Invest. **120**, 29–40 (2010). doi:10.1172/JCI40543.patients
133. Farris, W., Mansourian, S., Chang, Y., et al.: Insulin-degrading enzyme regulates the levels of insulin, amyloid beta-protein, and the beta-amyloid precursor protein intracellular domain in vivo. Proc. Natl. Acad. Sci. USA **100**, 4162–4167 (2003). doi:10.1073/pnas.0230450100
134. Marr, R.A., Rockenstein, E., Mukherjee, A., et al.: Neprilysin Gene Transfer Reduces Human Amyloid Pathology in Transgenic Mice. J. Neurosci. **23**, 1992–1996 (2003)
135. Yasojima, K., Akiyama, H., Mcgeer, E.G., Mcgeer, P.L.: Reduced neprilysin in high plaque areas of Alzheimer brain: a possible relationship to deficient degradation of b-amyloid peptide. Neurosci. Lett. **297**, 97–100 (2001)
136. Singer, O., Marr, R.A., Rockenstein, E., et al.: Targeting BACE1 with siRNAs ameliorates Alzheimer disease neuropathology in a transgenic model. Nat. Neurosci. **8**, 1343–1349 (2005). doi:10.1038/nn1531
137. Pickford, F., Masliah, E., Britschgi, M., et al.: The autophagy-related protein beclin 1 shows reduced expression in early Alzheimer disease and regulates amyloid beta accumulation in mice. J. Clin. Invest. **118**, 2190–2199 (2008). doi:10.1172/JCI33585
138. Vidal, R.L., Matus, S., Bargsted, L., Hetz, C.: Targeting autophagy in neurodegenerative diseases. Trends Pharmacol. Sci. **35**, 583–591 (2014). doi:10.1016/j.tips.2014.09.002
139. Rubinsztein, D.C.: The roles of intracellular protein-degradation pathways in neurodegeneration. Nature **443**, 780–786 (2006). doi:10.1038/nature05291
140. Kordower, J.H., Freeman, T.B., Snow, B.J., et al.: Neuropathological evidence of graft survival and striatal reinnervation after the transplantation of fetal mesencephalic tissue in a patient with Parkinson's disease. N. Engl. J. Med. **332**, 1118–1124 (1995). doi:10.1056/NEJM199504273321702
141. Piccini, P., Brooks, D., Björklund, A.: Dopamine release from nigral transplants visualized in vivo in a Parkinson's patient. Nature Neuroscience **2**, 1137–1140 (1999)
142. Barker, R.A., Barrett, J., Mason, S.L., Björklund, A.: Fetal dopaminergic transplantation trials and the future of neural grafting in Parkinson's disease. Lancet Neurol. **12**, 84–91 (2013). doi:10.1016/S1474-4422(12)70295-8
143. Piccini, P., Pavese, N., Hagell, P., et al.: Factors affecting the clinical outcome after neural transplantation in Parkinson's disease. Brain **128**, 2977–2986 (2005). doi:10.1093/brain/awh649

144. Bjorklund, L.M., Sa, R., Chung, S., et al.: Embryonic stem cells develop into functional dopaminergic neurons after transplantation in a Parkinson rat model. Proc. Natl. Acad. Sci. **99**, 2344–2349 (2002)
145. Kriks, S., Shim, J.-W., Piao, J., et al.: Dopamine neurons derived from human ES cells efficiently engraft in animal models of Parkinson's disease. Nature **480**, 547–551 (2011). doi:10.1038/nature10648
146. Kim, J., Auerbach, J.M.: Dopamine neurons derived from embryonic stem cells function in an animal model of Parkinson's disease. Nature **418**, 50–56 (2002)
147. Politis, M., Lindvall, O.: Clinical application of stem cell therapy in Parkinson's disease. BMC Med. **10**, 1 (2012). doi:10.1186/1741-7015-10-1
148. Wernig, M., Zhao, J.-P., Pruszak, J., et al.: Neurons derived from reprogrammed fibroblasts functionally integrate into the fetal brain and improve symptoms of rats with Parkinson's disease. Proc. Natl. Acad. Sci. USA **105**, 5856–5861 (2008). doi:10.1073/pnas.0801677105
149. Hargus, G., Cooper, O., Deleidi, M., et al.: Differentiated Parkinson patient-derived induced pluripotent stem cells grow in the adult rodent brain and reduce motor asymmetry in Parkinsonian rats. Proc. Natl. Acad. Sci. USA **107**, 15921–15926 (2010). doi:10.1073/pnas.1010209107
150. Kwon, Y.-W., Chung, Y.-J., Kim, J., et al.: Comparative study of efficacy of dopaminergic neuron differentiation between embryonic stem cell and protein-based induced pluripotent stem cell. PLoS One **9**, e85736 (2014). doi:10.1371/journal.pone.0085736
151. Hentze, H., Graichen, R., Colman, A.: Cell therapy and the safety of embryonic stem cell-derived grafts. Trends Biotechnol. **25**, 24–32 (2007). doi:10.1016/j.tibtech.2006.10.010
152. Caiazzo, M., Dell'Anno, M.T., Dvoretskova, E., et al.: Direct generation of functional dopaminergic neurons from mouse and human fibroblasts. Nature **476**, 224–227 (2011). doi:10.1038/nature10284
153. Pfisterer, U., Kirkeby, A., Torper, O., et al.: Direct conversion of human fibroblasts to dopaminergic neurons. Proc. Natl. Acad. Sci. USA **108**, 10343–10348 (2011). doi:10.1073/pnas.1105135108
154. Jarraya, B., Boulet, S., Ralph, G.S., et al.: Dopamine Gene Therapy for Parkinson's Disease in a Nonhuman Primate Without Associated Dyskinesia. Sci. Transl. Med. **1**, 1–11 (2009)
155. Azzouz, M., Martin-rendon, E., Barber, R.D., et al.: Multicistronic Lentiviral Vector-Mediated Striatal Gene Transfer of Aromatic L -Amino Acid Decarboxylase, Tyrosine Hydroxylase, and GTP Cyclohydrolase I Induces Sustained Transgene Expression, Dopamine Production, and Functional Improvement in a Rat M. J. Neurosci. **22**, 10302–10312 (2002)
156. Bankiewicz, K.S., Eberling, J.L., Kohutnicka, M., et al.: Convection-enhanced delivery of AAV vector in parkinsonian monkeys; in vivo detection of gene expression and restoration of dopaminergic function using pro-drug approach. Exp. Neurol. **164**, 2–14 (2000). doi:10.1006/exnr.2000.7408
157. Bankiewicz, K.S., Forsayeth, J., Eberling, J.L., et al.: Long-term clinical improvement in MPTP-lesioned primates after gene therapy with AAV-hAADC. Mol. Ther. **14**, 564–570 (2006). doi:10.1016/j.ymthe.2006.05.005

158. Muramatsu, S., Fujimoto, K., Kato, S., et al.: A phase I study of aromatic L-amino acid decarboxylase gene therapy for Parkinson's disease. Mol. Ther. J. Am. Soc. Gene. Ther. **18**, 1731–1735 (2010). doi:10.1038/mt.2010.135
159. Luo, J., Kaplitt, M.G., Fitzsimons, H.L., et al.: Subthalamic GAD Gene Therapy in a Parkinson's Disease Rat Model. Science **298**(80), 425–429 (2002)
160. LeWitt, P.A., Rezai, A.R., Leehey, M.A., et al.: AAV2-GAD gene therapy for advanced Parkinson's disease: a double-blind, sham-surgery controlled, randomised trial. Lancet Neurol. **10**, 309–319 (2011). doi:10.1016/S1474-4422(11)70039-4
161. Kordower, J.H., Emborg, M.E., Bloch, J., et al.: Neurodegeneration Prevented by Lentiviral Vector Delivery of GDNF in Primate Models of Parkinson's Disease. Science **290**(80), 767–773 (2000)
162. Gasmi, M., Herzog, C.D., Brandon, E.P., et al.: Striatal delivery of neurturin by CERE-120, an AAV2 vector for the treatment of dopaminergic neuron degeneration in Parkinson's disease. Mol. Ther. **15**, 62–68 (2007). doi:10.1038/sj.mt.6300010
163. Marks, W.J., Ostrem, J.L., Verhagen, L., et al.: Safety and tolerability of intraputaminal delivery of CERE-120 (adeno-associated virus serotype 2-neurturin) to patients with idiopathic Parkinson's disease: an open-label, phase I trial. Lancet Neurol. **7**, 400–408 (2008). doi:10.1016/S1474-4422(08)70065-6
164. Marks, W.J., Bartus, R.T., Siffert, J., et al.: Gene delivery of AAV2-neurturin for Parkinson's disease: a double-blind, randomised, controlled trial. Lancet Neurol. **9**, 1164–1172 (2010). doi:10.1016/S1474-4422(10)70254-4
165. Åkerud, P., Canals, J.M., Snyder, E.Y., Arenas, E.: Neuroprotection through Delivery of Glial Cell Line-Derived Neurotrophic Factor by Neural Stem Cells in a Mouse Model of Parkinson's Disease. J. Neurosci. **21**, 8108–8118 (2001)
166. Biju, K., Zhou, Q., Li, G., et al.: Macrophage-mediated GDNF delivery protects against dopaminergic neurodegeneration: a therapeutic strategy for Parkinson's disease. Mol. Ther. **18**, 1536–1544 (2010). doi:10.1038/mt.2010.107
167. Brody, D.L., Holtzman, D.M.: Active and passive immunotherapy for neurodegenerative disorders. Annu. Rev. Neurosci. **31**, 175–193 (2008). doi:10.1146/annurev.neuro.31.060407.125529
168. DeMattos, R.B., Bales, K.R., Cummins, D.J., et al.: Peripheral anti-A beta antibody alters CNS and plasma A beta clearance and decreases brain A beta burden in a mouse model of Alzheimer's disease. Proc. Natl. Acad. Sci. USA **98**, 8850–8855 (2001). doi:10.1073/pnas.151261398
169. Dodart, J.-C., Bales, K.R., Gannon, K.S., et al.: Immunization reverses memory deficits without reducing brain Abeta burden in Alzheimer's disease model. Nat. Neurosci. **5**, 452–457 (2002). doi:10.1038/nn842
170. Wilcock, D.M., Dicarlo, G., Henderson, D., et al.: Intracranially Administered Anti-AB Antibodies Reduce B-Amyloid Deposition by Mechanisms Both Independent of and Associated with Microglial Activation. J. Neurosci. **23**, 3745–3751 (2003)
171. Holmes, C., Boche, D., Wilkinson, D., et al.: Long-term effects of Aβ 42 immunisation in Alzheimer's disease: follow-up of a randomised, placebo-controlled phase I trial. Lancet **372**, 11–14 (2008)
172. Oddo, S., Billings, L., Kesslak, J.P., et al.: Abeta immunotherapy leads to clearance of early, but not late, hyperphosphorylated tau aggregates via the proteasome. Neuron **43**, 321–332 (2004). doi:10.1016/j.neuron.2004.07.003

173. Klyubin, I., Walsh, D.M., Lemere, C.A., et al.: Amyloid beta protein immunotherapy neutralizes Abeta oligomers that disrupt synaptic plasticity in vivo. Nat. Med. **11**, 556–561 (2005). doi:10.1038/nm1234
174. Asuni, A.A., Boutajangout, A., Quartermain, D., Sigurdsson, E.M.: Immunotherapy targeting pathological tau conformers in a tangle mouse model reduces brain pathology with associated functional improvements. J. Neurosci. **27**, 9115–9129 (2007). doi:10.1523/JNEUROSCI.2361-07.2007
175. Boutajangout, A., Quartermain, D., Sigurdsson, E.M.: Immunotherapy targeting pathological tau prevents cognitive decline in a new tangle mouse model. J. Neurosci. **30**, 16559–16566 (2010). doi:10.1523/JNEUROSCI.4363-10.2010
176. Boutajangout, A., Ingadottir, J., Davies, P., Sigurdsson, E.M.: Passive immunization targeting pathological phospho-tau protein in a mouse model reduces functional decline and clears tau aggregates from the brain. J. Neurochem. **118**, 658–667 (2011). doi:10.1111/j.1471-4159.2011.07337.x
177. Masliah, E., Rockenstein, E., Adame, A., et al.: Effects of alpha-synuclein immunization in a mouse model of Parkinson's disease. Neuron **46**, 857–868 (2005). doi:10.1016/j.neuron.2005.05.010
178. Benner, E.J., Mosley, R.L., Destache, C.J., et al.: Therapeutic immunization protects dopaminergic neurons in a mouse model of Parkinson's disease. Proc. Natl. Acad. Sci. USA **101**, 9435–9440 (2004). doi:10.1073/pnas.0400569101
179. Reynolds, A.D., Banerjee, R., Liu, J., et al.: Neuroprotective activities of CD4+CD25+ regulatory T cells in an animal model of Parkinson's disease. J. Leukoc. Biol. **82**, 1083–1094 (2007). doi:10.1189/jlb.0507296
180. Reynolds, A.D., Stone, D.K., Hutter, J.A.L., et al.: Regulatory T cells attenuate Th17 cell-mediated nigrostriatal dopaminergic neurodegeneration in a model of Parkinson's disease. J. Immunol. **184**, 2261–2271 (2010). doi:10.4049/jimmunol.0901852

Part V
Biomedical Devices and Imaging for Diagnostics and Therapeutics

Biomedical *in vivo* Optical Imaging for Disease Espying and Diagnosis

Abdul Mohaimen Safi and Euiheon Chung

Abstract Biomedical optical imaging is a rapidly emerging field providing non-invasive or minimally invasive means in the preclinical and clinical realm. At present, optical imaging can deliver structural and functional information in great detail, making it a contender for biopsy. *In vivo* optical imaging modalities can perform an 'optical biopsy' that is envisaged to have a substantial impact on the detection and diagnosis of a myriad of diseases. Here we introduce optical modalities ranging from the nanoscopic to macroscopic scale. We have illustrated their recent developments in preclinical areas and also highlighted clinical optical imaging technologies that have moved from 'benchtop to bedside'. Their perspectives and remaining challenges are also depicted. An abridged review, covering the applications of optical imaging for diagnosis of diseases and its future in guided treatment and monitoring therapies has been presented, which will be a suitable reference for the researchers who aspire to enter into the arena of biomedical optical imaging in vivo.

Keywords: Biomedical optics · In vivo imaging · Optical biopsy · Optical diagnosis · Preclinical · Clinical · Benchtop to bedside · Cancer · Neurodegenerative diseases

1 Introduction

The first thing that comes into one's mind when hearing the term "optical imaging" would be "microscope". Microscopy, an icon of modern science, allows us to

A.M. Safi
School of Mechatronics, Gwangju Institute of Science and Technology, Gwangju, South Korea

E. Chung(✉)
Department of Medical System Engineering & School of Mechatronics, Gwangju Institute of Science and Technology, Gwangju, South Korea
e-mail: ogong50@gist.ac.kr

© Springer International Publishing Switzerland 2016

H. Jo et al. (eds.), *Biomedical Engineering: Frontier Research and Converging Technologies*, Biosystems & Biorobotics 9,
DOI: 10.1007/978-3-319-21813-7_14

see tiny objects with our macroscopic eyes, hence not only reveals the complexity of living systems but also answers many fundamental questions of science [2] . Biomedical optical imaging, including modern microscopy, is experiencing rapid technological advances. The realm of biomedical optics is expanding, as it not only includes the light-based diagnostic techniques, but also includes light-meditated therapeutic potential by augmenting other medical imaging modalities such as magnetic resonance imaging (MRI), X-ray computed tomography (X-ray CT) or positron emission tomography (PET) when direct, portable, or continuous access to the patient is required [3].

1.1 An Overview of Biomedical Optical Imaging

The history of optics traces back to the early 11th century when a seven volume treatise, the "Book of Optics", (Kitab al-Manzir) by Ibn al-Haytham ("the father of modern optics") was published [4]. The era of optical imaging began around 1665 with the advent of Robert Hooke's microscope. Over the last 400 years optical imaging has undergone significant advancements. The 20th century, in particular, has seen the most remarkable progress in the light microscopy field. One pivotal advancement in light microscopy was the invention of the confocal microscope [5, 6] by Marvin Minsky in 1957, as it allowed three-dimensional high-resolution by optical sectioning. Confocal microscopy has been widely accepted in the biological community since the mid-1980s [3]. In 1990 Denk et al. introduced the two-photon fluorescence microscope - a new biophysical research tool based on nonlinear excitation. The advent of two-photon microscopy with fluorescence labeling has enabled the study of normal and diseased tissue at microscopic scale much deeper inside the living body than confocal microscopy. Two-photon microscopy has become a unique tool in developmental biology and in vivo pathophysiology as it allows for longitudinal observation over time in various organs such as the brain, kidney, skin, or heart [3, 7-9].

The journey into the ultra-detail in- and outside the cell began with the invention of super-resolution microscopy by breaking the Abbe's diffraction limit [10], coining the term 'nanoscopy'[11]. Super-resolution nanoscopes such as STED [12, 13], RESOLFT [14], SSIM [15, 16], STORM [17], PALM [18] and FPALM [19] have opened a new era of bioimaging by providing cellular and molecular details in biological processes [20-22]. Over the past decade, light-sheet microscopy has demonstrated its unique ability to image a whole biological sample for an extended period of time without damage [2, 23-29]. Despite substantial progress in the optical imaging field, only a few techniques are currently utilized in the clinical milieu, while many techniques are still in the preclinical phase.

Fluorescence-guided surgery and endoscopy promise improved outcomes in cancer patients and more accurate judgment of tissue visibility during operations. Intravascular imaging has the potential to probe the underlying biology in inflamed atherosclerotic plaques and stent-induced injury. Diffuse fluorescence and optical tomography may provide improved, simpler diagnosis of cancer, especially breast cancer or diseases of extremities such as arthritis as well as neurodegenerative diseases. Photoacoustic tomography or multispectral optoacoustics have applications in various targets including skin, sentinel lymph nodes, and blood vessels. Functional near–infrared spectroscopy (fNIRS) can provide monitoring of brain hemodynamics used for studying brain function and dysfunction. Two-photon microscopy techniques can play a role in dermatology and may, in the future, be a potential tool for the diagnosis of neurodegenerative diseases such as stroke or Alzheimer's disease [30, 31]

1.2 The Role of Biomedical Optical Imaging for in vivo Applications

Biomedical optical imaging allows us to non-invasively look at molecules, cells or tissues inside the body by using light. It is employed by scientists in their respective fields of research and by clinicians to diagnose or treat diseases. Traditional research of this kind has been centered around mainly *in vitro* or *ex vivo*, i.e. experimentation done on cells in a test tube or tissues outside an organism for more controllability at the expense of altering the natural microenvironment respectively. On the contrary, in vivo approaches can provide more realistic information for studying diseases while being costly and prone to individual variability. In vivo imaging approaches allow for monitoring and visualizing of disease progression with a more native environment. Optical imaging can provide various contrasts compared to other imaging modalities such as X-ray CT, MRI, or PET that can be more expensive and usually not accessible outside of the hospital [32, 33].

In addition, optical imaging typically employs light in the visible or near-infrared spectrum. Thus it is less harmful than some other imaging modalities such as X-ray CT or PET that utilize ionizing radiations. However optical imaging, like any other modalities, is not free from limitation. The main hindrance is the intrinsic scattering of light while interacting with biological tissue [3, 7, 34, 35]. Due to elastic scattering of photons within the tissue, the resolution of the image is degraded as light undergoes attenuation while travelling through the tissue. So, there exists a trade-off between imaging depth and resolution (Figure 1).

To overcome this limitation, high resolution in vivo optical images can be obtained often by creating an optically transparent window directly on the tissue of interest by removing skin or skull *via* surgical procedures. For instance, tumors growing in subcutaneous space can be imaged with a dorsal skin-fold chamber, brain tumors through a cranial window, breast tumors through a mammary fat pad window, and pancreatic or liver tumors through abdominal windows [36]. With its high resolution and relatively low cost, in vivo optical imaging provides an effective screening or diagnostic assays with high sensitivity and specificity [33].

2 Biomedical Optical Imaging Modalities for Detecting Disease

Optical imaging especially with live animal disease models is becoming an indispensible tool for biomedical research. Recent advances in optical imaging allow unprecedented imaging capability at greater depth with finer resolution. The field of optical imaging has expanded from the nanoscopic realm to a macroscopic one by enabling the imaging of small animals with a large field of view of millimeters or up to centimeters [28, 37] (Figure 1). The comparisons among different modalities with imaging agents are detailed in Table 1.

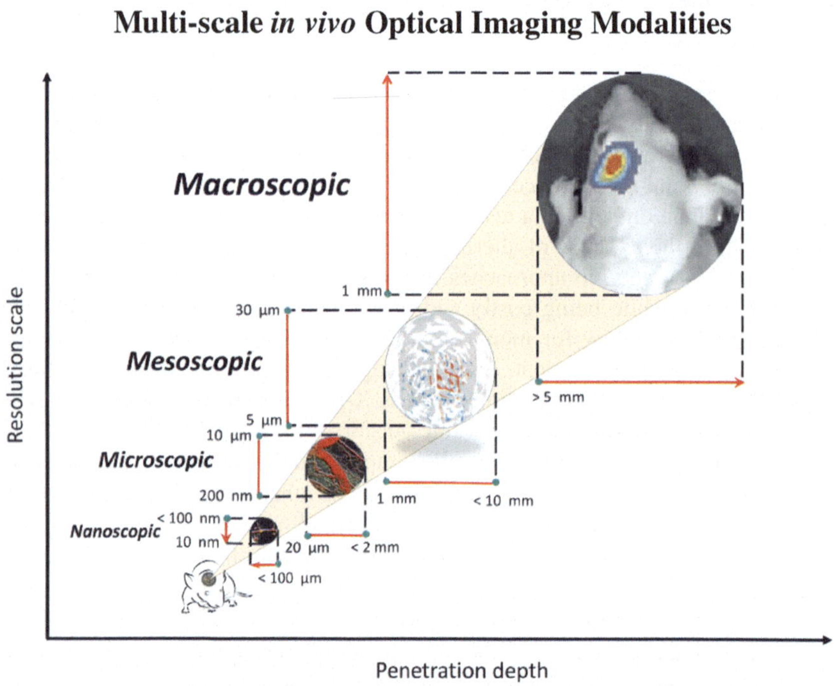

Fig. 1 Schematic diagram of *in vivo* optical imaging modalities from nanoscopic to macroscopic range. Nanoscopic figure is from Berning, S., et al., *Nanoscopy in a living mouse brain. Science*, 2012. 335(6068): p. 551.; with permission, microscopic figure is from Kodack, D.P., et al., *Combined targeting of HER2 and VEGFR2 for effective treatment of HER2-amplified breast cancer brain metastases*. Proceedings of the National Academy of Sciences, 2012. 109(45): p. E3119-E3127 [38]. Mesoscopic figure is from Wang, X., et al., *Noninvasive laser-induced photoacoustic tomography for structural and functional in vivo imaging of the brain*. Nature biotechnology, 2003. 21(7): p. 803-806; with permission [39]. Macroscopic figure is from Chung, E. et al. unpublished data

2.1 Nanoscopic Imaging Modalities

Nanoscopy, alternatively known as super-resolution microscopy, bridges the gap between electron microscopy and conventional microscopy by overcoming the diffraction limit of conventional light microscopy[10]. Thus, a nanoscope allows the visualization of ultra-structures of a biological sample such as cell organelles as well as protein identification in great detail. Super-resolution microscope techniques such as stimulated depletion emission (STED) microscopy employs reversible saturated fluorescence transitions [12]; Structured illumination microscopy (SIM) relies on the operating principle of the Moiré pattern to extract sub-diffraction limit spatial resolution [15]; Photo activation localization microscopy (PALM) [18] and stochastic optical reconstruction microscopy (STORM) [17] are based on the principle of single molecule localization [17, 18]. The rapid developments of nanoscopic techniques may allow in vivo nanoscopic imaging where exquisite spatial resolution is required in the foreseeable future.

2.2 Microscopic Imaging Modalities

Microscopy engenders images at resolutions beyond that which the naked eye can perceive, thus it has been a fundamental tool for biological observation. Here, we describe representative microscopic imaging modalities that are preferred for basic biology or preclinical in vivo studies.

Optical Coherence Tomography (OCT). OCT is analogous to ultrasound imaging except that it employs light instead of sound waves. In OCT, a light beam is scanned across the sample and backscattered photons are collected for reflection profiling in depth using low-coherence interferometry to provide an image as a function of depth and transverse position [40, 41]. In combination with fluorescence-based imaging, OCT can be a great tool to evaluate therapeutic strategies in preclinical animal disease models as it can provide images at cellular scale resolution ($< 10\ \mu m$) with relatively deep tissue penetration (< 2 mm) [42, 43]. Furthermore optical frequency domain imaging (OFDI) is an advanced form of the OCT technique with the capacity of rapid volumetric imaging of tumors at depths up to several millimeters with intrinsic contrast mechanisms. Combined with the principle of Doppler effect, Doppler-OFDI can reveal the network structures of tumors, blood and lymphatic vessels. Utilizing intrinsic contrast from molecular and cellular motion within circulation, OFDI requires no vascular tracer and thus allows for longitudinal imaging over prolonged periods while fluorescence-based vascular imaging has significant limitation from renal clearance and leakage of exogenous vascular tracer due to abnormal permeability of angiogenic tumor vasculatures [44, 45].

Confocal Laser Scanning Microscopy (CLSM). Prior to the invention of confocal microscopy, conventional wide-field microscopy has a fundamental limitation for providing three-dimensional imaging due to blurring of the image caused by the out-of-focus plane signal. Confocal laser scanning microscopy employs a focused laser beam to scan through a sample point by point, thus forming a three-dimensional image by reconstruction of signal. CLSM provides axial resolution by detecting signal only from the focal plane while rejecting the out-of-focus light with a pinhole in front of the photo detector, providing "optical sectioning" up to a depth of 100 - 150 μm [7, 35, 42, 46]. CLSM can be used with fluorescence imaging as well as reflection imaging.

Two-Photon/Multiphoton Microscopy (2P/MP). Multiphoton excitation phenomenon relies on the simultaneous absorption of two or multiple photons by a fluorophore, and provides an inherent three-dimensional optical sectioning. This phenomenon only occurs at the focal spot of a microscope objective lens due to its nonlinear dependence on the illumination intensity [9]. Localized excitation provides minimal photobleaching and phototoxicity compared with confocal microscopy. The use of near-infrared light excitation and the absence of a confocal pinhole provide an intrinsic advantage for deep tissue microscopic imaging down to several hundred microns [42]. Even further imaging depth over one millimeter can be achieved by employing a longer excitation wavelength and higher-order nonlinear excitation (i.e. three-photons) in an intact animal brain equipped with chronic window [1].

Selective Plane Illumination Microscopy (SPIM). SPIM is an approach where the sample is illuminated with a thin sheet of light formed by a cylindrical lens, while wide-field imaging is performed perpendicular to the illumination sheets [25, 27]. This orthogonal decoupling of illumination and detection axes allows fast data acquisition with less photobleaching compared to other imaging modalities [35, 47]. While the resolution of SPIM technology is limited to 0.5 - 10 μm with a penetration depth below 100 μm, this technology has rapidly evolved for in vivo and ex vivo imaging [2, 28].

Optical Projection Tomography (OPT). OPT is a novel imaging technique for three-dimensional imaging and visualization of biological specimens. OPT is analogous to X-ray computed tomography (X-ray CT) except it utilizes parallel ray projection rather than fan beam projection. To reconstruct the tomographic information, trans-illumination over multiple projections and analysis of data using inversion algorithms such as filtered-back projections are done [35, 47-50]. The specimens for OPT imaging should be either optically clear or transparent.

2.3 Mesoscopic Imaging Modalities

While microcopy gives a cellular picture of biological specimens, a comprehensive picture of cells or molecules within the context of tissue or whole organisms can be given by mesoscopic and macroscopic imaging modalities. Macroscopy deals with the diffusive regime at depths larger than 10 mm and microscopy deals with the ballistic regime at less than 1 mm depth. The mesoscopic regime, on the other hand, lies between the two length scales. Mesoscopes allow the visualization of tissue or small organisms at high resolutions, hence it is a useful tool to study neurodegenerative diseases or in studies of metastasis in small animals.

Mesoscopic Fluorescence Tomography (MFT). MFT is different from OPT as it takes light scattering into account using photon transport description through turbid tissue with multi-projection illumination. Here, the image is reconstructed by applying mathematical models and proper photon propagation for mesoscopic scale objects [35, 51]. MFT allows longitudinal study of in vivo specimens with higher resolution than macroscopy and penetration depth beyond that of microscopy. For instance, *in vivo* imaging of fluorescently labeled organs undergoing morphogenetic changes in *D. melanogaster* pupae has been demonstrated [51-53].

Functional Photoacoustic Microscopy (fPAM). fPAM is a hybrid technique to observe the local absorption of photons *via* photoacoustic effects or the thermoelastic expansion of tissue due to the illumination of short pulsed laser that can be detected by an ultrasonic transducer [3, 35]. By exploiting much lower ultrasonic scattering than light scattering, fPAM can improve tissue transparency by orders of magnitude and enables deep tissue functional and molecular imaging with high spatial resolution. Imaging is done with an ultrasonic signal allowing the light to be absorbed inside tissue [54]. By applying a raster scanning method, volumetric imaging can be implemented to provide functional parameters, such as oxygen saturation, for preclinical or clinical application [55, 56].

Multispectral Optoacoustic Tomography (MSOT). MSOT is similar to fPAM except that MSOT employs an expanded pulsed light beam to illuminate a large tissue area. Image formation is accomplished with the aid of mathematical inversion method with multiple light wavelengths. Interestingly the resolution of MSOT depends on the attenuation of acoustic frequency by the tissue and is not limited by the tissue scattering. Thus, light scattering and absorption of tissue govern the penetration depth. MSOT can provide a more accurate reconstruction of an image as a function of depth without the need of a baseline measurement compared to conventional optoacoustic imaging methods [35, 57-60].

2.4 Macroscopic Imaging Modalities

Macroscopic imaging has opened up a new avenue in biomedical sciences. With a large field of view, macroscopic techniques are suitable for studying cardiovascular disease, imaging the brain and visualizing tumors [35]. It could potentially be combined with other imaging modalities such as X-ray CT or MRI to provide more comprehensive anatomical or structural information.

Table 1 Comparison of *in vivo* optical imaging modalities

	Tech-nique	Resolution (µm)	Penetration depth (µm)	Imaging time	Contrast agents		Ref.
					Endogenous	Exogenous	
Nanoscopic	STED	< 70 nm	tens of µm	Minutes to hours	N / A	DY-480XL, KK114, ATTO620, EYFP	[67, 68]
Microscopic	OCT	2 - 10	up to 2 mm	Second to hours	label free (Reflection)	Gold nanoshells, gold nanoparticles	[35, 69, 70]
	Confocal	< 1	< 300	Second to hours	label free (Reflection), Autofluorescence	Gold nanoshells, gold nanoparticles	[35, 71, 72]
	2P / MP	< 1	< 400†	Second to hours	Autofluorescence	Florescent dyes, Fluorescent proteins	[11, 35]
	SPIM	0.5 - 10	< 100	Second to hours	N / A	Fluorescent dyes, Fluorescent proteins	[25, 28, 35]
	OPT	1 - 10	20	Second to minutes	Hb, HbO_2	Fluorescent dyes, Fluorescent proteins	[35, 55, 73]
Mesoscopic	MFT	> 20	~ 0.5 mm	Second to hours	N / A	Fluorescent proteins	[35, 52]
	fPAM	5 - 20	100 to few mm	Second to hours	Hb, HbO_2	Gold nanoparticle	[35, 74]
	MSOT	> 20	> 1 mm	Second to minutes	Hb, HbO_2	Fluorescent proteins	[35, 75]
Macroscopic	FMT	> 1mm	5 mm - 1cm	Minutes	N / A	Fluorescent proteins	[35, 66]
	BLI	Several mm	Few cm	Second to hours	Autofluorescence	Luciferien	[61, 72, 76, 77]
	NIRF	Several hundred µm	Few cm	Minutes	Hb, HbO_2	Near-infrared fluorophores	[72, 78]

Abbreviations:

OCT: Optical Coherence Tomography, 2P / MP: Two photon / Multiphoton, SPIM: Selective Plane Illuminative Microscopy, MFT : Mesoscopy Fluorescence Tomography, fPAM: Functional Photo Acoustic Microscopy, BLI : Bioluminescence imaging, NIRF: Near Infrared Fluorescence Imaging, FMT: Fluorescence Molecular Tomography, MSOT: Multispectral Optoacoustic Tomography, ,† with the use of 3P excitation microscopy of subcortical structures within an intact mouse brain depth of 1.135 mm has been achieved [1].

Bioluminescence Imaging (BLI). Bioluminescent light is generated from enzymatic reaction that converts chemical substrates into light emission. With a simple optical imager, BLI allows for sensitive detection of emitted light that contains the information of a reporter gene or its expression in cells or tissues. Thus BLI enables non-invasive real-time monitoring of disease progression or of the response to a treatment. BLI has various application areas including neuroscience, oncology, virology, and immunology [61-63].

Near-Infrared Fluorescence (NIRF) Imaging. NIR fluorescence provides a noninvasive and high-throughput imaging modality as NIR photons can penetrate deeper than visible photons through biological tissue, due to relatively low scattering and absorption. In addition the NIR regime generates much reduced tissue auto-fluorescence. Although NIRF achieves significant penetration depths up to centimeters, it comes with the cost of resolution degradation due to multiple scattering events within the turbid tissue [64, 65].

Fluorescence Molecular Tomography (FMT). To image the distribution of fluorescent reporters within small animals or tissues through macroscopic length scales over several transport mean free paths, one can use a diffuse optical tomographic technique. FMT utilizes point light illuminations over multiple-angles (projections) and a collection of excitation and emission photons to create quantitative images in three-dimension by use of mathematical processing of raw data. However the ill-posed nature of the image reconstruction due to scattering limits the resolution and accuracy of quantification [35, 66].

3 Preclinical Animal Models for Studying Disease – An Optical Approach

Optical imaging techniques can provide anatomical, molecular, and functional information in animal models of diseases, ranging from cancer to neurodegenerative diseases. With its scalable resolutions covering subcellular, cellular and tissue-organ scale, optical imaging takes a unique position to probe disease progression and treatment efficacy of new drugs in a living body that may not be easy with other imaging modalities.

3.1 Cancer Models

One of the emerging concepts in cancer research is the influence of tumor microenvironments for disease progression and responses to treatments. Tumor cell-oriented studies have been extended to understand the intricate interactions of tumor cells and surrounding host stromal tissue at various resolutions. Thus appropriate cancer models representing the distinct tumor microenvironment together with appropriate optical imaging techniques have been indispensable to deepen our understanding of cancer pathophysiology [42] (Figure 1).

Gastrointestinal Tract Cancer. The gastrointestinal (GI) tract is a major site for immunological challenge from food intake. *In situ* GI tract cancer detection requires an endoscopic imaging modality. Several endoscopic modalities have been reported including narrow-band imaging, autofluorescence imaging, chromoendoscopy and high-resolution confocal endomicroscopy which can reveal disease progression at a microscopic level [79-82]. Furthermore, florescence-based endoscopy with specific molecular targeting probes has emerged to screen suspicious lesions in the GI tract [83-85]. With fiber optic adaptation, endoscopic OCT has emerged as a potential tool for comprehensive analysis of GI tract cancer in murine models [86]. OCT is useful to demarcate different layered structures of the mouse GI tract, and thus disrupted regions with malignancies can be detected but it cannot provide molecular specificity by itself [87, 88]. On the other hand, endoscopic imaging of Cerenkov luminescence provides molecular imaging in GI cancer detection by capturing visible photons from Cerenkov radiation [89-91].

Brain Cancer. Despite the advances in cancer diagnosis, malignant brain cancer still remains a difficult disease with high mortality. Bioluminescence imaging allows for monitoring of brain tumor growth and metastatic spread over several weeks with various treatments [76, 92, 93]. With endogenously induced brain tumor cells, BLI can be used to evaluate the efficacy of drug treatment in vivo [93, 94]. NIRF imaging can also be used in brain tumor research. For instance NIRF flurophores can be used to analyze permeability of tumor vessels [78], and NIRF imaging with tumor targeting nanoparticles provides outstanding tumor specificity [95]. On the other hand, with higher resolution, OFDI provides tumor tissue viability in addition to tumor vascular and lymphatic network imaging, but this requires optically transparent cranial window [96]. Furthermore one- and two-photon fluorescence microendoscopy allows the time-lapse visualization of structural and functional changes of deep-lying brain tumors at microscopic resolution [97].

Breast Cancer. While optical imaging through bulky turbid tissue has limited spatial resolution, it plays a vital role in breast cancer diagnosis by assessing functional and molecular characteristics. As the main light absorbers in the NIR window are oxy- and deoxy-hemoglobin, optical imaging with multiple spectrum can provide a unique non-invasive imaging tool to quantifying tumor-related vascularization, especially oxygen saturation status surrounding breast tumors. Furthermore optical imaging can be combined with most other radiologic imaging modalities due to its compatibility for more comprehensive examination of breast tissue. Diffuse optical tomography (DOT) has proven to be a promising tool for breast cancer research, providing molecular information as cancer's molecular onset may precede anatomic changes [98]. The use of several optical modalities including multi-photon microscopy, bioluminescence with animal MRI and immunohistochemistry allowed preclinical evaluation of potential treatment strategies for breast cancer metastasis to the brain [38, 99]. Intravital optical imaging with cancer cells expressing novel photo-switchable fluorescent proteins can facilitate tracing of the invasion and intravastion of breast tumor cells within local microenvironments [100].

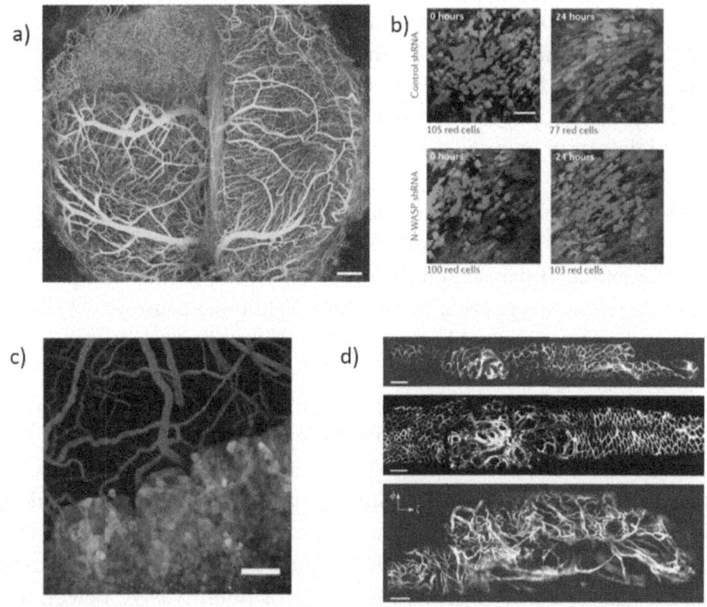

Fig. 2 Imaging cancer in preclinical animal models. a) The depth-projected vasculature within the first 2 mm of a mouse brain bearing a xenotransplanted U87 human glioblastoma multiforme tumor imaged with OFDI. Depth is denoted by color: yellow (superficial) to red (deep). Scale bar, 500 μm. (From Vakoc, B.J., et al., *Three-dimensional microscopy of the tumor microenvironment in vivo using optical frequency domain imaging*. Nature medicine, 2009. 15(10): p. 1219-1223; with permission) [96]. b) In a Dendra2-expressing breast tumor imaged through a mammary imaging window, the color of approximately 100 tumor cells was photo switched from green to red using a violet laser. The location (and therefore the migration) of these tumor cells was monitored on consecutive days. Scale bar represents 70 μm. (From Ellenbroek, S.I. and J. van Rheenen, *Imaging hallmarks of cancer in living mice*. Nature Reviews Cancer, 2014. 14(6): p. 406-418; with permission) [36]. c) Intravital multiphoton microscopy images illustrate the brain metastatic lesion boundary (From Kodack, D.P., et al., *Combined targeting of HER2 and VEGFR2 for effective treatment of HER2-amplified breast cancer brain metastases*. Proceedings of the National Academy of Sciences, 2012. 109(45): p. E3119-E3127) [38]. d) Fluorescence image of colorectal vasculature in a floxed Apc mouse at 11 weeks, 13 weeks and after adeno-Cre administration, and of a large lesion at week 17 in another adeno-Cre–treated mouse (From Kim, P., et al., *In vivo wide-area cellular imaging by side-view endomicroscopy*. Nature methods, 2010. 7(4): p. 303-305; with permission) [81].

3.2 Neurodegenerative Disease Models

Still in its infancy, optical imaging has emerged as an important tool for the study of neurodegenerative diseases. In addition to conventional techniques that use ex vivo tissue histopathology from post-mortem animal specimens, optical imaging permits longitudinal study of the neural tissue within the contextual influences of the intact microenvironment. Furthermore, in vivo optical imaging can trace reporter gene expression that is under the control of an activatable promoter which serves as a disease biomarker for studying neurodegenerative diseases. Alzheimer's disease, epilepsy and strokes are chosen for their significance and potential for rapid detection by optical imaging [33].

Alzheimer's Disease. As an age-dependent neurodegenerative disease, Alzheimer's disease (AD) is the most common cause of dementia. It can be epitomized by the extracellular accumulation of amyloid plaques composed of aggregated amyloid beta (Aβ) peptide in the brain. While the cause of AD is still vague, Aβ is believed to be responsible for neural inflammation the brain as well as disruption of cellular activities leading to neuronal death [33, 101]. In vivo optical imaging such as bioluminescence has proven to be useful in studying transgenic AD mouse models, producing robust Aβ deposition in the brain [77]. Using a cranial window, multiphoton microscopy can provide microscopic Aβ plaque detection in the mouse brain [102]. Studies on transgenic animal models showed the formation of Aβ plaques in the retina prior to the manifestation in the brain with non-invasive optical imaging [101] (Figure 3). To evaluate amyloid targeting reagents, the use of a NIR fluorescent probe for non-invasive detection of senile plaque has been reported [103].

Epilepsy. Epilepsy is a neurological disease characterized by recurrent and unpredictable seizures - a brief synchronous neural activity that is abnormally excessive [104]. To date, seizure detection has been based on the analysis of either surface or intracranial electroencephalography (EEG). As EEG suffers from low spatial and temporal resolution, currently no real-time EEG analysis can reliably predict seizure onset [105]. However, optical imaging may overcome these obstacles. In a recent study intrinsic optical signal (IOS) was used as a tool to study brain networks associated with epileptic seizures in living mice, demonstrating IOS's potential as a candidate for seizure detection [105]. In conjunction with IOS, laser speckle contrast imaging (LSCI) has been used to provide semi-quantitative information of hemodynamic changes in case of large neuronal discharges. [106] (Figure 4). To investigate epilepsy, various techniques such as spatial frequency domain imaging (SFDI), near-infrared fiber optic probe or OCT have been used [105]. Furthermore optical imaging in the primate visual cortex has shown the site of epileptic foci with high spatial resolution during seizure propagation [107]. Another functional neuroimaging modality, multispectral diffuse optical tomography (DOT) is a useful tool for mapping early seizure onset and propagation pathways with high spatiotemporal resolution [108].

Stroke. As a neurologic deficit of cerebrovascular cause, strokes account for the second most frequent cause of death after coronary artery disease, and are the leading cause of permanent physical and mental disability [109] (Figure 5). Non-invasive optical imaging techniques can prove to be an important asset as a diagnostic and monitoring tool with animal stroke models. LSCI has been used to

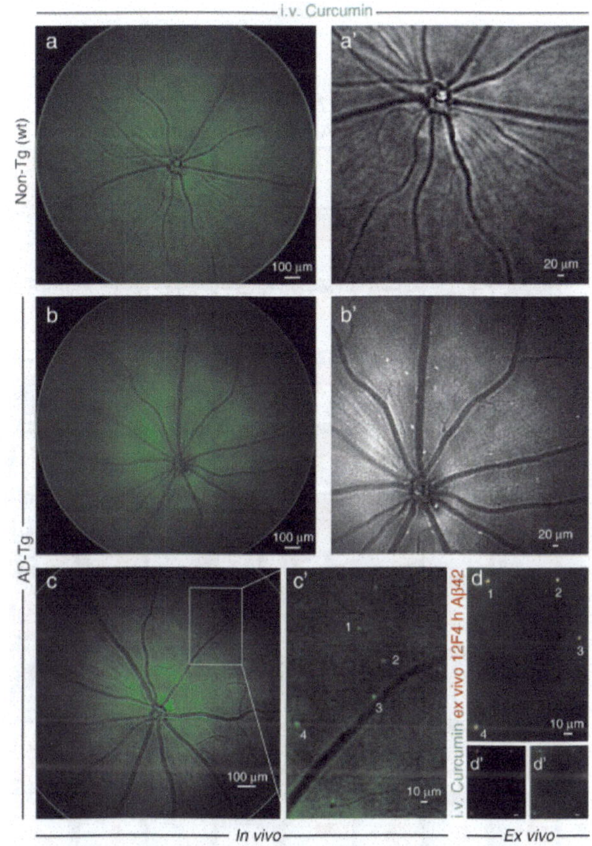

Fig. 3 Non-invasive *in vivo* optical imaging of curcumin-labeled Aβ plaques in AD-Tg mice retina. (a–c) Following i.v. administration of curcumin (7.5 mg/kg) for 5 consecutive days, retinas of 8–12 month-old AD-Tg and wild type mice were in vivo fluorescently imaged using Micron II rodent retinal microscope. Representative in vivo mouse fundus images: (a) no plaques could be visualized in wild type mouse; (a') higher magnification in grayscale. (b) Retinal curcumin-labeled plaques (green spots) were visible in live AD-Tg mouse; (b') higher magnification in grayscale. (b) Representative images demonstrate in vivo detection of curcumin-labeled plaques in the retinas of 8 month-old AD-Tg mice, following a single i.v. injection of curcumin (7.5 mg/kg) two hours prior to imaging. (c') Enlarged image of the selected area from (c) image, demonstrates that this imaging modality enables the identification of individual plaques or plaque clusters at a high spatial resolution. Note that blood vessels appear unstained (dark), possibly due to blood flow in the live mouse. (d) Whole-mount retina prepared from the same in vivo imaged mouse eye as in (c), following perfusion. An additional ex vivo staining with anti-human Aβ42 mAb (12F4; secondary Ab-Cy5 conjugate) further confirmed the specificity of curcumin signals (captured by Micron II) to Aβ plaques. (d',d'') separate channels for each staining. A similar pattern of curcumin-positive plaques imaged in vivo was identified ex vivo following antibody labeling (indicated by the numbers 1–4). (From M. Koronyo-Hamaoui et al., *Identification of amyloid plaques in retinas from Alzheimer's patients and noninvasive in vivo optical imaging of retinal plaques in a mouse model*. Neuroimage, 2011. 54: p. S204-S217; with permission) [101].

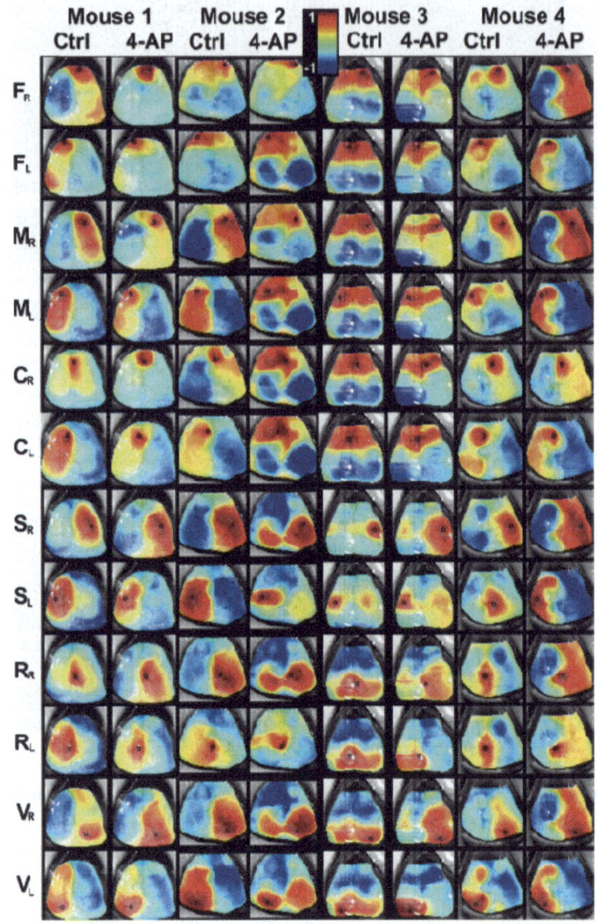

Fig. 4 Seed-based HbO_2 correlation maps for four mice. One control session and one post–4-AP injection session are displayed for each mouse. (F: frontal cortex, M: motor cortex, C: cingulate cortex, S: somatosensory cortex R: retrosplenial cortex, V: visual cortex; subscripts L and R refer to left or right hemisphere, respectively.) The scale for all correlation maps is from r = −1 to 1. Maps are shown overlaid on the anatomical image of the brain, acquired with green light. Seeds placement and sizes are indicated with black circles. (From Guevara, E., et al., *Optical imaging of acute epileptic networks in mice.* Journal of Biomedical Optics, 2013. 18(7): p. 076021-076021; with permission) [106].

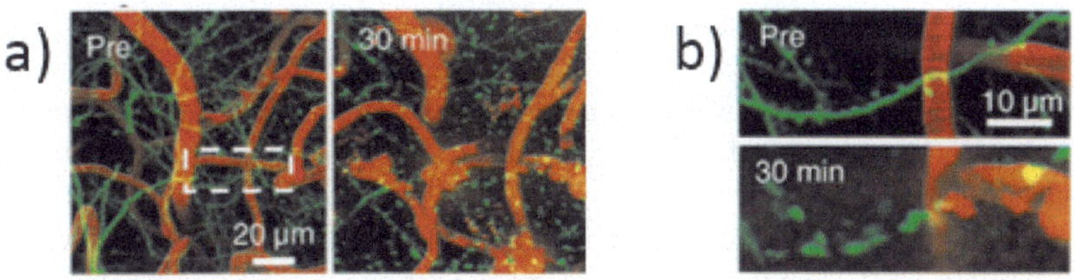

Fig. 5 Propagation of stroke (a) Images from a YFP mouse showing the vasculature (red) and dendrites (green) (maximal intensity projection of first 100 μm of the cortex) before and 30 min after photoactivation (1 min) of RB in an animal that resulted in severe ischemia and dendritic damage (>0.40 mm^2 surrounding area with clotted vessels). Vessels labeled with TR-dextran were all clotted in the imaged region. (b) A higher magnification view of the white-boxed region in (a) before and 30 min after photothrombosis. (From Zhang S, Murphy TH (2007) *Imaging the Impact of Cortical Microcirculation on Synaptic Structure and Sensory-Evoked Hemodynamic Responses In Vivo*. PLoS Biol 5(5): e119 [113].

study cerebral blood flow changes after a stroke. While LSCI allows high spatiotemporal resolution with inexpensive and simple analysis, it cannot provide absolute quantification of blood flow. Two-photon microscopy (TPM) allows precise quantification of direction and speed of blood flow. In stroke models, TPM showed rapid loss of dendritic microstructures in individual neurons and demonstrated functional rewiring of sensorimotor neurons in new regions of the cortex around the stroke area [110]. IOS is another tool for studying strokes. By using a photothrombotic stroke model, functional reorganization of sensory cortical maps for hindlimbs and forelimbs were revealed with long-term IOS imaging. Long-term structural and functional imaging modalities can provide mechanistic insight into strokes [111, 112]. For cortical microvasculature imaging, spectral Doppler optical coherence tomography (SDOCT), a special form of OCT, can provide quantified flow information of blood volume measurements [109].

3.3 Cardiovascular Disease Models

Cardiovascular disease is the leading cause of mortality in developed countries. To visualize its functional and molecular features, optical imaging is becoming an important preclinical tool. MSOT has been used to image cardiovascular systems – for example carotid arteries and aortas - of living mice in real-time [114]. Intravascular NIRF imaging has shown the delineation of the biological processes of atherosclerosis and stent-induced arterial injury [115]. At higher resolution, OCT can be a valuable tool for detecting cardiovascular diseases. The feasibility of OCT has been demonstrated by imaging coronary arteries of living pigs. With the capability of imaging coronary arterial stents, OCT can be used to aid cardiologists in measuring stent apposition [116].

3.4 Ocular Disease Models

Ocular diseases such as diabetic retinopathy can cause significant blindness. The past decade has witnessed OCT becoming a preferred imaging modality in the diagnosis and study of ocular disease. Recent developments in spectral domain OCT have enabled unprecedented imaging speed compared with conventional time domain OCT. In vivo imaging of mouse retinas can be a indispensable tool for studying various structural alterations in animal disease models[43, 117].

4 Towards Optical Diagnosis for Clinical Applications

While in vivo optical imaging has significantly contributed to preclinical research, it is relatively recent that optical imaging technologies for clinical applications have evolved to make a substantial impact. There is a wide range of clinical uses for optical imaging. With its relatively low cost and scalability in resolution and penetration depth, optical imaging holds the promise to visualize not only structural changes but also functional alterations of diseases for clinical applications. However signal attenuation, mainly due to light scattering, poses a big challenge for optical imaging in clinical applications especially if the target is located deep inside the tissue. Finding its niche area in clinical applications, such as adapting for endoscopic modality to screen pre-cancer on epithelial surface, optical imaging is likely to play a key role in fulfilling clinical needs [31, 118, 119]. Optical imaging is realizing the "optical biopsy" or the diagnosis based on light-tissue interaction without the excision of tissue under inspection in the clinic.

4.1 Fluorescence Imaging in the Clinic

Florescence endoscopy is emerging as an optical biopsy tool. Portable, inexpensive fluorescence microscopes have been developed for early detection of cancer which have proven to be useful for the detection of early neoplastic changes [120] such as with Barrett's esophagus, a common premalignant condition for upper GI tract cancer [121] (Figure 6). Clinical evidence of fluorescence imaging with aminolevulimic acid (ALA) in the case of bladder cancer has been demonstrated. Thus fluorescence with either intrinsic (i.e. autofluorescence) or extrinsic (i.e. indocyanine green) origin is emerging to provide optical contrast for colonoscopy, bronchoscopy or esophageal imaging [31] (Figure 7).

4.2 Optical Coherence Tomography (OCT) in the Clinic

Due to the rapid advancement of OCT in recent years, it has ensconced its position in the clinical realm. It has proven to be a powerful tool in ophthalmology because of its ability to detect early stages of ocular diseases and their progression. For example, OCT has been used to diagnose congenital X-linked retinoschisis [119].

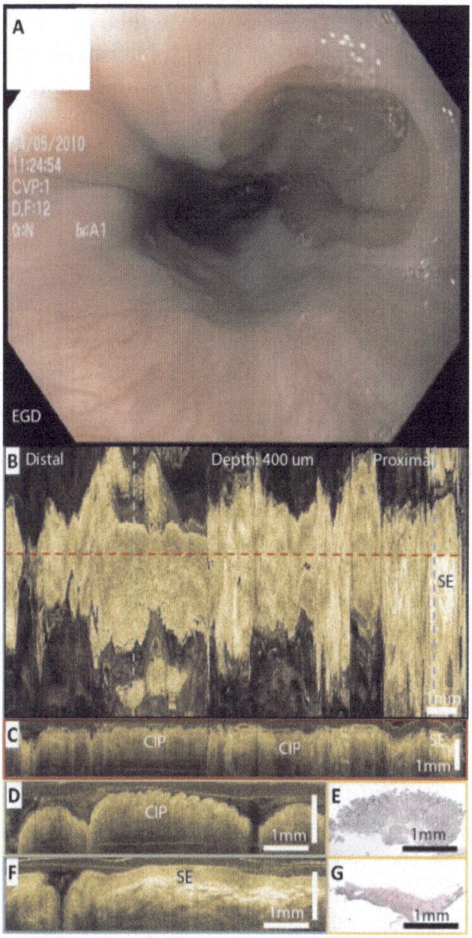

Fig. 6 Clinical OCT imaging of Barrett's esophagus. A 30-year-old white man with established Barrett's esophagus (BE) and continued symptoms of chronic severe heartburn, persistent cough, throat irritation, and asthma was referred for surveillance EGD at the VA Boston Healthcare System. During retraction of the endoscope, a pink circular lesion (A) was observed under white light endoscopy in the upper esophagus (spanning 20–22 cm from the incisors). Three-dimensional endoscopic optical coherence tomography (OCT) images were obtained of the region under direct visualization with white light by passing the probe through the standard accessory channel. An en face projection image (B) at 400-μm depth underneath the tissue surface showed columnar epithelium consistent with a cervical inlet patch (CIP) and surrounding normal squamous epithelium (SE). Cross-sectional OCT images along the probe pull-back direction (C) and the probe rotation direction (D and F) clearly demonstrated columnar and squamous epithelium in the CIP region and the surrounding esophagus, respectively. Biopsy specimens taken from the imaged lesion confirmed the finding of CIP. The OCT features matched representative hematoxylin and eosin histology (E and G). Both esophageal and extraesophageal symptoms responded to increased antacid therapy. (From Zhou, C., et al., Three-dimensional endoscopic optical coherence tomography imaging of cervical inlet patch. Gastrointestinal endoscopy, 2012. 75(3): p. 675-677 ; With permission) [128].

OCT also contributed to the diagnosis of neovascular age-related macular degeneration [122] and glaucoma progression [123]. In dermatology, it provides images of vascular systems in healthy and diseased skin with comprehensive details [124]. OCT combined with multi-photon microscopy has enabled wide-field imaging of skin lesion up to 2 mm penetration depth [125]. In oncology OCT can facilitate the delineation of tumor margins for resection by revealing the distinction between normal and invaded brain tissue as well as solid tumor tissue [126]. In gastroenterology, real-time OCT can be used to evaluate gastroesophageal reflux diseases in situ [127]. Endoscopic OCT can be used to diagnose GI tract cancers, particularly small lesions that are often missed [41]. In neurology OCT can help to study neurodegenerative diseases in the clinical field [129]. A recent study showed the use of OCT for the measurement of retinal nerve fiber layer (RNFL) to diagnose Alzheimer's disease [130]. OCT allows for the detection of changes in RNFL and macular volume with high resolution to enable the study of neurological diseases such as schizophrenia [129] and Parkinson's disease [131].

4.3 Confocal Imaging in the Clinic

With high-spatial resolution and three-dimensional capability, confocal imaging not only provides a reflection-based structural image but also offers specific fluorescence imaging. Thus confocal microscopes help clinicians to diagnose skin cancer at the bedside [132]. For example, an optical biopsy performed by confocal microscope revealed a lesion as an early in situ melanoma [43]. Furthermore endoscopy with confocal detection has been commercialized. As an example, confocal laser endomicroscopy (CLE) has expanded its applications from the upper and lower gastrointestinal tract to hepatobiliary and inter-abdominal functional imaging, significantly improving our understanding of GI diseases. Molecular imaging with CLE allows for the detection and characterization of lesions which may lead to the prediction of responses to targeted therapy [43, 133].

4.4 Two-Photon/ Multiphoton Imaging in the Clinic

Two-photon or multiphoton microscopy (TPM / MPM) is an emerging technique for clinical use as it enables fluorescence-based deep tissue imaging of potential lesions such as skin cancer. TPM has been used to study pigmented skin cancer melanoma in patients with high accuracy, sensitivity and specificity [134]. Another approach with TPM / MPM for disease diagnostics is the adaptation to microendoscopic imaging. Two-photon or confocal microendoscopy can be realized in two ways: rigid, needle-like lenses for imaging body surfaces and beneath, as well as flexible fiber optic probes with distal lenses for deep intra-cavity imaging [135]. Two-photon endomicroscopy provides reduced background fluorescence by use of near-infrared emission and becomes more powerful by resolving the issue of dispersion-induced pulse broadening through the use of optical fibers [136].

4.5 Diffuse Optical Imaging in the Clinic

Since near-infrared light can penetrate several centimeters into biological tissue in the diffusive regime, diffuse optical imaging (DOI) can image human brain and breast tissue. One example is to measure the hemodynamic response to brain activity. DOI has gained its popularity in functional brain imaging. Its unique temporal resolution allows for clear distinction between oxy- and deoxy-hemoglobin more effectively than other techniques thus making it well suitable for psychological studies [137]. Another area of DOI is breast imaging to predict breast cancer prognosis. Multimodal approaches such as DOT-MRI, PET-DOT and X-ray tomosynthesis have been reported for breast cancer screening and diagnosis [43, 137, 138]. Furthermore, DOI has been used to characterize pigmented skin lesions [139].

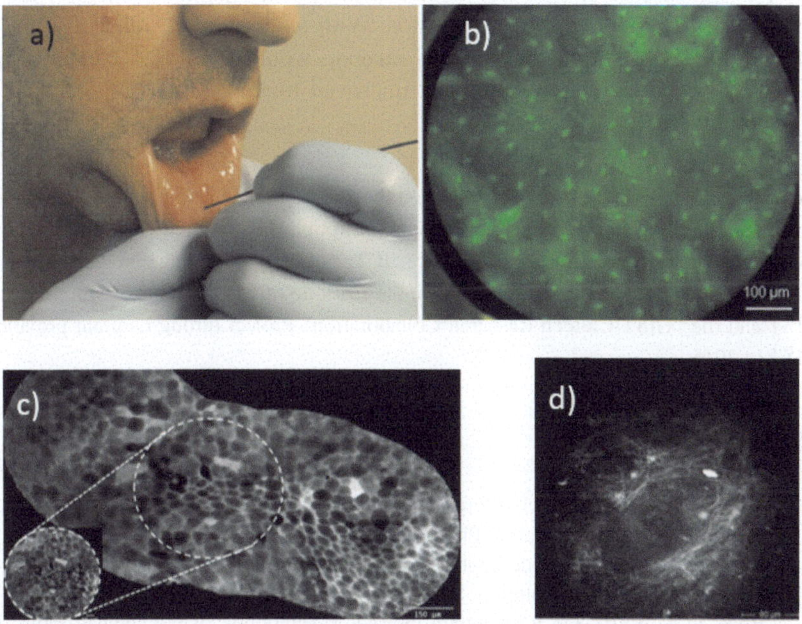

Fig. 7 Application of optical imaging modalities in clinical area. a), b) In vivo image of human volunteer. (From Shin, D., et al., *A fiber-optic fluorescence microscope using a consumer-grade digital camera for in vivo cellular imaging*. PloS one, 2010. 5(6): p. e11218) [120]. c) Confocal image processing of Normal bladder urothelium showing umbrel (From Wu, K., et al., *Dynamic real-time microscopy of the urinary tract using confocal laser endomicroscopy*. Urology, 2011. 78(1): p. 225-231; with permission) [140], d) Autofluorescence image of the two photon, 760 nm excited elastin network (From König, K., et al., *Clinical two-photon microendoscopy*. Microscopy research and technique, 2007. 70(5): p. 398-402; with permission) [141].

5 Summery and Outlook

Optical imaging is on the cusp of progressing from preclinical to clinical applications. In vivo optical imaging, in particular, is playing a major role as it can provide not only the structural or morphological information but also functional or molecular information, revealing accompanying disease processes and testing the efficacy of novel therapeutic approaches. For example, optical imaging in conjunction with fluorescently labeled molecular markers could potentially change our paradigm of cancer screening and improve the outcome of cancer surgery in both developed and developing countries [142]. Furthermore, the use of photons in biomedicine is not necessarily limited to passive imaging as the energy of photons can be exploited to actively modulate the tissue where disease is present. For example, localized tissue ablation for controlled ocular surgery or laser light therapy for skin disease could be combined with optical imaging to monitor the procedure and treatment of the target lesion. With its portability, relatively low cost and simple design, biomedical optical imaging would open new avenues to peek into the diseases as well as allow for improved treatments in the preclinical and clinical fields.

Acknowledgements The work is supported the Bio & Medical Technology Development Program and Basic Science Research Program through the National Research Foundation (NRF) funded by the Ministry of Science, ICT & Future Planning (No. 2011-0019633, 2012R1A1A1012853) and by a grant from the Institute of Medical System Engineering of GIST and the "GIST-Caltech Research Collaboration" Project through a grant provided by GIST in 2015.

References

1. Horton, N.G., et al.: In vivo three-photon microscopy of subcortical structures within an intact mouse brain. Nature Photonics **7**(3), 205–209 (2013)
2. Chen, B.-C., et al.: Lattice light-sheet microscopy: Imaging molecules to embryos at high spatiotemporal resolution. Science **346**(6208), 1257998 (2014)
3. Boas, D.A., Pitris, C., Ramanujam, N. (eds.): Handbook of biomedical optics. CRC Press, Boca Raton (2011)
4. Ellis, D.I., et al.: Illuminating disease and enlightening biomedicine: Raman spectroscopy as a diagnostic tool. Analyst **138**(14), 3871–3884 (2013)
5. Minsky, M.: Microscopy apparatus. Google Patents (1961)
6. Minsky, M.: Memoir on inventing the confocal scanning microscope. Scanning **10**(4), 128–138 (1988)
7. Pawley, J., Masters, B.R.: Handbook of biological confocal microscopy. Optical Engineering **35**(9), 2765–2766 (1996)
8. Denk, W., Strickler, J.H., Webb, W.W.: Two-photon laser scanning fluorescence microscopy. Science **248**(4951), 73–76 (1990)
9. Helmchen, F., Denk, W.: Deep tissue two-photon microscopy. Nature Methods **2**(12), 932–940 (2005)

10. Abbe, E.: Beiträge zur Theorie des Mikroskops und der mikroskopischen Wahrnehmung. Archiv für Mikroskopische Anatomie **9**(1), 413–418 (1873)
11. Park, Y.I., et al.: Upconverting nanoparticles: a versatile platform for wide-field two-photon microscopy and multi-modal in vivo imaging. Chemical Society Reviews **44**, 1302–1317 (2015)
12. Hell, S.W., Wichmann, J.: Breaking the diffraction resolution limit by stimulated emission: stimulated-emission-depletion fluorescence microscopy. Optics Letters **19**(11), 780–782 (1994)
13. Klar, T.A., Hell, S.W.: Subdiffraction resolution in far-field fluorescence microscopy. Optics Letters **24**(14), 954–956 (1999)
14. Hofmann, M., et al.: Breaking the diffraction barrier in fluorescence microscopy at low light intensities by using reversibly photoswitchable proteins. Proceedings of the National Academy of Sciences of the United States of America **102**(49), 17565–17569 (2005)
15. Gustafsson, M.G.: Nonlinear structured-illumination microscopy: wide-field fluorescence imaging with theoretically unlimited resolution. Proceedings of the National Academy of Sciences of the United States of America **102**(37), 13081–13086 (2005)
16. Heintzmann, R., Jovin, T.M., Cremer, C.: Saturated patterned excitation microscopy—a concept for optical resolution improvement. JOSA A **19**(8), 1599–1609 (2002)
17. Rust, M.J., Bates, M., Zhuang, X.: Sub-diffraction-limit imaging by stochastic optical reconstruction microscopy (STORM). Nature Methods **3**(10), 793–796 (2006)
18. Betzig, E., et al.: Imaging intracellular fluorescent proteins at nanometer resolution. Science **313**(5793), 1642–1645 (2006)
19. Hess, S.T., Girirajan, T.P., Mason, M.D.: Ultra-high resolution imaging by fluorescence photoactivation localization microscopy. Biophysical Journal **91**(11), 4258–4272 (2006)
20. Dickson, R.M., et al.: On/off blinking and switching behaviour of single molecules of green fluorescent protein. Nature **388**(6640), 355–358 (1997)
21. Moerner, W., Kador, L.: Optical detection and spectroscopy of single molecules in a solid. Physical Review Letters **62**(21), 2535 (1989)
22. Huang, B., Babcock, H., Zhuang, X.: Breaking the diffraction barrier: super-resolution imaging of cells. Cell **143**(7), 1047–1058 (2010)
23. Keller, P.J., Ahrens, M.B.: Visualizing Whole-Brain Activity and Development at the Single-Cell Level Using Light-Sheet Microscopy. Neuron **85**(3), 462–483 (2015)
24. Keller, P.J., Ahrens, M.B., Freeman, J.: Light-sheet imaging for systems neuroscience. Nature Methods **12**(1), 27–29 (2015)
25. Keller, P.J., Dodt, H.-U.: Light sheet microscopy of living or cleared specimens. Current Opinion in Neurobiology **22**(1), 138–143 (2012)
26. Krzic, U., et al.: Multiview light-sheet microscope for rapid in toto imaging. Nature Methods **9**(7), 730–733 (2012)
27. Stelzer, E.H.: Light-sheet fluorescence microscopy for quantitative biology. Nature Methods **12**(1), 23–26 (2015)
28. Dodt, H.-U., et al.: Ultramicroscopy: three-dimensional visualization of neuronal networks in the whole mouse brain. Nature Methods **4**(4), 331–336 (2007)
29. Bouchard, M.B., et al.: Swept confocally-aligned planar excitation (SCAPE) microscopy for high-speed volumetric imaging of behaving organisms. Nat. Photon. **9**(2), 113–119 (2015)
30. Chen, Y., et al.: Recent advances in two-photon imaging: technology developments and biomedical applications. Chinese Optics Letters **11**(1), 011703 (2013)

31. Taruttis, A., Ntziachristos, V.: Translational optical imaging. American Journal of Roentgenology **199**(2), 263–271 (2012)
32. Hillman, E.M., et al.: In vivo optical imaging and dynamic contrast methods for biomedical research. Philosophical Transactions of the Royal Society A: Mathematical, Physical and Engineering Sciences **2011**(369), 4620–4643 (1955)
33. Patterson, A.P., Booth, S.A., Saba, R.: The Emerging Use of In Vivo Optical Imaging in the Study of Neurodegenerative Diseases. BioMed Research International, 14 (2014)
34. Ryu, Y., et al.: Lensed fiber-optic probe design for efficient photon collection in scattering media. Biomedical Optics Express **6**(1), 191–210 (2015)
35. Ntziachristos, V.: Going deeper than microscopy: the optical imaging frontier in biology. Nature Methods **7**(8), 603–614 (2010)
36. Ellenbroek, S.I., van Rheenen, J.: Imaging hallmarks of cancer in living mice. Nature Reviews Cancer **14**(6), 406–418 (2014)
37. Ntziachristos, V., et al.: Looking and listening to light: the evolution of whole-body photonic imaging. Nature Biotechnology **23**(3), 313–320 (2005)
38. Kodack, D.P., et al.: Combined targeting of HER2 and VEGFR2 for effective treatment of HER2-amplified breast cancer brain metastases. Proceedings of the National Academy of Sciences **109**(45), E3119–E3127 (2012)
39. Wang, X., et al.: Noninvasive laser-induced photoacoustic tomography for structural and functional in vivo imaging of the brain. Nature Biotechnology **21**(7), 803–806 (2003)
40. Bouma, B.: Handbook of optical coherence tomography. Informa Health Care (2001)
41. Tsai, T.-H., Fujimoto, J.G., Mashimo, H.: Endoscopic optical coherence tomography for clinical gastroenterology. Diagnostics **4**(2), 57–93 (2014)
42. Chung, E., et al.: Uncovering tumor biology by intravital microscopy. In: Comprehensive Biomedical Physics, pp. 153–164. Elsevier, Oxford (2014)
43. Iftimia, N., Brugge, W.R., Hammer, D.X.: Advances in Optical Imaging for Clinical Medicine, vol. 6. John Wiley & Sons (2011)
44. Fujimoto, J.G., et al.: The development of OCT. In: Cardiovascular OCT Imaging, pp. 1–21. Springer (2015)
45. Li, J., et al.: Polarization sensitive optical frequency domain imaging system for endobronchial imaging. Optics Express **23**(3), 3390–3402 (2015)
46. Conchello, J.-A., Lichtman, J.W.: Optical sectioning microscopy. Nature Methods **2**(12), 920–931 (2005)
47. Gualda, E., et al.: Going "open" with Mesoscopy: a new dimension on multi-view imaging. Protoplasma **251**(2), 363–372 (2014)
48. Figueiras, E., et al.: Optical projection tomography as a tool for 3D imaging of hydrogels. Biomedical Optics Express **5**(10), 3443–3449 (2014)
49. Rieckher, M., et al.: Microscopic optical projection tomography in vivo. PloS One **6**(4), e18963 (2011)
50. Sharpe, J., et al.: Optical projection tomography as a tool for 3D microscopy and gene expression studies. Science **296**(5567), 541–545 (2002)
51. Vinegoni, C., et al.: In vivo imaging of Drosophila melanogaster pupae with mesoscopic fluorescence tomography. Nature Methods **5**(1), 45–47 (2008)
52. Ozturk, M.S., et al.: Mesoscopic Fluorescence Tomography of a Photosensitizer (HPPH) 3D Biodistribution in Skin Cancer. Academic Radiology **21**(2), 271–280 (2014)

53. Vinegoni, C., et al.: Mesoscopic fluorescence tomography for in-vivo imaging of developing Drosophila. Journal of Visualized Experiments: JoVE **30**, e1510 (2009)
54. Zhang, H.F., Maslov, K., Wang, L.V.: In vivo imaging of subcutaneous structures using functional photoacoustic microscopy. Nature Protocols **2**(4), 797–804 (2007)
55. Wang, L.V., Hu, S.: Photoacoustic tomography: in vivo imaging from organelles to organs. Science **335**(6075), 1458–1462 (2012)
56. Zhang, H.F., et al.: Functional photoacoustic microscopy for high-resolution and noninvasive in vivo imaging. Nature Biotechnology **24**(7), 848–851 (2006)
57. Ma, R., et al.: Multispectral optoacoustic tomography (MSOT) scanner for whole-body small animal imaging. Optics Express **17**(24), 21414–21426 (2009)
58. Ntziachristos, V., Razansky, D.: Molecular imaging by means of multispectral optoacoustic tomography (MSOT). Chemical Reviews **110**(5), 2783–2794 (2010)
59. Razansky, D., Baeten, J., Ntziachristos, V.: Sensitivity of molecular target detection by multispectral optoacoustic tomography (MSOT). Medical Physics **36**(3), 939–945 (2009)
60. Razansky, D., Buehler, A., Ntziachristos, V.: Volumetric real-time multispectral optoacoustic tomography of biomarkers. Nature Protocols **6**(8), 1121–1129 (2011)
61. Contag, C.H., Bachmann, M.H.: Advances in in vivo bioluminescence imaging of gene expression. Annual Review of Biomedical Engineering **4**(1), 235–260 (2002)
62. Rehemtulla, A., et al.: Rapid and quantitative assessment of cancer treatment response using in vivo bioluminescence imaging. Neoplasia **2**(6), 491–495 (2000)
63. Sato, A., Klaunberg, B., Tolwani, R.: In vivo bioluminescence imaging. Comparative Medicine **54**(6), 631–634 (2004)
64. Shah, K., Weissleder, R.: Molecular optical imaging: applications leading to the development of present day therapeutics. NeuroRx **2**(2), 215–225 (2005)
65. Amiot, C.L., et al.: Near-infrared fluorescent materials for sensing of biological targets. Sensors **8**(5), 3082–3105 (2008)
66. Ale, A., et al.: FMT-XCT: in vivo animal studies with hybrid fluorescence molecular tomography-X-ray computed tomography. Nature Methods **9**(6), 615–620 (2012)
67. Berning, S., et al.: Nanoscopy in a living mouse brain. Science **335**(6068), 551 (2012)
68. Pellett, P.A., et al.: Two-color STED microscopy in living cells. Biomedical Optics Express **2**(8), 2364–2371 (2011)
69. Cang, H., et al.: Gold nanocages as contrast agents for spectroscopic optical coherence tomography. Optics Letters **30**(22), 3048–3050 (2005)
70. Adler, D.C., et al.: Photothermal detection of gold nanoparticles using phase-sensitive optical coherence tomography. Optics Express **16**(7), 4376–4393 (2008)
71. Gratton, E.: Deeper tissue imaging with total detection. Science **331**(6020), 1016–1017 (2011)
72. Chong, K., et al.: Current Optical Imaging Techniques for Brain Tumor Research: Application of in vivo Laser Scanning Microscopy Imaging with a Cranial Window System, pp. 155–172. InTech, Rijeka (2011)
73. Yao, J., et al.: Label-free oxygen-metabolic photoacoustic microscopy in vivo. Journal of Biomedical Optics **16**(7), 076003-11 (2011)
74. Mallidi, S., et al.: Multiwavelength photoacoustic imaging and plasmon resonance coupling of gold nanoparticles for selective detection of cancer. Nano Letters **9**(8), 2825–2831 (2009)
75. Stiel, A.C., et al.: High contrast imaging of reversibly switchable fluorescent proteins via temporally unmixed Multispectral Optoacoustic Tomography (tuMSOT). arXiv preprint arXiv:1412.3241 (2014)

76. Contag, C.H., Ross, B.D.: It's not just about anatomy: in vivo bioluminescence imaging as an eyepiece into biology. Journal of Magnetic Resonance Imaging **16**(4), 378–387 (2002)
77. Watts, J.C., et al.: Bioluminescence imaging of Aβ deposition in bigenic mouse models of Alzheimer's disease. Proceedings of the National Academy of Sciences **108**(6), 2528–2533 (2011)
78. Choi, M., et al.: Minimally invasive molecular delivery into the brain using optical modulation of vascular permeability. Proceedings of the National Academy of Sciences **108**(22), 9256–9261 (2011)
79. Kuiper, T., et al.: Feasibility and accuracy of confocal endomicroscopy in comparison with narrow-band imaging and chromoendoscopy for the differentiation of colorectal lesions. The American Journal of Gastroenterology **107**(4), 543–550 (2012)
80. Shahid, M.W., et al.: Diagnostic accuracy of probe-based confocal laser endomicroscopy in detecting residual colorectal neoplasia after EMR: a prospective study.
Gastrointestinal Endoscopy **75**(3), 525–533 (2012). e1
81. Kim, P., et al.: In vivo wide-area cellular imaging by side-view endomicroscopy. Nature Methods **7**(4), 303–305 (2010)
82. Wallace, M., et al.: Miami classification for probe-based confocal laser endomicroscopy **43**(10), 882–891 (2011)
83. Oh, G., et al.: Intravital imaging of mouse colonic adenoma using MMP-based molecular probes with multi-channel fluorescence endoscopy. Biomedical Optics Express **5**(5), 1677–1689 (2014)
84. Habibollahi, P., et al.: Optical imaging with a cathepsin B activated probe for the enhanced detection of esophageal adenocarcinoma by dual channel fluorescent upper GI endoscopy. Theranostics **2**(2), 227 (2012)
85. Hsiung, P.-L., et al.: Detection of colonic dysplasia in vivo using a targeted heptapeptide and confocal microendoscopy. Nature Medicine **14**(4), 454–458 (2008)
86. Winkler, A.M., et al.: Quantitative tool for rapid disease mapping using optical coherence tomography images of azoxymethane-treated mouse colon. Journal of Biomedical Optics **15**(4), 041512-10 (2010)
87. Alex, A., et al.: Characterization of eosinophilic esophagitis murine models using optical coherence tomography. Biomedical Optics Express **5**(2), 609–620 (2014)
88. Iftimia, N., et al.: Fluorescence-guided optical coherence tomography imaging for colon cancer screening: a preliminary mouse study. Biomedical Optics Express **3**(1), 178–191 (2012)
89. Kothapalli, S.-R., et al.: Endoscopic imaging of Cerenkov luminescence. Biomedical Optics Express **3**(6), 1215–1225 (2012)
90. Liu, H., et al.: Intraoperative imaging of tumors using Cerenkov luminescence endoscopy: a feasibility experimental study. Journal of Nuclear Medicine **53**(10), 1579–1584 (2012)
91. Mitchell, G.S., et al.: In vivo Cerenkov luminescence imaging: a new tool for molecular imaging. Philosophical Transactions of the Royal Society of London A: Mathematical, Physical and Engineering Sciences **2011**(369), 4605–4619 (1955)
92. Liu, L., Mason, R.P., Gimi, B.: Dynamic bioluminescence and fluorescence imaging of the effects of the antivascular agent Combretastatin-A4P (CA4P) on brain tumor xenografts. Cancer Letters **356**(2), 462–469 (2015)
93. O'Neill, K., et al.: Bioluminescent imaging: a critical tool in pre-clinical oncology research. The Journal of Pathology **220**(3), 317–327 (2010)

94. Uhrbom, L., Nerio, E., Holland, E.C.: Dissecting tumor maintenance requirements using bioluminescence imaging of cell proliferation in a mouse glioma model. Nature Medicine **10**(11), 1257–1260 (2004)
95. Na, J.H., et al.: Real-time and non-invasive optical imaging of tumor-targeting glycol chitosan nanoparticles in various tumor models. Biomaterials **32**(22), 5252–5261 (2011)
96. Vakoc, B.J., et al.: Three-dimensional microscopy of the tumor microenvironment in vivo using optical frequency domain imaging. Nature Medicine **15**(10), 1219–1223 (2009)
97. Barretto, R.P., et al.: Time-lapse imaging of disease progression in deep brain areas using fluorescence microendoscopy. Nature Medicine **17**(2), 223–228 (2011)
98. Ntziachristos, V., Chance, B.: Breast imaging technology: Probing physiology and molecular function using optical imaging-applications to breast cancer. Breast Cancer Research **3**(1), 41 (2000)
99. Chung, E., et al.: Secreted Gaussia luciferase as a biomarker for monitoring tumor progression and treatment response of systemic metastases. PloS One **4**(12), e8316 (2009)
100. Kedrin, D., et al.: Intravital imaging of metastatic behavior through a mammary imaging window. Nature Methods **5**(12), 1019–1021 (2008)
101. Koronyo-Hamaoui, M., et al.: Identification of amyloid plaques in retinas from Alzheimer's patients and noninvasive in vivo optical imaging of retinal plaques in a mouse model. Neuroimage **54**, S204–S217 (2011)
102. Meyer-Luehmann, M., et al.: Rapid appearance and local toxicity of amyloid-β plaques in a mouse model of Alzheimer's disease. Nature **451**(7179), 720–724 (2008)
103. Skoch, J., et al.: Development of an optical approach for noninvasive imaging of Alzheimer's disease pathology. Journal of Biomedical Optics **10**(1), 011007–0110077 (2005)
104. Schwartz, T.H., Bonhoeffer, T.: In vivo optical mapping of epileptic foci and surround inhibition in ferret cerebral cortex. Nature Medicine **7**(9), 1063–1067 (2001)
105. Binder, D.K., Haut, S.R.: Toward new paradigms of seizure detection. Epilepsy & Behavior **26**(3), 247–252 (2013)
106. Guevara, E., et al.: Optical imaging of acute epileptic networks in mice. Journal of Biomedical Optics **18**(7), 076021 (2013)
107. Haglund, M.M.: Optical imaging of visual cortex epileptic foci and propagation pathways. Epilepsia **53**(s1), 87–97 (2012)
108. Zhang, T., et al.: Pre-seizure state identified by diffuse optical tomography. Scientific Reports **4**, 3798 (2014). doi:10.1038/srep03798110
109. Yu, L., et al.: Spectral Doppler optical coherence tomography imaging of localized ischemic stroke in a mouse model. Journal of Biomedical Optics **15**(6), 066006 (2010)
110. Winship, I.R., et al.: Augmenting collateral blood flow during ischemic stroke via transient aortic occlusion. Journal of Cerebral Blood Flow & Metabolism **34**(1), 61–71 (2014)
111. Clarkson, A.N., et al.: Multimodal examination of structural and functional remapping in the mouse photothrombotic stroke model. Journal of Cerebral Blood Flow & Metabolism **33**(5), 716–723 (2013)
112. Lay, C.C., et al.: Mild sensory stimulation protects the aged rodent from cortical ischemic stroke after permanent middle cerebral artery occlusion. Journal of the American Heart Association **1**(4), e001255 (2012)

113. Zhang, S., Murphy, T.H.: Imaging the impact of cortical microcirculation on synaptic structure and sensory-evoked hemodynamic responses in vivo. PLoS Biology **5**(5), e119 (2007)
114. Taruttis, A., et al.: Real-time imaging of cardiovascular dynamics and circulating gold nanorods with multispectral optoacoustic tomography. Optics Express **18**(19), 19592–19602 (2010)
115. Jaffer, F.A., et al.: Two-dimensional intravascular near-infrared fluorescence molecular imaging of inflammation in atherosclerosis and stent-induced vascular injury. Journal of the American College of Cardiology **57**(25), 2516–2526 (2011)
116. Guillermo, J.T., et al.: Optical coherence tomography in cardiology. In: Handbook of Optical Coherence Tomography, pp. 693–703. CRC Press (2001)
117. Chauhan, B.C., et al.: Longitudinal in vivo imaging of retinal ganglion cells and retinal thickness changes following optic nerve injury in mice. PloS One **7**(6), e40352 (2012)
118. Luker, G.D., Luker, K.E.: Optical imaging: current applications and future directions. Journal of Nuclear Medicine **49**(1), 1–4 (2008)
119. Napp, J., Mathejczyk, J.E., Alves, F.: Optical imaging in vivo with a focus on paediatric disease: technical progress, current preclinical and clinical applications and future perspectives. Pediatric Radiology **41**(2), 161–175 (2011)
120. Shin, D., et al.: A fiber-optic fluorescence microscope using a consumer-grade digital camera for in vivo cellular imaging. PloS One **5**(6), e11218 (2010)
121. Wang, T.D., Van Dam, J.: Optical biopsy: a new frontier in endoscopic detection and diagnosis. Clinical Gastroenterology and Hepatology **2**(9), 744–753 (2004)
122. Castillo, M., et al.: Optical coherence tomography for the diagnosis of neovascular age-related macular degeneration: a systematic review. Eye **28**(12), 1399–1406 (2014)
123. Mrugacz, M., Bakunowicz-Lazarczyk, A.: Optical coherence tomography measurement of the retinal nerve fiber layer in normal and juvenile glaucomatous eyes. Ophthalmologica. Journal International D'ophtalmologie. International Journal of Ophthalmology. Zeitschrift fur Augenheilkunde **219**(2), 80–85 (2004)
124. Blatter, C., et al.: In situ structural and microangiographic assessment of human skin lesions with high-speed OCT. Biomedical Optics Express **3**(10), 2636–2646 (2012)
125. Koenig, K., et al.: Clinical optical coherence tomography combined with multiphoton tomography of patients with skin diseases. Journal of Biophotonics **2**(6–7), 389–397 (2009)
126. Fujimoto, J.G., Farkas, D.: Biomedical optical imaging. Oxford University Press, New York (2009)
127. Zhou, C., et al.: Cervical inlet patch-optical coherence tomography imaging and clinical significance. World Journal of Gastroenterology **18**(20), 2502–2510 (2012)
128. Zhou, C., et al.: Three-dimensional endoscopic optical coherence tomography imaging of cervical inlet patch. Gastrointestinal Endoscopy **75**(3), 675–677 (2012)
129. Chu, E.M.-Y., et al.: A window into the brain: An in vivo study of the retina in schizophrenia using optical coherence tomography. Psychiatry Research: Neuroimaging **203**(1), 89–94 (2012)
130. Larrosa, J.M., et al.: Potential new diagnostic tool for Alzheimer's disease using a linear discriminant function for Fourier domain optical coherence tomography. Investigative Ophthalmology & Visual Science **55**(5), 3043–3051 (2014)
131. Satue, M., et al.: Use of Fourier-domain OCT to detect retinal nerve fiber layer degeneration in Parkinson's disease patients. Eye **27**(4), 507–514 (2013)

132. Halpern, A.C., Rajadhyaksha, M., Toledo-Crow, R.: Bringing histology to the bedside. Journal of Investigative Dermatology **124**(3), viii–x (2005)
133. Goetz, M.: Confocal Laser Endomicroscopy: Applications in Clinical and Translational Science—A Comprehensive Review. ISRN Pathology, pp. 1–13 (2012)
134. Dimitrow, E., et al.: Sensitivity and specificity of multiphoton laser tomography for in vivo and ex vivo diagnosis of malignant melanoma. Journal of Investigative Dermatology **129**(7), 1752–1758 (2009)
135. Perry, S.W., Burke, R.M., Brown, E.B.: Two-photon and second harmonic microscopy in clinical and translational cancer research. Annals of Biomedical Engineering **40**(2), 277–291 (2012)
136. Murari, K., et al.: Compensation-free, all-fiber-optic, two-photon endomicroscopy at 1.55 μm. Optics Letters **36**(7), 1299–1301 (2011)
137. Gibson, A., Hebden, J., Arridge, S.R.: Recent advances in diffuse optical imaging. Physics in Medicine and Biology **50**(4), 1–43 (2005)
138. Herranz, M., Ruibal, A.: Optical imaging in breast cancer diagnosis: the next evolution. Journal of Oncology (2012). Article ID 863747
139. Garcia-Uribe, A., et al.: In-vivo characterization of optical properties of pigmented skin lesions including melanoma using oblique incidence diffuse reflectance spectrometry. Journal of Biomedical Optics **16**(2), 020501 (2011)
140. Wu, K., et al.: Dynamic real-time microscopy of the urinary tract using confocal laser endomicroscopy. Urology **78**(1), 225–231 (2011)
141. König, K., et al.: Clinical two-photon microendoscopy. Microscopy Research and Technique **70**(5), 398–402 (2007)
142. Thekkek, N., Richards-Kortum, R.: Optical imaging for cervical cancer detection: solutions for a continuing global problem. Nature Reviews Cancer **8**(9), 725–731 (2008)

Therapeutic Uses of Atmospheric Pressure Plasma: Cancer and Wound

Bomi Gweon, Kijung Kim, Wonho Choe and Jennifer H. Shin

Abstract Atmospheric pressure plasma (APP), composed of multiple active components, including charged particles, reactive oxygen species, and radicals, recently emerged as a promising tool in cancer therapy and wound treatment. Driven by the increasingly growing interests on plasma medicine, numerous *in vivo* tests and *in vitro* studies on the applicability of the APP for clinical treatments have been performed. In particular, the APP was shown to be effective in removing the targeted cells by causing apoptosis or necrosis. More importantly, in some studies, these effects were demonstrated to be specific to cancer cells, which was further confirmed in animal models where the APP effectively suppressed the tumor growth. In addition to the cancer therapy, the APP treatment targeting the chronic wound has shown that the plasma could improve the healing process by accelerating blood coagulation, bacteria sterilization, and re-epithelialization around the wounded area. Interestingly, the plasma therapy on cutaneous wound also presented promising outcomes in an animal study with reduced scar formations. In this chapter, we report a general overview of the recent trends and progresses in the effects of the APP treatment on cancer and wound, and discuss the potential applications of the APP as a therapeutic tool.

B. Gweon
Department of Biomedical Engineering, Hanyang University, Wangsimni-ro 222, Seongdong-gu, Seoul 133-791, Republic of Korea

K. Kim · W. Choe
Department of Physics, Korea Advanced Institute of Science and Technology, 291 Daehak-ro, Yuseong-gu, Daejeon 305-701, Republic of Korea

J.H. Shin(✉)
Department of Mechanical Engineering, Korea Advanced Institute of Science and Technology, 291 Daehak-ro, Yuseong-gu, Daejeon 305-701, Republic of Korea

Graduate School of Medical Science and Engineering, Korea Advanced Institute of Science and Technology, 291 Daehak-ro, Yuseong-gu, Daejeon 305-701, Republic of Korea
e-mail: j_shin@kaist.ac.kr

© Springer International Publishing Switzerland 2016

H. Jo et al. (eds.), *Biomedical Engineering: Frontier Research and Converging Technologies*, Biosystems & Biorobotics 9,
DOI: 10.1007/978-3-319-21813-7_15

1 Atmospheric Pressure Plasma (APP) Devices

The main purpose of this chapter is to introduce and to provide an overview of recent advances in plasma medicine, especially in the fields of cancer therapy and wound treatment. Firstly, we would like to provide a short introduction on a few different types of plasma devices to allow better understanding of the applicability of atmospheric pressure plasma (APP).

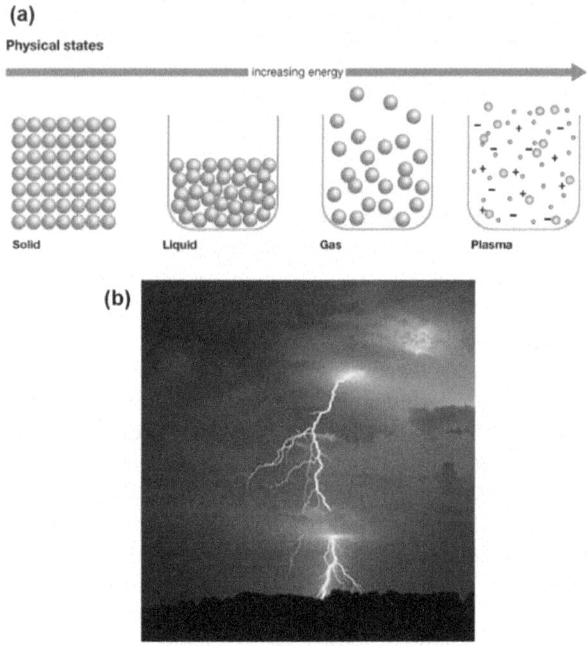

Fig. 1 (a) Four state of matters [1], (b) lightening [2]

In biology, the term "plasma" commonly denotes a non-cellular component of blood that exists in liquid form. There is another usage of the same word, plasma, which shares the same Greek origin, *πλάσμα* (thing formed or molded), in physics [3]. This "plasma" refers to a gaseous substance consisting of charged particles and neutrals, often described as the fourth state of matter alongside the other states, namely solid, liquid, and gas [Fig. 1(a)]. In addition to electrons and ions, the plasma contains UV photons and a wide range of highly reactive chemical species such as reactive oxygen species (ROS), nitric oxides (NO_X), and ozone (O_3). Interestingly, these reactive compounds generated by the plasma can interact with living targets and alter their physicochemical properties, which are proven to be useful in therapeutics. Lately, the number of studies on the application of APP targeting biological tissues and cells increases rapidly, building up a cutting-edge interdisciplinary field of plasma medicine [4-8].

Plasma sources applicable to the living objects are practically limited only to the APP because the living biological samples would require the discharged-environment to be at an ambient pressure. One representative APP that is naturally generated would be lightening [Fig. 1(b)], which obviously cannot be directly utilized for any practical biomedical uses. For medical applications, APP should be carefully optimized considering the material properties of its bio-target and the purpose of the treatments. During the plasma generation, energy is transferred to the gas, charging it by unfolding ions and electrons. The simplest way to ionize the gas is to apply an electric field, which can be generated in the discharge region by direct current (DC), alternating current (AC) from Hz to kHz, radiofrequency (RF) wave around MHz, or microwave aroud GHz into the electrodes. Unlike other states, each of electrons, ions and uncharged molecules in the plasma can have different temperatures from one another. The measured bulk gaseous temperature reflects mostly that of the ions, uncharged molecules or heavy particles, not that of the electrons. Depending on whether the electrons and the ions of the plasma coexist in thermal equilibrium or not, the APP can be categorized into two different types, namely the equilibrium thermal plasma, and the non-equilibrium cold plasma. In the equilibrium thermal plasma, the electrons and the ions have a similar yet high temperature whereas in the non-equilibrium cold plasma, the ions have a much lower temperature than the electrons, resulting in the cold bulk temperature. At an early stage of plasma medicine, the historically more common equilibrium thermal plasma, whose typical discharge temperature is well above 40°C, was utilized to enhance blood coagulation [9-11]. In most of those studies, the subjects suffered mild thermal damages or burns to the skin tissue by plasma treatment but still experienced complete wound healing [8, 12-15]. To avoid any complications associated with thermal damage or pain during the treatment, the plasma scientists began to suggest a utilization of the non-equilibrium cold plasma (non-thermal or cold plasma) in biological samples, which has now become the mainstream subject across the field of plasma medicine. Therefore, in this chapter, we limit our discussions to the non-equilibrium cold plasma.

The key merit of the APP, compared to other low-pressure plasmas, is the fact that it does not require vacuum systems. Attributed to this, the APP can be designed in many different forms, not having to worry about integrating a bulky vaccuumm system. One of the most commly used APP applicators in plasma medicine is a plasma jet [Fig. 2(a-d)]. Plasma jet has been popular thanks to its simple configuration and high biocompatibility based on its low temperature and electric current. The earliest model of the plasma jet consists of a rigid glass-type outer tube that encloses the gas flow, and a single electrode placed in the middle of the tube or wrapped around the tube [Fig. 2(a)]. Recently, an endoscope type plasma jet that consists of a flexible tubing has been developed to treat the interior lesions [Fig. 2(d)] [16]. As shown in Fig. 2, the typical plasma plume has jet sized from few micrometer to millimeter in diameter, determined by specific discharge conditions, which include, but not restricted to, the gas type, dimensions of the tubing, input voltage, frequency and, etc. The jet type allows easy access to the target substrate and controllable targeting of the specific small regions of interest.

When the treatment ought to expand over a substantially large coverage, an array of multiple jets can be utilized [12, 17, 18] to cover the entire area at once or a motorized stage can be employed to scan the area in a sequential manner [19].

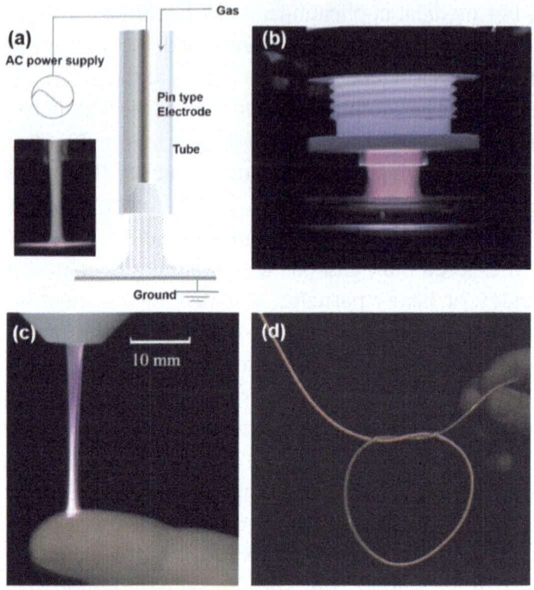

Fig. 2 (a) A schematic of plasma jet, (b) a plasma jet discharged with helium gas, (c) a plasma jet impinging to a human finger [20], (d) a plasma jet with a flexible dielectric tubing [21].

Another commonly used applicator of the APP in bio-application is the dielectric barrier discharge (DBD) plasma [Fig. 3(a-b)]. Conventional dielectric barrier discharge (DBD) consists of two parallel flat electrodes covered with a dielectric material, between which a carrier gas is filling the gap. Typically, one electrode is supplied by AC power with the frequency range of kHz and high voltage in the range of kV while the other electrode is grounded. In this case, the sample of interest should be placed between two electrodes, which is very narrow (few mm); this limits the applicability of the conventional DBD to intact biological tissues such as skin. To treat the intact tissues, a floating electrode DBD (FE-DBD) has been developed [Fig. 3(b)] [5]. In FE-DBD, any sample can replace the ground electrode, which allows greater flexibility and versatility in treating surfaces of bulky objects, including human tissues. Since all of the APPs with the dielectric barrier can be classified as the DBD type plasma, the shapes and designs of the DBD plasmas can vary depending on the conditions and geometries of the target samples. In this sense, the plasma jet with a dielectric material placed between electrodes can also be categorized as DBD.

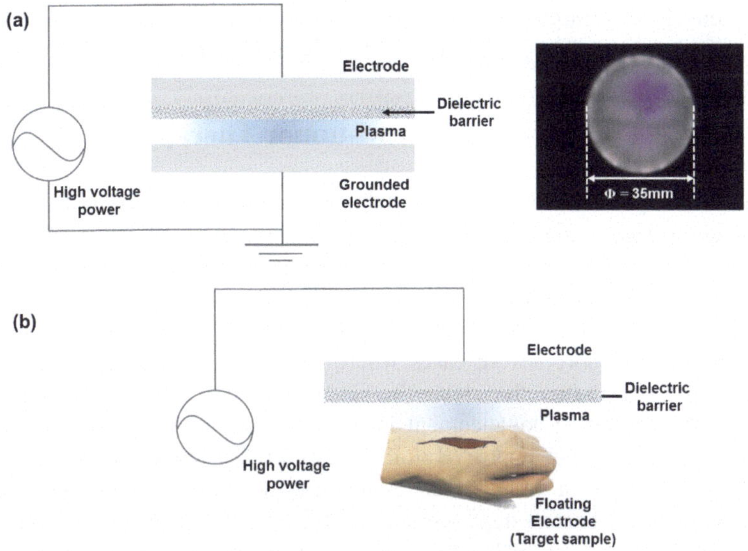

Fig. 3 (a) A schematic of dielectric barrier discharge (DBD) plasma and an example image of DBD discharge, (b) A schematic of floating electrode DBD (FE-DBD) applied on a human tissue

Some of the aforementioned APP types have been tested on animal subjects and employed even in clinical studies of wound and cancer treatments. Portrait Plasma® skin regeneration3 (PSR^3; RF, 1 - 4 J/pulse, N_2) is one of the commercialized devices approved by the United States Food and Drug Administration (FDA) [5, 22]. MicroPlaSter® (2.45 GHz, 86 W, Ar) [23], PlasmaDerms® VU-2010 (400 Hz DBD plasma, 120 mW/cm^2, air) [24, 25], kINPen 09 (DC, 3.5W, Ar) [26], Plasmafon [27], and Plazon [5, 28] are tested on living cells or tissues and have demonstrated their effectiveness in treatment outcomes. As such, the APP sources have been validated as a non-invasive and harmless tool for living organisms, especially with the pain-free feature thanks to its low discharge temperature [29]. Despite the advancement in the utilization of the APP as a therapeutic tool, active key components of the plasma species and the associated effects on the tissue microenvironment and the comprising cells have not been fully identified. The number of reports that characterize the active components increases very rapidly over the last decade, but no standard protocols have been established. Specific discharge conditions and the associated outcomes vary greatly depending on the type of devices and the samples of choice. In other words, any changes in the applied power, the frequency and waveform of input voltage, the treatment time, the gas components, or the physical setup of the plasma source will affect the anticipated outcomes. For that reason, more systematic approach to the underlying mechanisms, both physical and biochemical, involved in the plasma-induced effects would be crucial. Furthermore, on the engineering side, meticulously determined safety limit and improvement of the treatment efficacy would be necessary to establish the APP as a proper therapeutic tool.

2 APP Treatment as an Anti-tumor Therapy

When a tumor mass is detected during diagnosis, the first procedure is the physical removal of tumor mass by surgery, either open-chest or minimally invasive, followed by chemo- or radio-therapies to kill any remaining cancer cells in the neighboring tissue. In some cases when the tumor mass is too large to remove, the chemotherapy is used before the surgery to shrink the tumor to allow less extensive surgery. In more severe cases where cancer cells have spread or migrated deeply into the tissue, the chemotherapy can be used as the main treatment. Despite the exploding advancement in targeted chemotherapy, the survival rates for the tumor have not improved dramatically in past few decades. The working principle of the most standard chemotherapy that targets rapidly growing cells remains to be the best strategy while the selectivity continues to be a major issue. Targeting fast growing cells can compromise other fast growing non-cancerous cells to cause undesired complications, as well as the possible occurrence of drug resistance in cancer cells. With the needs, the APP emerged as an innovative surgical tool whose effects combine both physical removal of tumor mass and biochemical auxiliary effects. The jet type of the APP can be readily employed to treat a local region of interests to remove cancer cells [30-34]. While its selectivity against other normal cells in the surrounding tissue and the secondary effects on these cells have not been fully tested *in vivo*, the limited studies have demonstrated the feasibility of the APP as a cancer treatment at a cellular level *in vitro,* which will be discussed in the next section.

3 *In vitro* APP Studies in Anti-tumor Therapy

One of the earlier applications of the APP was the sterilization of the surfaces, including biological samples. Researchers demonstrated that the APP treatment can successfully exterminate bacteria cells without any apparent damage to the substrate properties [7, 38, 39]. Based on these results, one may easily speculate the applicability of the APP on killing mammalian cells and Stoffel et al. demonstrated successful cellular necrosis via membrane damages in Chinese hamster ovarian cells (CHO-K1) using helium plasma (RF) generated at a needle end with 1 W input power [35] [Fig. 4(a)]. Similarly, Friedman et al. (2007) showed that the treatment with the FE-DBD plasma can also induce deaths of melanoma cells (A2058) within few seconds by compromising cellular membranes [30]. Gweon et al. (2010) showed that the plasma induced cellular necrosis attributed to the disruption in cellular cytoskeleton in a human liver cell (THLE-2) [Fig. 4(b)]. Actin stress fibers and microtubules in cytosol were demolished after 2 minutes of direct exposure to helium plasma jet [36, 40]. In addition to this instantaneous cell death by sudden structural disturbance, cellular detachment induced by plasma treatment was also reported [31, 36, 41-43]. When the APP treatment condition was modified to a mild one (lower plasma discharge power, thicker liquid barrier, shorter treatment time), these abrupt cellular changes were not observed. Instead, after longer waiting time, some cells exhibited programmed cell death, in other words, apoptosis. APP induced apoptosis, distinctive from necrosis, was a new finding when first suggested by Kieft et al. (2006). Kieft et al. showed that APP can induce apoptosis in about 24 hours in 3T3 mouse fibroblasts [41]. The plasma-induced apoptosis is meaningful because the cancer cells normally do not activate their apoptosis process or cell death program spontaneously. Although 3T3 cells were not a malignant cell type, this study provoked the possibility of the usage of the APP in cancer treatment. In following years, a few other groups demonstrated the APP induced apoptosis in cancer cell types originated from various organs, such as skin, breast, bladder, brain, and pancreas, using various APP sources [30, 32, 44] [Fig. 4 (c)]. This notion of the APP induced cellular apoptosis opened up a new window for the APP application for the cancer treatment.

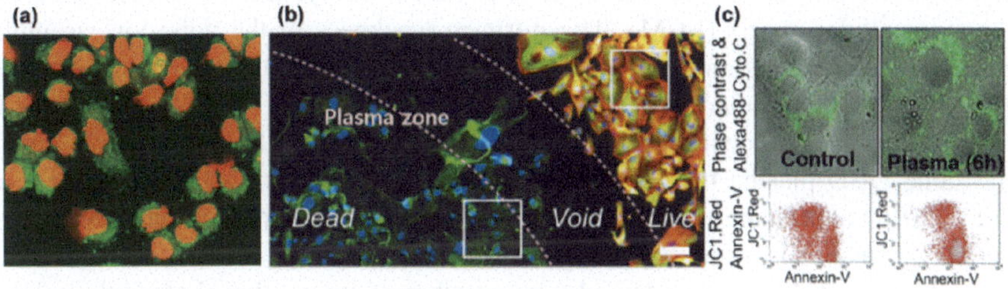

Fig. 4 (a) Cells necrotized by plasma [35], (b) cytoskeletal disruption by plasma; actin (red), microtubule (green), and nucleus (blue) [36], (c) cells apoptotized by plasma in 6 hours [37]

Melanoma cell, cutaneous skin cancer cells were frequently picked as APP target because the skin was considered to be the easiest target surface to approach with the APP [30, 31, 45, 46]. Fridman et al. (2007) used melanoma cells (A2058) to show that the 5-30s treatment with the FE-DBD plasma induced both apoptosis and necrosis, accessed by TUNEL staining assay. Apoptotic cell death appeared at the shortest treatment time of 5 seconds; however, at longer treatment times, instantaneous necrotic cell death became predominant [30]. Similarly, Kim, et al. (2010) showed that breast cancer cells (MCF7) became apoptotic by a short treatment (5 – 30 s) of a pulsed DC plasma jet where the apoptotic responses in cells were detected by Annexin V and PI (propidium iodide) staining [32]. They further observed that additional oxygen mixing in helium plasma increased the efficacy of apoptosis induction in cancer cells. According to Iseki et al. ovarian cancer cells (SKOV3 and HRA) treated by argon (Ar) plasma jet also showed dramatic suppression in proliferation due to apoptotic activities [47]. The APP was shown to induce apoptosis in various types of cancer cells such as skin cancer cells (mouse; B10-F16, B16-F10 and B16-F0) [33, 48, 49], brain cancer cells

(T98G) [50], nerve cancer cells (glioblastoma cell lines; U87MG, LN18, and LN229) [51, 52], pancreatic cancer cells (human cell lines; Colo-357, PaTu8988T and MIA PaCa2, mouse cell line; 6606PDA) [34, 53], bladder (EJ) [44], lung cancer cells (human cell line; A549, H460 and HCC1588, mouse cell line; TC-1) [54-56], oral cancer cells (SCC-25) [57]. Non-adherent typed leukemia cancer cells (THP-1 and CCRF-CEM) prepared as a media suspension in a flask was also tested with the jet plasma to confirm that cells exhibited apoptotic behavior [8, 58]. In other studies, researchers have also found that the specific APP treatment conditions could lead to cellular senescence, aging of cells [46].

More importantly, multiple studies have demonstrated that these plasma-induced effects can be highly selective for cancer cells over normal cells. Gweon et al. (2011) showed that metastatic liver cancer cells (SK-HEP-1) detached from the substrate more readily than the normal liver cells (THLE-2) upon plasma treatment [40]. The authors also confirmed the consistent result for another set of cancer/ normal cells (MDA-MB-231 vs. MCF10A) originated from the mammary gland epithelium. Biological assays showed that the cancer cells intrinsically have fewer and smaller adhesion complexes, thus the weaker biochemical bond to the extracellular matrix (ECM). The intrinsic weakness in the adhesion strength made the cancer cells more susceptible to the APP treatment. In the same context, Panngom et al. reported that lung cancer cell (H460 and HCC1588) presented higher apoptotic cell death compared to normal lung cell (MRC5 and L132) under the same DBD plasma treatment conditions [56]. These two studies by Gweon et al. (2011) and Panngom et al. (2013) are particularly valuable because the cancer and normal cells of the same origin were compared.

As listed in Table 1, there have been a number of reports that showed the selective effect of the APP for cancer cells as opposed to the normal cells of different origin. Lung cancer cells (SW900) exhibited easier detachment from the substrate when compared to the normal human bronchial epithelial cells (NHBE) [48]. From the same study, Keidar and co-workers also showed that in a condition where the primary mouse macrophages survived with no apparent apoptotic behavior, 40 - 50% of the melanoma cells (B16-F10) underwent apoptosis within 48 hours after the APP treatment. Ovarian cancer cells (SKOV3 and HRA), also showed higher sensitivity to the APP treatment than normal cells (human fetal lung fibroblast: WI-38 and MRC-5) [47]. In a separate study, the mouse lung melanomas (B16-F0) went into the apoptotic pathway more readily by the plasma treatment compared to the normal fibroblasts (CL.7) upon exposure to same dosage and duration of the plasma treatment [33]. When human colon cancer cells (COLO 320DM) and mouse lung cancer cells (B16-F10) were compared to normal mouse macrophage cells (RAW 264.7), both of the cancer cells exhibited apoptotic response while the normal cells remained intact [60]. As shown above, a number of findings indicate positive results in terms of the selectivity of the APP for the cancer cells, which can make the APP a promising therapeutic tool for cancer treatment. However, most of the comparisons discussed above were made between normal and cancer cells of different origins. What turned out to be very intriguing was an article by Volotskova, O., et al. in which the authors showed that the APP treatment mostly impacted cells during DNA synthesis (S phase) to induce a transient cell cycle arrest at G2/M phase, demonstrated this phenomenon only in three different keratinocytes (wild type, 308 (benign), and PAM212 (carcinogenic)) [61]. For this reason, the APP treatment appeared to impact more upon cancer cells than normal cells because actively dividing cancer cells have a higher portion of cells in the S phase [61]. If this were to be true for all cell types, the APP treatment would be very powerful in selective killing of fast-dividing cancer cells over normal cells in the surrounding tissue. What separates the APP from the conventional chemotherapy is that the APP treatment is applied locally affecting only the tumor and the neighbors whereas the chemotherapy affects all body parts, leading to undesired killing of fast-dividing normal cells.

Table 1 Summary of the APP effects on different types of cells *in vitro*

Cancer cell type	Normal cell type	Selectivity: Cancer cell viability Normal cell viability	Plasma type	Ref.
Human liver cancer (SK-HEP-1)	Human liver epithelial (THLE-2)	Detached / intact	Plasma jet (He, 50 kHz)	[40]
Human mammary gland (MCF10A)	Human mammary gland (MDA-MB-231)	Detached / intact	Plasma jet (He, 50 kHz)	[40]
Human lung cancer (SW900)	Human bronchial epithelial (NHBE)	Apoptosis / intact	Plasma jet (He, 30 kHz)	[48]
Mouse melanoma cells (B16-F10)	Mouse macrophages	Apoptosis/ viable	Plasma jet (He, 30 kHz)	[48]
Ovarian cancer cells (SKOV3 and HRA)	Human lung fibroblast (WI-38 and MRC-5)	Apoptosis/ less apoptosis	Plasma jet (Ar, 60 Hz)	[47]
Mouse lung melanomas (B16-F0)	Mouse fibroblast (CL.7)	Apoptosis/ viable	Plasma jet (He, 32 kHz)	[33]
Human colon cancer cells (COLO 320DM) and Mouse lung cancer cells (B16-F10)	Mouse macrophage cells (RAW 264.7)	Apoptosis/ viable	Pulsed plasma jet (He, Ar, and O_2)	[60]
Human lung cancer cell (H460 and HCC1588)	Human lung epithelial cells (MRC5 and L132)	Apoptosis/ less apoptosis	DBD (Air, 5.7 W)	[56]
Mouse keratinocytes 308 (benign) and PAM212 (carcinogenic)	Mouse keratinocytes (wild type)	More cell cycle arrest in more tumorigenic cells	Plasma jet (He, 13 kHz)	[61]

To increase the selectivity of the plasma treatment for cancer cell, Kim G.C. et al. suggested to conjugate gold nanoparticles (gNPs) to melanoma cells [45]. They found that, when melanoma cells (G361) were incubated with gold nanoparticles (gNP), gNPs got phagocytized into the cytosol and aggravated the plasma-induced killing of the cells by two folds. Furthermore, when gNPs were conjugated specifically to focal adhesion kinase (FAK) in cells, the number of cellular deaths increased even further by five folds. FAK protein was chosen because G361 cells were shown to overexpress in FAK proteins compared to the normal cells. The authors utilized a combination of air plasma and gNPs conjugation to specific target proteins and demonstrated the enhanced melanoma deactivation by five-fold compared to the case when only the air plasma was used. In separate studies, the same group tried two different types of complexes, anti-epidermal growth receptor (EGFR) antibody-conjugated gNPs (EGFR-gNPs) and anti- transferring receptor (TFR) antibody-conjugated gNPs (TFR-gNPs) in combination with air plasma to treat oral cancer (SCC25) cells. Similarly to the case of FAK, both EGFR and TFR were overexpressed in oral cancer cells, and thus the use of these complexes with air plasma treatment led to much higher cellular death [62]. These results imply that the plasma therapy can be expanded over any types of cancers even when the plasma did not show selectivity upon cancer cells over the surrounding cells. Thus far, we discussed the potential usage of the APP as a surgical tool to selectively remove cancer cells by causing apoptosis, necrosis, or detachment. However, the effect of APP on the neighboring tissues of tumors remains largely unknown.

Tumor is structurally very complex, consisting of many different cell types. In tumor progression, two very important processes occur, namely epithelial-mesenchymal transition (EMT) of the cancer cells and the angiogenesis by endothelial cells in the neighboring tissue. For the efficacy of the APP for the cancer treatment, its effects on other cell types must be investigated. For example, it would be ideal if APP could suppress both EMT and angiogenesis to minimize the potential for metastasis. First, in the context of cancer metastasis, the migratory capacity of the cells is crucial for the cancer cells to be disseminated from the original tumor mass. In this perspective, the effects of the APP treatment on cellular migration should be considered, and some reports successfully showed the suppressing effect of the APP on cellular migration [Fig. 5]. Volotskova and coworkers have measured the distribution of cellular velocity and observed the migration rate of plasma treated skin cancer cells (PAM212) was about 20% less than that of the untreated cells. Kim et al. (2010 a, b) used colorectal cancer cells (SW480) to show that the APP treatment inhibited cell migration and, further, invasion activity [59, 63].

Secondly, as much as the targeting the cancer cells, targeting the environment of the tumor is also important to control cancer metastasis. On the notion that cancer metastasis requires angiogenesis, effective suppression of angiogenesis around the tumor mass would also be important for cancer management. Interestingly, Gweon et al. (2014) observed that the APP can impede the angiogenesis process by affecting endothelial cells [64]. They used a helium plasma jet upon human aortic endothelial cells (HAECs) to demonstrate that the APP treatment affected the proliferation by causing cell cycle arrest. Along with the cycle arrest, migration of the cells slowed down and the ability to form new vessels was suppressed by the APP treatment.

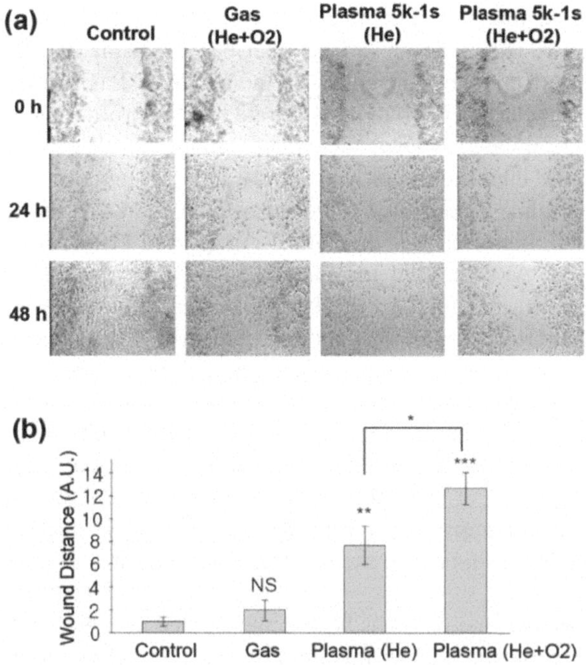

Fig. 5 Wound healing assay with colorectal cancer cells (SW480). (a) SW480 cells at 0, 24 and 48 hours with and without plasma treatment, (b) wound distance quantified from wound healing assay. (NS: not significant, AU: arbitrary unit) [59]

4 *In vivo* APP Studies in Anti-tumor Therapy

Compared to the in vitro studies, much fewer studies have been reported on the APP based cancer treatments on animal models. Early studies of the APP application on animal tissue examined dosage-dependent damages on living tis-sues, accessed by either histological evaluation or 3D two-photon microscopy, to validate the treatment conditions of the APP for cancer therapy [5, 25, 65-67].

Supported by numerous in vitro studies and the growth of the related field, multiple in vivo studies have been reported in recent years [33, 53, 68-71]. In these studies, many different types of mouse tumor models were tested, which include glioblastoma, melanoma, pancreatic tumor, and neuroblastoma. These in vivo tumor models have been prepared by injecting cancer cells subcutaneously. Vandamme et al. injected U87 glioma cells into mice to develop tumor model and treated the tumor with the FE-DBD (100 Hz) plasma for about 6 minutes every day for 5 consecutive days. Bioluminescence imaging (BLI) indicated that the measured tumor volume was reduced by 33% [68]. Kim et al. (2011), created melanoma with a size of 40 mm^3, and then treated the tumorous regime with the plasma jet [33]. To obtain the tumor growth rate, volume of the tumor was monitored every 2 - 3 days to find that suppression of tumor was significant in the group treated four times per day. In addition, the feasibility of the plasma cancer endoscopy was tested using an optical fiber-based plasma jet. On a pancreatic cancer model with human pancreatic carcinoma (MIA PaCa2-luc), Brulle et al. used a plasma jet (Plasma Gun) to demonstrate that the proliferations of pancreatic carcinoma were inhibited *in vivo*, and, when combined with gemcitabine (chemotherapy drug), the inhibition improved significantly [53]. Walk et al. showed that the APP significantly decelerated tumor growth, so that the survival rate of the mouse with neuroblastoma was dramatically improved [70]. All of these studies demonstrated the APP induced impediment in tumor growth in many different tumor models. Furthermore, the study by Walk et al. additionally emphasized that the APP induced effects are enough to prolong the life of the subject animals [70].

While there have been several promising research outcomes reported on animal studies, it would require much more systematic studies on the consequences and side effects of the APP before adopting the non-thermal APP treatment to actual clinical studies. As a side note, the classical argon plasma coagulation (APC) device, which is a type of thermal plasma of a high temperature, has been used as a surgical tool as a comparable tool to the laser ablator or electrical scalpel [10, 72, 73], but we leave this out to the references because it is beyond the scope of this chapter.

Taken all together, the APP has been demonstrated as a potent therapeutic tool for the cancer treatment, whether it is utilized as a surgical tool to directly remove the tissue or it induces biochemical reactions that leads to cell deaths and detachment. Then, the next question is what the underlying mechanism involved in these cell death, senescence, and suppressed cellular motility is. To understand the basic mechanism of the plasma therapy, it is important to correlate the plasma dosage with the phenomenological outcomes of the cellular responses. For example, at low input power or short treatment time, cells undergo apoptosis or become detached from the substrate whereas at higher input power or treatment time over a certain threshold, the cells become necrotized [5, 40, 74, 75]. Yet, again, the users of the APP need to be cautious as these specific treatment conditions are the device specific and thus cannot be generalized in any simple manner. Also, the APP generates a handful of many different active species that could influence cellular behaviors in an orchestrated manner. Therefore, identifying and correlating the specific component to the observed phenomenon would be very much desired by isolating these individual components in the plasma. Alternatively, we can find the dominant player amongst many components and formulate a quantitative relationship between the dosage and the effect. In recent studies, the amount of the APP induced reactive species was suggested as a direct parameter to represent the plasma dosage instead of the device specifications and experimental conditions. To quantify the reactive species, some of the researchers monitored the emission spectrum of the plasma discharge to correlate the reactive species with the APP effect. Optical emission spectroscopy (OES) is a popular and powerful method to monitor the active species such as OH, O, N_2, and N_2^+ in the plasma where the peaks at the spectrum represent the relative amount of the existing molecules in the discharged plasma. The OES measurements from a few different types of the APP devices featured the peaks around 200 - 400 nm, from which UV photons were suggested as another key player in plasma [76]. This observation was consistent with previous findings on the effect of UV irradiation around 260 nm that was known to alter DNA structure, leading to cellular apoptosis. However, the significance of UV photon would also depend on the type of APP devices [6, 76]. In addition to UV, the generated radicals in the plasma have been considered to be responsible for the plasma-induced cellular behaviors. Among many radicals, reactive oxygen species (ROS) has been recognized as one of the key players in plasma-induced apoptosis, senescence or suppressed migration of cells [51, 77]. Multiple studies showed that the plasma-induced elevation in ROS in the media increased the efficacy of the APP to kill target cells; addition of oxygen gas at discharged plasma led to higher number of apoptotic cells [32, 44, 59]. Unfortunately, however, the amount of radicals cannot be measured by the OES method, requiring to finding other methods to quantify the radicals.

As described, in plasma-side mechanism, there still remains lots of missing pieces and controversies. Then, what about the mechanism of the target cells? – i.e. what upstream factors in the cell provoke these changes inside cells or which signaling pathways will be triggered by plasma treatment? In recent years, there have been a number of molecular level studies to reveal the involved molecular signaling pathways. For example, DNA damages by the APP were shown to trigger cellular apoptosis, and cell cycle arrest via p21 CDK inhibition [51, 78], p53 phosphorylation [79] or p53/p21 dependent pathway [51, 61, 64, 78]. Furthermore, other studies revealed that the APP induced mitochondria dysfunction leading to cellular apoptosis [49, 56]. Overall, there has not been any single answer as to how the APP affects any specific cellular functions. Because the kinds of active components in the discharged plasma vary depending on particular types of the devices and specific experimental conditions, again we cannot emphasize enough how important it is to identify and pair the effectors and corresponding responses.

5 APP Treatment for Wound Healing

Wounds may be considered as a less life-threatening disease than cancer. However, chronic wound such as ulcer can cause persistent pain and discomfort and, sometimes, the wounds of a large affected area or with severe infection can become life threatening. As much as the complex tumor progression, the wound healing is a very complicated physiological process that involves a series of cellular and molecular events with a handful of active players. The major phases of the healing starts with the coagulation, followed by vasoconstriction, vasodilation, inflammation, re-epithelialization, contraction, angiogenesis, and finally maturation and remodeling. At early stage of the APP application, the hot thermal plasma was first employed to enhance the coagulation at the wound site. As a pain-free and damage-free tool on the living tissues, cold APP was revealed to be effective on these multiple physiological steps of wound healing. Thanks to the proven efficacy of the APP in killing bacteria, the cold APP was used as a sterilization tool to minimize infections. In addition, the APP was shown to improve the quality of the wound healing. Unlike the research path on APP cancer therapy, the investigations on wound healing started from the *in vivo* studies first and then expanded to the *in vitro* cellular and molecular studies. Although the underlying mechanism is still largely unknown, the applications for different types of wounds have shown promising outcomes. In the following sections, we will discuss the reported *in vitro* studies followed by *in vivo* studies.

6 *In vitro* APP studies in wound healing

The key phases that are crucial in cellular processes of wound healing are inflammation, angiogenesis, fibroplasia, epithelialization, and dermal remodeling [80, 81]. Each phase requires collaborative contribution of various types of proteins, cells and tissues in the microenvironment of the wound, and the APP appears to contribute in multiple phases of the wound healing process [82].

Among these complicated processes, blood coagulation of the APP was recognized first by many plasma physicists. Since early 1990s, the argon plasma coagulator (APC) discharged at 40 - 100 W in the MHz range frequency has been used as a potent blood coagulator [9, 72] by thermally ablating and desiccating the wounded tissue [9, 83]. More recent studies have demonstrated that the cold plasma could also trigger and accelerate blood coagulation. In studies by Kalghatgi et al., it was demonstrated that blood collected from a healthy human coagulated in less than a minute when treated by the FE-DBD plasma (1.5 W/cm^2). Normally, the coagulation would take up to 15 minutes without any treatment. After treatment with the FE-DBD, significant alterations were observed in fibrinogen composition and other important coagulation factors in the blood [75, 84]. Choi et al. also reported that the treatment with the plasma jet (3 W of input power) could accelerate blood coagulation when they tested on blood from a healthy mouse (CBA/N strain) pre-treated with an anti-coagulant (EDTA; ethylenediaminetetraacetic acid, 5 mmol/l). Coagulation of the plasma treated blood initiated within 10 seconds and was completed by 20 seconds whereas the untreated sample did not show any sign of coagulation [85]. Similarly, Janani et al. and co-workers found the treatment with the cold argon plasma jet around room temperature successfully decreased blood coagulation time down to 10 seconds, which was 8 minutes without treatment [86]. Because the utilized plasma sources in these studies were non-thermal cold plasma whose temperature was maintained below 38.3°C during the treatment time, the action mechanism must be different from what was initially believed as thermal dehydration [85]. This requires further investigation.

Inflammation is another key phase in the wound healing process. Inflammation in the tissue is caused by bacterial infection around the wounded tissue. When the inflammatory response becomes excessive, it can drastically delay the wound repair process [87]. For example, if the level of infectious bacteria was not controlled well, elevated pro-inflammatory cytokines would be prolonged, initiating the chronicity of wounds [88]. When the bacterial loads larger than 10^5 CFU/tissue, the number of bacteria species at the site more than four, or the presence of resistant pathogens (*Staphylococcus aureus, Pseudomonas aeruginosa*, and etc.) can all lead to wound infection, which will delay the healing process [7, 89-91]. Therefore, controlling infection in and around the wound bed would be crucial in proper healing of the wound, making the germicidal properties of the APP plasma useful in chronic wounds with serious infection along with its blood coagulation ability. Bacterial decontamination studies using the APP in vitro dated back to late 1990s [38, 92, 93], and since then, many research groups started to evaluate the sterilization efficacy of the APP for its wound healing capability [5, 20, 82, 94, 95]. Friedman et al. (2007) established an *in vitro* wound model using the block of agarose gel containing 5% sheep blood to culture bacteria (*Staphylococci, Streptococci*, and *Candida* species of yeast) on it, and demonstrated that their plasma (FE-DBD) treatment drastically deactivated bacteria grown on the agarose gel surface [96]. Weltmann et al. grew *Escherichia coli* (*E. coli*) on agar gel, treated the surface using the APP to demonstrate the effectiveness of the plasma on decontamination [29]. On agarose gel mixed with blood, 11 strains of bacteria were routinely cultivated, and 8 of them were sterilized by the plasma treatment [94]. Daeschlein et al. reported that the RF plasma treatment was effective for most of the typical wound pathogens found on acute wound surfaces (*Staphylococcus aureus*, *Pseudomonas aeruginosa*, and *b-hemolyzing Streptococci*) *in vitro* on agar plate. For two pathogens (*Enterococcus faecium and Candida albicans*), the RF plasma failed to reach the antiseptic level within 6 minutes of treatment [97].

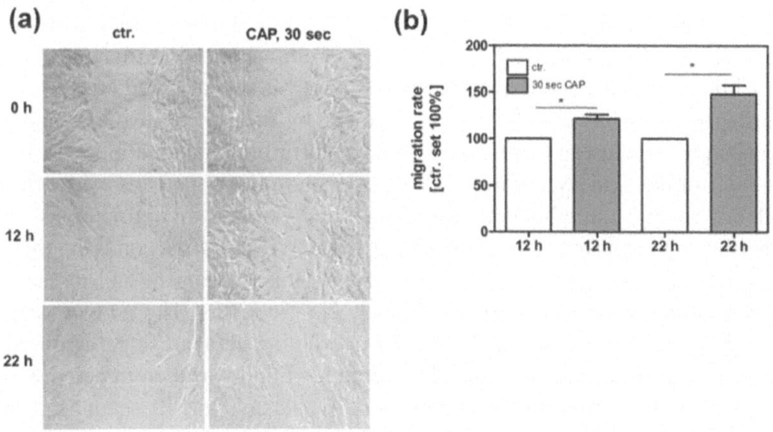

Fig. 6 Wound healing assay of fibroblast cells (w/ and w/o plasma treatment), (a) images of scratch wound at 0, 12 and 22 hours: APP treated (for 30 seconds) wound showed faster closure compared with untreated control, (b) the migration rate for APP treated wound ■ and untreated control □ (at 12 and 22 hours after APP treatment) [23]

Next important step in wound healing would be angiogenesis and fibroplasia in which endothelial cells and dermal fibroblasts are representative cells for these processes, respectively. Interestingly, some researchers have found that the APP can directly affect endothelial cells and dermal fibroblast cells to influence vessel formation and fibroplasia processes [Fig. 6] [23, 98]. To investigate the plasma effect on the physiological functions of the vessels, Kalghatgi and colleagues designed an *in vitro* experiment using porcine aortic endothelial cells [98]. They first established the optimal low dose plasma treatment condition (below 30 seconds treatment time or 4 J/cm^2 energy density for 30 seconds), and showed that the proliferation of the APP treated cells was twice as high as the untreated control five days after the plasma treatment. They also demonstrated that the plasma treated endothelial cell proliferated more and secreted fibroblast growth factor-2 (FGF2), which diminished when they inhibited FGF2 activity or ROS. Based on these, they suggested that the plasma-induced endothelial cell proliferation was due to ROS induced FGF2 release. The same group also demonstrated the enhanced angiogenesis using porcine aortic endothelial cells and DBD plasma (Arjunan et al. 2012). Endothelial cells treated by plasma (4.2 J/cm^2) showed increased migration by 40 percent in 2D surface and 15 percent in 3D matrix compared to untreated control. Tube formation efficiency measured by total tube length was also enhanced by 2.6 times in the plasma treated cells compared to the untreated control [99]. Arndt et al. investigated the plasma effect in fibroplasia examining how plasma improved wound healing processed by measuring the molecular composition changes in human fibroblasts after 2 minute treatment using a commercialized plasma source called the MicroPlaSter®. Their *in vitro* analysis unveiled that the plasma treatment elevated expression of key factors for wound healing (TGF-ß1, TGF-ß2, IL-6, etc.) and increased production of extracellular matrix (collagen type I) in fibroblast cells. Also, as shown in Figure 6, *in vitro* wound healing assays showed an increase in cellular motility after plasma treatment while neither the proliferation nor the apoptosis was affected [100].

Furthermore, it has also been demonstrated by the *in vitro* wound model that the plasma treatment can be effective without perturbing the keratinocytes in outer most layer of the skin. According to Wende et al., microorganisms were effectively killed while the epithelial cells were intact and viable at an optimal plasma condition (RF argon plasma jet, 2 - 6 kV, $P \approx 10$ W, and 40 seconds of treatment time) [101]. To evaluate the efficacy in killing microorganism and the safety in normal healthy cells, they cultured human keratinocytes (HaCaT) *in vitro* and made a modified scratch wound model by overlaying an agarose gel upon the cell layer to mimic the wounded tissue. The scratch wound was then infected with 10^4 of *Staphylococcus epidermidis* and treated by plasma for 40 seconds. HaCaTs were able to grow and migrate to close the scratch wound 24 hours after the plasma treatment while the non-treated ones were not able to close the wound. This result indicated that the APP could act selectively on prokaryotic cells while promoting the migratory capacity of the eukaryotic cells. Haertel et al. investigated whether an argon plasma jet compromised the integrity of the adhesion molecules of HaCaT by the APP treatment that exterminated the bacteria (*Staphylococcus aureus*) [26]. Expressions of $\alpha 2$ and $\beta 1$ integrin, e-cadherin (transmembrane protein forming adherent junction) and the EGFR did not change much by 10 seconds treatment but the longer treatment for 30 seconds increased $\beta 1$ integrin and decreased e-cadherin and EGFR expression, measured by flow cytometry 24 hours after the plasma treatment. These results indicated that the 10 seconds plasma treatment did not affect the adhesion molecule or epithelial cell function. Similar standards would be useful in determining the treatment conditions for the safe clinical usage.

7 *In vivo* APP Studies in Wound Healing

In many *in vitro* studies, it was shown that APP could enhance the healing process by participating in multiple wound healing phases. Intriguingly, recent *in vivo* plasma researches on animal models or human subjects started to turn this potential into an optimistic outcome. As previously mentioned, there exist a number of different types of commercialized plasma devices which are available for clinical usage [24, 28, 95, 102]. To this extent, *in vivo* studies have already achieved great progress in clinical applications, and, now, researchers start to take a step back to investigate the detailed mechanism on how APPs could affect the living tissues.

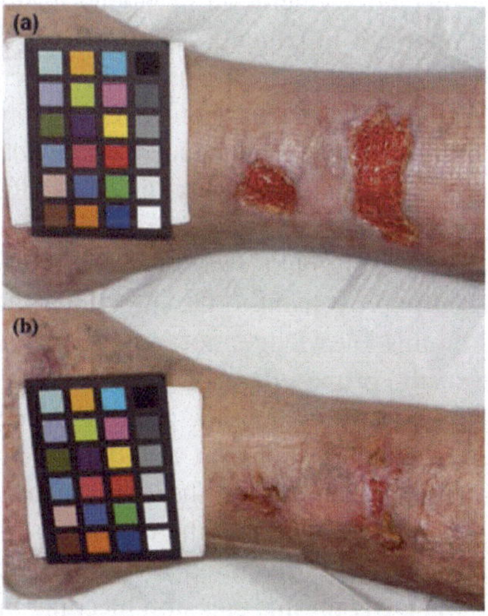

Fig. 7 Venous ulcers treated by plasma (61-year-old patient): (a) wounds before plasma treatment and (b) after 11 treatments. On daily basis, MicroPlaSter® was treated for 2 minutes. [76]

Many researchers highlighted that the APP impacted preferentially on prokaryotic cells (bacteria) over the surrounding eukaryotic tissues to control the inflammation process, which then influenced the neighboring cells indirectly. The first clinical study of the APP on the wound was initiated using the 'Plazon' [5, 28]. Trophic ulcers of 318 patients were treated by air plasma with exogenous NO, and the total healing time was reduced by 2.5 - 4 times compared to the non-treated ones. In this study, the key therapeutic factor was plasma-generated

NO rather than the plasma itself, but in more recent studies, the direct contact with the plasma plume was shown to be effective in the healing of the wound, showing promising outcomes. Isbary and co-workers performed a randomized controlled phase II clinical trial using the MicroPlaSter®; 38 infected chronic wounds in 36 patients were exposed to this argon plasma for 5 minutes on daily basis as an add-on therapy [Fig. 7] [102]. To quantify the bacteria species, standard bacterial swabs were used, and, as a result, bacterial counts were significantly reduced ($< 10^{-6}$) in the plasma-treated wounds compared to the non-treated controls. Similarly, Brehmer et al. conducted a randomized clinical trial with PlasmaDerms® VU-2010 first to determine the safety standard for the add-on therapeutic tool against the chronic venous ulcers [24].

Among 14 patients, half of them received standard wound care (control group) and the other half received the plasma treatment in addition to standard care (plasma group; 3 times per week for 8 weeks). Compared to control group, the plasma group patients showed greater wound reduction on average (control: -3.4 cm^2, plasma: -5.3 cm^2). According to Daeschlein et al., a single argon plasma jet treatment reduced multidrug-resistant (MDR) bacteria in all wounds [95]. Fetykov et al. also reported the marked acceleration in wound healing as well as pain reduction by the plasma treatment in patients suffering from diabetic foot syndrome [27]. Mertens et al. tested the plasma treatment on the patients with chronic atopic dermatitis caused by pathogenic microorganism colonization to show that the DBD plasma treatment (90 – 700 Hz) reduced *Staphylococcus aureus* colony forming unit (CFU) by more than ten-fold only in 2 days [76, 103].

Although the detailed physiological basis for the plasma-assisted acceleration in wound healing remains largely unknown, many researchers demonstrated that the APPs induced marked improvement to the infection of the chronic wounds by clinical studies. Based on this knowledge, not only are the germicidal characteristics considered as one of the prime effect of the plasma treatment, but there are also other factors assisting the wound healing process. As demonstrated in a number of *in vitro* studies, fast blood coagulation would be one of them. In addition to the thermal plasma-based coagulation tool, the APC, many *in* and *ex vivo* studies as well as *in vitro* studies have proven that the treatments using non-thermal cold plasma devices can also induce accelerated blood coagulation. Fridman et al. (2006) showed that the FE-DBD plasma can stop bleeding in 30 seconds on the ex vivo model of a human spleen where the authors demonstrated that the temperature remained at room temperature and the tissue was kept moist indicating negligible tissue damage and desiccation [75]. Janani and co-workers tested the efficacy of the APP in coagulation in rats with live injuries [86]. When the total bleeding time was measured on the incised liver until complete coagulation, it took about 4.5 minutes in control group as opposed to 12 seconds in the plasma treatment group.

Interestingly, according to Nastuta et al., the plasma was shown to directly affect the cells or tissues physiology related to the curing process. Re-epithelialization, a part of the skin regeneration process, in which cellular components are recruited by epidermal and dermal interaction, was shown to be accelerated by the APP treatment [20]. After helium plasma jet (4 - 5 kV, 2 kHz) was applied on burned wounds of Wistar rat skin for 40 seconds for 21days, the re-epithelization rate was compared to those treated by the conventional type of dressing (polyurethane) as well as the untreated control. The observed hematological, biochemical and histological parameters, in both of the classic dressing and plasma-assisted samples, indicated positive physiological outcome corresponds to the accelerated recovery compared to those in non-treated wounds. Surprisingly, on the plasma treated tissue, the re-epithelization of epidermis was accelerated while the recovery of superficial dermis was decelerated [20, 105].

Some studies revealed that the APP not only accelerated the healing process but also improved the appearance of skin. In a recent report by Lee et al., their APP treatment showed promising outcome in the context of perfect wound repair by reduced scarring of the skin in a mouse model. They applied the helium plasma (3 slpm mixed with 50 sccm of air, 4 kHz) on C57/BL6 mice (aged 8 weeks) 3 times for 3 minutes per week over 6 weeks. In plasma group, the hypertrophic scar was suppressed [Fig. 8] [104]. Indeed, in regards to the aesthetic and functional aspects, reducing cutaneous scar or suppressing scar formation process would be considered very crucial in dermatology and plastic surgery.

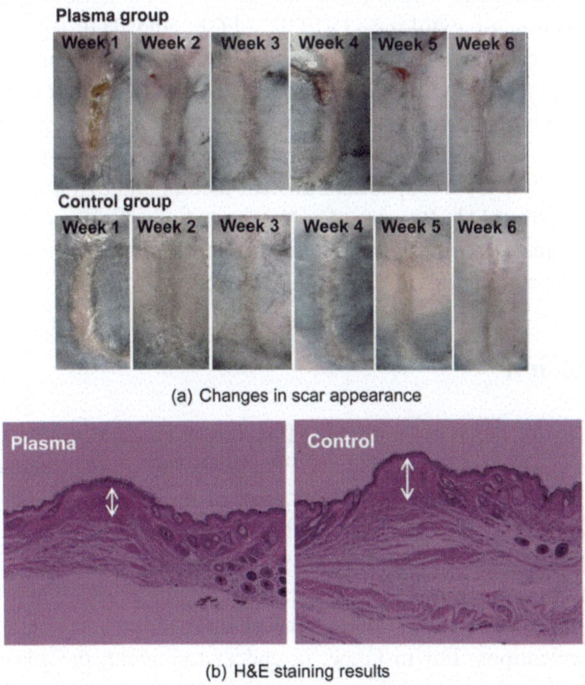

Fig. 8 (a) Scar appearance in the plasma and control group. (The additional wound appeared in week 4 in the plasma group is a scratch from another mouse.) (b) Hematoxylin and eosin (H&E) stained scar tissue. Slightly more fibrosis in the control tissue compared with the plasma treated tissue [104].

Furthermore, Bogle et al. showed that the nitrogen plasma treatment could induce skin regeneration (PSR[3], RF, 1.2 - 1.8 J). Eight volunteers received full-face plasma therapy every 3 weeks in total 3 treatments, and were evaluated 4 days after each treatment as well as the 30 and 90 days after the last (3rd) treatment. During the whole time course of the treatment, the skin underwent partial damages and re-epithelization. A histologic evaluation demonstrated a band of new collagen in the epidermal junction and less elastin in the upper dermis. At 3 months after the last treatment, a 37% improvement in facial rhytides was observed [22]. During the period of plasma treatment, the researchers also scaled the pain and monitored any signs of side effects. Intriguingly, in most clinical studies, the patients felt much less pain compared to the conventional wound treatment. As well as human subjects, animal models employed in these studies also showed good tolerance against the plasma treatment with no sign of side effects or complications [24, 102].

The underlying mechanism of the APP-induced wound healing still remains unclear. The NO molecule, for example, can be one of the key stimulators in the plasma therapy. Based on the generally known NO biochemistry, the specific roles of the radical to the wound healing process may include the enhanced phagocytosis of bacterial pathogens, vasodilator, stabilization of microcirculation, vascular growth by secretion of cytokines (VEGF, bFGF, and TGF-β), proliferation of fibroblast and keratinocytes, and increased collagen synthesis and deposition [106-109]. Additionally, multiple researchers observed the intra- and the extracellular abundance of the reactive oxygen species (ROS) induced by the plasma treatment. ROS is known to play a crucial role in the antibacterial activity to be involved in activation of the upstream growth factors and cytokines in the signaling pathways [110, 111]. Besides the NO and direct ROS activities from the plasma sources, the UV irradiation may also contribute in the inactivation of microorganism or in turning on the intracellular ROS [7, 76].

8 Conclusion

The applicability of the non-thermal APP has evolved rapidly in last few decades in the field of plasma medicine. The APP clearly has a huge potential in cancer treatment and wound healing. Especially, one of the most appreciable properties of plasma treatment would be "one applicator for multiple functions" where it controls or influences multiple processes in the complex tumor or wound microenvironment at the same time [Fig 9 (a) and (b)]. There exist a handful of different key effectors of the plasma that lead to the phenomenological outcomes useful in therapeutic applications. For instance, in cancer treatment, the plasma treatment can effectively deactivate or kill cancer cells, suppress vascularization near the tumor, and inhibit cellular metastasis. For the wound therapy, the plasma is capable of promoting coagulation, controlling the infection and excess inflammation, inducing angiogenesis, and accelerating the re-epithelialization for better healing of the tissue.

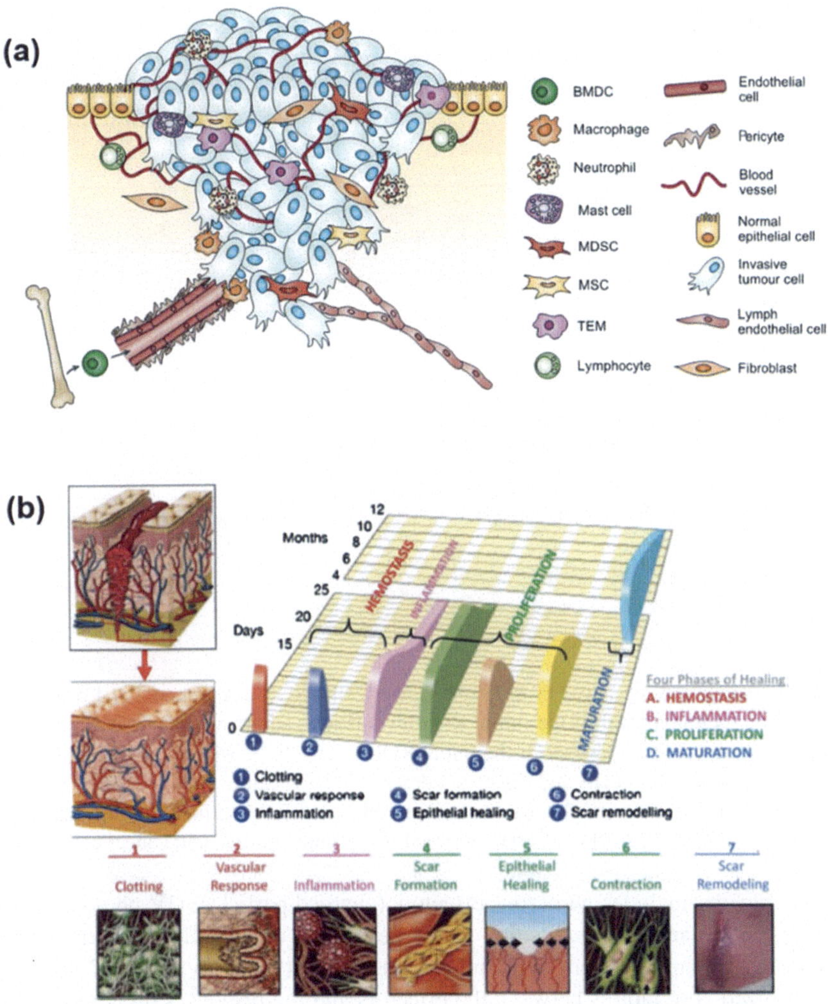

Fig. 9 (a) Schematic of the tumor microenvironment; numerous of cells involved in a multiplex structure [112], (b) complicated wound healing process involving multiple phases [82]

As such, the plasma medicine, as an innovative therapeutic concept, already penetrated into the field of medicine, but there still remain some open questions that await solutions; what is the key player or effector of the plasma? How should the dosage of the APP be optimized to enhance the outcomes while minimizing the toxicity? Would the APP therapy cause any latent side effects? In addition to these questions, the questions on the fundamental understanding of the interaction between the APP and the living cells and tissues remained unanswered, requiring further investigations in physical, biological as well as physiological perspectives.

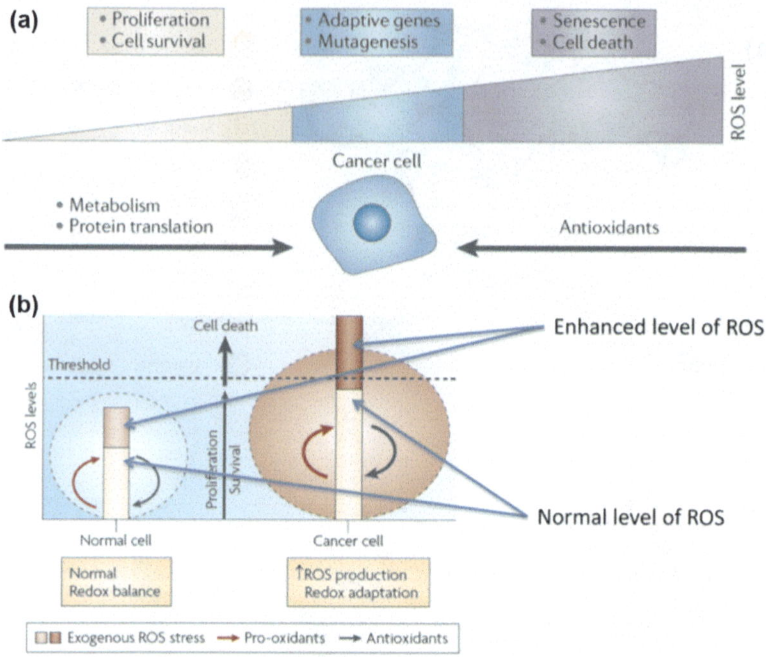

Fig. 10 (a) Relationship between ROS and cancer cells: ROS can be beneficial or harmful depends on its initial level. If the initial ROS level is low, additional ROS can support survival pathway of cells. However, if the initial ROS level is high, with additional ROS, ROS will become too high in cells and cause cell death [113], (b) cancer redox biology: since the initial level of ROS of normal cells and cancer cells are different, the redox mechanism or tolerance against ROS can be different between two groups (normal and cancer cells) [114]

The plasma includes numerous effectors (charged particles, UV photons, ROS, NO_X, O_3, etc.) and its composition varies depending on specific discharge conditions such as input power, gas composition, humidity and so on. Until now, the standardized correlations between the effectors and the responses do not exist and the plasma-induced effects are believed to be the device-specific consequences. What has been most studied amongst many factors is the plasma induced ROS. ROS has been identified as a predominant key factor in the plasma-induced cellular processes, and the majority of the research has shown the elevated level of intracellular ROS in cells or tissues (mostly epithelium) upon treatment by the APP. Surprisingly, our recent study on the stromal cells indicated that the intracellular ROS level was indeed reduced in the APP treated cells by the elevated antioxidant activities. Our result supports that cells of different types not only have different level of initial ROS but also have different level of ROS managing capacity via ROS scavenging process [Fig. 10]. Based on this, we concluded that the APP-cell interactions could be cell type specific (this cell type specificity, indeed, could be related to the selectivity of APP for cancer cells). Additional to the cell type sensitive properties, there are other variables, which make the plasma therapy more complicated. Multiple studies have demonstrated that, depending on the plasma dosage (treatment time, input power) or the type (input power, voltage, frequency, and supplied gas), the plasma effects on cells or the cellular responses would be entirely different. For example, a particular experimental condition of the plasma treatment can be pro-angiogenic whereas another condition can be anti-angiogenic, requiring the researchers to establish more rigorous operation criteria specific to the targeted diseases for consistent outcomes. At last, as questioned earlier, toxicity and potential side effects still remain unsolved. So far, in most *in vivo* or clinical trials, the maximum acceptable dose was defined just by excluding the dosage that led to short term phenotypic damages. However, the toxicity of plasma should be addressed more precisely and the potential side effects, including mutagenic capabilities, should be carefully accessed over a long period of time.

In conclusion, despite the unsolved mysteries on the working principles and possible complications, the APP still has a great potential to become a groundbreaking therapeutic tool for cancer treatment and wound healing. This can be realized if and only if the APP treatment becomes safe and reliable. To establish this basic knowledge, it is important to foster broader multidisciplinary research filed across different disciplines of plasma physics, biomedical engineering, biology, biochemistry, oncology, and dermatology. Reciprocal collaborative efforts across these diverse fields will allow the scientists to gain better understanding of the working mechanisms and principles, the minimum effective threshold, the optimal parameters, as well as other risk factors, based on which the proper therapeutic applications of the APP would be realized.

Acknowledgments Researches in the GDPL and SoftBM lab was supported by R&D Program through the National Fusion Research Institute of Korea (NFRI) funded by the Korean government.

References

1. States of matter, Art. Encyclopædia Britannica Online. Web, April 2015. http://global.britannica.com/EBchecked/media/148660/States-of-matter
2. Griffinstorm (Own work) [CC BY-SA 4.0]. http://commons.wikimedia.org/wiki/File:Staccoto_Lightning.jpg
3. http://biblehub.com/greek/4110.htm
4. Laroussi, M.: Low-Temperature Plasmas for Medicine? IEEE Transactions on Plasma Science **37**(6), 714–725 (2009)
5. Fridman, G., et al.: Applied plasma medicine. Plasma Processes and Polymers **5**(6), 503–533 (2008)
6. Kong, M.G., et al.: Plasma medicine: an introductory review. New Journal of Physics **11**, 115012 (2009)
7. Heinlin, J., et al.: Plasma applications in medicine with a special focus on dermatology. Journal of the European Academy of Dermatology and Venereology **25**(1), 1–11 (2011)

8. Hoffmann, C., Berganza, C., Zhang, J.: Cold Atmospheric Plasma: methods of production and application in dentistry and oncology. Medical Gas Research **3**, 21 (2013)
9. Waye, J.D., Grund, K.E., Farin, G.: Argon plasma coagulation (APC) - Clinical usefulness in flexible endoscopy. Gastrointestinal Endoscopy **43**(4), 64 (1996)
10. Raiser, J., Zenker, M.: Argon plasma coagulation for open surgical and endoscopic applications: state of the art. Journal of Physics D: Applied Physics **39**(16), 3520–3523 (2006)
11. Kaassis, M., et al.: Argon Plasma Coagulation for the Treatment of Hemorrhagic Radiation Proctitis. Endoscopy **32**(09), 673–676 (2000)
12. Laroussi, M., Akan, T.: Arc-free atmospheric pressure cold plasma jets: A review. Plasma Processes and Polymers **4**(9), 777–788 (2007)
13. Ehlbeck, J., et al.: Low temperature atmospheric pressure plasma sources for microbial decontamination. Journal of Physics D: Applied Physics **44**(1), 013002 (2011)
14. Laroussi, M.: Nonthermal decontamination of biological media by atmospheric-pressure plasmas: Review, analysis, and prospects. IEEE Transactions on Plasma Science **30**(4), 1409–1415 (2002)
15. Tendero, C., et al.: Atmospheric pressure plasmas: A review. Spectrochimica Acta Part B: Atomic Spectroscopy **61**(1), 2–30 (2006)
16. Kim, J.Y., et al.: Single-Cell-Level Cancer Therapy Using a Hollow Optical Fiber-Based Microplasma. Small **6**(14), 1474–1478 (2010)
17. Weltmann, K.D., et al.: Antimicrobial treatment of heat sensitive products by miniaturized atmospheric pressure plasma jets (APPJs). Journal of Physics D: Applied Physics **41**(19), 194008 (2008)
18. Park, G.Y., et al.: Atmospheric-pressure plasma sources for biomedical applications. Plasma Sources Science & Technology **21**(4), 043001 (2012)
19. Kim, J.Y., et al.: 15 mu m-sized single-cellular-level and cell-manipulatable microplasma jet in cancer therapies. Biosensors & Bioelectronics **26**(2), 555–559 (2010)
20. Nastuta, A.V., et al.: Stimulation of wound healing by helium atmospheric pressure plasma treatment. Journal of Physics D: Applied Physics **44**(10), 105204 (2011)
21. Robert, E., et al.: Experimental Study of a Compact Nanosecond Plasma Gun. Plasma Processes and Polymers **6**(12), 795–802 (2009)
22. Bogle, M.A., Arndt, K.A., Dover, J.S.: Evaluation of plasma skin regeneration technology in low-energy full-facial rejuvenation. Archives of Dermatology **143**(2), 168–174 (2007)
23. Arndt, S., et al.: Cold Atmospheric Plasma (CAP) Changes Gene Expression of Key Molecules of the Wound Healing Machinery and Improves Wound Healing In Vitro and In Vivo. Plos One **8**(11), e79325 (2013)
24. Brehmer, F., et al.: Alleviation of chronic venous leg ulcers with a hand-held dielectric barrier discharge plasma generator (PlasmaDerm (R) VU-2010): results of a monocentric, two-armed, open, prospective, randomized and controlled trial (NCT01415622). Journal of the European Academy of Dermatology and Venereology **29**(1), 148–155 (2015)
25. Emmert, S., et al.: Atmospheric pressure plasma in dermatology: Ulcus treatment and much more. Clinical Plasma Medicine **1**(1), 24–29 (2013)
26. Haertel, B., et al.: Non-thermal atmospheric-pressure plasma can influence cell adhesion molecules on HaCaT-keratinocytes. Experimental Dermatology **20**(3), 282–284 (2011)

27. Fetykov, A.I., et al.: The effectiveness of cold plasma treatment of diabetic feet syndrome, complicated by purulo-necrotic process. In: 2nd International Conference on Plasma Medicine, San Antonio, TX, USA (2009)
28. Shekhter, A.B., et al.: Experimental and clinical validation of plasmadynamic therapy of wounds with nitric oxide. Bulletin of Experimental Biology and Medicine **126**, 829–834 (1998)
29. Weltmann, K.D., et al.: Atmospheric Pressure Plasma Jet for Medical Therapy: Plasma Parameters and Risk Estimation. Contributions to Plasma Physics **49**(9), 631–640 (2009)
30. Fridman, G., et al.: Floating electrode dielectric barrier discharge plasma in air promoting apoptotic behavior in melanoma skin cancer cell lines. Plasma Chemistry and Plasma Processing **27**(2), 163–176 (2007)
31. Lee, H.J., et al.: Degradation of adhesion molecules of G361 melanoma cells by a non-thermal atmospheric pressure microplasma. New Journal of Physics **11**, 115026 (2009)
32. Kim, S.J., et al.: Induction of apoptosis in human breast cancer cells by a pulsed atmospheric pressure plasma jet. Applied Physics Letters **97**(2), 023702 (2010)
33. Kim, J.Y., et al.: Single-Cell-Level Microplasma Cancer Therapy. Small **7**(16), 2291–2295 (2011)
34. Partecke, L.I., et al.: Tissue Tolerable Plasma (TTP) induces apoptosis in pancreatic cancer cells in vitro and in vivo. BMC Cancer **12**, 473 (2012)
35. Stoffels, E., Kieft, I.E., Sladek, R.E.J.: Superficial treatment of mammalian cells using plasma needle. Journal of Physics D-Applied Physics **36**(23), 2908–2913 (2003)
36. Gweon, B., et al.: Plasma effects on subcellular structures. Applied Physics Letters **96**(10), 101501 (2010)
37. Kim, G.J., et al.: DNA damage and mitochondria dysfunction in cell apoptosis induced by nonthermal air plasma. Applied Physics Letters **96**(2), 021502 (2010)
38. Laroussi, M.: Sterilization of contaminated matter with an atmospheric pressure plasma. IEEE Transactions on Plasma Science **24**(3), 1188–1191 (1996)
39. Gweon, B., et al.: Escherichia coli deactivation study controlling the atmospheric pressure plasma discharge conditions. Current Applied Physics **9**(3), 625–628 (2009)
40. Gweon, B., et al.: Differential responses of human liver cancer and normal cells to atmospheric pressure plasma. Applied Physics Letters **99**(6), 063701 (2011)
41. Kieft, I.E., Kurdi, M., Stoffels, E.: Reattachment and apoptosis after plasma-needle treatment of cultured cells. IEEE Transactions on Plasma Science **34**(4), 1331–1336 (2006)
42. Shashurin, A., et al.: Living tissue under treatment of cold plasma atmospheric jet. Applied Physics Letters **93**(18), 181501 (2008)
43. Kieft, I.E., et al.: Plasma treatment of mammalian vascular cells: A quantitative description. IEEE Transactions on Plasma Science **33**(2), 771–775 (2005)
44. Joh, H.M., et al.: Reactive oxygen species-related plasma effects on the apoptosis of human bladder cancer cells in atmospheric pressure pulsed plasma jets. Applied Physics Letters **101**(5), 053703 (2012)
45. Kim, G.C., et al.: Air plasma coupled with antibody-conjugated nanoparticles: a new weapon against cancer. Journal of Physics D: Applied Physics **42**(3), 032005 (2009)
46. Arndt, S., et al.: Cold atmospheric plasma, a new strategy to induce senescence in melanoma cells. Experimental Dermatology **22**(4), 284–289 (2013)

47. Iseki, S., et al.: Selective killing of ovarian cancer cells through induction of apoptosis by nonequilibrium atmospheric pressure plasma. Applied Physics Letters **100**(11), 113702 (2012)
48. Keidar, M., et al.: Cold plasma selectivity and the possibility of a paradigm shift in cancer therapy. British Journal of Cancer **105**(9), 1295–1301 (2011)
49. Kim, G.J., et al.: DNA damage and mitochondria dysfunction in cell apoptosis induced by nonthermal air plasma. Applied Physics Letters **96**(2) (2010)
50. Kaushik, N.K., et al.: Effect of jet plasma on T98G human brain cancer cells. Current Applied Physics **13**(1), 176–180 (2013)
51. Vandamme, M., et al.: ROS implication in a new antitumor strategy based on non-thermal plasma. International Journal of Cancer **130**(9), 2185–2194 (2012)
52. Koritzer, J., et al.: Restoration of Sensitivity in Chemo - Resistant Glioma Cells by Cold Atmospheric Plasma. Plos One **8**(5), e64498 (2013)
53. Brulle, L., et al.: Effects of a Non Thermal Plasma Treatment Alone or in Combination with Gemcitabine in a MIA PaCa2-luc Orthotopic Pancreatic Carcinoma Model. Plos One **7**(12), e52653 (2012)
54. Huang, J., et al.: Deactivation of A549 cancer cells in vitro by a dielectric barrier discharge plasma needle. Journal of Applied Physics **109**(5), 053305 (2011)
55. Kim, J.Y., et al.: Apoptosis of lung carcinoma cells induced by a flexible optical fiber-based cold microplasma. Biosensors & Bioelectronics **28**(1), 333–338 (2011)
56. Panngom, K., et al.: Preferential killing of human lung cancer cell lines with mitochondrial dysfunction by nonthermal dielectric barrier discharge plasma. Cell Death & Disease **4**, e642 (2013)
57. Han, X., et al.: DNA damage in oral cancer cells induced by nitrogen atmospheric pressure plasma jets. Applied Physics Letters **102**(23), 233703 (2013)
58. Thiyagarajan, M., Waldbeser, L., Whitmill, A.: THP-1 leukemia cancer treatment using a portable plasma device. Studies in Health Technology and Informatics **173**, 515–517 (2012)
59. Kim, C.H., et al.: Effects of atmospheric nonthermal plasma on invasion of colorectal cancer cells. Applied Physics Letters **96**(24), 243701 (2010)
60. Georgescu, N., Lupu, A.R.: Tumoral and Normal Cells Treatment With High-Voltage Pulsed Cold Atmospheric Plasma Jets. IEEE Transactions on Plasma Science **38**(8), 1949–1955 (2010)
61. Volotskova, O., et al.: Targeting the cancer cell cycle by cold atmospheric plasma. Scientific Reports **2**, 636 (2012)
62. Lee, J.K., et al.: Biomedical Applications of Low Temperature Atmospheric Pressure Plasmas to Cancerous Cell Treatment and Tooth Bleaching. Japanese Journal of Applied Physics **50**(8), JF01 (2011)
63. Kim, C.H., et al.: Induction of cell growth arrest by atmospheric non-thermal plasma in colorectal cancer cells. Journal of Biotechnology **150**(4), 530–538 (2010)
64. Gweon, B., et al.: Suppression of angiogenesis by atmospheric pressure plasma in human aortic endothelial cells. Applied Physics Letters **104**(13), 133701 (2014)
65. Moon, S.Y., et al.: Feasibility study of the sterilization of pork and human skin surfaces by atmospheric pressure plasmas. Thin Solid Films **517**(14), 4272–4275 (2009)
66. Awakowicz, P., et al.: Biological Stimulation of the Human Skin Applying Health-Promoting Light and Plasma Sources. Contributions to Plasma Physics **49**(9), 641–647 (2009)

67. Rajasekaran, P., et al.: Characterization of Dielectric Barrier Discharge (DBD) on Mouse and Histological Evaluation of the Plasma-Treated Tissue. Plasma Processes and Polymers **8**(3), 246–255 (2011)
68. Vandamme, M., et al.: Antitumor Effect of Plasma Treatment on U87 Glioma Xenografts: Preliminary Results. Plasma Processes and Polymers **7**(3–4), 264–273 (2010)
69. Keidar, M., et al.: Cold atmospheric plasma in cancer therapy. Physics of Plasmas **20**(5), 057101 (2013)
70. Walk, R.M., et al.: Cold atmospheric plasma for the ablative treatment of neuroblastoma. Journal of Pediatric Surgery **48**(1), 67–73 (2013)
71. Yajima, I., et al.: Non-equilibrium atmospheric pressure plasmas modulate cell cycle-related gene expressions in melanocytic tumors of RET-transgenic mice. Experimental Dermatology **23**(6), 424–425 (2014)
72. Vargo, J.J.: Clinical applications of the argon plasma coagulator. Gastrointestinal Endoscopy **59**(1), 81–88 (2004)
73. Canady, J., et al.: Characterization of Plasma Parameters and Tissue Injury Produced by Plasma Electrosurgical Systems. Plasma Medicine **3**(4), 279–289
74. Stoffels, E., Kieft, I.E., Sladek, R.E.J.: Superficial treatment of mammalian cells using plasma needle. Journal of Physics D: Applied Physics **36**(23), 2908–2913 (2003)
75. Fridman, G., et al.: Blood coagulation and living tissue sterilization by floating-electrode dielectric barrier discharge in air. Plasma Chemistry and Plasma Processing **26**(4), 425–442 (2006)
76. Heinlin, J., et al.: Plasma medicine: possible applications in dermatology. Journal Der Deutschen Dermatologischen Gesellschaft **8**(12), 968–976 (2010)
77. Ahn, H.J., et al.: Atmospheric-Pressure Plasma Jet Induces Apoptosis Involving Mitochondria via Generation of Free Radicals. Plos One **6**(11), e28154 (2011)
78. Yan, X., et al.: On the Mechanism of Plasma Inducing Cell Apoptosis. IEEE Transactions on Plasma Science **38**(9), 2451–2457 (2010)
79. Tuhvatulin, A.I., et al.: Non-thermal Plasma Causes p53-Dependent Apoptosis in Human Colon Carcinoma Cells. Acta Naturae **4**(3), 82–87 (2012)
80. Martin, P.: Wound healing - Aiming for perfect skin regeneration. Science **276**(5309), 75–81 (1997)
81. Guo, S., DiPietro, L.A.: Factors Affecting Wound Healing. Journal of Dental Research **89**(3), 219–229 (2010)
82. Lloyd, G., et al.: Gas Plasma: Medical Uses and Developments in Wound Care. Plasma Processes and Polymers **7**(3–4), 194–211 (2010)
83. Tirakotai, W., et al.: Argon plasma coagulation (APC) in brain tumor surgery: experimental study and clinical experiences. Clinical Neuropathology **23**(6), 257–261 (2004)
84. Kalghatgi, S.U., et al.: Mechanism of blood coagulation by nonthermal atmospheric pressure dielectric barrier discharge plasma. IEEE Transactions on Plasma Science **35**(5), 1559–1566 (2007)
85. Choi, J., et al.: 900-MHz Nonthermal Atmospheric Pressure Plasma Jet for Biomedical Applications. Plasma Processes and Polymers **7**(3–4), 258–263 (2010)
86. Janani, E., Ale-Ebrahim, M., Mortazavi, P.: In Vitro and in Vivo studies of the Effects of Cold Argon Plasma on Decreasing the Coagulation Time. Iranian Journal of Medical Physics **2013**(10), 31–36 (2013)
87. Eming, S.A., Krieg, T., Davidson, J.M.: Inflammation in wound repair: Molecular and cellular mechanisms. Journal of Investigative Dermatology **127**(3), 514–525 (2007)

88. Harding, K.G., Moore, K., Phillips, T.J.: Wound chronicity and fibroblast senescence–implications for treatment. Int. Wound. J. **2**(4), 364–368 (2005)
89. Edwards, R., Harding, K.G.: Bacteria and wound healing. Current Opinion in Infectious Diseases **17**, 91–96 (2004)
90. Robson, M.C., Heggers, J.P.: Delayed wound closures based on bacterial counts. Journal of Surgical Oncology **2**(4), 379–383 (1970)
91. Bowler, P.G., Duerden, B.I., Armstrong, D.G.: Wound microbiology and associated approaches to wound management. Clinical Microbiology Reviews **14**(2), 244–269 (2001)
92. Laroussi, M., Alexeff, I., Kang, W.L.: Biological decontamination by nonthermal plasmas. IEEE Transactions on Plasma Science **28**(1), 184–188 (2000)
93. Herrmann, H.W., et al.: Decontamination of chemical and biological warfare, (CBW) agents using an atmospheric pressure plasma jet (APPJ). Physics of Plasmas **6**(5), 2284–2289 (1999)
94. Ermolaeva, S.A., et al.: Bactericidal effects of non-thermal argon plasma in vitro, in biofilms and in the animal model of infected wounds. Journal of Medical Microbiology **60**(1), 75–83 (2011)
95. Daeschlein, G., et al.: Skin and wound decontamination of multidrug-resistant bacteria by cold atmospheric plasma coagulation. Journal Der Deutschen Dermatologischen Gesellschaft **13**(2), 143–150 (2015)
96. Fridman, G., et al.: Comparison of direct and indirect effects of non-thermal atmospheric-pressure plasma on bacteria. Plasma Processes and Polymers **4**(4), 370–375 (2007)
97. Daeschlein, G., et al.: Antibacterial Activity of an Atmospheric Pressure Plasma Jet Against Relevant Wound Pathogens in vitro on a Simulated Wound Environment. Plasma Processes and Polymers **7**(3–4), 224–230 (2010)
98. Kalghatgi, S., et al.: Endothelial Cell Proliferation is Enhanced by Low Dose Non-Thermal Plasma Through Fibroblast Growth Factor-2 Release. Annals of Biomedical Engineering **38**(3), 748–757 (2010)
99. Arjunan, K.P., et al.: Non-thermal dielectric barrier discharge plasma induces angiogenesis through reactive oxygen species. Journal of the Royal Society Interface **9**(66), 147–157 (2012)
100. Arndt, S., et al.: Cold Atmospheric Plasma (CAP) Changes Gene Expression of Key Molecules of the Wound Healing Machinery and Improves Wound Healing In Vitro and In Vivo. Plos One **8**(11) (2013)
101. Wende, K., et al.: Distinctive Activity of a Nonthermal Atmospheric-Pressure Plasma Jet on Eukaryotic and Prokaryotic Cells in a Cocultivation Approach of Keratinocytes and Microorganisms. IEEE Transactions on Plasma Science **38**(9), 2479–2485 (2010)
102. Isbary, G., et al.: A first prospective randomized controlled trial to decrease bacterial load using cold atmospheric argon plasma on chronic wounds in patients. British Journal of Dermatology **163**(1), 78–82 (2010)
103. Mertens, N., et al.: Low temperature plasma treatment of human tissue. In: 2nd International Conference on Plasma Medicine, San Antonio, TX, USA (2009)
104. Lee, D.H., et al.: Suppression of scar formation in a murine burn wound model by the application of non-thermal plasma. Applied Physics Letters **99**(20), 203701 (2011)
105. Grigoras, C., et al.: Influence of Atmospheric Pressure Plasma Treatment on Epithelial Regeneration Process. Romanian Journal of Physics **56**, 54–61 (2011)

106. Luo, J.D., Chen, A.F.: Nitric oxide: a newly discovered function on wound healing. Acta Pharmacologica Sinica **26**(3), 259–264 (2005)
107. Shekhter, A.B., et al.: Beneficial effect of gaseous nitric oxide on the healing of skin wounds. Nitric Oxide-Biology and Chemistry **12**(4), 210–219 (2005)
108. Soneja, A., Drews, M., Malinski, T.: Role of nitric oxide, nitroxidative and oxidative stress in wound healing. Pharmacological Reports **57**, 108–119 (2005)
109. Witte, M.B., Barbul, A.: Role of nitric oxide in wound repair. American Journal of Surgery **183**(4), 406–412 (2002)
110. Gamou, S., Shimizu, N.: Hydrogen-Peroxide Preferentially Enhances the Tyrosine Phosphorylation of Epidermal Growth-Factor Receptor. FEBS Letters **357**(2), 161–164 (1995)
111. Gordillo, G.M., Sen, C.K.: Revisiting the essential role of oxygen in wound healing. American Journal of Surgery **186**(3), 259–263 (2003)
112. Joyce, J.A., Pollard, J.W.: Microenvironmental regulation of metastasis. Nature Reviews Cancer **9**(4), 239–252 (2009)
113. Cairns, R.A., Harris, I.S., Mak, T.W.: Regulation of cancer cell metabolism. Nature Reviews Cancer **11**(2), 85–95 (2011)
114. Trachootham, D., Alexandre, J., Huang, P.: Targeting cancer cells by ROS-mediated mechanisms: a radical therapeutic approach? Nature Reviews Drug Discovery **8**(7), 579–591 (2009)

Cancer Imaging for Therapy Assessment

Hyunki Kim

Abstract Monitoring tumor response following treatment is essential to tailor therapeutic strategy to achieve the most favorable clinical outcomes for cancer patients. Imaging is a non-invasive approach to assess tumor response quantitatively, and it has been clinically validated as a reliable tool. RECIST (Response Evaluation Criteria in Solid Tumors) is a published protocol to determine tumor response mainly based on change in tumor size as a response to therapy. The high-resolution anatomical information of a tumor can be obtained using computer tomography (CT), ultrasound imaging or magnetic resonance imaging (MRI). However, it is often insufficient to evaluate tumor viability and aggressiveness using only anatomical information, thus a new protocol named PERCIST (Positron Emission Tomography (PET) Response Criteria in Solid Tumors) has been recently proposed. PERCIST evaluates therapy based on the change of glycolic metabolism using ^{18}F-FDG PET, and its effectiveness has been compared with RECIST. In addition, many other physiologic and molecular imaging modalities have been tested for more objective, rapid, and reproducible measurement of tumor response. In this chapter, both conventional and experimental non-invasive imaging modalities to evaluate therapy for cancer patients having solid tumors are reviewed.

Keywords: CT · MRI · PET · SPECT · Ultrasound imaging · Personalized medicine · Precision medicine · Physiologic imaging · Molecular imaging

1 Introduction

Cancer is a disease involving uncontrolled growth of cells that commonly create abnormal tissue masses, known as malignant tumors. Unlike benign tumors, malignant tumors are capable of migrating to other sites, called metastasis, and

H. Kim(✉)
Department of Radiology, The University of Alabama at Birmingham,
Birmingham, AL 35244, USA
e-mail: hyunki@uab.edu

© Springer International Publishing Switzerland 2016

H. Jo et al. (eds.), *Biomedical Engineering: Frontier Research and Converging Technologies*,
Biosystems & Biorobotics 9,
DOI: 10.1007/978-3-319-21813-7_16

invading tissue boundaries. In the United States, cancer is the leading cause of human death, second only to heart diseases (1). Signs and symptoms of many cancers are usually non-specific and often appear at the late stage, so the early diagnosis of cancer is typically difficult, which is the primary reason of the high mortality rate.

When a tumor is localized and resectable, maximal safe resection is the primary treatment option in most cancer types. Additionally, various chemical and radiographic treatment strategies may be considered according to the type of cancer and cancer stage. Radiotherapy uses ionizing radiation to damage DNA of cells, leading to the cell death. Radiation is in general delivered externally to the region of a localized target tissue. Chemotherapy utilizes a chemical substance to interrupt DNA replication and cell division to eliminate actively proliferating cells. Chemotherapy is not entirely specific to cancer cells, thus it induces adverse side effects against actively proliferating normal cells like blood cells generated from bone marrow. To reduce toxicity against normal cells, targeted therapy has been developed and many drugs are being introduced to significantly increase therapeutic efficacy in cancer treatment, while also reducing off-target effects. Targeted therapy uses a drug engineered to target cancer cells and/or extracellular matrix in tumors. For example, cetuximab is a commercialized drug to target endothelial growth factor receptor (EGFR) that is overexpressed on cancer cell membrane, and functions as an antagonist to block the internal cellular signaling to trigger uncontrolled cell division. Cetuximab has been utilized to treat several cancers including colorectal, lung, and head and neck cancers (2-4). As another example, tamoxifen is a hormonal antagonist targeting estrogen receptor alpha (ERα) overexpressed on the cancer cell membrane, and suppresses the cell growth stimulated by ERα (5). Radiotherapy, chemotherapy and targeted therapy can also be used in combination either simultaneously or sequentially. For example, there are two standard regimens to treat advanced pancreatic cancer; the first regimen is to use four different chemotherapeutic agents such as oxaliplatin, leucovorin, irinotecan and fluorouracil sequentially (6), and the second regimen is the dual combination therapy with abraxane and gemcitabine (7).

The therapeutic efficacy of a drug traditionally varies among patient populations, which has led to the development of tailored strategies to select the optimal drug (or drug combination) for each individual patient. This approach is called personalized medicine, which is a concept often confused with precision medicine (8). Precision medicine is a strategy to select a drug, which has the highest likelihood of effective therapy for a certain group of patients. To enable precision medicine, a predictive biomarker of drug response is identified. For example, tamoxifen is prescribed only for ERα positive breast cancer patients; this strategy is precision medicine, and ERα plays a role as a predictive biomarker. Similarly, EGFR, c-KIT, HER2, and p53 serve as predictive biomarkers for cetuximab, sunitinib, trastuzumab, and cisplatin, respectively (9-12). The presence of a predictive biomarker is usually confirmed with histologic analysis, flow cytometry or western blotting analysis, but non-invasive molecular imaging can be also utilized. To enable personalized medicine, the therapeutic efficacy of a drug should be verified for each individual patient using a surrogate biomarker. For example, if tumor volume is monitored to evaluate therapy, the tumor volume change serves as a surrogate biomarker. The change of intratumoral glycolysis, perfusion and diffusion during therapy can also be utilized as surrogate biomarkers, which are commonly measured by non-invasive imaging techniques.

Figure 1 illustrates four different steps of therapy response in cancer, and three different steps where imaging can be applied to measure the therapy response. The first response against a drug appears in molecular level like DNA transcription or RNA translation. The cellular change can affect the tumor physiology that can lead to the change of tumor morphology or size, and eventually clinical outcomes. The drug response should be assessed at the earliest time point to allow tailoring therapeutic strategy, however this often requires several weeks or months to observe changes in tumor anatomy. Thus molecular or physiologic imaging has been perceived as more ideal tools to achieve personalized medicine. In fact, the molecular imaging of glycolysis was successfully established as a non-invasive tool in cancer therapy evaluation. In this chapter, various types of anatomic, physiologic and molecular imaging and the clinical application of these strategies for cancer therapy assessment will be reviewed.

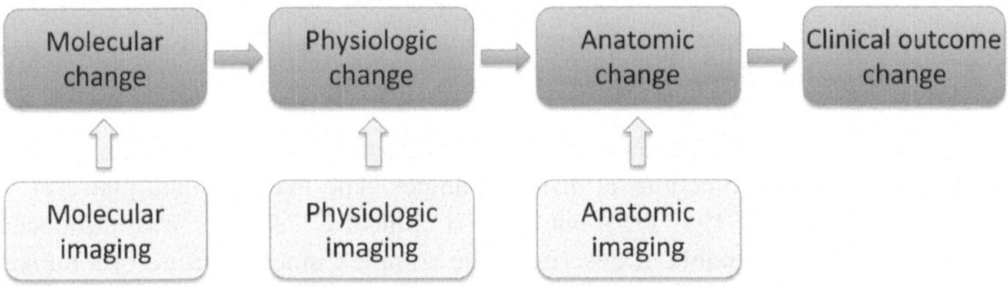

Fig. 1 Four steps of therapy response in cancer, and imaging method to measure therapy response in each step

2 Anatomic Imaging

Anatomic imaging has been employed for routine clinical therapy assessment in solid tumors. Response evaluation criteria in solid tumors (RECIST) was initially published in the year, 2000 (version 1.0), and updated in 2009 (version 1.1) (13). RECIST defines tumor response based on the change of tumor dimensions measured by anatomic imaging including radiography, computed tomography (CT), magnetic resonance imaging (MRI) and ultrasound imaging.

2.1 Radiography

Projection radiography is the creation of 2-dimensional (2D) X-ray projection images using a radiation source and a detector. An X-ray is electromagnetic radiation that has 0.01~10 nm wavelength. An X-ray can pass through the human body and reach an X-ray film (or X-ray detector arrays) placed behind the subject. X-ray is attenuated by the tissue when it travels through. Since the higher density tissue attenuates the more X-rays, radiography produces image contrast according to the tissue density. Radiography is useful to visualize higher density tissues like bone or fiber surrounded by lower density tissue like fat or muscle. The spatial resolution of a current clinical radiography is typically about 150 μm. In oncologic application, radiography has been routinely used for breast cancer screening. Breast is composed of about 90% fat and 10% fibroglandular tissue, and the density of tumors are typically higher than that of fat. Thus tumors can be readily visualized in breast X-ray images. Radiography specialized for breast examination is called mammography, and it can detect a 10-mm size tumor in 91% sensitivity (14). However, the limitation of radiography is obvious; the data loss due to 2D projection. For example, a tumor located behind a higher density tissue like fiber or bone cannot be visualized by radiography. Therefore cross-sectional imaging (tomography) modalities were soon developed.

2.2 Computed Tomography

Computed tomography (CT) is a method to create cross-sectional images using a series of 2D X-ray projections at different angles. The mathematical platform of CT was established in 1917 (15), but the first clinical CT scanner was introduced in 1972 (16). CT is capable of describing the complex inner structure of a human body in high resolution; the current whole body CT scanner provides about 0.5 mm spatial resolution and about 200 ms temporal resolution. CT is a standard clinical tool to measure tumor response based on RECIST in many cancers such as lung cancer and most abdominal cancers. Since CT provides poor contrast between soft tissues, contrast agents are commonly used together. Table 1 shows various CT contrast agents used clinically. CT contrast agents brighten tissues having higher vasculature, and tumor boundary can be delineated based on the contrast difference between tumor and the surrounding tissue. One major concern of CT imaging is the high dose of radiation. The single radiation dose of CT is in the range of 5~10 mSv, while the maximum annual radiation dose is 50 mSv for an adult and 5 mSv for children under the age of 18 years. As a consequence, CT has been used in limited capacity for pediatric cancer, although radiation dose can be reduced at the cost of image quality (17). Another concern of CT imaging is metal artifact, which occurs when a metal implant blocks most projected X-rays. Several methods have been introduced to suppress the metal artifact, however these attempts were not able to entirely eliminate it (18).

Table 1 CT contrast agents used for common clinical use

CT contrast agents		
Iodinated (i.v.)		**Barium (oral)**
Ionic	**Non-ionic**	
Diatrizoate	Iodixanol	
Loxaglate	Iohexol	Barium sulfate (gastro-intestinal imaging)
Metrizoate	Iopamidol	
	Iopromide	
	Ioxilan	

2.3 *Magnetic Resonance Imaging*

Magnetic resonance imaging (MRI) is to generate cross-sectional images using nuclear magnetic resonance (NMR) principle. NMR technique was originally utilized to examine chemical compounds, but has been employed for imaging since 1980s for medical use. MR signal is generated from the magnetic moment of a certain element, most commonly hydrogen, in a biological entity. To generate MR images, three different types of external magnetic fields are required: the main magnetic field (B_0), the radiofrequency (RF) field, and the gradient fields (x, y, z directions). MR image quality is in proportion to the field strength of B_0; 1.5T and 3.0T are most commonly used for human imaging. When B_0 is present through the central bore of an MR scanner (z direction), the axes of the magnetic moments in a human body are likely to be aligned with the direction of B_0. All magnetic moments spin at a constant rate proportional to the field strength of B_0, and the spinning frequency is called Larmor frequency. If a RF field is externally applied at the Larmor frequency, the axes of the magnetic moments precess down to the x-y plane, and MR signal can be induced, based on Faraday's Law, to a receiver coil placed nearby the subject. When the RF field is turned off, the axes of the magnetic moments precess back up to the z direction, and it is called longitudinal relaxation. Before longitudinal relaxation occurs, the spins of magnetic moments are gradually out of phase due to spin-spin interaction, and therefore the total sum of the magnetic moments is gradually reduced over time, which is called transverse relaxation. Time to reach 63% of full longitudinal relaxation is called T1 value, and time to reach 63% of signal reduction due to transverse relaxation is called T2 value. Each tissue has unique T1 and T2 values. When image contrast between tissues is achieved by T1 value difference, it is called T1 weighted MRI (T1W MRI), and when T2 value is used to create image contrast, it is called T2 weighted MRI (T2W MRI). If the image contrast is created by the difference in hydrogen density, it is called proton density weighted MRI (PDW MRI). T2W MRI and contrast-enhanced T1W MRI are commonly used to achieve the anatomic details of cancer in clinical MRI. The gradient fields are used to encode spatial information to the MR signals (19). One distinguished feature of MRI is that MR signals are obtained in frequency domain first, known as K-space, and then transformed to spatial domain. As a result, MRI is vulnerable to subject motion, and thereby MRI use has been limited for abdominal cancers severely influenced by breathing and peristaltic motions. Another concern of MRI is susceptibility artifact. When two different substances are closely located, MR images may experience signal drop out, which is one of the reasons why there are few MRI applications for pulmonary use. MRI is also vulnerable to metal implant, because metal disturbs the uniformity of the main magnetic field resulting in signal drop out.

Regardless of those concerns described above, MRI has been extensively used for various medical examinations including cancer, primarily due to excellent contrast between soft tissues. MR image contrast, however, can be further enhanced by using contrast agents. Table 2 summarizes the US FDA approved contrast agents for common clinical use. Gadolinium compounds are usually employed in T1W MRI, which brighten hypervasuclar tissues, while iron oxide compounds are used in T2W MRI, making hypervascular tissues dark. Several iron oxide compounds have been clinically used for a while, but most of them, with the exception of lumirem, are currently not produced. One well-known concern in the use of gadolinium compounds is nephrogenic systemic fibrosis (NSF) (20). NSF is a disease associated with free gadolinium that has detached from the contrast agents becomes toxic in patients with abnormal renal function. NSF appears like a skin disease, and can cause even death. Thus gadolinium compounds should only be used for patients without chronic kidney diseases. MRI has been intensively and most successfully used for brain cancer assessment, although it has been widely investigated for the others. Since MRI requires no ionizing radiation, multiple MRI examinations can be readily implemented even for children. The spatial resolution of MR images widely vary according to the imaging sequence, but typically about 0.5~1.5 mm. MRI acquisition time varies from several seconds to several minutes per scan. However multiple scans are commonly implemented sequentially at the imaging site, so the total acquisition time is in the range of 30~45 minutes.

Table 2 MRI contrast agents used for common clinical use

MRI contrast agents	
Gadolinium based	**Iron Oxide based**
Gadopentetate (Magnevist)	
Gadoteridol (Prohance)	
Gadobenate (MultiHance)	
Gadodiamide (Omniscan)	
Gadoterate (Dotarem)	Lumirem (Gastromark)
Gadoversetamide (OptiMARK)	
Gadoxetate (Eovist)	
Gadobutrol (Gadavist)	
Gadofosveset (Ablavar)	

2.4 *Ultrasound Imaging*

Ultrasound imaging utilizes high frequency acoustic waves to generate 2D cross-sectional images (or 3D images) of a target region. The frequency of acoustic waves used in medical application is in the range of 1~20 MHz. An ultrasound scanner utilizes a transducer to transform electrical signals generated from a computer to mechanical vibration, and vice versa. The transducer transmits a short pulse of ultrasound wave through the target region, and some of the pulse, called echo, is reflected back from the boundary between two tissues having different acoustic impedances. Based on the returned time and remaining strength of the echo, the distance to the tissue boundary and the stiffness of the tissue that the pulse traveled through can be estimated, producing a cross-sectional image. In general, the higher frequency ultrasound waves produce greater spatial resolution, but the high frequency ultrasound waves cannot propagate in deep tissue due to scattering. So in order to image a tumor located deep inside tissue, lower frequency ultrasound waves should be utilized at the cost of spatial resolution. Ultrasound imaging has been extensively used to delineate the boundary of various tumors in breast and abdomen. However, ultrasound waves cannot penetrate bone, thus brain tumors cannot be assessed using ultrasound imaging non-invasively. Ultrasound imaging is a safe, easily accessible and relative inexpensive modality. Spatial resolution varies according to the ultrasound wave frequency, but typically about 0.2~1.0 mm for human patients, and temporal resolution is about 10~30 ms. Ultrasound microbubbles, which are gas-filled lipid or polymer micelles ranging 1-10um in diameter, has been used as a contrast agent for ultrasound imaging, and approved for cancer diagnosis by regulatory agencies in both Asia and Europe, but not in the United States. Ultrasound waves bounce off microbubbles prominently, brightening the region where microbubbles are present during imaging. However, unlike CT or MRI contrast agents, microbubbles are not small enough to leak out to the extravascular space, so only vessel regions are brightened.

3 Physiologic Imaging

Physiologic change in tumors occurs earlier than the anatomical change, thus physiologic imaging has been considered as a more appropriate tool to enable personalized medicine. However, the physiologic change does not always lead to the anatomic change in tumors. So in general, physiologic imaging evaluates the therapeutic efficacy of a drug in the earlier time period at the cost of accuracy. Various physiologic imaging modalities have been investigated for early therapy assessment in cancer.

3.1 *Perfusion Imaging*

Perfusion imaging is performed by measuring the vascular perfusion in the region of interest (ROI) by monitoring the signal change during and after intravenous injection of a contrast agent. Most malignant tumors are hypervascular and hyperperfused to support the rapidly growing cancer cells (21). In general, tumor microvasculature is regressed after effective killing of cancer cells, so the reduction in vascular perfusion has been utilized as a surrogate biomarker to indicate the therapy effectiveness. However, the magnitude of the reduction is highly related to the therapeutic mechanism of a drug. For example, antiangiogenic drug like bevacizumab can decrease vascular perfusion more effectively than regular chemotherapeutic or biologic agents, but not necessarily induce better antitumor effects (22). Therefore the reduction in vascular perfusion should be interpreted differently according to the type of cancer, the initial vascularity of a tumor, and the therapeutic mechanism of a drug. Since most contrast agents are rapidly washed out from the human body, the modality for perfusion imaging should allow rapid and repeated imaging within a short period of time. Thus CT is not appropriate tool for perfusion imaging of human patients. Here we introduce two perfusion imaging modalities, dynamic contrast-enhanced MRI (DCE-MRI) and dynamic contrast-enhanced ultrasound imaging (DCE-US), which have great potential to be translated for routine clinical use.

3.1.1 Dynamic Contrast-Enhanced MRI

DCE-MRI has been evaluated as a minimally invasive tool to analyze tumor vascularity for many years. DCE-MRI employs gadolinium compounds, which are small enough to leak out of the vessels freely, so the signal enhancement in the target tissue is dominated by vascular perfusion. Vascular permeability can also be measured using gadolinium labeled macromolecule like albumin, although its clinical use has not been approved (23). DCE-MRI allows quantitating several vascular parameters including wash-in rate (K^{trans}; volume transfer constant), washout rate (k_{ep}; reverse reflex rate constant), v_e (K^{trans}/k_{ep}; extravascular extracellular space), and v_p (blood volume). Since most gadolinium based contrast agents are non-targeting small molecules, one tissue compartment model is commonly employed to quantitate the vascular parameters. The change of contrast concentration in artery during and after intravenous contrast injection is called arterial input function (AIF), and reliable AIF is essential for accurate quantification of vascular parameters. When a large artery like abdominal aorta is seen within the field of view (FOV), AIF can be directly measured from the images. If only a small artery is available like the case of brain imaging, AIF may experience severe partial volume effect, which can be fixed at the cost of SNR (signal-to-noise ratio) (24). When no artery is present like the case of breast imaging, a population-averaged AIF can be used (25). But since the population-averaged AIF does not reflect the variance across individual patients and different contrast agents, it may result in significant quantification error. The early change of K^{trans} (or k_{ep}) value has been proposed as a surrogate biomarker for therapy assessment. In hypervascular tumors, K^{trans} and k_{ep} values are typically reduced by an effective therapy. But in hypovascular tumors like pancreatic ductal adenocarcinomas, those parameters can be rather increased after an effective therapy, so care must be taken when interpreting tumor response. One major concern of quantitative DCE-MRI is the variability of perfusion parameters across different scanners and quantification software (26), which limits the modality for multicentral clinical trials. Therefore the standardization of DCE-MRI protocol and quantification method must be introduced to establish global criteria for therapy evaluation based on DCE-MRI.

3.1.2 Dynamic Contrast-Enhanced Ultrasound Imaging

DCE-US is an approach to examine vascularity in a target tissue based on ultrasound imaging, and has been investigated in both preclinical and clinical studies. DCE-US monitors the change of the signal intensity during and after the injection of ultrasound contrast agent (microbubbles) intravenously, and yields semi-quantitative vascular parameters including area under curve (AUC; the area under the time intensity curve) and the peak intensity (PI; the highest intensity subtracted by baseline intensity in the time-intensity curve), which are proportional to the vascular perfusion and blood volume, respectively, based on the assumption that the signal intensity is in direct proportion to the microbubble density. Both PI and AUC were significantly correlated with microvessel density (MVD) assessed by histologic analysis in a preclinical study (27), and the early change of AUC was verified as a reliable predictive biomarker for the response of various types of tumors against antiangiogenic therapy in a multicenter clinical trial (28).

3.2 Diffusion Imaging

Apoptotic volume decrease (AVD) occurs in the early phase of apoptosis, pushing out the intracellular water molecules to extracellular space. The extracellular water molecules are not spatially confined within the cell membrane, thus they diffuse more freely. Therefore the increase of extracellular water molecules, which is the early indicator of apoptosis, can be estimated by the increase of diffusion of water molecules. Diffusion weighted imaging (DWI) is a physiologic MRI technique to quantitate the diffusion of water molecules noninvasively; apparent diffusion coefficient (ADC) represents the magnitude of the diffusion, and the early change of ADC has been evaluated as a surrogate biomarker of therapeutic efficacy in various cancers. Extracellular water molecules are also increased by necrosis; when necrosis occurs, cell membrane ruptures, and the intracellular water molecules diffuse out. Thus the increase of ADC values in tumors during therapy can be an indicator of effective killing of cells regardless of apoptosis or necrosis. However, DWI may not be an appropriate tool to measure the therapeutic efficacy of a cytostatic drug like antiangiogenic agents or anti-EMMPRIN (CD147) antibody regulating extracellular matrix (29). DWI is very sensitive to motion, so fast imaging sequences like echo planar imaging (EPI) are commonly employed especially for abdominal cancers (30), but EPI may cause geometric distortion and signal loss. The increased extracellular water molecules produced by a therapy are not accumulated within the tumor boundary, but can be cleared out via blood vessels or simply diffused out of the tumor region. Therefore in DWI, the interpretation of therapeutic response can vary according to the evaluation time point. For example, mean ADC value in breast tumor xenografts increased about 30% for 3 days of treatment using monoclonal antibodies targeting death receptor 5, but decreased thereafter, although an additional dosing was applied (31). The optimal imaging time point in DWI may vary according to the type of cancer, the physiologic properties of a tumor, and the therapeutic mechanism of a drug. Typically, DWI is applied before therapy (baseline imaging) and after one or two cycles of treatment, and the ADC change during the period is used as a surrogate biomarker for therapy assessment.

4 Molecular Imaging

Imaging intracellular molecular change responding to a therapy allows tailoring therapeutic strategy at the earliest possible time for each individual patient. Molecular imaging has been implemented using various imaging modalities, but most successfully with nuclear medicine imaging tools such as positron emission tomography (PET) and single photon emission computed tomography (SPECT). PET has been extensively used in molecular imaging due to its high sensitivity. For example, MRI can detect contrast agents in the range of micromolar, but PET does in picomolar range. PET employs a tracer conjugated with a radioisotope, emitting a positron. When a radioactive tracer is delivered to a target tissue, the emitted positron travels a short distance (1~2 mm) until it bumps into a free electron. Then the positron and electron annihilate each other, creating two photons (gamma rays) having 511 KeV, traveling in opposite directions (180±0.25°). A PET system has a ring of detector array, and the two photons are detected by two facing detector elements simultaneously. After many events, the location where the radiotracers are present is estimated. Among many radioisotopes used in PET imaging, fluorine-18 (^{18}F) has been most extensively employed. The half-life of ^{18}F is about 110 minutes, appropriate for routine clinical imaging, and most commonly, ^{18}F is labeled with fludeoxyglucose (FDG) to image glucose metabolism or fluorothymidine (FLT) to image cell proliferation in various cancers. The spatial resolution of a clinical PET image is about 5 mm.

SPECT also uses radiotracers for imaging, however these radioisotopes emit a single photon at a time. ^{99m}Tc (energy: 140 KeV; half-life: 6 hours) and ^{111}In (energy: 245 KeV; half-life: 2.8 days) are most commonly employed radioisotopes for clinical SPECT imaging. The radioisotope is conjugated with a targeting agent, typically antibody or peptide that targets molecules expressed on disease-specific cell membrane or intracellular space. Photons emitted from the target area are collected by gamma cameras. However, a photon can travel to any directions, and gamma rays cannot be converged by regular optics to create images. So, in SPECT imaging, photons traveling through a particular direction are allowed to reach the camera for imaging, and the others are all rejected using a collimator. As a result, the photon collection efficiency of SPECT is typically about 100 times lower than that of PET. The spatial resolution of a clinical SPECT image is about 10 mm.

4.1 FDG PET Imaging

A glucose analog, ^{18}F-FDG, is commonly used in PET imaging for cancer. Since ^{18}F-FDG appears as a regular glucose to cells, it accumulates inside cells at levels proportionate to the glucose metabolism. Since cancer cells actively proliferate, requiring high level of glycolysis, ^{18}F-FDG is preferentially taken up by cancer cells, which creates contrast between a malignant tumor and background tissue. Since ^{18}F-FDG competes with regular glucose in blood, patients are usually fasted for 6~12 hours before ^{18}F-FDG injection. The radiotracer uptake in the target tissue is semi-quantitatively represented by standard uptake value (SUV). SUV is the radiotracer concentration normalized to the patient body weight and the initial injected dose. The higher SUV of ^{18}F-FDG is corresponding to the higher glycolysis. The change of SUV has been used as a surrogate biomarker to evaluate the therapeutic efficacy in many cancers. In cases where tumor size or morphology may not be altered as a result of therapy, intratumoral SUV may be decreased significantly, representing the therapy effectiveness. Therefore, ^{18}F-FDG PET imaging has been perceived as a more reliable and accurate method to measure the therapeutic response in cancer. In fact, PET Response Criteria In Solid Tumors (PERCIST), version 1.0, was published in the year 2009 (32). PERCIST defines tumor response to four different categories like progressive metabolic disease (PMD), stable metabolic disease (SMD), partial metabolic response (PMR) and complete metabolic response (CMR) based on the change of SUV in tumor region during therapy. Since SUVs in a tumor region are often heterogeneous, the maximum SUV in the tumor region may evaluate the therapy more accurately than the mean SUV. One major concern of ^{18}F-FDG PET imaging is, however, the non-specificity when targeting cancer. That is, ^{18}F-FDG targets glycolysis, not cell proliferation, thus it may be also accumulated in inflammatory cells, actively consuming glucose. Therefore the increase of SUV can be induced by the increase of inflammation, not by cancer cell proliferation, so care must be taken in the image interpretation. ^{18}F-FDG PET imaging for therapy assessment in cancer is approved in most countries including the United Sates.

4.2 FLT PET Imaging

^{18}F-FLT, a pyrimidine analogue, targets S-phase cells specifically, thus ^{18}F-FLT PET imaging yields contrast between proliferating cells and background tissue.

Since no uptake of ^{18}F-FLT was observed in inflammatory tissues (33), ^{18}F-FLT has been perceived as an agent to be delivered to cancer cells more specifically than ^{18}F-FDG. Intratumoral SUV of ^{18}F-FLT is typically lower than that of ^{18}F-FDG, but the SUV in the background tissue is also lower, so the contrast between the target region and the background tissue is comparable in both agents (34). Of interest, it was reported that ^{18}F-FDG is more effective to assess the efficacy of cytotoxic agents, while ^{18}F-FLT is better for cytostatic agents, but this theory has not been established yet (35, 36). ^{18}F-FLT was evaluated in many cancers (37), and the significant decrease of SUV was demonstrated in most studies. However, McKinley *et al* recently reported that ^{18}F-FLT is not as sensitive as Ki-67 expression to measure the magnitude of cell proliferation, thus may not be practical to predict therapy effectiveness in oncology (38). In fact, the correlation between the SUV of ^{18}F-FLT and Ki-67 expression was significant in lymphoma and lung cancer (39, 40), but not in sarcoma or gastric cancer (41, 42). Therefore, more studies will need to be followed to verify the usefulness of ^{18}F-FLT PET imaging for early cancer prognosis.

4.3 Other PET Based Molecular Imaging

Biomarkers overly expressed on cancer cell membrane can be utilized as imaging probes for cancer diagnosis and prognosis. Overexpressed biomarkers used in cancer imaging include EGFR, HER2, VEGFR, integrin, etc. When cancer cells are effectively killed, the tumor uptake of tracers targeting those biomarkers will be reduced. Thus the change of SUV in the tumor region can be used as a surrogate biomarker. Furthermore the baseline image can also be used to identify the targeting biomarkers before therapy initiation, so the baseline SUV in the tumor region can serve as a predictive biomarker. ^{11}C labeled PD153035 is a PET imaging radiotracer, targeting EGFR. ^{11}C is a positron-emitting radioisotope with a half-life of about 20 minutes, and PD153035 is a tyrosine kinase inhibitor specifically targeting EGFR (43). ^{11}C-PD153035 has been studied in multiple preclinical cancer models including neuroblastoma, glioma and breast (44-46), and tested in human subjects having lung cancer and glioma (47, 48); the intratumoral SUV in the baseline image was positively correlated with clinical outcomes when patients were treated with an EGFR inhibitor, erlotinib. HER2, another human epidermal growth factor receptor, is overexpressed in about 30% of breast cancer, and plays an important role in tumor progression (49). Trastuzumab is a monoclonal antibody targeting HER2, and ^{89}Zr labeled trastuzumab has been employed to image HER2 (50). The half-life of ^{89}Zr is 78 hours, which is markedly longer than those of typical radioisotopes used in PET, so it allows longitudinal examination for several days after a single dosing. ^{64}Cu-DOTA labeled VEGF (vascular endothelial growth factor) is to image VEGFRs (vascular endothelial growth factor receptors). ^{64}Cu is another PET radioisotope with the half-life of about 13 hours, and conjugated with VEGF via a chelator, DOTA. VEGF binds to VEGFR-1 and VEGFR-2. VEGFR-2 stimulates neoangiogenesis via binding to VEGF, while VEGFR-1 is known to function as a modulator of VEGFR-2 intracellular signaling (51). ^{64}Cu-DOTA-VEGF was used to evaluate the early tumor response against a novel antiangiogenic agent, KR-31831, in an ovarian tumor mouse model (52), but not tested in human subjects yet. Integrins are glycoprotein receptors mediating interaction between cells, and to date, a total of 24 integrins were identified (53). While the functions of integrins have been widely studied in cancer, integrin αVβ3 was particularly identified as a marker for cancer progression. Because the increase of integrin αVβ3 is correlated with the malignancy of tumors, integrin αVβ3 has been considered as a target for both therapy and imaging. One of the most successful PET imaging agents targeting integrin αVβ3 is ^{18}F-galacto labeled arginylglycylaspartic acid (RGD), which has been clinically tested in breast cancer and glioma (54, 55).

4.4 SPECT Based Molecular Imaging

Various radiotracers are under investigation for therapy assessment based on SPECT imaging or scintigraphy, which is 2-dimensional (2D) single photon projection imaging. Due to the low photon-collection efficiency of SPECT, it takes long time to acquire robust SPECT images, so scintigraphy is often used instead. When scintigraphy is used for breast examination, it is called scintimammography. Since breast is a relatively simple organ composed of mainly fat and fibroglandular, 2D imaging is usually sufficient to measure the tumor uptake of a radiotracer. ^{99m}Tc labeled trastuzumab has been tested to image HER2 positive breast cancer (56) and was recently evaluated using a whole-body scintigraphy (57); the tumor uptake of ^{99m}Tc-trastuzumab was significantly reduced during the early therapeutic period of trastuzumab for HER2 positive metastatic breast cancer patients. ^{99m}Tc labeled NC100692 targeting integrin αVβ3 was also clinically tested using scintigraphy (58, 59); ^{99m}Tc-NC100692 was efficient to detect brain and lung metastases, but not for liver and bone metastases. VEGFR can also be imaged using the single photon imaging technique; ^{99m}Tc labeled single-change VEGF was tested in a breast cancer murine model, and the tumor uptake was significantly reduced after treatment with sunitinib, a multi-targeted RTK (receptor tyrosine kinase) inhibitor (60), although it has not been translated to clinical studies yet.

4.5 MR Based Molecular Imaging

Efforts to develop tracers used in MR based molecular imaging are relatively modest compared with those in nuclear medicine, although several investigators have achieved substantial success. MRI probe targeting estrogen receptor (ER) was developed by conjugating gadolinium with tamoxifen (PTA-Gd) or 17β-estradiol (EPTA-Gd) (61), and the specific uptake of those agents to ER positive breast tumor xenografts was confirmed in an *in vivo* animal experiment (62). In another example, Iron oxide can be conjugated with trastuzumab to image HER2.

Typically, iron oxide cannot be used alone (63), so it is coated with polymer that serves as the foundation for antibody conjugation (64-66). The specific delivery of trastuzumab labeled iron oxide nanoparticles (IONPs) to HER2 positive breast tumors have also been confirmed in a murine model (67, 68). In a similar study, IONPs were conjugated with EGFR targeting antibody, and specific uptake to orthotopic brain tumor xenografts was demonstrated in a preclinical model (69). However, IONP based MRI tracers can be easily trapped in liver or spleen, which needs to be resolved for clinical translation.

4.6 Ultrasound Based Molecular Imaging

Microbubbles have the size of 1~10 μm, thus cannot leak out through the gap between endothelial cells. Therefore microbubble based molecular imaging targets biomarkers expressed on the endothelial cells. For example, anti-VEGFR2 antibody binding to the surface of microbubbles was developed (70), and the higher retention rate in tumors was observed when compared with unlabeled microbubbles in a breast cancer murine model. Similarly, microbubbles conjugated with monoclonal antibodies targeting endoglin (CD105) or VEGFR2 were introduced, and the significantly higher tumor uptakes of those were confirmed in pancreatic tumor murine models (71). Microbubbles were also conjugated with RGD to target αVβ3, and its improved tumor uptake was demonstrated (72). Furthermore, multiple targeting agents can be attached on microbubbles; microbubbles conjugated with antibodies targeting anti-VEGFR2 antibody, αVβ3, and P-selectin was tested to measure the early tumor response following an antiangiogenic therapy in a breast tumor murine model (73). However, microbubbles cannot target biomarkers presented on cancer cell membrane or extracellular matrix, which limits the scope of this application for cancer diagnosis or prognosis.

References

1. Murphy, S.L., Xu, J., Kochanek, K.D.: Deaths: final data for 2010. Natl. Vital. Stat. Rep. **61**(4), 1–117 (2013)
2. Ang, K.K., Zhang, Q., Rosenthal, D.I., Nguyen-Tan, P.F., Sherman, E.J., Weber, R.S., et al.: Randomized phase III trial of concurrent accelerated radiation plus cisplatin with or without cetuximab for stage III to IV head and neck carcinoma: RTOG 0522. Journal of Clinical Oncology : Official Journal of the American Society of Clinical Oncology **32**(27), 2940–2950 (2014). PMCID: 4162493
3. Heymach, J.V., Nilsson, M., Blumenschein, G., Papadimitrakopoulou, V., Herbst, R.: Epidermal growth factor receptor inhibitors in development for the treatment of non-small cell lung cancer. Clinical Cancer Research : an Official Journal of the American Association for Cancer Research **12**(14 Pt 2), 4441s–4445s (2006)
4. Maiello, E., Giuliani, F., Gebbia, V., Piano, A., Agueli, R., Colucci, G.: Cetuximab: clinical results in colorectal cancer. Ann. Oncol. **18**(suppl 6), vi8–vi10 (2007)

5. Osborne, C.K.: Tamoxifen in the treatment of breast cancer. The New England Journal of Medicine **339**(22), 1609–1618 (1998)
6. Conroy, T., Desseigne, F., Ychou, M., Bouche, O., Guimbaud, R., Becouarn, Y., et al.: FOLFIRINOX versus gemcitabine for metastatic pancreatic cancer. N. Engl. J. Med. **364**(19), 1817–1825 (2011)
7. Von Hoff, D.D., Ervin, T., Arena, F.P., Chiorean, E.G., Infante, J., Moore, M., et al.: Increased survival in pancreatic cancer with nab-paclitaxel plus gemcitabine. The New England Journal of Medicine **369**(18), 1691–1703 (2013)
8. Marquet, P., Longeray, P.H., Barlesi, F., Ameye, V., Auge, P., Cazeneuve, B., et al.: Translational research: precision medicine, personalized medicine, targeted therapies: marketing or science? Therapie **70**(1), 11–19 (2015)
9. Abrams, T.J., Lee, L.B., Murray, L.J., Pryer, N.K., Cherrington, J.M.: SU11248 inhibits KIT and platelet-derived growth factor receptor beta in preclinical models of human small cell lung cancer. Molecular Cancer Therapeutics **2**(5), 471–478 (2003)
10. Baselga, J.: The EGFR as a target for anticancer therapy–focus on cetuximab. European Journal of Cancer **37**(Suppl 4), S16–S22 (2001)
11. Baselga, J., Norton, L., Albanell, J., Kim, Y.M., Mendelsohn, J.: Recombinant humanized anti-HER2 antibody (Herceptin) enhances the antitumor activity of paclitaxel and doxorubicin against HER2/neu overexpressing human breast cancer xenografts. Cancer Research **58**(13), 2825–2831 (1998)
12. Fritsche, M., Haessler, C., Brandner, G.: Induction of nuclear accumulation of the tumor-suppressor protein p53 by DNA-damaging agents. Oncogene **8**(2), 307–318 (1993)
13. Eisenhauer, E.A., Therasse, P., Bogaerts, J., Schwartz, L.H., Sargent, D., Ford, R., et al.: New response evaluation criteria in solid tumours: revised RECIST guideline (version 1.1). European Journal of Cancer **45**(2), 228–247 (2009)
14. Weedon-Fekjaer, H., Lindqvist, B.H., Vatten, L.J., Aalen, O.O., Tretli, S.: Breast cancer tumor growth estimated through mammography screening data. Breast Cancer Res. **10**(3), R41 (2008). PMCID: 2481488
15. Radon, J.: On the Determination of Functions from Their Integral Values along Certain Manifolds. IEEE. Trans. Med. Imaging **5**(4), 170–176 (1986)
16. Hounsfield, G.N.: Computerized transverse axial scanning (tomography). 1. Description of system. The British Journal of Radiology **46**(552), 1016–1022 (1973)
17. Zacharias, C., Alessio, A.M., Otto, R.K., Iyer, R.S., Philips, G.S., Swanson, J.O., et al.: Pediatric CT: strategies to lower radiation dose. AJR American Journal of Roentgenology **200**(5), 950–956 (2013)
18. Axente, M., Paidi, A., Von Eyben, R., Zeng, C., Bani-Hashemi, A., Krauss, A., et al.: Clinical evaluation of the iterative metal artifact reduction algorithm for CT simulation in radiotherapy. Med. Phys. **42**(3), 1170 (2015)
19. Lauterbur, P.C.: Image Formation by Induced Local Interactions: Examples Employing Nuclear Magnetic Resonance. Nature **242**, 190–191 (1973)
20. Marckmann, P., Skov, L., Rossen, K., Dupont, A., Damholt, M.B., Heaf, J.G., et al.: Nephrogenic systemic fibrosis: suspected causative role of gadodiamide used for contrast-enhanced magnetic resonance imaging. J. Am. Soc. Nephrol. **17**(9), 2359–2362 (2006)
21. Jain, R.K.: Normalizing tumor vasculature with anti-angiogenic therapy: a new paradigm for combination therapy. Nat. Med. **7**(9), 987–989 (2001)

22. Kim, H., Folks, K.D., Guo, L., Stockard, C.R., Fineberg, N.S., Grizzle, W.E., et al.: DCE-MRI detects early vascular response in breast tumor xenografts following anti-DR5 therapy. Molecular Imaging and Biology: MIB: the Official Publication of the Academy of Molecular Imaging **13**(1), 94–103 (2011). PMCID: 4138021
23. Feng, Y., Jeong, E.K., Mohs, A.M., Emerson, L., Lu, Z.R.: Characterization of tumor angiogenesis with dynamic contrast-enhanced MRI and biodegradable macromolecular contrast agents in mice. Magnetic Resonance in Medicine: Official Journal of the Society of Magnetic Resonance in Medicine / Society of Magnetic Resonance in Medicine **60**(6), 1347–1352 (2008). PMCID: 2649794
24. Strul, D., Bendriem, B.: Robustness of anatomically guided pixel-by-pixel algorithms for partial volume effect correction in positron emission tomography. J. Cereb. Blood Flow Metab. **19**(5), 547–559 (1999)
25. Parker, G.J., Roberts, C., Macdonald, A., Buonaccorsi, G.A., Cheung, S., Buckley, D.L., et al.: Experimentally-derived functional form for a population-averaged high-temporal-resolution arterial input function for dynamic contrast-enhanced MRI. Magnetic Resonance in Medicine : Official Journal of the Society of Magnetic Resonance in Medicine / Society of Magnetic Resonance in Medicine **56**(5), 993–1000 (2006)
26. Heye, T., Davenport, M.S., Horvath, J.J., Feuerlein, S., Breault, S.R., Bashir, M.R., et al.: Reproducibility of dynamic contrast-enhanced MR imaging. Part I. Perfusion characteristics in the female pelvis by using multiple computer-aided diagnosis perfusion analysis solutions. Radiology **266**(3), 801–811 (2013)
27. Pitre-Champagnat, S., Leguerney, I., Bosq, J., Peronneau, P., Kiessling, F., Calmels, L., et al.: Dynamic contrast-enhanced ultrasound parametric maps to evaluate intratumoral vascularization. Investigative Radiology **50**(4), 212–217 (2015)
28. Lassau, N., Bonastre, J., Kind, M., Vilgrain, V., Lacroix, J., Cuinet, M., et al.: Validation of dynamic contrast-enhanced ultrasound in predicting outcomes of antiangiogenic therapy for solid tumors: the French multicenter support for innovative and expensive techniques study. Investigative Radiology **49**(12), 794–800 (2014). PMCID: 4222794
29. Kim, H., Zhai, G., Liu, Z., Samuel, S., Shah, N., Helman, E.E., et al.: Extracelluar matrix metalloproteinase as a novel target for pancreatic cancer therapy. Anti-Cancer Drugs **22**(9), 864–874 (2011)
30. Kim, H., Arnoletti, P.J., Christein, J., Heslin, M.J., Posey, J.A., Pednekar, A., et al.: Pancreatic adenocarcinoma: a pilot study of quantitative perfusion and diffusion-weighted breath-hold magnetic resonance imaging. Abdom Imaging (2014)
31. Kim, H., Morgan, D.E., Zeng, H., Grizzle, W.E., Warram, J.M., Stockard, C.R., et al.: Breast tumor xenografts: diffusion-weighted MR imaging to assess early therapy with novel apoptosis-inducing anti-DR5 antibody. Radiology **248**(3), 844–851 (2008)
32. Wahl, R.L., Jacene, H., Kasamon, Y., Lodge, M.A.: From RECIST to PERCIST: Evolving Considerations for PET response criteria in solid tumors. Journal of nuclear medicine: official publication, Society of Nuclear Medicine **50**(Suppl 1), 122S–150S (2009). PMCID: 2755245
33. van Waarde, A., Jager, P.L., Ishiwata, K., Dierckx, R.A., Elsinga, P.H.: Comparison of sigma-ligands and metabolic PET tracers for differentiating tumor from inflammation. J. Nucl. Med. **47**(1), 150–154 (2006)

34. Smyczek-Gargya, B., Fersis, N., Dittmann, H., Vogel, U., Reischl, G., Machulla, H.J., et al.: PET with [18F]fluorothymidine for imaging of primary breast cancer: a pilot study. European Journal of Nuclear Medicine and Molecular Imaging **31**(5), 720–724 (2004)
35. Schwaiger, M., Wester, H.J.: How many PET tracers do we need? Journal of Nuclear Medicine: Official Publication, Society of Nuclear Medicine. **52**(Suppl 2), 36S–41S (2011)
36. De Saint-Hubert, M., Brepoels, L., Mottaghy, F.M.: Can evaluation of targeted therapy in oncology be improved by means of 18F-FLT? Journal of nuclear medicine: official publication, Society of Nuclear Medicine. **51**(10), 1499–1500 (2010)
37. Sanghera, B., Wong, W.L., Sonoda, L.I., Beynon, G., Makris, A., Woolf, D., et al.: FLT PET-CT in evaluation of treatment response. Indian J. Nucl. Med. **29**(2), 65–73 (2014). PMCID: 3996774
38. McKinley, E.T., Ayers, G.D., Smith, R.A., Saleh, S.A., Zhao, P., Washington, M.K., et al.: Limits of [18F]-FLT PET as a biomarker of proliferation in oncology. PLoS One **8**(3), e58938 (2013). PMCID: 3598948
39. Yang, W., Zhang, Y., Fu, Z., Sun, X., Mu, D., Yu, J.: Imaging proliferation of (1)(8)F-FLT PET/CT correlated with the expression of microvessel density of tumour tissue in non-small-cell lung cancer. European Journal of Nuclear Medicine and Molecular Imaging **39**(8), 1289–1296 (2012)
40. Herrmann, K., Wieder, H.A., Buck, A.K., Schoffel, M., Krause, B.J., Fend, F., et al.: Early response assessment using 3'-deoxy-3'-[18F]fluorothymidine-positron emission tomography in high-grade non-Hodgkin's lymphoma. Clinical Cancer Research: an Official Journal of the American Association for Cancer Research. **13**(12), 3552–3558 (2007)
41. Benz, M.R., Czernin, J., Allen-Auerbach, M.S., Dry, S.M., Sutthiruangwong, P., Spick, C., et al.: 3'-deoxy-3'-[18F]fluorothymidine positron emission tomography for response assessment in soft tissue sarcoma: a pilot study to correlate imaging findings with tissue thymidine kinase 1 and Ki-67 activity and histopathologic response. Cancer **118**(12), 3135–3144 (2012). PMCID: 3436595
42. Ott, K., Herrmann, K., Schuster, T., Langer, R., Becker, K., Wieder, H.A., et al.: Molecular imaging of proliferation and glucose utilization: utility for monitoring response and prognosis after neoadjuvant therapy in locally advanced gastric cancer. Annals of surgical oncology **18**(12), 3316–3323 (2011)
43. Bos, M., Mendelsohn, J., Kim, Y.M., Albanell, J., Fry, D.W., Baselga, J.: PD153035, a tyrosine kinase inhibitor, prevents epidermal growth factor receptor activation and inhibits growth of cancer cells in a receptor number-dependent manner. Clinical Cancer Research: an Official Journal of the American Association for Cancer Research **3**(11), 2099–2106 (1997)
44. Fredriksson, A., Johnstrom, P., Thorell, J.O., von Heijne, G., Hassan, M., Eksborg, S., et al.: In vivo evaluation of the biodistribution of 11C-labeled PD153035 in rats without and with neuroblastoma implants. Life Sci. **65**(2), 165–174 (1999)
45. Wang, H., Yu, J., Yang, G., Song, X., Sun, X., Zhao, S., et al.: Assessment of 11C-labeled-4-N-(3-bromoanilino)-6,7-dimethoxyquinazoline as a positron emission tomography agent to monitor epidermal growth factor receptor expression. Cancer Science **98**(9), 1413–1416 (2007)
46. Wang, H., Yu, J.M., Song, X.R., Yang, G.R., Mu, D.B., Zhao, S.Q., et al.: Molecular imaging of epidermal growth factor receptor in glioma-bearing rats. Zhonghua Zhong Liu Za Zhi. **30**(5), 343–346 (2008)

47. Meng, X., Loo Jr, B.W., Ma, L., Murphy, J.D., Sun, X., Yu, J.: Molecular imaging with 11C-PD153035 PET/CT predicts survival in non-small cell lung cancer treated with EGFR-TKI: a pilot study. Journal Of Nuclear Medicine : Official Publication, Society of Nuclear Medicine **52**(10), 1573–1579 (2011)
48. Sun, J., Cai, L., Zhang, K., Zhang, A., Pu, P., Yang, W., et al.: A pilot study on EGFR-targeted molecular imaging of PET/CT With 11C-PD153035 in human gliomas. Clin. Nucl. Med. **39**(1), e20–e26 (2014)
49. Mitri, Z., Constantine, T., O'Regan, R.: The HER2 Receptor in Breast Cancer: Pathophysiology, Clinical Use, and New Advances in Therapy. Chemother. Res. Pract. 743193 (2012). PMCID: 3539433
50. Dijkers, E.C., Kosterink, J.G., Rademaker, A.P., Perk, L.R., van Dongen, G.A., Bart, J., et al.: Development and characterization of clinical-grade 89Zr-trastuzumab for HER2/neu immunoPET imaging. Journal of Nuclear Medicine : Official Publication, Society of Nuclear Medicine **50**(6), 974–981 (2009)
51. Holmes, K., Roberts, O.L., Thomas, A.M., Cross, M.J.: Vascular endothelial growth factor receptor-2: structure, function, intracellular signalling and therapeutic inhibition. Cell. Signal. **19**(10), 2003–2012 (2007)
52. Lee, I., Yoon, K.Y., Kang, C.M., Lin, X., Chen, X., Kim, J.Y., et al.: Evaluation of the angiogenesis inhibitor KR-31831 in SKOV-3 tumor-bearing mice using (64)Cu-DOTA-VEGF(121) and microPET. Nucl. Med. Biol. **39**(6), 840–846 (2012). PMCID: 3629961
53. Ruoslahti, E.: Integrins. J. Clin. Invest. **87**(1), 1–5 (1991). PMCID: 294975
54. Beer, A.J., Niemeyer, M., Carlsen, J., Sarbia, M., Nahrig, J., Watzlowik, P., et al.: Patterns of alphavbeta3 expression in primary and metastatic human breast cancer as shown by 18F-Galacto-RGD PET. Journal of Nuclear Medicine: Official Publication, Society of Nuclear Medicine **49**(2), 255–259 (2008)
55. Schnell, O., Krebs, B., Carlsen, J., Miederer, I., Goetz, C., Goldbrunner, R.H., et al.: Imaging of integrin alpha(v)beta(3) expression in patients with malignant glioma by [18F] Galacto-RGD positron emission tomography. Neuro Oncol. **11**(6), 861–870 (2009). PMCID: 2802406
56. Perik, P.J., Lub-De Hooge, M.N., Gietema, J.A., van der Graaf, W.T., de Korte, M.A., Jonkman, S., et al.: Indium-111-labeled trastuzumab scintigraphy in patients with human epidermal growth factor receptor 2-positive metastatic breast cancer. Journal of Clinical Oncology: Official Journal of the American Society of Clinical Oncology. **24**(15), 2276–2282 (2006)
57. Gaykema, S.B, de Jong J.R., Perik P..J, Brouwers, A.H., Schroder, C.P., Oude Munnink, T.H, et al.: (111)In-trastuzumab scintigraphy in HER2-positive metastatic breast cancer patients remains feasible during trastuzumab treatment. Molecular Imaging **13** (2014)
58. Axelsson, R., Bach-Gansmo, T., Castell-Conesa, J., McParland, B.J.: An open-label, multicenter, phase 2a study to assess the feasibility of imaging metastases in late-stage cancer patients with the alpha v beta 3-selective angiogenesis imaging agent 99mTc-NC100692. Acta. Radiol. **51**(1), 40–46 (2010)
59. Bach-Gansmo, T., Danielsson, R., Saracco, A., Wilczek, B., Bogsrud, T.V., Fangberget, A., et al.: Integrin receptor imaging of breast cancer: a proof-of-concept study to evaluate 99mTc-NC100692. Journal of Nuclear Medicine: Official Publication, Society of Nuclear Medicine **47**(9), 1434–1439 (2006)

60. Levashova, Z., Backer, M., Hamby, C.V., Pizzonia, J., Backer, J.M., Blankenberg, F.G.: Molecular imaging of changes in the prevalence of vascular endothelial growth factor receptor in sunitinib-treated murine mammary tumors. Journal of Nuclear Medicine: Official Publication, Society of Nuclear Medicine **51**(6), 959–966 (2010)
61. Gunanathan, C., Pais, A., Furman-Haran, E., Seger, D., Eyal, E., Mukhopadhyay, S., et al.: Water-soluble contrast agents targeted at the estrogen receptor for molecular magnetic resonance imaging. Bioconjug Chem. **18**(5), 1361–1365 (2007)
62. Pais, A., Gunanathan, C., Margalit, R., Biton, I.E., Yosepovich, A., Milstein, D., et al.: In vivo magnetic resonance imaging of the estrogen receptor in an orthotopic model of human breast cancer. Cancer Research **71**(24), 7387–7397 (2011). PMCID: 3242887
63. Moghimi, S.M., Hunter, A.C., Murray, J.C.: Long-circulating and target-specific nanoparticles: theory to practice. Pharmacol. Rev. **53**(2), 283–318 (2001)
64. Kohler, N., Sun, C., Fichtenholtz, A., Gunn, J., Fang, C., Zhang, M.: Methotrexate-immobilized poly(ethylene glycol) magnetic nanoparticles for MR imaging and drug delivery. Small **2**(6), 785–792 (2006)
65. Lacava, L.M., Lacava, Z.G., Da Silva, M.F., Silva, O., Chaves, S.B., Azevedo, R.B., et al.: Magnetic resonance of a dextran-coated magnetic fluid intravenously administered in mice. Biophys. J. **80**(5), 2483–2486 (2001). PMCID: 1301436
66. Quaglia, F., Ostacolo, L., De Rosa, G., La Rotonda, M.I., Ammendola, M., Nese, G., et al.: Nanoscopic core-shell drug carriers made of amphiphilic triblock and star-diblock copolymers. Int. J. Pharm. **324**(1), 56–66 (2006)
67. Chen, H., Wang, L., Yu, Q., Qian, W., Tiwari, D., Yi, H., et al.: Anti-HER2 antibody and ScFvEGFR-conjugated antifouling magnetic iron oxide nanoparticles for targeting and magnetic resonance imaging of breast cancer. Int. J. Nanomedicine **8**, 3781–3794 (2013). PMCID: 3794963
68. Yang, H.M., Park, C.W., Woo, M.A., Kim, M.I., Jo, Y.M., Park, H.G., et al.: HER2/neu Antibody Conjugated Poly(amino acid)-Coated Iron Oxide Nanoparticles for Breast Cancer MR Imaging. Biomacromolecules **11**(11), 2866–2872 (2010)
69. Hadjipanayis, C.G., Machaidze, R., Kaluzova, M., Wang, L., Schuette, A.J., Chen, H., et al.: EGFRvIII antibody-conjugated iron oxide nanoparticles for magnetic resonance imaging-guided convection-enhanced delivery and targeted therapy of glioblastoma. Cancer Research **70**(15), 6303–6312 (2010). PMCID: 2912981
70. Lyshchik, A., Fleischer, A.C., Huamani, J., Hallahan, D.E., Brissova, M., Gore, J.C.: Molecular imaging of vascular endothelial growth factor receptor 2 expression using targeted contrast-enhanced high-frequency ultrasonography. Journal of Ultrasound in Medicine: Official Journal of the American Institute of Ultrasound in Medicine **26**(11), 1575–1586 (2007). PMCID: 2634836
71. Korpanty, G., Carbon, J.G., Grayburn, P.A., Fleming, J.B., Brekken, R.A.: Monitoring response to anticancer therapy by targeting microbubbles to tumor vasculature. Clinical Cancer Research: an Official Journal of the American Association for Cancer Research **13**(1), 323–330 (2007)
72. Anderson, C.R., Hu, X., Zhang, H., Tlaxca, J., Decleves, A.E., Houghtaling, R., et al.: Ultrasound molecular imaging of tumor angiogenesis with an integrin targeted microbubble contrast agent. Investigative Radiology **46**(4), 215–224 (2011). PMCID: 3075480
73. Sorace, A.G., Saini, R., Mahoney, M., Hoyt, K.: Molecular ultrasound imaging using a targeted contrast agent for assessing early tumor response to antiangiogenic therapy. J. Ultrasound Med. **31**(10), 1543–1550 (2012). PMCID: PMC3464103

Advanced Stents for Cardiovascular Applications

Mahmoud A. Elnaggar, Yoon Ki Joung and Dong Keun Han

Abstract Over the past decades, there has been significant evolution in coronary stents used in percutaneous coronary intervention. These coronary stents were developed to prevent the recoil and acute closure that limited the effectiveness of angioplasty. New problems of in-segment restenosis and stent thrombosis emerged despite the advantages of bare metal stents (BMS) for these issues. Antiplatelet therapy and better stent ballooning strategies decreased complications; however, restenosis remained a problem until the development of drug-eluting stents (DES). First generation sirolimus- and paclitaxel-eluting stents were very successful in lowering restenosis, but still other safety concerns emerged, in particular, higher rates of late stent thrombosis. This led to the continuous research and development of new types of DES, and the birth of fully biodegradable stent platforms, which is foreseen to truly overcome the obstacle of having a permanent polymer and metallic stent remaining in the vessel wall that may precipitate sustained inflammation, persistent vasomotor dysfunction, and in-stent neo-atherosclerosis. In this chapter, we will cover the wide spectrum of advancements in stent technologies, and the approaches that were undertaken to surmount the issues that arose with the use of the various devices *in vivo*, such as material, design, drugs etc., until the most recent ones, an overview of the biological background for the use of percutaneous coronary intervention, and the complications that appear through the use of the different generations of stents.

Keywords: Coronary stent · Bare metal stent · Drug-eluting stent · Restenosis · Late thrombosis · Bioabsorbable stent · Future generation stent

---

M.A. Elnaggar · Y.K. Joung · D.K. Han(✉)
Center for Biomaterials, Biomedical Research Institute,
Korea Institute of Science and Technology, Seoul 136-791, Republic of Korea
e-mail: dkh@kist.re.kr

M.A. Elnaggar · Y.K. Joung · D.K. Han
Department of Biomedical Engineering, Korea University of Science and Technology,
113 Gwahangno, Yuseong-gu, Daejeon 305-333, Republic of Korea

© Springer International Publishing Switzerland 2016
H. Jo et al. (eds.), *Biomedical Engineering: Frontier Research and Converging Technologies*,
Biosystems & Biorobotics 9,
DOI: 10.1007/978-3-319-21813-7_17

1 Coronary Angioplasty

Coronary arteries are the vessels that supply the heart with oxygenated blood. If the artery becomes narrow or get blocked by plaque, the supply of oxygen to cardiac muscles decreases leading to coronary artery disease (CAD) which is the most common type of heart disease and cause of heart attacks [1]. This narrowing is known as stenosis. As of 2012, CAD became the most common cause of death in the world [2], which emphasizes how important finding a treatment for such disease is and why it is the subject of interest worldwide. According to the American Heart Association, CAD has a number of risk factors. The most common risk factors include smoking, family history of CAD, hypertension, obesity, diabetes, lack of exercise, stress, and hyperlipidemia.

1.1 Ballooning

Since the first report of coronary angioplasty by Gruentzig, *et al.* in 1987, percutaneous coronary intervention (PCI) has become the most common revascularization procedure for occluded coronary artery disease [3]. The procedure provided a novel technique to expand the stenotic coronary lesion with a balloon catheter and thereby relieve the myocardial ischemia, which was only treated by coronary artery bypass graft surgery prior to the discovery of this technique.

A balloon catheter is inserted surgically in the arteries until it reaches the lesion; then water pressure is added into the catheter, which adds pressure onto the walls of the vessel, thus forcing the vessels to open (Figure 1A). However, the rate of restenosis for patients undergoing balloon angioplasty was as high as 30-50% within 6 months of the operation [4,5].

1.2 Coronary Stents

Coronary stents, specifically bare metal stents (BMS), were developed to provide an intracoronary scaffolding device to reduce elastic recoil and remodeling, thus reducing the rate of restenosis. The BMS is a small mesh tube which is pre-loaded on a catheter and that is guided forward from the aorta to the blocked area. At the lesion site it is expanded within the vessel. After expansion, the BMS serves as a scaffold to leave the artery open. The stents did reduce the rate of restenosis; however, the rate of restenosis remained unacceptably high (at about 10-30% of patients receiving PCI). This resulted in recurrent angina, impaired quality of life and the need for repeat PCI [4-6] (Figure 1A & B).

The narrowing of the artery after stent deployment is called in-stent restenosis (ISR). ISR is mainly due to neointimal hyperplasia, caused by increase in the vascular smooth muscle cells' (VSMCs) proliferation, migration, extracellular matrix and collagen synthesis, and triggered by vessel injury during implantation [7].

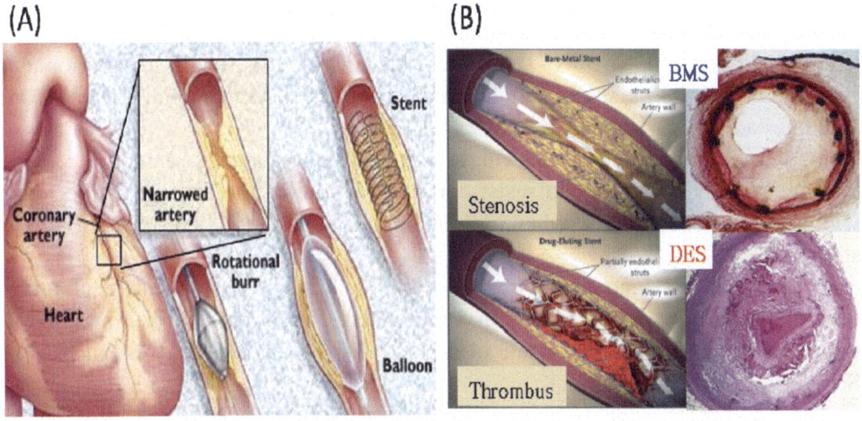

Fig. 1 (A) Three general methods to deal with artery lumen narrowing and (B) clinical problems in bare metal stent vs. drug eluting stent

2 Bare Metal Stent (BMS): The Beginning of an Era

2.1 History

The drawbacks of balloon angioplasty, despite its initial success, led to the development of a scaffold which would prevent restenosis (Figure 1A).Early stent development used a stainless steel scaffold due to its strength and corrosion resistance. In addition to these characteristics, flexibility of the mesh is very important so that it can be easily guided through the blood vessels, but to obtain a flexible material, the thickness of the struts must be decreased, making tracking using X-ray attenuation difficult; thus, guiding will be compromised. In addition to that, there was a hypothesis during that time that thinner struts would decrease the recurrence of restenosis.

It wasn't until the ISAR-STEREO clinical trial results in 2001 which compared two stents with difference in thickness (50 and 140 μm) [8], along with the similar study published in 2002 which used a larger variety of struts and in smaller vessels [9], that this hypothesis was proved, and it was concluded that thinner struts resulted in lower restenosis rates.

This led to the introduction of a stronger metal alloy that has better X-ray attenuation properties, thus enabling the development of balloon-expanding metallic stents with thinner struts. The metal alloy was a cobalt-chromium alloy (Co-Cr) that was already used for stents [10].

The high elastic modulus of the metal alloy allowed the creation of the self-expandable Wallstent® (Boston Scientific Corp., Natick, MA), while one of the first balloon-expandable stents made of Co-Cr alloy was the Multi-Link Vision™ coronary stent (Abbott Laboratories, Abbott Park, IL) made of L605 allot (Co-20Cr-15W-10Ni) that provided both strength and increased X-ray attenuation compared to stainless steel, thus allowing for thinner struts [11].

After Vision™, a similar strut thickness coronary stent was introduced to the market, the Driver® (Medtronic Inc., Minneapolis, MN), which was made of MP35N alloy (Co-20Cr-35Ni-10Mo) and had increased strength and radiopacity compared to stainless steel. Initial clinical trials showed that both – Driver and Vision - had comparable performances [12].

It is worth mentioning that in addition to the added strength of cobalt-chromium alloy, its ability to resist corrosion is one of the driving factors for it wide-spread use, especially when molybdenum is added, as it contributes to higher stability [13].

Other metal-amalgamates have been explored, such as stainless steel mixed with platinum, for added strength and radiopacity. Early prototypes contained up to 10% platinum [14]. A candidate of such strategy is the Element™ stent (Boston Scientific Corp., Natick, MA).

The TriMaxx™ stent (Abbott Laboratories, Abbott Park, IL), a stent with a unique design comprised of a thin layer of tantalum in between two layers of 316L stainless steel, provides high strength and radiopacity, thus allowing thinner struts [15].

Some strategies that were used to increase the radiopacity of stents involved coating a layer of gold on conventional stainless steel stents. Examples of such efforts are the Inflow™ stents (InFlow Dynamics AG, Munich, Germany) and NIR™ stent design (Medinol Ltd., Jerusalem, Israel). However, due to very poor clinical outcomes, interest in gold coatings declined, keeping cobalt-chrome and other alloys on top [16,17].

2.2 Pros and Cons of BMS

Despite the initial success of BMS to decrease the 30-50% incidence of restenosis from conventional ballooning strategies, bare metal stent technology was only successful in decreasing the probability of restenosis to 10-30%; this led to further advancements in bare metal stent design to make them more biocompatible, thus reducing restenosis rates.

Diamond-like Carbon (DLC) was investigated as a potential coating on stainless steel stents as an effort to reduce restenosis. BioDiamond DLC-coated stainless steel stents (BioDiamond, Mainz, Germany) exhibited reduced metal ion release and lower platelet activation when compared to uncoated stents; [18] however, *in vivo* studies didn't demonstrate this according to a clinical study reported by Airoldi *et al.*, comparing Diamond Flex AS stent (Phytis Medical Devices GmbH, Berlin, Germany) in its DLC-coated form and its uncoated form. No difference in restenosis rate was detected [19]. Similar results were shown by Meieles *et al.* when they compared Phytis™ DLC-coated stent (Phytis Medical devices GmbH, Berlin, Germany) with the Multi-Link Penta™ stainless steel stent (Abbott Laboratories, Abbott Park, IL) [20].

Another approach to increase the biocompatibility of bare metal stents was to replace diamond-like carbon with Pyrolytic carbon as it was used successfully for artificial heart valves, with one of the first carbon-coated stents to be released in the market called Carbostent™ (Sorin Biomedica, Saluggia, Italy). However, according to clinical trials conducted, it showed more typical results as with bare metal stents [21]. Another variation of carbon coated stents is the ARTHOS^{inert} stent (Amg International GmbH, Raesfeld-Erle, Germany) that has a surface with ion-implanted carbon, as the theory behind it suggests that it would adhere much better to the devices. However, clinical trials failed to show any difference when compared with an equivalent stainless steel stent [22].

Silicon carbide was an early approach to the high levels of restenosis caused by BMS. The idea was that interaction between fibrinogen and the stent surface would lead to fibrin deposition, leading to platelet adhesion and thrombus formation; thus adding a semiconductor may decrease these electron interactions. [23] Unfortunately the clinical studies of stents coated by amorphous, phosphorous-doped, hydrogen rich silicon carbide showed no difference in restenosis rate when compared to the typical BMS [24].

Based on the fact that nitric oxide is important in regulating endothelium functions, inhibiting smooth muscle cell proliferation, suppressing platelet aggregation, and cellular adhesion [25], the idea of adding a layer of titanium by physical vapor deposition in an oxygen and nitrogen gas mixture [26] has proven to have promising *in vivo* results when compared to bare metal stents, proving that TiNOX could be a successful future competitor in the market [27].

Restenosis after PCI is not well understood, although it is widely accepted that it is a maladaptive response to angioplasty-induced trauma, resulting in inflammation , while thrombosis, cellular proliferation (peaks at about 7 days) and extracellular matrix production (formation from 1 week onwards) [28].

Negative arterial remodeling, resulting from collagen deposition in the matrix and adventitia thickening, plays an important role in restenosis after balloon angioplasty. This is decreased by the use of BMS; however, neointimal hyperplasia that results from the proliferation and migration of smooth muscle cells and extracellular matrix formation can still cause in-stent stenosis [29].

A study involving 56 patients with 116 stents showed that neointimal thickness, neointimal vascular channel density and inflammatory cell density were much higher when there was contact between the struts and ruptured arterial media compared with the fibrous plaque or intact fibrous capsule [30]. This study suggests that coronary stenting accompanied by damage in the arterial media increased the induction of arterial inflammation, leading to neointimal growth, and thus in-stent stenosis.

3 Drug-Eluting Stents (DES)

Due to the high restenosis rate that accompany BMS (Figure 2), many efforts have been made to decrease the inflammation that is caused by stenting, such as by systemically administering anti-inflammatory drugs; this was unsuccessful in humans due to the short period of time the compounds remained within the vascular wall.

The problem was addressed by adding the drugs into a polymer matrix that is coated on the stent. This provided continuous local drug release to the injured lesion.

A handful of drugs have been studied in DES, such a paclitaxel, sirolimus, and its related family members (zotarolimus, biolimus A9, and everolimus). Many polymeric materials such as polylactic acid (PLA), polyethylene-co-vinyl acetate (PEVA), poly(styrene-b-isobutylene-b-styrene) (SIBBS), poly-L-lactic acid (PLLA), phosphorylcholine (PC), and poly-n-butyl methacrylate (PBMA), were used for drug coatings, all of which are unique to each class of DES.

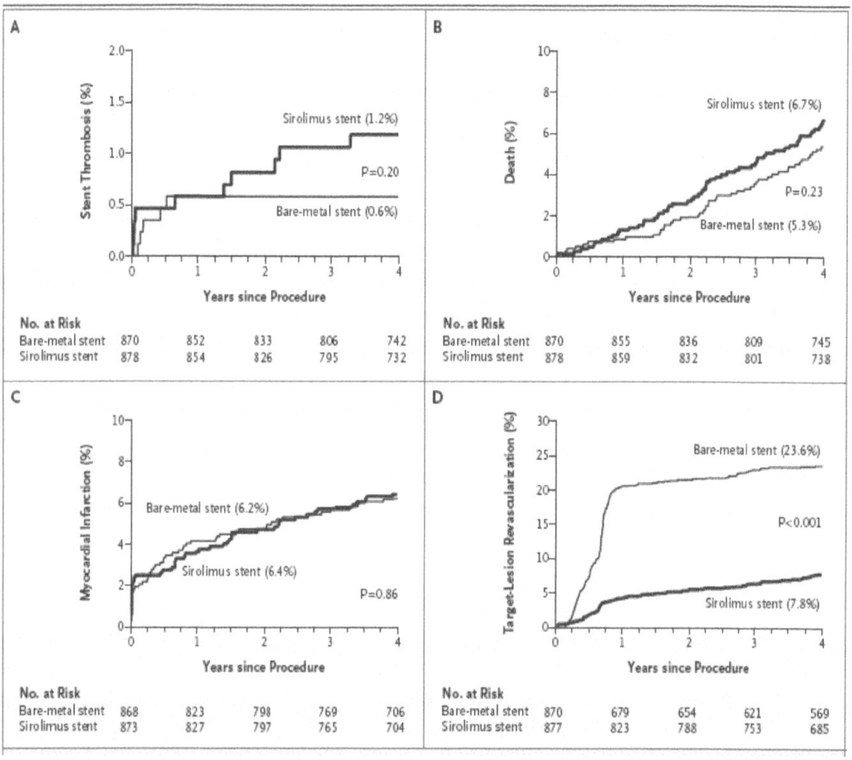

Fig. 2 Four years clinical outcome of BMS when compared to DES; (A) in-stent thrombosis reached 1.2% using DES compared to 0.6% using BMS, (B) death risk of patients with BMS was at 5.3% after 4 years while DES was at 6.7% but the difference was not considered to be significant, (C) myocardial infarction of both was same all over the time period of 4 years, and (D) Restenosis levels were drastically higher for BMS than DES, with a 23% probability over 4 years [93].

3.1 First Generation

Two main US FDA-approved DESs are classified as first generation. The first is the Cypher (Cordis Corp., Miami lakes, FL), which is a sirolimus releasing stent covered with a PEVA and PBMA polymer layer (5 micron thickness) coated on stainless steel struts 140 microns thick. The polymer layer modulates the release of the sirolimus to a release profile of 80% of total drug (140 microgram per square centimeter) within 30 days [31].

The sirolimus is effective in inhibition of neointimal hyperplasia via FKBP 12 binding protein, which subsequently binds to mTOR, thus blocking the cell cycle (mainly of smooth muscle cells) from G1 to S phase [32-34].

The first clinical study was the RAVEL trial, which showed that no cases of restenosis were observed with the DES, but 26.6% of BMS cases suffered from restenosis [31]. A 3 year follow-up of the RAVEL study showed sustained clinical benefits [35], during which another set of trials called the SIRIUS study compared the safety and efficacy of the sirolimus eluting stent with two smaller studies that were performed in Europe (E-SIRIUS) and Canada (C-SIRIUS) [36-39]. The combined results showed a reduction in in-stent restenosis from 38.5% (BMS) to 3.1% (sirolimus eluting stent).

Taxus™ (Boston Scientific, Natick, MA, USA) is another FDA-approved stent that is categorized as first generation. In this stent the drug is paclitaxel (1 microgram/mm^2 per unit area of stent surface) incorporated in a SIBBS layer coated on a stainless steel stent with 132 micron-thick struts. Only about 5% of the drug is released over the course of the first two months (Figure 3).

Paclitaxel stabilizes the microtubules and inhibits cell division in the G0/G1 and G2/M phases, thus inhibiting neointimal hyperplasia [40]. It was first evident that paclitaxel eluting stents were effective when the TAXUS I results came out: after six months, angiographic restenosis rates were 0% for paclitaxel group and 10% for the bare metal stent group [41]. TAXUS II trials, in which more patients were tested, also tested the difference between slow-releasing paclitaxel and moderate-releasing paclitaxel stents. The results showed lower restenosis rates for the slow release stents (2.3% in slow release vs. 4.7% in moderate release) after 6 month follow-up [42].

Then came the TAXUS IV, V, and VI, which tested more complicated lesions with more diabetic patients included in the studies [43-45]. These showed that the paclitaxel groups always had a much lower rate of restenosis than bare metal stents groups, proving that paclitaxel releasing stents are much more effective than conventional bare metal stents.

Despite the success of Cypher™ and Taxus™, histopathological results showed incomplete endothelial coverage of the struts with persistent fibrin deposition and focal platelet aggregation at 16 months of Cypher implantation [46] and with DES in general when compared to BMS [47]. Moreover, late stent thrombosis (LST) was observed in more than half of the patients that received Cypher or Taxus 3-4 months after implantation [48].

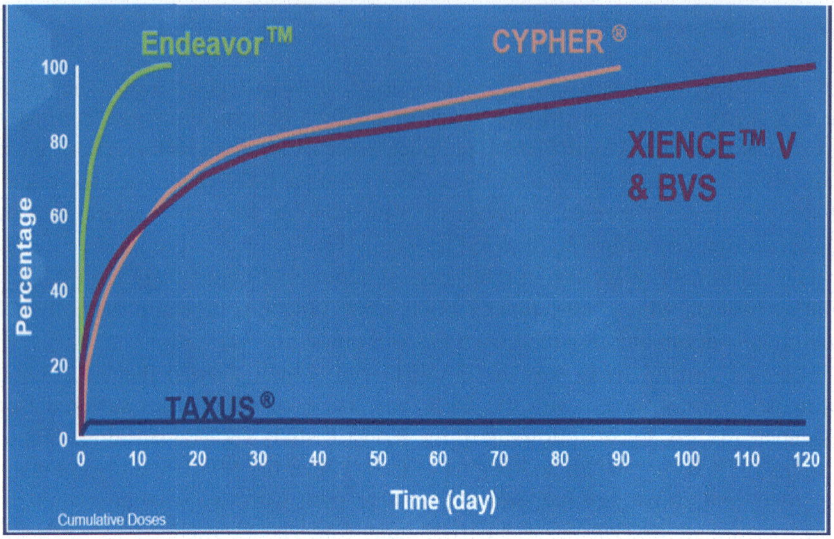

Fig. 3 Drug release patterns of 5 DESs available in the market. Taxus drug release was the lowest amongst all types with a plateau at around 5% for more than 120 days; Endeavor had the fastest drug release profile with 100% of drug released within 20 days, while Cypher, Xience and BVS had a moderate release rate with 100% release achieved between 90–120 days.

Many findings correlated LST to low vascular healing, which is most likely due to underlying patient problems and/or lesion morphology, [49-52] although the devices also affect healing as hypersensitivity reactions have been reported following DES implantations [53]. The most important complication is thrombosis as most cases had LST, [54] but this observation was limited to Cypher and not to Taxus. As most of the reactions occurred after 1 year, it suggested that they were due to the non-biodegradable polymer layer, as the drug was already 100% eluted by that time. This was proved in other studies [55,56].

On the other hand, because the fibrin deposition is dose related, it could be due to paclitaxel cytotoxicity, and may or may not be related to LST [57].

3.2 Second Generation

In this section, second generation drug eluting stents are described, the first of which is the Endeavor™ (Medtronic Vascular, Santa Rosa, CA). This stent's metallic mesh is made up of cobalt-chromium with strut thickness of 91 microns, coated with zotarolimus drug (10 μg zotarolimus per 1 mm stent length) with 70% released after 30 days, blended with phosphorylcholine (a synthetic copy of the most common phospholipid in the outer membrane of RBCs and has high biological compatibility). In animal studies, phosphorylcholine coated stents have significantly less platelet adhesion when compared with uncoated stents [58, 59].

The one arm study ENDEAVOR I trial showed that Endeavor had low incidence of restenosis: 2.1% (4 months) and 5.4% (12 months). [60] This was confirmed by ENDEAVOR II trials that compared Endeavor with BMS [59].

On the other hand, ENDEAVOR III trials showed that Endeavor is associated with higher rates of restenosis when compared with Cypher [61]; however, no very late stent thrombosis has been reported after 4 years of Endeavor use, showing long-term safety of Endeavor [60].

The other stent approved by the FDA is the Xience™ V (Abbott Vascular, Santa Rosa, CA) with everolimus, which is another drug in the limus family and is a powerful anti-proliferative agent: by forming a complex with the cytoplasmic protein FKBP12, it can inhibit growth factor-simulated phosphorylation of p70 S6 kinase and 4E-BPI. Moreover, by interfering with the function of FKBP12-rapamycin associated protein, it can cause cell-cycle arrest during the late G1 stage. For the Xience V, the drug is mixed with a fluoropolymer coated on a cobalt-chromium stent with 81 micron-thick struts.

The main clinical data for everolimus-releasing stents comes from the SPIRIT clinical trial program. This was the first-in-man randomized and controlled multicenter study. Everolimus-releasing stents showed significant reduction in angiographic late loss without an increase of stent thrombosis when compared to Taxus [62].

Despite the fact that this generation of DES reduced in-stent restenosis greatly, the delayed endothelialization or its dysfunction is a big issue [48, 54, 63]. This is most likely due to hypersensitivity caused by the non-degradable polymers used as a coating and drug release medium, [47] which could be solved by the birth of the next generation DESs.

3.3 Next Generation

After the initial great success that was subsequently faced with many challenges in the second generation DES, many new designs of DES are now taking into consideration the arterial healing process.

The first of these approaches involves stents that have biodegradable polymer coating layers with newer drugs. These polymers include PLA, PLLA, poly-lactic-co-glycolic acid (PLGA), and polyglycolic acid (PGA), that hydrolyze to their basic monomers by water hydrolysis of the ester bonds through a process called bulk degradation. New drugs are also being used in these new generation stents, such as the more lipophilic rapamycin analogue, biolimus A9, which has similar smooth muscle cell suppressing power to sirolimus [32-34, 64-66]. There is also tacrolimus and pimecrolimus that act by the calcium/calmodulin-dependent protein phosphatase calcineurin pathway [67].

Stents of such types include the Sparrow™ NiTi (CardioMind Inc., Sunnyvale, CA) that uses SynBiosys™ biodegradable PLGA polymer (Surmodics Inc., Eden Prairie, MN) which releases rapamycin [32-34, 68]. Another is the BioMatrix 316L device (Biosensors International, Singapore) which elutes biolimus A9 from a PLA coating on a stainless steel 112 micron-thick strut stent, but it is still too early to judge its long term effectiveness. However, according to STEALTH and LEADERS studies, the results are promising [66, 69].

3.3.1 Recent Trends in Developments

Just recently, 2013, the Promus Premier everolimus-eluting (Boston Scientific, Natick, MA) stent was FDA approved, which has the same composition of Xience V, everolimus as the drug, same fluoropolymer, same drug concentration of 100 g/cm^2, same strut thickness of 82 micons, with the only difference being in the substrate, where they replaced Co-Cr with platinum-chromium. In the 3 year PLATINUM trial, Promus showed comparable clinical results as Xience [72]. Despite the comparable results with all the second generation DESs, bench studies showed that the stent had lower longitudinal strength, which means that by time the stent shortens [73] which was solved by adding extra connectors at the proximal end to provide longitudinal strength, and this was confirmed by point-force compression bench testing [74].

Resolute Integrity (Medtronic, Minneapolis, MN) is a 2012 FDA approved stent, that uses zotarolimus (160 g/cm^2) that is 85% eluted in the first 2 months, blended with Biolinx polymer coated on a cobalt-chromium 91 micron strut. Despite its similarity with Endeavour, it was redesigned to have higher biocompatibility and controlled drug release. It was shown that Resolute had comparable results with everolimus eluting stents in the RESOLUTE All-comers trial [75-77], with other trials such as the TWENTE trial having comparable results [78]. Albeit Resolute is the only DES to be indicated specifically for diabetic patients, data from EXCELLENT and RESOLUTE-Korea show that the same could go for everolimus eluting stents [79].

3.3.2 Fully Biodegradable DES

Despite the success of DES, still better prevention of restenosis and in stent thrombosis is searched for by scientists. In addition to that, the metallic stent matrix that remains in the vessels is still considered to be a foreign body, thus having potential disadvantages.

Reduced or abolished late stent thrombosis, facilitation of multiple surgeries in the lesion area, better imaging all these could be potential advantages of having a stent that actually disappears completely after it does its job. This led scientists to start developing fully biodegradable stents (FBS) (Figure 4).

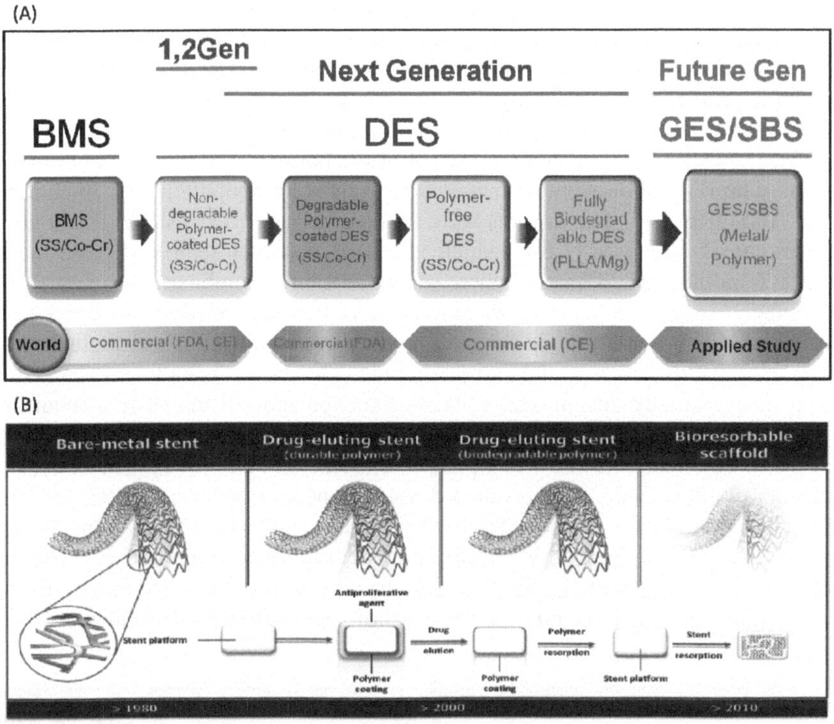

Fig. 4 (A) and (B) showing evolution of the stent over the years [89]

3.3.2.1 Polymer-Based Stents

The Igaki-Tamai stent (Igaki Medical Planning Co., Kyoto, Japan) is the first FBS to be used in humans; it is composed of PLLA [78]. It is a zigzag helical coil with straight bridges and strut thickness of 170 microns (Figure 5A). It is mounted using a balloon, but self-expands due to dilatation by the warm surface of the blood vessel. Gold markers are added at both ends of the stent, as PLLA lacks radiopacity. No drug is blended with the PLLA, and the stent is degraded due by bulk erosion of the ester bonds; the degradation products are metabolized in the Krebs cycle into water and carbon dioxide. The first clinical trials that were conducted with 50 patients were followed up for 4 years, with all patients showing low complication rate with only 1 person-in hospital stent thrombosis, 18% repeat PCI, and no surgical revascularization [80].

Next, the BVS everolimus-eluting stent (Abbott Vascular, Santa Clara, CA) was the first fully biodegradable DES to have similar clinical results to metallic DES after implantation for 2 years, with the added benefit of being fully biodegradable [79, 80]. It is made up of a PLLA backbone, with a poly-D,L-lactide coating that controls the everolimus drug release. Although the PLLA used in the Igaki-Tamai stent is the same as in the BVS stent, the processing method is different, thus giving it other physical properties. 80% of the drug is released within 30 days making it very similar to the metallic DES. Like the Igaki-Tamai, it is also absorbed by bulk erosion.

The BVS has two designs: the 1.0 revision and the 1.1 revision (Figure 5B). Both designs are made of the same material, but with different processing methods, giving increased radial vessel wall support but the same degradation period (around 2 years) [80]. Both designs have platinum radio-opaque markers at each end, and are easily differentiated from calcium to during tracking using fluoroscopy, while tracking the implantation. The strut thickness is 150 microns in both the designs. The BVS underwent first-in-patient trials with a population of 30 patients called ABSORB [79]. The patients were followed up at 6, 12, 18, and 24 months. The restenosis rate was 11.5%, which is lower than that seen with bare metal stents at 6 months. At 24 months, there was no in-stent thrombosis.

Another example of a biodegradable stent is the REVA (Reva Medical Inc., San Diego, CA), a tyrosine-derived polycarbonate, which after metabolism produces amino acids, ethanol, and carbon dioxide. The REVA is balloon expandable with a slide and lock design which allows the expansion of the stent without deformation (Figure 5C). Iodine is its source of radiopacity. It has thick struts of 200 microns. Preclinical data show complete re-endothelialization [81]. Currently, the paclitaxel-eluting REVA stent is under development.

Finally, we have the Bioabsorbable therapeutics stent (Bioreabsorbable Therapeutics Inc., Menlo Park, CA), a sirolimus-eluting FBS composed of salicylic acid. It is also coated with a polymer layer responsible for controlled release of the drugs (Figure 5D). The coating is made up of repeated salicylic acid molecules

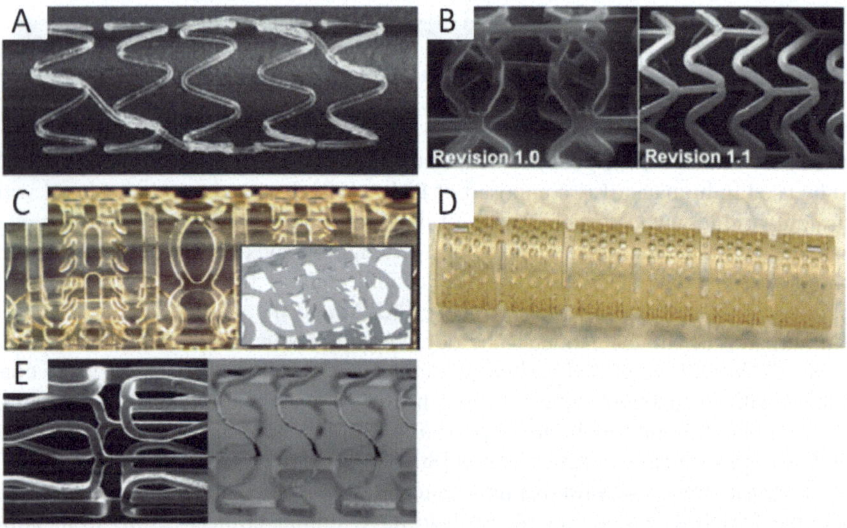

Fig. 5 (A) The expanded Igaki-Tamai bioabsorbable stent showing the zigzag helical coil with straight bridges, (B) the BVS stent with the two different designs, (C) the REVA stent in its expanded form, (D) the bioabsorbable salicylic acid-based stent, and (E) a bioabsorbable magnesium stent unexpanded on the left and expanded on the right with different magnifications [88].

linked by adipic acid molecules. The backbone is the same, but with a different linker. The sirolimus has similar dosing to Cypher stents. The absorption of the stent occurs by surface erosion [82]. After clinical trials, it was found out that the stents caused higher than expected intimal hyperplasia; thus, a subsequent design change will have a higher dose of sirolimus with thinner struts. However, it still had a significant reduction in in-stent restenosis rates [81].

3.3.2.2 Metal-Based Stents

The first metal-based fully bioabsorbable stent (figure 5E) is made up of 93% magnesium with 7% rare earth metals: the WE-43 (Biotronik, Berlin, Germany), which degrades within 2 months into inorganic salts [83]. It is balloon expandable and has 165 micron-thick struts. The coverage of the stent is similar to metallic stents [9]. The radial strength is similar to stainless steel stents [84]. The absorption is similar to the Bioreabsorbable Therapeutics polymer stent, and occurs by surface erosion. The clinical results were similar to those of metallic stents [8]. One of the problems of this stent was that it lost its radial strength after days from implantation, which made it hard to counteract the negative remodeling [83]; also, perhaps due to the lack of drugs, it had high restenosis rates [83].

One of the issues with absorbable stents is that they lose – as we saw – their radial strength long before they are actually fully degraded. This begs the question: how long does the stent need to stay strong?

In a clinical study, it was shown that there was early positive remodeling up to 1 month, and then negative remodeling took place from 1 until 6 months [85]. With balloon angioplasty, restenosis appeared between the first and third months and is rare after that [86, 87]. Taking all this information into consideration, the optimal fully biodegradable stent should retain its radial strength throughout the first and third months, and after that they can start to degrade and weaken, as stents are not really needed after 6 months post-angioplasty.

4 Future Generation Stent

4.1 Gene-eluting Stent

The basic concept of gene-eluting stents is that stents will be covered with genes responsible for the production of growth factors. These genes will leave the stent and will enter the injured vascular cells; consequently, the cell will secrete growth factors that will bind to C-MET, thus activating the C-MET pathway, leading to increase in mitogenesis, more cell proliferation, and suppression of cell apoptosis. This will lead to the proliferation and migration of endothelial cells in addition to smooth muscle cell migration to the injured site [90, 91].

4.2 *Stem Cell-Based Stent*

Metallic stents covered with stem cells were proposed as a plausible solution for preventing in-stent thrombosis and restenosis. Raina *et al.* published in 2014 *in vivo* results of 152 implanted hTEC covered stents and compared the outcomes with BMS. The preliminary results showed earlier strut coverage with endothelial cells and no increased neointimal proliferation compared to BMS [92].

5 Summary and Perspective

Innovation in stent technologies has improved massively over the years when compared to balloon angioplasty. This started with BMSs, passing through DESs, and arriving at fully biodegradable, gene-eluting, and stem cell-based stents. As our understanding of the biological complications of PCI – restenosis and in-stent thrombosis – increases, better options are provided as solutions for the underlying diseases. While the ideal stent has not yet been produced, fully biodegradable stents along with future generation stents are very promising approaches to solve the problems of restenosis and late stent thrombosis.

Acknowledgments This work was supported by Cell Regeneration Program (2014M3A9C6049717), Pioneer Research Center Program (2014M3C1A3056052), and the KIST Program (2E24680) through the National Research Foundation of Korea funded by the Ministry of Science, ICT & Future Planning (MSIP), and Core Materials Program (10048019) funded by the Ministry of Trade, Industry and Energy (MOTIE), Republic of Korea.

References

1. Bedair, T.M., Cho, Y.J., Park, B.J., Joung, Y.K., Han, D.K.: Coating defects in polymer-coated drug-eluting stents. Biomater. Biomed. Eng. **1**(3), 121–141 (2014)
2. Finegold, J.A., Asaria, P., Francis, D.P.: Mortality from ischaemic heart disease by country, region, and age: Statistics from World Health Organisation and United Nations. Int. J. Card. **168**(2), 934–945 (2012)
3. Gruentzig, A.R., King III, S.B., Schlumpf, M., Siegenthaler, W.: Long-term follow-up after percutaneous transluminal coronary angioplasty: the early Zurich experience. N. Engl. J. Med. **316**, 1127–1132 (1987)
4. Holmes Jr, D.R., Vliestra, R.E., Smith, H.C., et al.: Restenosis after percutaneous transluminal coronary angioplasty (PTCA): a report from the PTCA Registry of the National Heart, lung, and blood Institute. Am. J. Cardiol. **53**, 77–81 (1984)
5. Fischman, D.L., Leon, M.B., Baim, D.S., et al.: A randomized comparison of coronary-stent placement and balloon angioplasty in treatment of coronary artery disease. Stent study investigators. N. Engl. J. Med. **331**, 496–501 (1994)

6. Serruys, P.W., De Jaegere, P., Kiemeneij, F., et al.: A comparison of Balloon-expandable-stent implantation with Balloon angioplasty in patients with coronary artery disease. Benestent Study Group. N. Engl. J. Med. **349**, 1315–1323 (1994)
7. Tesfamarium, B.: Local vascular toxicokinetics of stent-based drug delivery. Toxicol. Lett. **168**, 93–102 (2007)
8. Kastrati, A., Mehilli, J., Dirschinger, J., Dotzer, F., Schuehlen, H., Neumann, F.J., et al.: Intracoronary stenting and angiographic results: strut thickness effect on restenosis outcome (ISAR-STEREO) trial. Circulation **103**, 2816–2821 (2001)
9. Briguori, C., Sarais, C., Pagnotta, P., Liistro, F., Montorfano, M., Chieffo, A., et al.: In-stent restenosis in small coronary arteries: impact of strut thickness. J. Am. Coll. Cardiol. **40**, 403–409 (2002)
10. Clerc, C.O., Jedwab, M.R., Mayer, D.W., Thompson, P.J., Stinson, J.S.: Assessment of wrought ASTM F1058 cobalt alloy properties for permanent surgical implants. J. Biomed. Mater. Res. Appl. Biomater. **38**, 229–234 (1997)
11. Kereiakes, D.J., Cox, D.A., Hermiller, J.B., Midei, M.G., Bachinsky, W.B., Nujta, E.D., et al.: Usefulness of a cobalt chromium coronary stent alloy. Am. J. Cardiol. **92**, 463–466 (2003)
12. Sketch, M.H., Ball, M., Rutherford, B., Pompa, J.J., Russell, C., Kereiakes, D.J.: Evaluation of the Medtronic (Driver) cobalt-chromium alloy coronary stent system. Am. J. Cardiol. **95**, 8–12 (2005)
13. Metikos-Hukovic, M., Pilic, Z., Babie, R., Omanovic, D.: Influence of alloying elements on the corrosion stability of CoCrMo implant alloy in Hank's solution. Acta Biomater. **2**, 693–700 (2006)
14. Craig, C.H., Friend, C.M., Edwards, M.R., Cornish, L.A., Gokcen, N.A.: Mechanical Properties and microstructure of platinum enhanced radiopaque stainless steel (PERSS) alloys. J. Alloy. Comp. **361**, 187–199 (2003)
15. Azibad, A., Popma, J.J., Tanajura, L.F., Hattori, K., Solberg, B., Larracas, C., et al.: Clinical and angiographic results of percutaneous coronary revascularization using a trilayer stainless steel-tantalum-stainless steel phosphorylcholine-coated stent: The TriMaxx trial. Catherter Cardiovasc. Interv. **70**, 914–919 (2007)
16. Kastrati, A., Schoemig, A., Dirschinger, J., Mehilli, J., von Welser, N., Pache, J., et al.: Increased risk of restenosis after placement of gold-coated stents. Circulation **101**, 2478–2483 (2000)
17. Reifart, N., Morice, M.C., Silber, S., Benit, E., Hauptmann, K.E., de Sousa, E., et al.: The NUGGET study: NIR ultra gold-gilded equivalency trial. Catheter Cardiovasc. Interv. **62**, 18–25 (2004)
18. Gutensohn, K., Beythien, C., Bau, J., Fenner, T., Grewe, P., Koester, R., et al.: In vitro analysis of diamond-like carbon coated stents: reduction of metal ion release, platelet activation, and thrombogenecity. Thromb. Res. **99**, 577–585 (2000)
19. Airoldi, F., Colombo, A., Tavano, D., Stankovic, G., Klugmann, S., Paolillo, V., et al.: Comparison of Diamond-like carbon-coated stents versus uncoated stainless steel stents in coronary artery disease. Am. J. Cardiol. **93**, 474–477 (2004)
20. Meirles, G.C.X., de Abreu, L.M., da Cruz Forte, A.A., Sumita, M.K., Sumita, J.H., Aliaga, J.: Randomized comparative study of diamond-like carbon coated stainless steel stent versus uncoated stent implantation in patients with coronary artery disease. Arc. Bras. Cardiol. **88**(4), 343–347 (2007)
21. Antoniucci, D., Valenti, R., Migliorini, A., Moschi, G., Trapani, M., Bolognese, L., et al.: Clinical and angiographic outcomes following elective implantation of the carbostent in patients at high risk of restenosis and target vessel failure. Catheter Cardiovasc. Interv. **54**, 420–426 (2001)

22. Kim, Y.H., Lee, C.W., Hong, M.K., Park, S.W., Tahk, S.J., Yang, J.Y., et al.: Randomized comparison of carbon ion-implanted stent versus bare metal stent in coronary artery disease: the asian pacific multicancer arthos stent study (PASS) trial. Am. Heart J. **149**, 336–341 (2005)
23. Rzany, A., Schaldach, M.: Smart material silicon carbide: reduced activation of cells and proteins on a-SiC:H-coated stainless steel. Prog. Biomed. Res. **3**, 182–194 (2001)
24. Heublein, B., Pethig, K., Ozbek, C., Elsayed, M., Bolz, A., Schaldach, M.: Silicon carbide coating-a new hybrid design of coronary stents. Prog. Biomed. Res. **1**, 33–39 (1998)
25. Vallence, P., Chan, N.: Endothelial fuction and nitric oxide: clinical relevance. Heart **85**, 342–350 (2001)
26. Windecker, S., Mayer, I., De Pasquale, G., Maier, W., Dirsch, O., De Groot, P., et al.: Stent coating with titanium-nitride-oxide for reduction of neointimal hyperplasia. Circulation **104**, 928–933 (2001)
27. Windecker, S., Simon, R., Lins, M., Klauss, V., Eberli, F.R., Roffi, M., et al.: Randomized comparison of a titanium-nitride-oxide-coated stent with a stainless steel stent for coronary revascularization The TiNOX Trial. Circulation **111**, 2617–2622 (2005)
28. Nikol, S., Huehns, T.Y., Hofling, B.: Molecular biology and post-angioplasty restenosis. Atherosclerosis **123**, 17–31 (1996)
29. Rajagopal, V., Rockson, S.G.: Coronary restenosis: a review of mechanisms and management. Am. J. Med. **115**, 547–553 (2003)
30. Farb, A., Weber, D.K., Kolodgie, F.D., Burke, A.P., Virmani, R.: Morphological predictors of restenosis after coronary stenting in humans. Circulation **105**, 2974–2980 (2002)
31. Morice, M.C., Serruys, P.W., Sousa, J.E., et al.: A randomized comparison of a sirolimus-eluting stent with a standard stent for coronary revascularization. N. Engl. J. Med. **346**(23), 1773–1780 (2002)
32. Mark, S.O., Mark, A.R.: Bench to bedside: the development of rapamycin and its application to stent stenosis. Circulation **104**, 853–858 (2001)
33. Burke, S.E., Lubbers, N.L., Chen, Y.W., et al.: Neointimal formation after balloon-induced vascular injury in yucaran minipigs is reduced by oral rapamycin. J. Cardiovasc. Pharmacol. **33**, 829–835 (1999)
34. Gallo, R., Padurean, A., Jayaraman, T., et al.: Inhibition of intimal thickening after balloon angioplasty in porcine coronary arteries by targeting regulators of the cell cycle. Circulation **99**, 2164–2170 (1999)
35. Fajadet, J., Morice, M.C., Bode, C., et al.: Maintenance of long-term clinical benefit with sirolimus-eluting coronary stents: three-year results of the RAVEL trial. Circulation **111**, 1040–1044 (2005)
36. Weisz, G., Leon, M.B., Holmes, D.R., et al.: Two-year outcomes after sirolimus-Eluting stent implantation: results from the sirolimus-Eluting in de Novo Native Coronary Lesions (SIRIUS) trial. J. Am. Coll. Cardiol. **47**, 1350–1355 (2006)
37. Schofer, J., Schluter, M., Gershlick, A.H., et al.: Sirolimus-eluting stents for treatment of patients with long atherosclerotic lesions in small coronary arteries: double-blinded, randomized controlled trial (E-SIRIUS). Lancet **362**, 1093–1099 (2003)
38. Schampaert, E., Cohen, E.A., Schluter, M., et al.: The Canadian study of the sirolimus-eluting stent in the treatment of patients with long *de novo* lesions in small native coronary arteries (C-SIRIUS). J. Am. Coll. Cardiol. **43**, 1110–1115 (2004)

39. Schampaert, E., Moses, J.W., Schofer, J., et al.: Sirolimus-eluting stents at two years: a pooled analysis of SIRIUS, E-SIRIUS, and C-SIRIUS with emphasis on late revascularization and stent thrombosis. Am. J. Cardiol. **98**, 36–41 (2006)
40. Sollott, S.J., Cheng, L., Pauly, R.R., et al.: Taxol inhibits neointimal smooth muscle cell accumulation after angioplasty in the rat. J. Clin. Invest. **95**, 1869–1876 (1995)
41. Grube, E., Silber, S., Hauptmann, K.E., et al.: TAXUS I: six- and twelve- month results from a randomized, double blind trial on a slow-release paclitaxel –eluting stent for de novo coronary artery lesions. Circulation **107**, 38–42 (2003)
42. Colombo, A., Drzewiecki, J., Banning, A., et al.: Randomized study to assess the effectiveness of slow- and moderate- release polymer-based paclitaxel-eluting stents for coronary artery lesions. Circulation **108**, 788–794 (2003)
43. Stone, G.W., Ellis, S.G., Cox, D.A., et al.: one-year clinical results with the slow-release, polymer-based, paclitaxel-eluting TAXUS stent: the TAXUS IV trial. Circulation **109**, 1942–1947 (2004)
44. Stone, G.W., Ellis, S.G., Cannon, L., et al.: Comparison of a polymer-based paclitaxel-eluting stent with a bare-metal stent in patients with complex coronary artery disease: a randomized control trial. JAMA **294**, 1215–1223 (2005)
45. Dawkins, K.D., Grube, E., Guagliumi, G., et al.: Clinical efficacy of polymer-based paclitaxel-eluting stents in the treatment of complex, long coronary artery lesions from a multicenter, randomized trial: support for the use of drug-eluting stents in contemporary clinical practice. Circulation **112**, 3306–3313 (2005)
46. Guagliumi, G., Farb, A., Musumeci, G., et al.: Images in cardiovascular medicine, Sirolimus-eluting stent implanted in human coronary artery for 16 months: pathological findings. Circulation **107**(9), 1340–1341 (2003)
47. Joner, M., Finn, A.V., Farb, A., et al.: Pathology of drug eluting stents in humans: delayed healing and late thrombotic risk. J. Am. Coll. Cardiol. **48**(1), 193–202 (2006)
48. Finn, A.V., Joner, M., Nakazawa, G., et al.: Pathological correlates of late drug-eluting stent thrombosis: strut coverage as a marker of endothelialization. Circulation **115**(18), 2435–2441 (2007)
49. Nakazawa, G., Finn, A.V., Joner, M., et al.: Delayed arterial healing and increased late stent thrombosis at culprit sites after drug-eluting stent placement for acute myocardial infarction patients: an autopsy study. Circulation **118**(11), 1138–1145 (2008)
50. Awata, M., Kotani, J., Uematsu, M., et al.: Serial angioscopic evidence of incomplete neointimal coverage after sirolimus-eluting stent implantation: comparison with bare-metal stents. Circulation **116**(8), 910–916 (2007)
51. Matsumoto, D., Shite, J., Shinke, T., et al.: Neointimal coverage of sirolimus-eluting stents at 6 month follow-up: evaluated by optical coherence tomography. Eur. Heart J. **28**(8), 961–967 (2007)
52. Kubo, T., Imanishi, T., Kitabata, H., et al.: Comparison of vascular response after sirolimus-eluting stent implantation between patients with unstable and stable angina pectoris a serial optical coherence tomography study. J. Am. Coll. Cardiol. Img. **1**(4), 475–484 (2008)
53. Virmani, R., Guagliumi, G., Farb, A., et al.: Localized hypersensitivity and late coronary thrombosis secondary to a sirolimus-eluting stent: should we be cautious? Circulation **109**(6), 701–705 (2004)
54. Finn, A.V., Nakazawa, G., Joner, M., et al.: Vascular responses to drug-eluting stents: importance of delayed healing. Arterioscler. Thromb. Vasc. Biol. **27**(7), 1500–1510 (2007)

55. Ahmed, D.D., Sobczak, S.C., Yunginger, J.W.: Occupational allergies caused by latex. Immunol. Allergy Clin. North. Am. **23**(2), 205–219 (2003)
56. Leggat, P.A., Kedjarune, U.: U. Toxicity of methyl methacrylate in dentistry. Int. Dent. J. **53**(3), 126–131 (2003)
57. Farb, A., Heller, P.F., Shroff, S., et al.: Pathological analysis of local delivery of paclitaxel via a polymer-coated stent. Circulation **104**(4), 473–479 (2001)
58. Whelan, D.M., van der Giessen, W.J., Krabbendam, S.C., van Vliet, E.A., Verdouw, P.D., Serruys, P.W., van Beusekom, H.M.: Biocompatibility of phosphorylcholine coated stents in normal porcine coronary arteries. Heart **83**, 338–345 (2000)
59. Fajadet, J., Wijns, W., Laarman, G.J., et al.: Randomized, Double-Blind, Multicenter Study of the Endeavor Zotarolimus-Eluting Phosphorylcholine-Encapsulated Stent for Treatment of Native Coronary Artery Lesions: clinical and angiographic results of the ENDEAVOR II trial. Circulation **114**, 798–806 (2006)
60. Meredith, I.T., Ormiston, J., Whitbourn, R., et al.: Four-year clinical follow-up after implantation of the endeavor zotarolimus-eluting stent: ENDEAVOR I, the first-in-human study. Am. J. Cardiol. **100**, 56–61 (2007)
61. Kandzari, D.E., Leon, M.B., Popma, J.J., et al.: Comparison of zotarolimus-eluting and sirolimus-eluting stents in patients with native coronary artery disease: a randomized controlled trial. J. Am. Coll. Cardiol. **48**(12), 2440–2447 (2006)
62. Stone, G.W., Midei, M., Newman, W., et al.: Comparison of an everolimus-eluting stent and a paclitaxel-eluting stent in patients with coronary artery disease: a randomized trial. JAMA **299**(16), 1903–1913 (2008)
63. Mauri, L., Hsieh, W.H., Massaro, J.M., et al.: Stent thrombosis in randomized clinical trialsof drug-eluting stents. N. Engl. J. Med. **356**(10), 1020–1029 (2007)
64. Ostojic, M., et al.: First clinical comparison of Nobori-Biolimus A9 eluting stents with Cypher-SIROLIMUS eluting stents: NOBORI CORE nine months angiographic and one year clinical outcomes. Eurointervention **3**, 574–579 (2008)
65. Grube, E., Buellesfeld, L.: BioMatrix Biolimus A9-eluting coronary stent: a next-generation drug drug-eluting stent for coronary artery disease. Expert Rev. Med. Dev. **3**, 731–741 (2006)
66. Grube, E., et al.: Six-month results of a randomized study to evaluate safety and efficacy of a biolimus A9 eluting stent with a biodegradable polymer coating. Eurointervention **1**, 53–57 (2005)
67. Donners, M.M., et al.: Inflammation and restenosis: implications for therapy. Ann. Med. **35**(7), 523–531 (2003)
68. Ac, Abizaid, et al.: The CardioMind coronary stent delivery system: stent delivery on a 0.014" guidewire platform. EuroInterv **3**, 154–157 (2007)
69. Windecker, S., Serruys, P.W., Wandel, S., et al.: Biolimus-eluting stent with biodegradable polymer versus sirolimus-eluting stent with durable polymer for coronary revascularization (LEADERS): a randomized non-inferiority trial. Lancet **372**(9644), 1163–1173 (2008)
70. Meredith, I.T., Teirstein, P.S., Bouchard, A., Carrié, D., Möllmann, H., Oldroyd, K.G., et al.: Three-year results comparing platinum-chromium PROMUS elementand cobalt chromium XIENCEV everolimus-eluting stents in denovo coronary artery narrowing (from the PLATINUM Trial). Am. J. Cardiol. **113**, 1117–1123 (2014)
71. Ormiston, J.A., Webber, B., Webster, M.W.: Stent longitudinal integrity bench insights into a clinical problem. JACC Cardiovasc. Interv. **4**, 1310–1317 (2011)

72. Ormiston, J.A., Webber, B., Ubod, B., White, J., Webster, M.W.: Stent longitudinal strength assessed using point compression: insights from a second-generation, clinically related bench test. JACC Cardiovasc. Interv. **7**, 62–69 (2014)
73. Taniwaki, M., Stefanini, G.G., Silber, S., Richardt, G., Vranckx, P., Serruys, P.W., et al.: 4-year clinical outcomes and predictors of repeat revascularizationin patients treated withnew- generation drug-eluting stents: are port from the RESOLUTE All-Comers trial (a randomized comparison of a zotarolimus-eluting stent with an everolimus-eluting stent for percutaneous coronary intervention). J. Am. Coll. Cardiol. **63**, 1617–1625 (2014)
74. Camenzind, E., Wijns, W., Mauri, L., KurowskiV, Parikh K., Gao, R., et al.: Stent thrombosis and major clinical events at 3 years after zotarolimus-eluting or sirolimus-eluting coronary stent implantation: a randomised, multicentre, open- label, controlled trial. Lancet **380**, 1396–1405 (2012)
75. Park, D.W., Kim, Y.H., Yun, S.C., Kang, S.J., Lee, S.W., Lee, C.W., et al.: Comparison of zotarolimus-eluting stents with sirolimus- and paclitaxel-eluting stents for coronary revascularization: the ZEST (comparison of the efficacy and safety of zotarolimus-eluting stent with sirolimus-eluting and paclitaxel-eluting stent for coronary lesions) randomized trial. J. Am. Coll. Cardiol. **56**, 1187–1195 (2010)
76. Tandjung, K., SenH, Lam M.K., Basalus, M.W., Louwerenburg, J.H., Stoel, M.G., et al.: Clinical outcome following stringent dis- continuation of dual antiplatelet therapy after 12 months in real-world patients treated with second-generation zotarolimus-eluting resolute and everolimus-eluting Xience V stents: 2-yearfollow-up of the randomized TWENTE trial. J. Am. Coll. Cardiol. **61**, 2406–2416 (2013)
77. Park, K.W., Lee, J.M., Kang, S.H., Ahn, H.S., Kang, H.J., Koo, B.K., et al.: Everolimus-eluting xiencev/promus versus zotarolimus-eluting resolute e stents in patients with diabetes mellitus. JACC Cardiovasc. Interv. **7**, 471–481 (2014)
78. Tamai, H., et al.: Initial and 6-month results of biodegradable poly-L-Lactic acid coronary stents in humans. Circulation **102**, 399–404 (2000)
79. Ormiston, J.A., et al.: A bioabsorbable everolimus-eluting coronary stent system for patients with single de-novo coronary artery lesions (ABSORB): a prospective open-label trial. Lancet **371**(9616), 899–907 (2008)
80. Serruys, P.W., et al.: Absorb trial first-in-man evaluation of a bioabsorbable everolimus-eluting coronary stent system: two-year outcomes and results from multiple imaging modalities. Lancet **373**, 897–910 (2009)
81. Ramcharitar, S., Serruys, P.W.: Fully biodegradable coronary stents: progress to date. Am. J. Cardiovasc. Drugs **8**(5), 305–314 (2008)
82. Jabara, R., et al.: Evaluation of a novel fully biodegradable salicylate-based sirolimus-eluting stent: histologic analysis in pig coronary artery implants. Am. J. Cardiol. **102**, 187i (2008)
83. Erbel, R., et al.: Temporary scaffolding of coronary arteries with bioabsorbable magnesium stents: a prospective, non-randomized multicentre trial. Lancet **369**, 1869–1875 (2007)
84. Ormiston, J., et al.: First-in-human implantation of a fully bioabsorbable drug-eluting stent: the BVS poly-L-Lactic acid everolimus-eluting coronary stent. Catheter Cardiovasc. Interv. **69**, 128–131 (2007)
85. Kimura, T., et al.: Remodeling responses of human coronary arteries undergoing coronary angioplasty and atherectomy. Circulation **96**, 475–483 (1997)

86. Nobuyoshi, M., et al.: Restenosis after successful percutaneous transluminal coronary angioplasty: serial angiographic follow-up of 229 patients. J. Am. Coll. Cardiol. **12**, 616–623 (1988)
87. Serruys, P.W., et al.: Incidence of restenosis after successful coronary angioplasty: a time-related phenomenon. A quantitative angiographic study in 342 consecutive patients at 1, 2, 3, and 4 months. Circulation **77**, 361–371 (1988)
88. Ormiston, J.A., Serruys, P.W.S.: Bioabsorbable coronary stents. Circ. Cardiovasc. Intervent. **2**, 255–260 (2009)
89. Gogas, B.D., et al.: Novel drug-eluting stents for coronary revascularization. Trends Cardiovascul. Med. **24**, 305–313 (2014)
90. Brito, L.A., Chandrasekhar, S., Little, S.R., Amiji, M.M.: In vitro and in vivo studies of local arterial gene delivery and transfection using lipopolyplexes-embeded stents. J. Biomed. Mater. Res. **93A**, 325–336 (2009)
91. Yang, J., Zeng, Y., Zhang, C., et al.: The prevention of restenoss in vivo with a VEGF gene and paclitaxel co-eluting stent. Biomaterials **34**, 1635–1643 (2013)
92. Raina, T., Arnold, N., Moore, H., et al.: 152 Stem cell coated metallic coronary stents show accelerated strut coverage without excessive neointimal proliferation in a porcine model. Heart **100**, A88–A89 (2014)
93. Stone, G.W., Moses, J.W.: Ellis GS Safety and efficacy of sirolimus and paclitaxel-eluting coronary stents. N. Engl. J. Med. **356**, 998–1008 (2007)

Recent Advances in Regenerative Approaches to Intervertebral Disc Degeneration

Gilson Khang, Jeong Eun Song, Nirmalya Tripathy, Eun Young Kim and Dongwon Lee

Abstract Intervertebral disc (IVD) degeneration is one of the major neurodegenerative diseases throughout the world. Development of new and advanced treatment methods are highly warranted due to the limitations of conventional protocols. Very recently, regenerative medicine techniques based on biomaterials, stem cell and bioactive molecules have been demonstrated as the most powerful concept to regenerate IVD. In this chapter, we will review the recent progress in (1) biomaterial scaffolds for annulus fibrosus (AF), nucleus pulposus (NP) and cartilaginous end plates (CEP), (2) stem cells as cell sources, and finally introduce (3) our recent research studies focusing on the fabrication of natural/synthetic hybrid scaffolds for various purposes during the last 5 years.

1 Introduction

One of the major causes of disability is low back pain which affects millions of people worldwide. In USA, it is found that approximately 20.7% of patients suffered from low back pain in 1997 which further increased to 24.7% in 2005. The economic burden of low back problems in USA has been estimated to be $85.9 billion. As well, this trend is found to be similar phenomenon all over the world. Furthermore, studies have reported that the low back pain might be associated with intervertebral disc (IVD) degenerations [1].

IVD is the cornerstone of the joint complex comprising the spinal motion segment. IVD permits only a limited motion and flexibility, while maintaining segmental stability by absorbing and distributing external loads [2,3]. IVD is a complex structure comprised of an outer ring of collagen-rich fibrous annulus

G. Khang(✉) · J.E. Song · N. Tripathy · E.Y. Kim · D. Lee
Department of Polymer Nano Science and Technology, Department of BIN Fusion Technology, Chonbuk National University, Jeonju 561-756, Republic of Korea
e-mail: gskhang@jbnu.ac.kr

© Springer International Publishing Switzerland 2016

H. Jo et al. (eds.), *Biomedical Engineering: Frontier Research and Converging Technologies*, Biosystems & Biorobotics 9,
DOI: 10.1007/978-3-319-21813-7_18

fibrosus (AF), a gelatinous core nucleus pulposus (NP) and joined to the adjacent vertebrae by cartilaginous end plates (CEP). The structure of the normal IVD includes the NP, composed primarily of proteoglycans and collagen type II, with a capacity to absorb and distribute load, whereas the outer AF is well-organized layers consisting of collagen type I and collagen type II serving as stabilizers of motion segment [4,5].

The structure and functions of the IVD may be altered by processes including normal physiological aging, mechanical factors, i.e., trauma and repetitive stress, segmental instability of the spine, inflammatory and biochemical factors. Patients with degenerative disc disease (DDD) constitute majority of patients specifically with spinal origin back pain. The IVD undergoes extrinsic morphological changes during their lifetime. In fact, most of the adult population faces degenerative disc problem by the age of 50. The current treatment options range from NSAIDs to invasive procedures including discectomy and spinal fusion. Unfortunately, these treatments not only retain the original problem, which is the degeneration of IVD itself, but also increase the load of intact discs and joints located adjacently to the degenerated disc [6,7].

Spurred by the progresses in the tissue engineering and regenerative research, studies were now more focused to understand the causes of these diseases and regeneration of IVD [8,9]. Many cellular and molecular mechanisms have shown relation with the degenerative trends, strongly suggesting possibilities to regenerate from DDD based cell therapy, combined with seeding cells, bioactive substances, stimulation of cells and biocompatible materials [10,11]. To realize a successful therapy using regenerative medicinal technique for IVD, triad components are prerequisites [12-16], that is: (i) natural or artificial extracellular matrix (ECM) known as scaffolds which provides a favorable environment for cells adhesion, growth resulting in desired number of healthy cells with specific functions; (ii) specific/desired cells as harvested and dissociated primary cells, adult stem cell, embryonic stem cell (ESC) and induced pluripotent cell (iPS); (iii) biomolecules including growth factors, cytokines and phospholipids which furnishes the nourishment for cell proliferation and/or differentiation by up-regulation or down-regulation of specific proteins, receptors, etc.

In this chapter, we will discuss about the major contribution of the researchers towards the progress of IVD and DDD.

2 Biomaterials for Disc Regenerations

Scaffolds biomaterials play a critical role in IVD regeneration. The main function of scaffolds is to direct the growth of cells, seeded within the porous structure of the scaffold or the cells migrating from surrounding tissue. In general, the AF and NP cells are anchorage-dependent, which can lead to cell death in the absence of suitable adhesion substrate. The scaffold matrices can be successfully used to achieve desired cell delivery to the specific sites with high loading efficiency. Therefore, the scaffold must provide a suitable substrate for cell attachment, proliferation, differentiation and cell migration. The prerequisite physicochemical properties of scaffolds are (i) to support and delivery of cell at the target site, (ii) to induce, differentiate and conduct tissue growth, (iii) to achieve cellular adhesion to the substrate, (iv) to stimulate cellular response, (v) wound healing barrier, (vi) biocompatible and biodegradable, (vii) relatively easy processability and malleability into desired shapes, (viii) highly porous with large surface/volume, (ix) mechanical strength and dimensional stability, and so on [12–17]. Generally, 3-dimensional porous scaffolds can be fabricated from natural and synthetic polymers. Some of the recent works were summarized below.

Brona *et al* [18] have tested the relationship between alginate concentration, scaffold stiffness and viscoelastic behavior mimics that of the NP. The effect of variations in scaffold stiffness had been investigated on the expression of ECM molecules specific to NP. The sample discs were prepared of various concentrations of alginate (1~6%) by two different methods i.e. diffusion and in situ gelation. They found that the alginate can mimic the viscoelastic properties of the NP and capable of preserving the biosynthetic phenotype of NP cells.

Zhuang *et al* [19] have studied tissue engineered composite IVD consisting of demineralized bone matrix gelatin (DBMG) and collagenII/hyaluronate/chondroitin-6-sulfate (CII/HyACS) scaffolds seeded with AF and NP cells, respectively, followed by its evaluation using gross, histological morphology and biochemical properties. They showed the feasibility of creating a tissue engineered composite IVD with similar morphological and biochemical properties to the native tissue.

Strange *et al* [20] investigated the time-dependent mechanical properties of gelatin, agar hydrogels and gelatin-agar composites. Several gel formulations displayed comparable equilibrium elastic behavior to the NP, but with much greater permeability values than those of the native tissue. Gorapalli *et al* [21] have developed a novel deacetylated poly-N-acetyl glucosamine hydrogel and characterized its biochemical effect as well as biomechanical properties. This work showed that sulphated deacetylated glycosaminoglycan derivative hydrogel possesses promising characteristics, motivating for further evaluation as a potential therapy for NP degeneration.

Chen *et al* [22] have developed an injectable oxidized hyaluronic acid (high molecular weight, 1900 kDa)-gelatin-adipic acid dihydrazide (oxi-HAG–ADH) hydrogel and analyzed their viscoelastic properties. An increased NP cell expression as COL2A1, AGN, SOX-9, and HIF-1A was observed for the designed hydrogels. Further, Li *et al* [23] have tried to synthesize using fibrinogen (FBG) and hyaluronic acid (235kDa) (FBG/HA w/w ratio of 17:1) which have shown superior gel stability and mechanical properties. The results showed pronounced increment in glycosaminoglycan synthesis compared with a FBG/HA mixture-based hydrogels or fibrin gels. Also, Jeong *et al* [24] have tried to make a HA derivatives as eight different hydrogels from thiolated HA (HA-SH) and PEG vinylsulfone (PEG-VS) macromers. Results showed that NP and AF cell numbers were highest upon the HA-PEG hydrogels formed from the lower-molecular-weight HA, with evidence of higher sulfated glycosaminoglycan production upon lower-HA-molecular-weight composite gels. Fransisco *et al* [25] have developed a photo-crosslinkable poly(ethylene glycol)-laminin 111 (PEG-LM111) hydrogel and were revealed to promote cell clustering and increased levels of sGAG production as compared to stiffer LM111-presenting and only PEG gels.

Feng *et al* [26] reported the regeneration of NP tissue using a biodegradable nanofibrous scaffold. The NP cells on the nanofibrous scaffolds proliferated faster than those on control solid-walled scaffolds in vitro. The tissue-engineered NP survived, produced functional ECM, remained in place, and maintained the disc height, which is similar to the native NP tissue. Kim *et al* [27] developed porous PLGA scaffolds with various pore sizes for NP cell cultures and examined the effects of pore size on cell ingrowth and ECM synthesis. Also, they have investigated a novel, biodegradable, synthetic/natural hybrid scaffold composed of PLGA and small intestinal submucosa (SIS) (PLGA-SIS) with suitable pore size for NP regeneration. This study showed the criticality of scaffolds with proper pore size for NP regeneration. As well, it emphasized the usability of PLGA-SIS scaffolds with suitable pores as substrates for tissue-engineered bio-discs.

Pereira *et al* [28] have been developed gellan gum microparticles (GGMPs)-based hydrogels for acellular and cellular tissue-engineered biodiscs in order to reinforcement of the hydrogel matrix with biocompatible and biodegradable GGMPs. This study showed that GG hydrogel matrices reinforced with cell-loaded MPs could be an efficient candidate for NP regeneration. Nagae *et al* [29] have evaluated the regenerative effects of platelet-rich plasma (PRP) for the degenerated IVD *in vivo*. This work showed that the combined administration of PRP and gelatin hydrogel microspheres into the IVD could be a promising therapeutic modality for IVD degeneration. Le Visage *et al* [30] have studied the possibility of porcine SIS to support the in vitro proliferation of human disc cells and the synthesis of ECM. This work showed that toluidine blue staining detected large areas of proteoglycans as well as positive gene expression for collagens I, II, and X, aggrecan, and Sox-9 confirmed deposition of new ECM components.

As previously emphasized, scaffold materials for biodisc tissue engineering must contain the active site of cellular and molecular induction and adhesion with an appropriate mechanical strength, and must allow cell migration and proliferation through its porosity. It should also maintain strength, flexibility, biostability and biocompatibility to mimic a more natural, 3-D environment. With the advances of more sophisticate CAD/CAM techniques with 3-D printer, there is a great possibility of mimicking human body's structure. Also, the combination of the cells and redesigned bioactive scaffolds has attempted to expand to the tissue level of hierarchy. From this point of view, novel, advanced scaffold biomaterials with various fabrication techniques and translational clinical protocol should be developed especially for the biodisc regeneration.

3 Cell Sources

Right selection of the cells source is highly important for the regeneration for IVD and DDD. Generally, three types of stem cells have been widely considered for organ regeneration i.e. adult stem cells, ESC and iPS including direct reprogramed cells. Compared to ESC and iPS (leading to easy teratoma formation), adult stem cells were extensively employed in pilot researches as promising cell based treatment owing to direct harvest from patients, rapid cell proliferation, ease of manipulation for replacement of nonfunctional genes, homing ability and to interact with surrounding tissues. Moreover, some lineage-specific adult stem cells were reported to possess greater plasticity and thus regarded as multipotent – ability to differentiate into more than one cell type regardless of the parent cell [15]. Depending on the origin type, adult stem cells can be further classified as mesenchymal stem cells (MSCs) and hemopoetic stem cells (HSCs). MSCs were derived from a variety of tissues such as bone marrow stem cells (BMSCc), limbal stem cells, blood derived (BSC), adipose derived (ADSC), hepatic stem cells, dermal stem cells, adipose derived (ADSC) whereas HSCs are found in cord/peripheral blood. MSCs encompassing multipotent cells are sufficiently capable of carrying out various cell types.

3.1 BMSCs-Mediated Therapies

Among all, BMSCc have received intense attention for regenerative therapeutics due to its efficient differentiation capability, easy optimization of isolation and ease of marker profile [2–4,9]. BMSCs are mostly pluripotent in nature, thereby behave as supportive cells as well as can generate all types of cells in our body except cells of embryonic membrane. When implanted in in vivo, BMSCs differentiates into multiple tissue (for example, tissues of heart, liver, blood vessel, etc.) in order to replenish the body with healthy cells. This restoration was realized through a series of dynamic processes such as signaling, homing, incorporation, inflammation, proliferation, differentiation and so on, where differentiation plays a key role for development of regenerative medicine [15]. In this section, we review the case of application of IVD and DDD treatment.

Chen *et al* [31] have studied the possibility of MSC derived from eGFP-transgenic porcine (MSC-GFP), platelet-rich plasma (PRP) and MSC-GFP/PRP combined treatment, and confirmed in *in vivo* animal model. This work has been successfully demonstrated chondrogenesis of MSC for the IVD. Cai *et al* [32] have been investigated the curative effects of transplantation of BMSCs on IVD regeneration and tested the feasibility of the quantitative T2 mapping method for evaluating repair of the NP after implantation of BMSCs. They showed good regeneration of degenerative discs by the transplantation of BMSCs and non-invasive T2 mapping was very useful with good sensitivity. Kumar *et al* [33] have been demonstrated the potential of a novel, photo-curable, injectable, synthetic polymer hydrogel (pHEMA-co-APMA grafted with polyamidoamine (PAA)) to encapsulate and differentiate human MSCs towards a NP phenotype under hypoxic conditions which could be used to restore NP tissue function and mechanical properties.

Another possibility is the application of mesenchymal precursor cells (MPCs) loaded in hydrogel with soluble factor as pentosan polysulphate [34]. Frith *et al* showed evidence that this system has significant potential as both a delivery vehicle for MPCs and as a matrix for fibrocartilage tissue engineering applications.

Wang *et al* [35] have been compared the regenerative potentials of the AF, NP and CEP-derived stem cells with that of the classic BMSCs in a rabbit disc degeneration model. They found cartilage endplate-derived stem cells (CESCs) with superior capacity compared with the annulus fibrosus-derived stem cells (AFSCs), nucleus pulposus-derived stem cells (NPSCs), and BM-MSCs ($p<0.05$). This information provides good direction for the application of disc-derived stem cells.

Huang *et al* [36] summarized the current status of cell-based treatments for IVD degeneration and identify the additional research needed before ADSCs can be evaluated in a one-step procedure for regenerative treatment of the IVD. Liang *et al* [37] investigated the viability, proliferation and the expression of main matrix proteins of ADSCs in the chemical microenvironment of IVD under normal and degeneration conditions as low glucose, acidity, high osmolarity, and combined conditions. They found that IVD-like low glucose is a positive factor but IVD-like high osmolarity and low pH are deleterious factors affecting the survival and biological behaviors of ADSCs.

Potier *et al* [38] have been suggested that notochordal cells (NCs) might be involved in IVD homeostasis. Using alginate beads, bovine NP was exposed for 28 days to porcine NC conditioned medium (NCCM); direct co-culture with porcine NCs or bovine BMSCs; or the combination of BMSCs and NCCM. This study showed that secretomes by NCs constitute a promising alternative to the proposed "standard" injection of BMSCs for disc repair.

Shoukry *et al* [39] have reviewed synovium-derived stem cells (SDSCs) as a new candidate for the potential application of IVD regeneration. With the evidence for the usage of SDSCs for NP regeneration, they have studied the adaptive mechanisms of NP cells in order to gain some insights about the factors important for constructing a tissue-specific in vitro niche. Also, they suggested the key information on producing sufficient numbers of high-quality NP-specific population that is able to regenerate deteriorated NP *in vivo*. Liu *et al* [40] have been suggested a new method to use porcine NP matrix to direct notochordal differentiation of human iPSCs (hiPSCs). The results showed that hiPSCs successfully differentiated into NC-like cells under the influence of devitalized porcine NP matrix.

Co-culture systems of NP, AF and CEP cells with adult stem cell system have been also tried. Tao *et al* [41] have tested the possibility of NP mesenchymal stem cells (NPMSCs) for IVD regeneration. This study carried to investigate the viability, proliferation and protein biosynthesis of NP cells, NPMSCs and co-cultured NPMSCs-NPCs under IVD-like high osmolarity conditions. This suggests that coculture of NPMSCs-NPCs has potential application for IVD regeneration. Feng *et al* [42] have compared transplanted MSCs with NPCs in a IVD model (rabbits). Chen *et al* [43] have tested a coculture system of SDSCs and NPCs in a serum-free pellet system. Gaetani *et al* [44] have developed the co-culture system of NP cells and adipose tissue non-adipocyte fraction cells (nAFs). Results indicated that the above co-culture systems capable of improving the quality of the *in vitro* reconstructed tissue in term of ECM production and 3D cell organization.

4 Hybrid Natural/Synthetic Scaffold for the IVD Regenerations

PLGA is a member of a group of poly(α-hydroxy acid) [12–17] that is among the few synthetic polymers approved for human clinical use by FDA. Consequently, it has been extensively used and tested for scaffold materials for the regeneration of IVD and DDD as a bioerodible material due to good biocompatibility, relatively good mechanical property, lower toxicity and controllable biodegradability. It has been clinically utilized for three decades as sutures, bone plates, screws and drug delivery vehicles and its safety has been proved in many medical applications. PLGA degrades by nonspecific hydrolytic scission of their ester bonds into their original monomer, lactic acid and glycolic acid. During these processes, there is very minimal systemic toxicity, however, in some cases, their acidic degradation products can decrease the pH in the surrounding tissue that result in local inflammatory reaction and potentially poor tissue development as shown in Figure 1. Also, its poor mechanical strength, small pore size and hydrophobic surface properties for cell seeding have limited its usage.

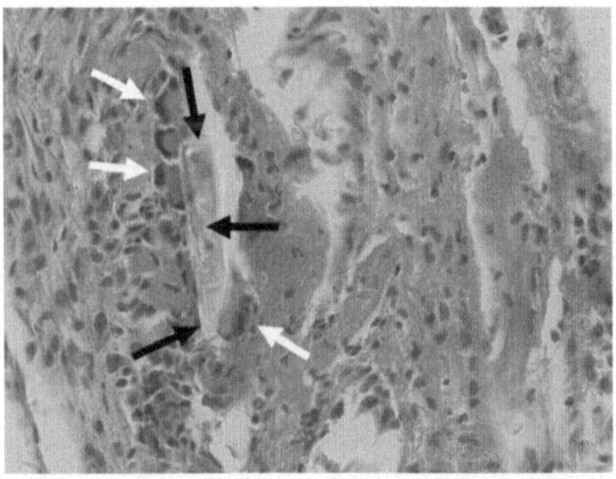

Fig. 1 Foreign body granuloma. Small PLGA debris (black arrows) broken off from the PLGA film, surrounded by macrophages and multinucleated giant cells (white arrows). These induced macrophages and multinucleated giant cells were remaining over 2 months.

In our laboratory, the natural/synthetic hybrid scaffolds have been investigated during the last 15 years such as SIS, DBP, DBP gel, fibrin, keratin, hyaluronic acid, collagen gel, silk and a 2-methacryloyloxyethyl phosphorylcholine (MPC) polymer (PMEH) with PLGA to reduce cellular inflammatory response and improve biocompatibility [17]. Also, the synthesis of hydrogel and modification of natural gel, application of BMSC and differentiation of BMSC to disc cell have been developed to study the their feasibility for the IVD regeneration. In this section, we have reviewed the research progress over last 15 years regarding the regeneration of IVD.

Ko *et al* [45,46] have investigated the feasibility of methoxy polyethyleneglycol (mPEG) and PLGA or poly(ε-caprolactone) (PCL) diblock copolymers for the application of IVD regenerations. Results showed that optimized copolymer composition and molecular weight can provide an appropriate biomechanical properties and drug delivery carrier for IVD. Bae *et al* [47] have assessed purified and non-purified alginate microcapsules for AF regenerationcell viability and proliferation via MTT assay, RT-PCR. Further, some of the results were shown in Figure 2 (H&E, MTS, and Safranin-O staining). AF cells in purified alginate microcapsules displayed higher cell survival rate and more activated phenotype expression than those in non-purified alginate microcapsules (Figure 3). The results suggest that purified alginate microcapsule might be useful for AF tissue regeneration

Choi *et al* [48] have evaluated 3-dimensional DBP sponge and collagen sponge for proliferation and phenotype maintenance of AF cells. DBP sponges were prepared by freeze-drying method after addition 2 wt% DBP solution and crosslinked with 1-ethyl-(3-3-dimethyl aminopropyl) carbodiimide hydrochloride (EDC) solution with 50 mM concentration for 24 h and lyophilized. In results, DBP sponges showed were higher cell viability, specific mRNA and uniform porosity, indicating that DBP sponges can be a better platform for intervertebral disc regeneration than collagen. Ha *et al* [49] have prepared PLGA, PLGA/SIS(20%), PLGA/DBP(20%) and PLGA/SIS(10%)/DBP(10%) scaffold by solvent casting/salt leaching method.

Rabbit AF cells were seeded into experimental groups. Biomechanical property, hydroxyproline production and DNA quantity of AF cells on each scaffold were measured. Cell-scaffold composites were implanted subcutaneously into athymic mice and then analyzed H&E, Safranin-O and type I collagen staining. This study showed the best disc cell proliferation capacity of PLGA/DBP(20%)

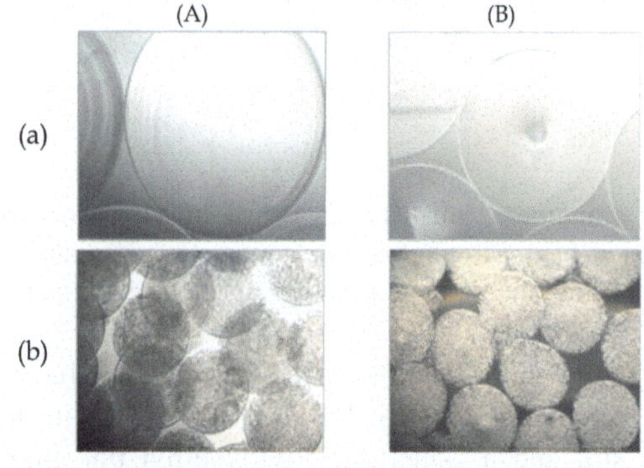

Fig. 2 (A) non-purified and (B) purified alginate microcapsules, (a) alginate microcapsule without cell (×100 magnification) and (b) alginate microcapsule with cell (×40 magnification) to confirm by inverted microscope [47].

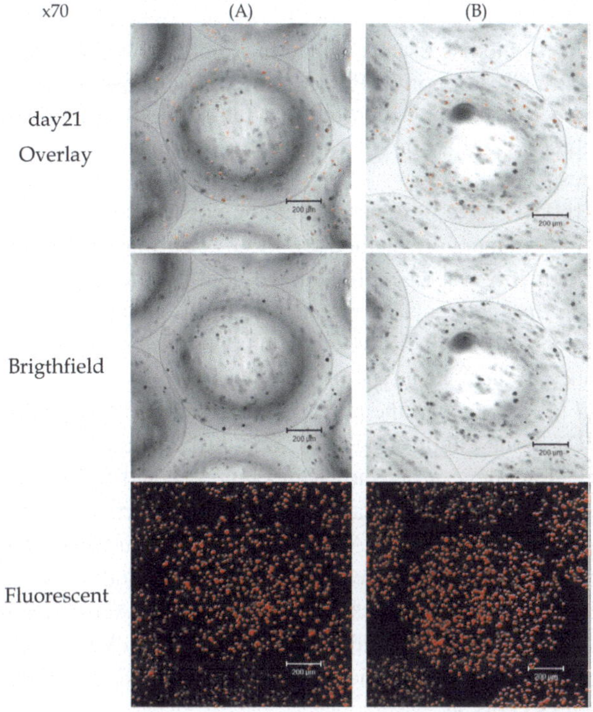

Fig. 3 Microcapsules confirmation by confocal microscope, the AF were labeled with PKH26, (A) non-purified and (B) purified alginate microcapsule including AF [47]

Ko *et al* [50–56] have designed DBP/PLGA scaffold for tissue engineered disc regeneration. They fabricated dual-structured scaffold with inner cylinder and outer doughnut shaped similar to disc tissue and further cultured NP and AF cells from rabbit intervertebral disc on them. The DBP/PLGA scaffold was characterized by porosity, wettability, and water uptake ability, MTT test followed by in vivo evaluations. Formation of disc-like tissue was found to be better DBP/PLGA hybrid scaffold than control ones. Specially, it showed that scaffold impregnated 20 and 40% DBP can affect proliferation of disc cell and formation of disc-like tissue.

So *et al* [57] have investigated the usability of PLGA microspheres (180~425 μm) prepared using oil-in-water (O/W) solvent evaporation method, as microcarriers for IVD cells. Human disc cells were used in a prospective study of attachment, proliferation, and gene expression within various sizes of PLGA microspheres. This study showed the human disc cells cultured PLGA microsphere injectable system might be very good for disc regeneration. Lee *et al* [58] designed the align-coated PLGA microspheres to control initial burst release and to improve the lag time by PLGA bulk degradation. This study showed that bovine serum albumin (BSA)-loaded alginate/PLGA microspheres can be used as an injectable system for biodisc tissue engineering application. Also, So *et al* [59] have assessed the pore sizes of porous PLGA scaffolds with five different pore sizes for AF tissue regeneration. They reported that the scaffold with 180~250 μm pores displayed high cell viability in vitro and produced higher ECM than scaffold with other pore sizes *in vivo*.

Kim *et al* [60,61] have designed novel composite scaffolds of PLGA/SIS prepared by simple immersion method of PLGA scaffolds in SIS solution under vacuum and then SIS was cross-linked. To confirm the effects of adhesion and proliferation on disc cells, cells were seeded on PLGA and SIS/PLGA according to the SIS contents. SIS with 10 wt% contents positively affected the proliferation and maintenance of disc phenotype in SIS/PLGA.

Saban *et al* [62] evaluated the potential of fibrin/PLGA scaffold for intervertebral disc (IVD) tissue engineering using annulus AF and NP cells in relation to potential clinical application as shown in Figures 4 and 5. PLGA scaffolds were soaked in cells–fibrin suspension and polymerized by dropping thrombin–sodium chloride ($CaCl_2$) solution. A PLGA–cell complex without fibrin was used as control. In conclusion, fibrin promotes cell proliferation, stable in vitro tissue morphology, superior cartilaginous tissue formation and sGAG production of AF and NP cells cultured in PLGA scaffold. The 3D porous PLGA scaffold–cell complexes using fibrin can provide an efficient vehicle for cell delivery to regenerate tissue-engineered IVD tissue.

Choi *et al* [63] have carried out studies on 3-dimensional HA loaded PLGA scaffold on proliferation and phenotype maintenance of AF cells. HA/PLGA scaffolds were prepared by emulsion freeze-drying method and then AF cells were seeded. AF cell viability in HA/PLGA scaffolds were higher than PLGA scaffolds in vitro. Song *et al* [64] have performed in vivo application of HA/PLGA scaffolds. AF strongly expressed their specific mRNA and produced well sGAG in HA/PLGA scaffold in *in vitro* and *in vivo*. These results indicated that HA/PLGA scaffold is useful for intervertebral disc regeneration.

Lee et al [65] have been developed laminated cylindrical scaffolds composed of PLGA and keratin, and investigated their potential for tissue engineering and disc regeneration. The scaffold was designed to have two parts, i.e. inner cylinder and outer disk, to mimic a natural disk. The outer disk was composed of PLGA and the inner cylinder was prepared using PLGA film or PLGA/keratin hybrid film. Scaffolds containing PLGA/keratin films showed a significantly higher AF cell proliferation and expression of collagen I and II than the counterpart with PLGA films. Keratin containing scaffolds also exhibited an excellent mechanical strength, demonstrating that keratin influences the proliferation of annulus fibrous cells. The results provide valuable information on PLGA/keratin films for tissue engineered disc regeneration.

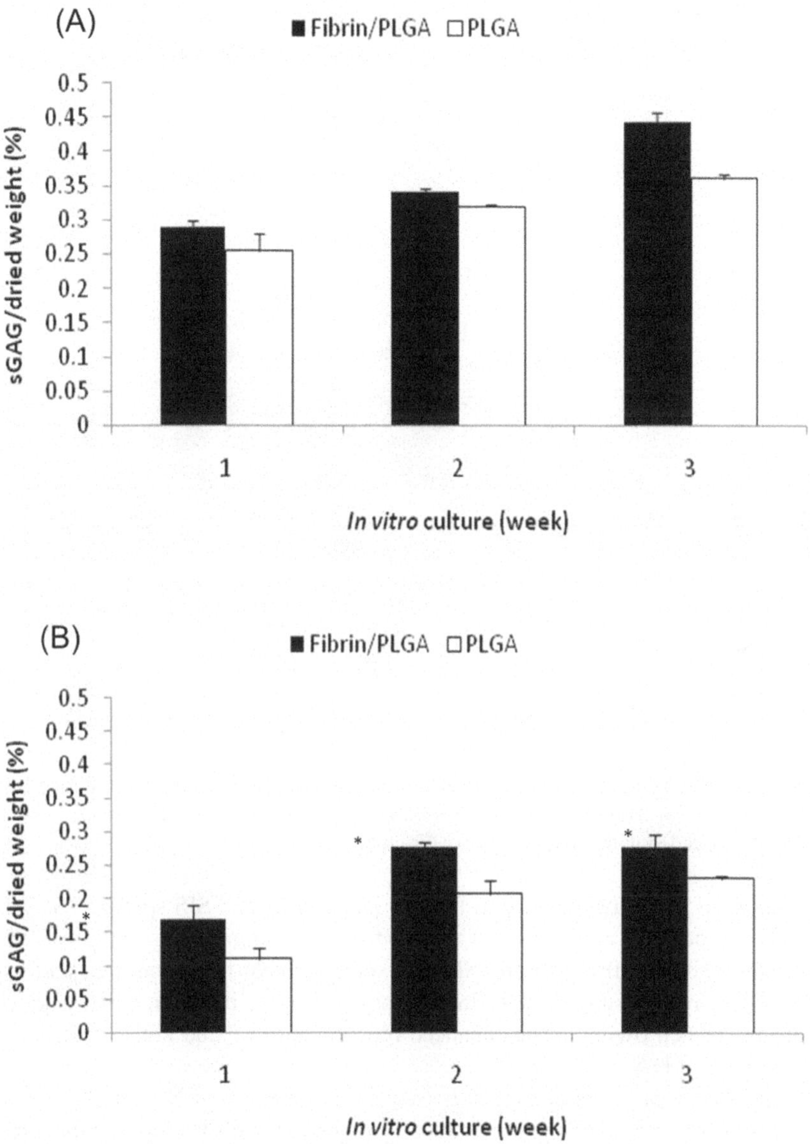

Fig. 4 Total sGAG production was normalized by the dried weight of each sample and represented as relative sGAG content in percentage (%). The sGAG production was higher in fibrin/PLGA seeded with AF cells (A) and NP cells (B) than in the PLGA group.

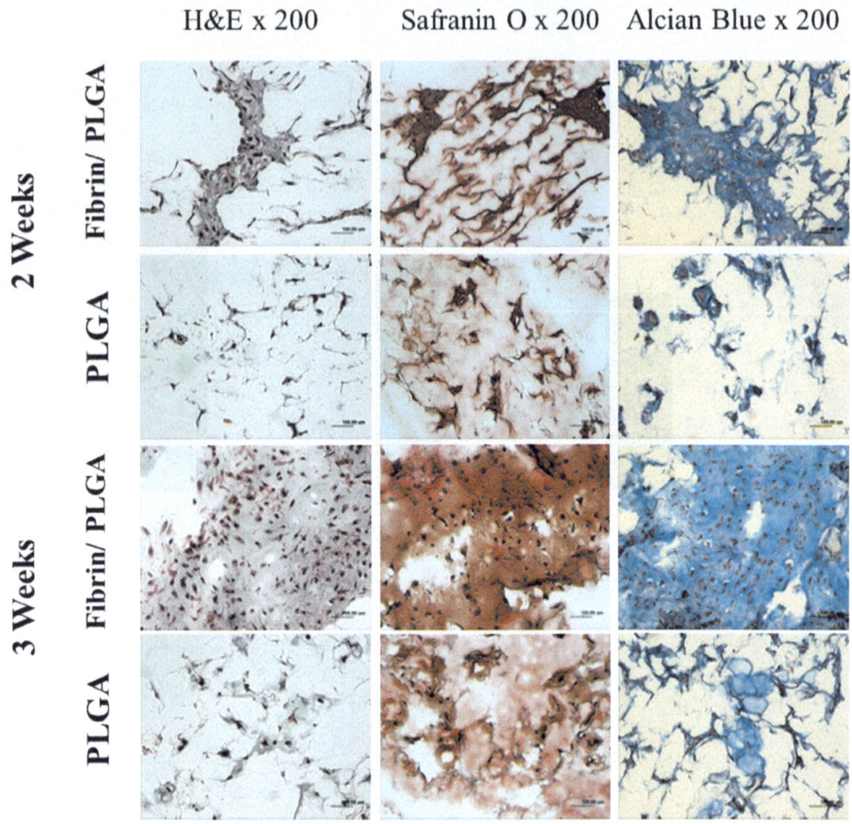

Fig. 5 Safranin O indicating presence of the proteoglycan-rich matrix corroborated with positive and Alcian Blue staining confirming GAG accumulation of the fibrin/PLGA and PLGA group seeded with AF cells at 2 weeks culture.

Hong *et al* [66] have developed the SIS and HA loaded SIS sponges (SIS/HA) to study the possible application of the biodisc tissue engineering using AF. SEM observation showed that SIS and SIS/HA sponges have interconnected and open pores. These results suggest that SIS/HA sponge may be useful to tissue engineering using AF cell owing to the enhanced biocompatibility and higher water retention capacity of HA.

Ha *et al* [67] have characterized the sequential passages of NP and AF. NP and AF cells were separated by enzymatically digestion with 0.25wt% collagenase, respectively. Morphological changes were observed by phase contrast microscope and cell proliferation was counted by hemacytometer. To analyze the biosynthesis of glycosaminoglycan and gene expression of Type I and Type II collagen, safranin-O staining and RT-PCR have been carried out, respectively. Proliferation of NP and AF cells increased up to passage 3 and 6. Morphology of NP cells was maintained to passage 5 and AF cells was maintained to passage 4. In the result of Safranin-O staining, NP and AF cells were stained positively. Type II collagen gene of NP cells was not expressed after passage 6 and expression of Type I collagen gene of AF cells was maintained up to passage 9. It can be expected that these results provide the important information for the application of tissue engineered biodisc.

Kang *et al* [68] have designed by monolayer direct co-culture with varying NP and BMSC contents in order to optimize the ratio of NP and BMSC in mixed culture. Collagen type I and II were evaluated RT-PCR and immunofluorescence staining. This study indicated that the optimal ratio of 25:75 presented on maintaining the function of cells and inducing cell-proliferation can be applied to clinical application for tissue engineering.

5 Conclusions and Future Prospects

Tissue engineering including regenerative medicine based on the biomaterials and stem cell for the IVD regeneration shows tremendous potential as a revolutionary research push. As previously emphasized, scaffold materials for the biodisc construction must contain the site of cellular and molecular induction and adhesion and must allow for the migration and proliferation of cell through porosity. It should also maintain strength, flexibility, biostability and biocompatibility to mimic natural, three dimensional environments. From this point of view, the control over precise biochemical signal must be needed by the combination of scaffold matrix and bioactive molecules including genes, peptide molecules and cytokines. Moreover, the combination of the cells and redesigned bioactive scaffolds has attempted to expand to a tissue level of hierarchy. Also, the understanding of stem cell must be needed as homing, differentiation, dedifferentiation, trans-differentiation as well as direct conversion toward AF and NP cells.

Moreover, future work in this area based on the current understanding should be performed to establish the causes of disc degeneration and mechanisms leading to disc degeneration. AsImproved understanding of these critical parameters will provide a platform for the emergence of more advanced regenerative treatments for this debilitating condition. In order to achieve this goal, the novel hybrid scaffold biomaterials, the novel scaffolds fabrication methods and the novel characterization methods in the respect of "Scaffold Biomaterials" must be developed.

Acknowledgments This work was supported by grants from KMOWH and KMOST.

References

1. Sakai, D., Anderson, G.B.J.: Stem cell therapy for intervertebral disc regeneration: obstacles and solution. Nat. Rev. Rheumatol. (2015). doi:10.1038/nrrheum.2015.13
2. Maidhof, R., Alipui, D.O., Rafiuddin, A., Levine, M., Grande, D.A., Chahine, N.O.: Emerging trends in biological therapy for intervertebral disc degeneration. Discov. Med. **14**, 401–411 (2012)

3. Kalson, N.S., Richardson, S., Hoyland, J.A.: Strategies for regeneration of the intervertebral disc. Regen. Medicine **3**, 717–729 (2008)
4. Wei, A., Shen, B., Williams, L., Diwan, A.: Mesenchymal stem cells: potential application in intervertebral disc regeneration. Transl. Pediatr. **3**, 71–90 (2014)
5. Huang, Y.C., Urban, J.P., Luk, K.D.: Intervertebral disc regeneration: do nutrients lead the way? Nat. Rev. Rheumatol. **10**, 561–566 (2014)
6. Whatley, B.R., Wen, X.: Intervertebral disc (IVD): Structure, degeneration, repair and regeneration. Mater. Sci. Eng. **C32**, 61–77 (2012)
7. Iatridis, J.C., Nicoll, S.B., Michalek, A.J., Walter, B.A., Gupta, M.S.: Role of biomechanics in intervertebral disc degeneration and regenerative therapies: what needs repairing in the disc and what are promising biomaterials for its repair? The Spine J. **13**, 243–262 (2013)
8. Nerurkar, N.L., Elliott, D.M., Mauck, R.L.: Mechanical design criteria for intervertebral disc tissue engineering. J Biomechan. **43**, 1017–1030 (2010)
9. Kepler, C.K., Ponnappan, R.K., Tannoury, C.A., Risbud, M.V., Anderson, D.G.: The molecular basis of intervertebral disc degeneration. The Spine J. **13**, 318–330 (2013)
10. Broek, P.R., Huyghe, J.M., Wilson, W., Ito, K.: Design of next generation total disk replacements. J. Biomechan. **45**, 134–140 (2012)
11. Sharifi, S., Bulstra, S.K., Grijpma, D.W., Kuijer, R.: Treatment of the degenerated intervertebral disc; closure, repair and regeneration of the annulus fibrosus. J. Tissue Eng. Regen. Med. (2014) doi:10.1002/term.1866
12. Khang, G., Kim, M.S., Lee, H.B.: Manuals in Biomedical Research, vol. 4. A Manual for Biomaterials/Scaffold Fabrication Technology. World Scientific, Singapore (2007)
13. Khang, G., Kim, S.H., Kim, M.S., Lee, H.B.: Hybrid, composite, and complex biomaterials for scaffolds. In: Lanza, R., Atala, A., Nerem, R., Thomson, W. (eds) Principles of Regenerative Medicine, pp 636–655. Academic Press. New York (2008)
14. Khang, G.: chap 1, introduction. In: Khang, G. (ed.) Handbook of Intelligent Scaffold for Tissue Engineering & Regenerative Medicine, pp 3–40. Pan Stanford Pub., New York (2012)
15. Khang, G., Shimizu, T:. Manuals for Differentiation of Bone Marrow-Derived Stem Cell to Specific Cell Types. World Scientific, Singapore (2014)
16. Khang, G., Kim, S.H., Rhee, J.M., Sha'Ban, M., Idrus, R.B.H.: Chap. synthetic/natural hybrid scaffold for cartilage and DIsc regenerations. In: Tateishi, T. (ed.) Biomaterials in Asia in Commemoration of the 1st Asian Biomaterials Congress World Scientific, Singapore (2008)
17. Khang, G.: Importance of inflammation reaction of scaffold for the application of regenerative medicine. Inflammation Regeneration **32**(5), 178–196 (2012)
18. Brona, J.L., Vonkb, L.A., Smita, T.H., Koenderinkd, G.H.: Engineering alginate for intervertebral disc repair. J. Mechan. Behavior. Biomed. Mater. **4**, 1196–1205 (2011)
19. Zhuang, Y., Huang, B., Li, C.Q., Liu, L.T., Pan, Y., Zheng, W.J., Luo, G., Zhou, Y.: Construction of tissue-engineered composite intervertebral disc and preliminary morphological and biochemical evaluation. Biochem. Biophys. Res. Comm. **407**, 327–332 (2011)
20. Strange, D.G., Oyen, M.L.: Composite hydrogels for nucleus pulposus tissue engineering. J. Mechan. Behavior Biomed. Mater. **11**, 16–26 (2012)
21. Gorapalli, D., Seth, A., Vournakis, J., Whyne, C., Akens, M., Zhang, A., Demcheva, M., Qamirani, E., Yee, A.: Evaluation of a novel poly N-acetyl glucosamine (pGlcNAc) hydrogel for treatment of the degenerating intervertebral disc. Life Sci. **91**, 1328–1335 (2012)

22. Chen, Y.C., Su, W.Y., Yang, S.H., Gefen, A., Lin, F.H.: In situ forming hydrogels composed of oxidized high molecular weight hyaluronic acid and gelatin for nucleus pulposus regeneration. Acta Biomaterialia **9**, 5181–5193 (2013)
23. Li, Z., Kaplan, K.M., Wertzel, A., Peroglio, M., Amit, B., Alini, M., Grad, S., Yayon, A.: Biomimetic fibrin-hyaluronan hydrogels for nucleus pulposus regeneration. Regen. Med. **9**(3), 309–326 (2014)
24. Jeong, C.G., Francisco, A.T., Niu, Z., Mancino, R.L., Craig, S.L., Setton, L.A.: Screening of hyaluronic acid-poly(ethylene glycol) composite hydrogels to support intervertebral disc cell biosynthesis using artificial neural network analysis. Acta Biomater. **10**(8), 3421–3430 (2014)
25. Francisco, A.T., Hwang, P.Y., Jeong, C.G., Jing, L., Chen, J., Setton, L.A.: Photocrosslinkable laminin-functionalized polyethylene glycol hydrogel for intervertebral disc regeneration. Acta Biomater. **10**(3), 1102–1111 (2014)
26. Feng, G., Zhang, Z., Jin, X., Hu, J., Gupte, M.J., Holzwarth, J.M., Ma, P.X.: Regenerating nucleus pulposus of the intervertebral disc using biodegradable nanofibrous polymer scaffolds. Tissue Eng. Part A **18**(21–22), 2231–2238 (2012)
27. Kim, S.H., Song, J.E., Lee, D., Khang, G.: Development of poly(lactide-co-glycolide) scaffold-impregnated small intestinal submucosa with pores that stimulate extracellular matrix production in disc regeneration. J. Tissue Eng. Regen. Med. **8**(4), 279–290 (2014)
28. Pereira, D.R., Silva-Correia, J., Caridade, S.G., Oliveira, J.T., Sousa, R.A., Salgado, A.J., Oliveira, J.M., Mano, J.F., Sousa, N., Reis, R.L.: Development of gellan gum-based microparticles/hydrogel matrices for application in the intervertebral disc regeneration. Tissue Eng. Part C Methods **17**(10), 961–972 (2011)
29. Nagae, M., Ikeda, T., Mikami, Y., Hase, H., Ozawa, H., Matsuda, K., Sakamoto, H., Tabata, Y., Kawata, M., Kubo, T.: Intervertebral disc regeneration using platelet-rich plasma and biodegradable gelatin hydrogel microspheres. Tissue Eng. **13**(1), 147–158 (2007)
30. Le Visage, C., Yang, S.H., Kadakia, L., Sieber, A.N., Kostuik, J.P., Leong, K.W.: Small intestinal submucosa as a potential bioscaffold for intervertebral disc regeneration. Spine(Phila Pa 1976) **31**(21), 2423–2430 (2006)
31. Chen, W.H., Liu, H.Y., Lo, W.C., Wu, S.C., Chi, C.H., Chang, H.Y., Hsiao, S.H., Wu, C.H., Chiu, W.T., Chen, B.J., Deng, W.P.: Intervertebral disc regeneration in an ex vivo culture system using mesenchymal stem cells and platelet-rich plasma. Biomaterials **30**(29), 5523–5533 (2009)
32. Cai, F., Wu, X.T., Xie, X.H., Wang, F., Hong, X., Zhuang, S.Y., Zhu, L., Rui, Y.F., Shi, R.: Evaluation of intervertebral disc regeneration with implantation of bone marrow mesenchymal stem cells (BMSCs) using quantitative T2 mapping: a study in rabbits. Int. Orthop., August 13 2014 (Epub ahead of print)
33. Kumar, D., Gerges, I., Tamplenizza, M., Lenardi, C., Forsyth, N.R., Liu, Y.: Three-dimensional hypoxic culture of human mesenchymal stem cells encapsulated in a photocurable, biodegradable polymer hydrogel: a potential injectable cellular product for nucleus pulposus regeneration. Acta Biomater. **10**(8), 3463–3474 (2014)
34. Frith, J.E., Cameron, A.R., Menzies, D.J., Ghosh, P., Whitehead, D.L., Gronthos, S., Zannettino, A.C., Cooper-White, J.J.: An injectable hydrogel incorporating mesenchymal precursor cells and pentosan polysulphate for intervertebral disc regeneration. Biomaterials **34**(37), 9430–9440 (2013)

35. Wang, H., Zhou, Y., Huang, B., Liu, L.T., Liu, M.H., Wang, J., Li, C.Q., Zhang, Z.F., Chu, T.W., Xiong, C.J.: Utilization of stem cells in alginate for nucleus pulposus tissue engineering. Tissue Eng. Part A. **20**(5–6), 908–920 (2014)
36. Huang, Y.C., Leung, V.Y., Lu, W.W., Luk, K.D.: Adipose stem cells for intervertebral disc regeneration: current status and concepts for the future. J. Cell. Mol. Med. **12**(6A), 2205–2216 (2008)
37. Liang, C., Li, H., Tao, Y., Zhou, X., Li, F., Chen, G., Chen, Q.: Responses of human adipose-derived mesenchymal stem cells to chemical microenvironment of the intervertebral disc. J. Translational. Med. **10**, 49–61 (2012)
38. Potier, E., de Vries, S., van Doeselaar, M., Ito, K.: Potential application of notochordal cells for intervertebral disc regeneration: an in vitro assessment. Eur. Cell. Mater. **25**, 68–80 (2014)
39. Shoukry, M., Li, J., Pei, M.: Reconstruction of an in vitro niche for the transition from intervertebral disc development to nucleus pulposus regeneration. Stem Cells Dev. **22**(8), 1162–1176 (2013)
40. Liu, Y., Fu, S., Rahaman, M.N., Mao, J.J., Bal, B.S.: Native nucleus pulposus tissue matrix promotes notochordal differentiation of human induced pluripotent stem cells with potential for treating intervertebral disc degeneration. J. Biomed. Mater. Res. A (2014). doi:10.1002/jbm.a.35243
41. Tao, Y.Q., Liang, C.Z., Li, H., Zhang, Y.J., Li, F.C., Chen, G., Chen, Q.X.: Potential of co-culture of nucleus pulposus mesenchymal stem cells and nucleus pulposus cells in hyperosmotic microenvironment for intervertebral disc regeneration. Cell Biol. Int. **37**(8), 826–834 (2013)
42. McCann, M.R., Bacher, C.A., Séguin, C.A.: Exploiting notochord cells for stem cell-based regeneration of the intervertebral disc. J. Cell Commun. Signal. **18**, 39–43 (2011)
43. Feng, G., Zhao, X., Liu, H., Zhang, H., Chen, X., Shi, R., Liu, X., Zhao, X., Zhang, W., Wang, B.: Transplantation of mesenchymal stem cells and nucleus pulposus cells in a degenerative disc model in rabbits: a comparison of 2 cell types as potential candidates for disc regeneration. J. Neurosurg. Spine **14**(3), 322–329 (2011)
44. Chen, S., Emery, S.E., Pei, M.: Coculture of synovium-derived stem cells and nucleus pulposus cells in serum-free defined medium with supplementation of transforming growth factor-beta1: a potential application of tissue-specific stem cells in disc regeneration. Spine(Phila Pa 1976) **34**(12), 1272–1280 (2009)
45. Ko, J.T., Lee, J.H., Kim, J.M., Kim, M.S., Rhee, J.M., Lee, H.B., Khang, G.: Development of biodegradable methoxy poly(ethyleneglycol)-polyesters diblock copolymers as drug carrier for implantable disc. Tissue Eng. Regen. Medicine. **3**, 158–168 (2006)
46. Ko, Y.K., Kim, S.H., Ha, H.J., Yoon, S.J., Rhee, J.M., Kim, M.S., Lee, H.B., Khang, G.: Emerging trends in biological therapy for intervertebral disc degeneration. Key Eng. Mater. **342–343**, 173–176 (2007)
47. Bae, J.Y., Shin, Y.N., Hwang, J.H., Kim, E.Y., Yoo, H., Lee, D., Yoon, K.H., Khang, G.: Effect of purified alginate microcapsules on the regeneration of annulus fibrosus cells. Internat J. Tissue Regen. **2**(3), 103–110 (2011)
48. Choi, J.H., Jang, J.W., Kim, S.H., Hong, H.K., Min, B.H., Son, Y., Rhee, J.M., Khang, G.: The effect of BMP-2 in DBP sponge for annulus fibrosus regeneration. Tissue Eng. Regen. Medicine. **5**(4–6), 877–884 (2008)
49. Ha, H.J., Kim, S.H., Yoon, S.J., Park, S.W., So, J.W., Rhee, J.M., Kim, M.S., Khang, G., Lee, H.B.: Evaluation of various scaffolds for tissue engineered biodisc using annulus fibrcosus cells. Polymer (Korea) **32**, 26–30 (2008)

50. Ko, Y.K., Kim, S.H., Jeong, J.S., Ha, H.J., Yoon, S.J., Rhee, J.M., Kim, K.S., Lee, H.B., Khang, G.: Biodisc tissue-engineered using PLGA/DBP hybrid scaffold. Polymer(Korea) **31**, 14–19 (2007)
51. Ko, Y.K., Kim, S.H., Ha, H.J., Kim, M.S., Han, C.W., Rhee, J.M., Son, Y., Lee, H.B., Khang, G.: Tissue engineered biodisc using human intervertebral disc cells and PLGA/DBP scaffold. Tissue Eng. Regen. Medicine **4**, 67–72 (2007)
52. Ko, Y.K., Jeong, J.S., Kim, S.H., Lim, J.Y., Rhee, J.M., Kim, M.S., Lee, H.B., Khang, G.: Effect of 2-D DBP/PLGA hybrid films on attachment and proliferation of intervertebral disc cells. Polymer(Korea) **32**, 109–115 (2008)
53. Lim, J.Y., Kim, S.H., Park, S.W., So, J.W., Back, M.O., Kim, M.S., Khang, G., Rhee, J.M., Lee, H.B.: Regeneration of biodisc using annulus fibrosus and nucleus pulposus cell on PLGA/DBP hybrid scaffold. Tissue Eng. Regen. Medicine **5**, 96–102 (2008)
54. Lee, S.K., Hong, H.K., Kim, S.J., Kim, Y.K., Song, Y.S., Ha, Y., Lee, D., Khang, G.: The comparison of sponges and PLGA scaffolds impregnated with DBP on growth behaviors of human intervertebral disc cells. Polymer(Korea) **34**, 398–404 (2010)
55. Park, S.M., Kim, H.M., Kim, E.Y., Lee, H.G., Lee, C.J., Lee, D., Khang, G.: Influence of cell-based demineralized bone particle gels and sponges on nucleus pulposus therapy for intervertebral disc regeneration; In vivo study. Inter. J. Tissue Regen. **5**, 96–102 (2013)
56. Jang, J.E., Kim, H.M., Song, J.E., Kee, D., Kwon, S.Y., Chung, J.W., Khang, G.: Regeneration of intervertebral disc using poly(lactic-co-glycolic acid) scaffolds included demineralized bone particle in vivo. Polymer(Korea) **37**, 669–676 (2013)
57. So, J.W., Kim, S.H., Baek, M.O., Lim, J.Y., Roh, H.W., Lee, N.R., Kim, M.S., Ryu, G.H., Cho, Y.H., Lee, S.J., Min, B.H., Khang, G., Lee, H.B.: Effect of size of PLGA microsphere on proliferation and phenotype for human intervertebral disc cells. Tissue Eng. Regen. Medicine **4**, 577–582 (2007)
58. Lee, J.H., Jang, J.W., So, J.W., Choi, J.H., Park, J.H., Ahn, S.I., Son, Y., Min, B.H., Khang, G.: Adhesion of human intervertebral disk cells on alginate/PLGA microspheres. Polymer(Korea) **33**, 7–12 (2009)
59. So, J.W., Jang, J.W., Kim, S.H., Choi, J.H., Rhee, J.M., Min, B.H., Khang, G.: The effect of pore sizes on poly(L-lactide-co-glycolide) scaffolds for annulus fibrosus tissue regeneration. Polymer(Korea) **32**, 516–522 (2008)
60. Kim, S.H., Choi, B.S., Ko, Y.K., Ha, H.J., Yoon, S.J., Rhee, J.M., Kim, M.S., Lee, H.B., Khang, G.: The characterization of PLGA/small intestinal submucosa composites as scaffolds for intervertebral disc. Key Eng. Mater. **342–343**, 112–116 (2007)
61. Collier, J.P., Mayor, M.B., Jensen, R.E., Surprenant, V.A., Surprenant, H.P., McNamar, J.L. Belec, L.: Mechanisms of failure of modular prostheses. Clin. Orthop. Relat. Res., 129–39 (1992)
62. Shaban, M., Yoon, S.J., Ko, Y.K., Ha, H.J., Kim, S.H., So, J.W., Idrus, R.B.H., Khang, G.: Fibrin promotes proliferation and matrix production of intervertebral disc cells cultured in three-dimensional poly(lactic-co-glycolic acid) scaffold. J. Biomater. Sci. Polymer. Edn. **19**, 1219–1237 (2008)
63. Choi, J.H., Kim, S.H., Hong, H.K., Kim, S.J., Lee, S.K., Kim, O.Y., Park, J.H., Rhee, J.M., Khang, G.: Regeneration of biodisc using annulus fibrosus cell on hyaluronic acid loaded PLGA scaffold. Tissue Eng. Regen. Medicine **6**, 83–88 (2009)
64. Jang, J.E., Kim, H.M., Song, J.E., Kee, D., Kwon, S.Y., Chung, J.W., Khang, G.: Regeneration of intervertebral disc using poly(lactic-co-glycolic acid) scaffolds included demineralized bone particle in vivo. Polymer(Korea) **37**, 669–676 (2013)

65. Lee, S.K., Hong, H.K., Kim, S.J., Kim, Y.K., Lee, D., Khang, G.: Effects of laminated cylindrical scaffolds of keratin/poly(lactic-co-glycolic acid) hybrid film on annulus fibrous tissue regeneration. Polymer (Korea) **34**, 474–479 (2010)
66. Hong, H.K., Lee, S.K., Song, Y.S., Kim, D.S., Eom, S., Kim, E.Y., Lee, D., Khang, G.: Biodisc regeneration using annulus fibrosus cell with hyaluronic acid impregnated small intestinal submucosa sponge. Polymer(Korea) **34**, 282–288 (2010)
67. Ha, H.J., Kim, S.H., Yoon, S.J., Ko, Y.K., Lee, E.K., Son, Y., Kim, M.S., Rhee, J.M., Khang, G., Lee, H.B.: Characterization of annulus fibrosus and nucleus pulposus with sequential passage. Tissue Eng. Regen. Medicine **3**, 416–422 (2006)
68. Kang, Y., Song, Y.S., Kim, E.Y., Oh, S.C., Song, J.E., Lee, D., Khang, G.: Differentiation of bone marrow stromal clls to nucleus pulposus cells by mixed co-culture. Intern. J. Tissue Regen. **3**, 21–27 (2012)

Pacing the Heart with Genes: Biological Pacing as an Alternative to Electronic Devices

Hee Cheol Cho and Jung Hoon Sung

Abstract The mammalian heart beats spontaneously without conscious input from the brain. Each heartbeat starts from a minuscule region, called the sinoatrial node (SA node or SAN). The SA node is a small highly-specialized structure containing just a few thousand genuine pacemaker cells. In contrast, the vast majority of the myocardium is populated by the ~5 billion working cardiomyocytes which remain quiescent until the electrical signal propagated from the SAN stimulates them. When the SAN fails, it could lead to circulatory collapse, requiring implantation of electronic pacemaker devices. These electronic devices generally work quite well. However, problems such as lead failure/repositioning, pediatric patients outgrowing the device, finite battery life, and infection call for biologics that are free from all hardware. Toward this goal, we and others have tested the concept of biological pacing. This article focuses on recent breakthroughs in the engineering of biological pacemakers. Combined with efforts to create clinically-relevant, large animal models of biological pacing, the field is moving beyond a conceptual novelty toward a future with clinical reality.

Keywords: Bradycardia · Gene therapy · Biological pacemakers

1 Introduction

Electronic pacemaker devices can sustain heart rate, or deliver shocks to terminate life-threatening tachycardias. However, the high cost of devices and complications

H.C. Cho(✉)
Urowsky-Sahr Scholar in Pediatric Bioengineering,
Department of Biomedical Engineering and Pediatrics Emory University,
HSRB E-184, 1760 Haygood Drive, Atlanta, GA 30322, USA
e-mail: heecheol.cho@emory.edu

J.H. Sung
Department of Cardiology CHA Bundang Medical Center, CHA University Seongnam,
Pocheon-si, Kyung-ki Do, Republic of Korea

© Springer International Publishing Switzerland 2016

H. Jo et al. (eds.), *Biomedical Engineering: Frontier Research and Converging Technologies*,
Biosystems & Biorobotics 9,
DOI: 10.1007/978-3-319-21813-7_19

such as pulmonary collapse, hemorrhage, bacterial infection, and lead/generator failure [1] represent limitations of the technology.

Biological pacing seeks to convert a focal, normally-quiescent region of myocardium to one that beats on its own by creating a spontaneous, electrical activity in that focal region. Earlier approaches to creating biological pacing exploited the existing knowledge on the molecular correlates that give rise to the unique electrophysiology of the SA nodal pacemaker cells. Such effort demonstrated the astonishing simplicity with which automaticity could be elicited from ordinary cardiomyocytes. It has also exposed, however, limitations to this approach; the cardiomyocytes with nascent automaticity lacked many of the defining features of the native SA nodal pacemaker cells, which are important for impulse propagation as well as impulse generation.

A recent breakthrough presents a new paradigm that transcends the previous approaches, creating induced pacemaker cells with striking semblance to the native SA nodal counterparts. A parallel approach is to derive standalone, spontaneously-beating cardiac tissues from pluripotent stem cells [2-4]. Beyond the natural formation of spontaneously beating embryoid bodies, however, the stem cell-based approach has gained rather incremental advances and is not discussed here. In the next sections, the gene-based approaches to creating biological pacing are discussed in detail, including the ongoing efforts to translate the basic findings to clinically-relevant, large-animal models.

2 Biological Pacing by Ion Channel Modulation

The first proof-of-concept study to trigger spontaneous action potentials (APs) in ventricular myocytes by modulating ionic currents was provided by Marbán and colleagues [5]. They posited that the cardiac myocytes in the ventricular and atrial chamber myocardium have the capacity to beat on their own, but are suppressed from doing so. This is because the hyperpolarizing ion current, I_{K1}, conducted by Kir2 channels, establishes a stable and strong negative resting membrane potential, thereby, functioning as an "electrical break" against spontaneous activity [6]. They sought to destabilize I_{K1} by adding a dominant-negative, mutant ion channel Kir2.1AAA. Dominant-negative Kir2 ion channels have been shown to indiscriminately co-assemble with all sub-families of Kir2 channels such as Kir2.1, 2.2, and 2.3; and down-regulate I_{K1} selectively and effectively in ventricular cardiomyocytes [7, 8]. A complementary approach taken by us and others was to overexpress an ion channel, HCN4, known for its role for spontaneous phase-4 depolarization in the SA and AV nodal pacemaker cells [9]. HCN4-mediated ionic current, I_f, could also provide rate responsiveness to neurohumoral input owing to its gating dependence on cAMP [9]. The ion channel-based rationale led to multiple approaches to engineering biological pacemakers in both small and large animal models. The striking simplicity with which spontaneous APs could be generated is demonstrated by the self-oscillating APs created by exogenous expression of three sarcolemmal ionic currents in a non-excitable, kidney cell line [10]. These approaches may represent a fast method to generate automaticity since the heterologously-expressed ion channels should reshape the AP profile instantly as the nascent proteins are delivered to the target cardiomyocytes' sarcolemma. However, these approaches mimic only a single ionic profile of the native SA nodal cells, and thus, fall short of replicating the coupled mechanisms of intracellular Ca^{2+} cycling and sarcolemmal ionic flow, which give rise to the remarkable robustness of automaticity observed in native nodal pacemaker cells [11]. The readers are referred to the previous reviews for detailed description of the ion channel-based approaches [10].

3 Induced Pacemaker Cells by Transcription Factor-Mediated Reprogramming

Reconfiguring the ventricular myocytes' sarcolemmal ion channel repertoire elegantly demonstrates the minimal complement of ionic currents for ordinary cardiomyocytes to pace themselves. Still, the repurposed myocytes remain largely as chamber cardiomyocytes with the characteristically long AP duration [12]. And the cell-to-cell electrical coupling in the ion channel-modified myocardium would continue to be mediated by Cx43, a predominant gap-junction protein in the chamber myocardium. In contrast, a low-conductance gap junction protein, Cx45, mediates the cell-to-cell electrical coupling in the SA node, and is thought to be a major component for overcoming the source-to-sink mismatch in the nodal structure [13, 14]. The ion channel-based approach also lacks a design element for spontaneous intracellular Ca^{2+} oscillations that are one of the hallmarks of genuine SA nodal pacemaker cells [15].

To engineer faithful biological replicas of SAN cells as an alternative therapeutic strategy, we reasoned that a gene/genes critical for early SAN specification may be able to convert working cardiomyocytes to the nodal pacemaker cells. During cardiogenesis, cardiomyocytes become specialized to exhibit ventricular, atrial, or pacemaker properties [16]. Pacemaker cells are exceedingly rare, comprising <10,000 of the ~10 billion cells (including myocytes and non-myocytes) in the adult mammalian heart [17], and yet the initiation of the heartbeat depends critically upon this diminutive subpopulation. Studies in mouse development suggest that embryonic SAN development is heavily dictated by transcriptional regulators such as Shox2, Tbx3, Tbx5, and Tbx18 [18]. Shox2 negatively regulates Nkx2.5 in the sinus venosus, and mice and zebrafish embryos lacking Shox2 exhibit bradycardia [19, 20]. Tbx3 strongly regulates SAN specification, either deficiency or ectopic expression of Tbx3 are associated with developmental defects [21]. Tbx5, which affects abnormal cardiac morphogenesis in Holt-Oram syndrome, potentiates Shox2 and Tbx3 expression [22]. Tbx18 seems to function upstream of Tbx3 and Tbx5, and Tbx18-positive mesenchymal progenitor cells define the sinus venosus, and differentiate *de novo* into SA nodal cells. Tbx18 is required for embryonic development of the SAN head area [23], but becomes minimally expressed by birth and in adulthood [24]. The following chapters describe our work in which re-expression of Tbx18 in postnatal cardiomyocytes could convert quiescent ventricular myocytes to pacemaker cells.

3.1 *Tbx18 Re-Expression Leads to Higher Automaticity in Neonatal Cardiomyocytes*

Freshly-isolated neonatal rat ventricular myocytes (NRVMs) were transduced individually with selected transcription factors in bicistronic adenoviral vectors. Tbx18-transduced NRVMs (Tbx18-NRVMs) exhibited an increased percentage of spontaneously-beating monolayer cultures compared to control and to the other transcription factors (Shox2, Tbx3, Tbx5 and Tbx20). Multiple, spontaneously-beating foci could often be observed in individual Tbx18-NRVM monolayers. This is likely due to the significant down regulation of Cx43, but not Cx45 or Cx40, by Tbx18 [25].

NRVMs are immature cardiomyocytes, and often exhibit spontaneous, syncytial contractions when cultured as monolayers. However the number of autonomously-beating cells are relatively small compared to the 'slave' cells that are driven by the spontaneous activity of the small subset of the cardiomyocytes [25]. When sparsely plated at a low density, a given cell rarely makes physical contact with neighboring cells. Under this condition a majority of the control ventricular myocytes, transduced singly with GFP, did not beat. In contrast, most Tbx18-NRVMs beat autonomously and spontaneously under the same condition, plated at a low density. The AP parameters of Tbx18-NRVMs replicated the features of SA nodal pacemaker cells: relatively depolarized maximum diastolic potential (MDP), accompanied by reduced I_{K1} density but elevated I_f. Native pacemaker cells exhibit spontaneous and rhythmic intracellular Ca^{2+} releases, which complement sarcolemmal ionic currents to generate robust automaticity [26]. A six-fold higher number of Tbx18-NRVMs exhibited spontaneous intracellular Ca^{2+} oscillationscompared to control. These sub-sarcolemmal, spontaneous local Ca^{2+} release events (LCRs) are indeed a hallmark of native SAN pacemaker cells [26]. During late diastole, LCRs activate sarcolemmal Na^+-Ca^{2+} exchanger currents (I_{NCX}), which helps the membrane potential to slowly depolarize (phase-4 depolarization) [26]. When Tbx18 was expressed in NRVMs, rhythmic LCRs were observed preceding each whole-cell Ca^{2+} transient, and those LCRs had larger amplitude compared to the occasional, LCRs observed in GFP-NRVMs. Larger Ca^{2+} stores in the sarcoplasmic reticulum (SR) would favor automaticity [26]. Thus, Tbx18-NRVMs exhibited spontaneous Ca^{2+} cycling which is one of the key features of native SAN pacemaker cells.

3.2 *Tbx18-Reprogrammed, de Novo Pacemaker Cells Could Function as Biological Pacemakers in Vivo*

The Tbx18-induced automaticity could represent a new way to provide pace the heart. Thus, Tbx18-mediated conversion of adult ventricular myocytes into pacemaker cells was examined *in vivo*. To accomplish this, an bi-cistronic adenoviral vector expressing Tbx18 and eGFP was directly and focally injected into the heart apex in guinea-pigs. Direction intra-myocardial injection led to robust expression of the transgene at the injection site in less than a day of the gene delivery (Figure 1). Electrocardiography was used to detect any ectopic ventricular activity in these animals two to four days after the initial gene delivery. Guinea pigs' resting heart rate is high at about 280 bpm, so the animals' native sinus rhythm was suppressed with methacholine. [27, 28]. In control animals, only a slow junctional rhythm was observed with antegrade polarity and narrow QRS duration (n=5). This is anticipated because slow, ectopic ventricular activity arises from the atrioventricular junctional area when the sinus rhythm is substantially slowed. Under the same conditions, five of seven Tbx18-injected guinea pigs exhibited a faster ectopic ventricular rhythm with wide QRS complexes and retrograde polarity (Figure 2). Since Tbx18 had been injected at the apex of the guinea pig heart, the retrograde polarity of the QRS complexes correlates with the site of gene injection. Furthermore, the wide QRS is likely the result of cardiac conduction through ventricular myocardium without the traveling through the secondary cardiac conduction system.

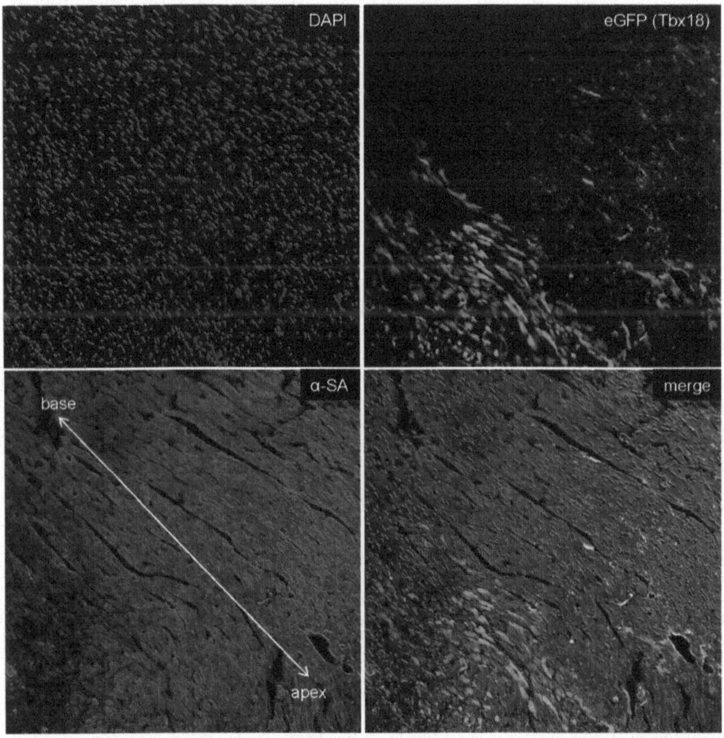

Fig. 1 Direct intra-myocardial injection of Tbx18 leads to robust expression of the transgene at the injection site. An adenoviral vector expressing Tbx18 and eGFP (in a bicstronic manner after IRES) was directly injected in the guinea pig ventricular myocardium *in vivo*. Twenty-four hours later, the animal was sacrificed and the heart was harvested for histology followed by immunostaining. Heterologous expression of Tbx18 is reported by eGFP expression (green myocytes), and the cardiomyocytes are marked by α-sarcomeric actinin staining. DAPI: nuclei.

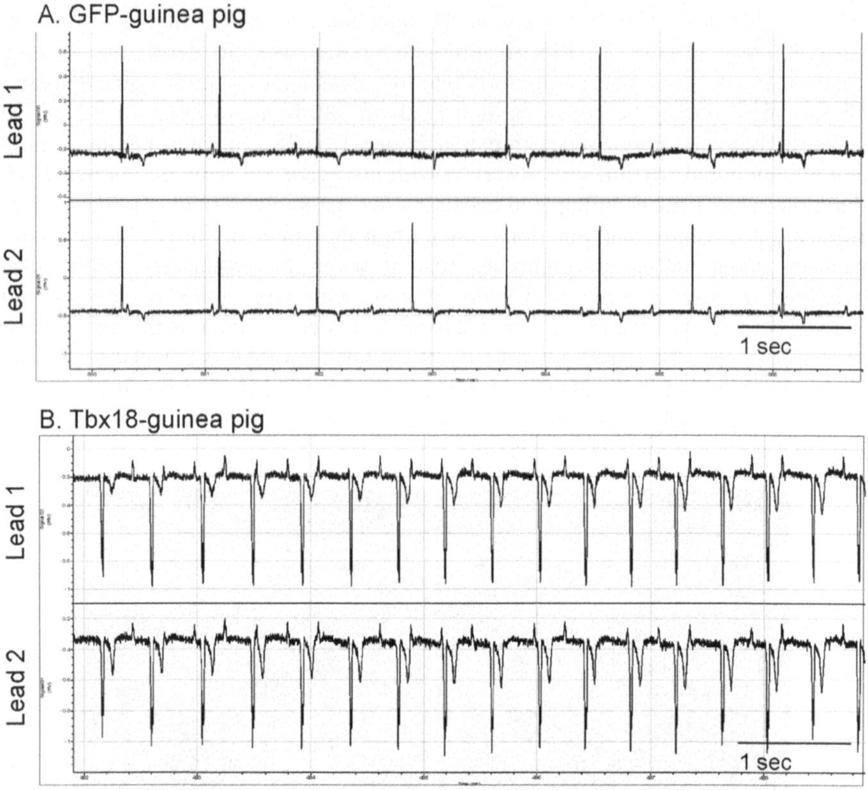

Fig. 2 Focal expression of Tbx18 in the apex of guinea pig hearts *in vivo* created ectopic ventricular beats. Representative electrocardiograms of GFP- or Tbx18- injected guinea pig (A and B, respectively). Downward (negative) deflection in ECG recording on lead II after methacholine injection (0.05mg/ml) revealed ectopic pacemaker activity originating from the ventricle with wide QRS complexes in Tbx18-injected guinea pigs (B). In contrast, GFP-injected guinea pigs (n=7) exhibited antegrade, junctional rhythm with narrow QRS complexes (A). No retrograde, ectopic beats with wide QRS complexes were observed in GFP-injected guinea pigs.

At the injection site, single cardiomyocytes isolated and examined 4-6 days after of the initial *in vivo* Tbx18 (or GFP) delivery. As a comparison, freshly-isolated, native SAN cells served as comparison 'gold standard' for native pacemaker cells. GFP-VMs were quiescent, displaying a stable resting potential near -80 mV, and firing an action potentials (AP) only upon electrical stimulation. In contrast, Tbx18-VMs fired spontaneous, rhythmic APs with prominent phase-4 diastolic depolarization similar to those in native SAN cells (Figure 3). Native SAN cells are smaller and narrower than ordinary ventricular myocytes. Remarkably, some of the Tbx18-VMs faithfully reproduced the tapering morphology of SAN cells; in contrast, non-transduced and GFP-VMs retained the bricklike appearance typical of ventricular myocytes (Figure 4).

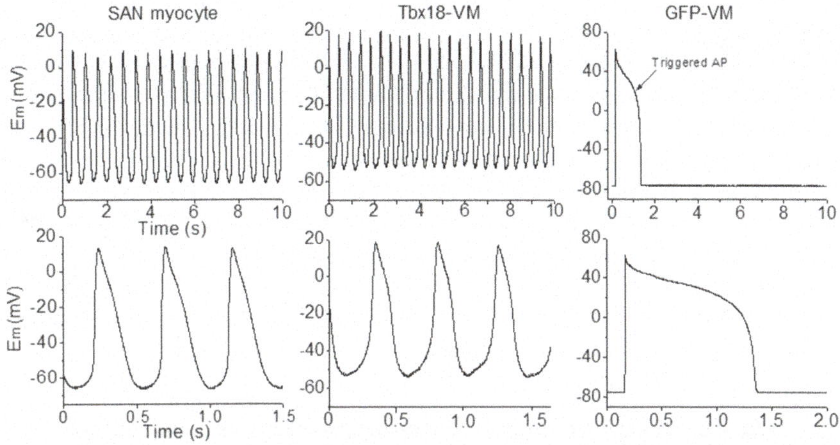

Fig. 3 Spontaneous action potentials recorded from freshly-isolated single Tbx18-VMs (n=5, middle panel) with perforated-patch current-clamp technique, showing robust and rhythmic APs with prominent diastolic depolarization recapitulating the electrophysiological hallmark of native SAN myocytes (left panel). The same recordings are expanded in the lower panel to show prominent diastolic depolarization in Tbx18-VMs and native SAN myocytes. Right panel: GFP-VMs displayed stable resting membrane potential and fired single action potential only upon electrical stimulation.

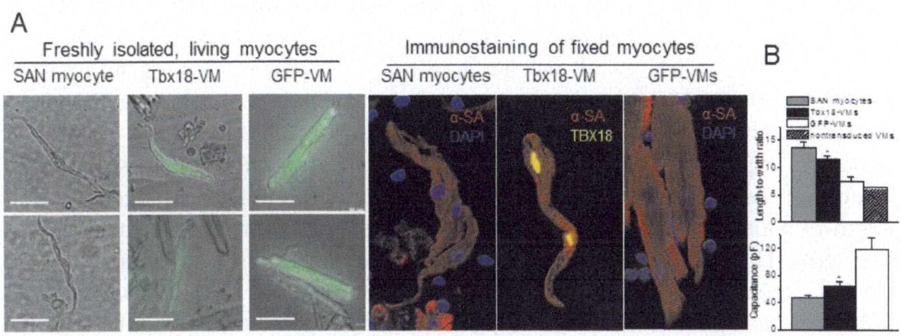

Fig. 4 Tbx18 converts adult guinea pig ventricular myocytes into pacemaker cells *in vivo*. **A**. Left panel: representative images of freshly-isolated SAN myocytes, Tbx18-VMs (reported by GFP expression), and GFP-VMs. Right panel: immunostaining against α-SA reveals disorganized myofibrillar structure in Tbx18-VMs akin to what is observed in native SAN myocytes. Scale bar = 50 μm. **B**. Analyses of myocyte length-to-width ratio and whole-cell capacitance as a measure of cell shape and size from freshly-isolated, living myocytes. Tbx18-VMs are smaller in size and spindle-shaped compared to GFP-VMs (n=53 for Tbx18-VMs and 80 for GFP-VMs and non-transduced VMs, $p<0.01$), but similar to native SAN myocytes (n=24).

A disease model of atrioventricular (AV) block is created in guinea pigs to simulate a common indication for the placement of an electronic pacemaker in humans. Electrocardiographic recordings of the beating hearts, perfused *ex vivo*, revealed ectopic ventricular beats in eight of eight Tbx18-injected hearts at a rate of 154 ± 6 bpm. The polarity and morphology of the ectopic beats was identical to that of electrode-paced beats at the site of Tbx18 injection, juxtaposing the origin of the ectopic beats to the site of gene delivery. Conversely, the hearts from the control animals exhibited a narrow-QRS junctional escape rhythm at an average rate of 120 ± 7 bpm, originating from the opposite end of the ventricle. Furthermore, Tbx18-injected hearts responded to autonomic regulation in an expected fashion; β-adrenergic stimulation followed by cholinergic suppression accelerated and then decelerated the heart rate to 235 ± 19 and then to 150 ± 23 bpm, respectively (n=7). Thus, the Tbx18-induced biological pacemaker activity responds appropriately to autonomic regulation in the intact heart.

3.3 The Induced Pacemaker Phenotype Persists Even After Tbx18 Expression Faded Away

Permanently reprogrammed cells, in principle, are thought to retain the target phenotype without sustained presence of the reprogramming factor [29, 30]. To ask if the *de novo* SAN phenotype persists beyond transient Tbx18 expression, Tbx18-VMs were isolated 6-8 weeks after the initial *in vivo* gene transfer. Then, the transcript levels of Tbx18 were quantified by single-cell, quantitative RT-PCR. When isolated just three days after *in vivo* gene transfer, the level of Tbx18 transcripts displayed a wide range of expression level from one Tbx18-VM to another. Spontaneously-beating ventricular myocytes were identified by live-cell video recording or whole-cell patch-clamp 6-8 weeks after *in vivo* gene transfer. The levels of Tbx18 transcript in most of the spontaneously-beating myocytes were negligible, similar to the negative level observed in control ventricular myocytes expressing GFP alone. A significant proportion of these induced pacemaker cells still exhibited SAN-like, lean morphology as judged by cell length-to-width (L-to-W) ratio up to six weeks after initial gene transfer. The long-term, Tbx18-injected animals demonstrated ectopic idioventricular rhythm at 165±14 bpm, with electrocardiograms parameters that are consistent with biological pacing from the Tbx18 injection site. Furthermore, the hearts from the long-term animals responded to autonomic regulation similar to the short-term, Tbx18-injected hearts. Thus, SAN-like cells retain pacemaker cell phenotype even after exogenous Tbx18 expression has ceased.

This study provides the first evidence that a singular expression of a transcription factor can convert adult ventricular myocytes to replicas of SAN pacemaker cells *in vitro* and *in vivo*. The new, induced pacemaker cells exhibit automaticity, and pace the heart *in vivo*. These pacemaker cell-specific phenotype persists even after the expression of exogenous Tbx18 has waned, indicating durable conversion. In another study, constitutive overexpression of Tbx3 in transgenic mouse model led to spontaneous atrial rhythm but failed to create ectopic ventricular beats [31]. A logical next step is to combine the two factors and look for synergistic effects for better reprogramming.

4 Biological Pacemaker Created by Minimally Invasive Gene Transfer in Pigs with Complete Heart Block

Electronic pacemakers have been successfully used for more than five decades with continuous refinement [32]. But, they are limited in some instances by complications ranging from inadequate autonomic support to lead fracture, infections, and adverse cardiac remodeling [33, 10]. Proof-of-concept studies of biological pacemakers have been performed in large animals, using delivery methods that are transferrable to clinical procedures [34, 10, 35, 33, 5, 36, 37]. However, all previous large animal studies have used ion channel-based approaches to elicit spontaneously oscillating APs from normal myocardium [37, 34, 35, 5, 36]. Motivated by *in vitro* and small animal *in vivo* reprogramming with *TBX18* [24], we have employed the minimally-invasive, porcine model of complete heart block [35] to achieve *in vivo* somatic reprogramming with *TBX18* in pigs [38].

Twelve pigs were given percutaneous injections of either *TBX18* (*n*=7) or green fluorescent protein (*GFP*; control) (*n*= 5) and monitored for 14 days. The baseline clinical, electrocardiographic (ECG), and laboratory data showed no significant differences between animals. The model involved first implanting an electronic pacemaker device with the lead placed at the right ventricular apex, and then creating complete heart block by ablating the AV node with a radiofrequency catheter. The electronic device served as a backup pacing device for the duration between the AV node ablation and until the biological pacing is achieved after the transgene expression. Upon generating complete heart block, the transgene, TBX18, is then injected in the AV junctional area, downstream of the ablated site.

After gene delivery, mean heart rate was higher in *TBX18*-transduced animals compared to *GFP*-transduced control animals starting at day 2 and persisting for 2 weeks. *TBX18*-transduced pigs achieved a higher maximal HR, and relied less on a backup electronic pacing device (<1% from day 5 to day 11) compared to controls (8 to 40% throughout the 2 weeks of follow-up). Thus, *TBX18* gene delivery successfully creates robust biological pacemaker activity, minimizing the need for backup electronic pacing.

The heart rhythm is known to correlate with the circadian rhythm [39]. Over the 14 days after the gene injection, both daytime and nighttime heart rates were higher in the *TBX18*-transduced group than in controls. The diurnal change, the difference between daytime and nighttime heart rate, was also greater in the *TBX18* group, resembling the behavior of the native SAN [39]. As a surrogate for the sinus node recovery time (SNRT) measurements [40], the pig's heart was paced electronically at a heart rate of 120 bpm for 1 min, and then electronic pacing was abruptly terminated to allow the biological pacemaker to emerge from overdrive suppression. Corrected recovery time after burst ventricular pacing was 10-fold shorter in the *TBX18* group at day 8 compared to control animals whose pacing was presumably coming from the endogenous and slow junctional escape rhythm.

4.1 Autonomic Regulation Controls the Rate of the *TBX18* Biological Pacemaker in Pigs

Distinct, yet unproven, advantages of biological pacemakers are their potential capacity for heart rate variability (HRV) as observed under normal sinus rhythm [41, 42] and response to β-adrenergic stimulation. Normalized low frequency component of the HRV tended to be higher while normalized high frequency tended to be lower, in the *TBX18* group, which are compatible with sympathetic predominance of the SAN [41, 42] and in the *TBX18*-transduced group. Isoproterenol infusion increased heart rate by 71% in *TBX18* animals, with less pronounced changes in control animals. The HRV analysis and the response to β-adrenergic agonist are indicative of the *TBX18* biological pacemaker's responsiveness to endogenous and pharmacologic autonomic stimuli.

The mean physical activity, measured by the implanted accelerometer, and the duration of bursts of activity were significantly greater in the *TBX18* group than in controls. The post-burst latency (time between bursts of activity) was shorter in *TBX18*-transduced animals than in controls. During maximal activity, animals in the *TBX18* group demonstrated a higher heart rate compared to controls, indicating superior accommodation to hemodynamic and metabolic demands (i.e., chronotropic adaptability). These data highlight another perceived advantage of biological pacing: unlike the electronic backup pacing device, the *TBX18* biological pacemaker provides autonomically sensitive chronotropic adaptability appropriate to support physical activity.

4.2 *TBX18* Biological Pacing Originates at the Focal Injection Site, and Does Not Increase Arrhythmic Risk

In order to derive anatomical origin of ectopic ventricular pacing activity *in vivo* electroanatomic mapping of the right ventricle (RV) and pace-mapping from the injection site were performed. On the day of device implantation and gene injection, the earliest activation originated, in both groups, from the right ventricular apex at which the electronic device's lead had been positioned to pace the heart. At day 14 after gene delivery, activation arose in the high septal region, anatomically close to the transgene injection site in *TBX18*-transduced animals but not in control animals.

Transient pacing of the gene injection site with a stimulation catheter generated electrocardiographs that showed a 100% match on all 12 of the 12 ECG leads with the *TBX18*-induced biological pacemaker rhythm. In contrast, there was no match between the catheter-paced ECGs with the nonspecific escape rhythms observed in *GFP*-transduced controls. The QRS duration of the *TBX18*-induced rhythm was significantly shorter than that in controls (55.4 ± 3.1 ms versus 67.4 ± 1.1 ms, $p= 0.01$), consistent with fast conduction from the high septal region [43]. These findings suggest that *TBX18*-induced rhythms originated from the gene delivery site, whereas spontaneous rhythms in control animals rarely did.

During the 2-week follow-up period, the cumulative incidence of sustained ventricular arrhythmias was low and did not differ significantly between the two experimental groups. Corrected QT interval and dispersion from 12-lead ECG and APD_{90} dispersion and duration did not differ in the two treatment groups. Programmed ventricular stimulation with and without isoproterenol infusion elicited no episodes of inducible ventricular tachycardia in either control or *TBX18*-transduced animals. These data suggest that TBX18 biological pacemaker does not increase arrhythmic risk.

4.3 Prospects for Clinical Translation of Biological Pacemakers

This study demonstrates that in situ somatic reprogramming can stimulate proper chronotropic heart function in a large-animal model of complete heart block. The gene delivery techniques were consistent with routine clinical practice in which venous access is considered the safest route for electronic pacemaker implantation [44]. The *TBX18* biological pacemaker was superior to an electronic pacemaker in its responsiveness to the circadian rhythm and autonomic regulation, showing features of the native SAN.

A major concern of biologic–driven automaticity is the potential occurrence of ventricular arrhythmias [10, 45]. Here, no evidence of increased ventricular arrhythmias was seen during 2 weeks of continuous monitoring. Repolarization parameters (QT and APD) and gradients (QT and APD_{90}dispersion) that can promote arrhythmogenesis [46, 47] were indistinguishable between *TBX18*-transduced animals and in the control animals. Moreover, no inducible ventricular tachycardia was seen by programmed ventricular stimulation two weeks after *TBX18* gene delivery. Together, these findings support the absence of pro-arrhythmic risk of the *TBX18*-induced biological pacemaker, at least in the short term. Biodistribution the *TBX18* adenoviral vector indicated that the viral vector was primarily localized at the injection site, with minimal systemic distribution. Nevertheless, the arrhythmic and systemic safety profiles should be rigorously sought out with a longer-term study.

The gene delivery technique is readily transferrable to a clinical setting and exhibits many advantages: low invasiveness, minimal pain/distress, minimal blood loss, lower risk of stroke by advancing the catheter to the right-sided circulation, and the ability to deliver the transgene focally into the AV junction area, resulting in antegrade conduction through the specialized His-Purkinje system. In addition, the transvenous injection catheter (NOGA MyoStar) has been extensively used in multiple human biological therapy trials [48-50].

5 Conclusions

A distinct advantage of biological pacemakers to gene therapies for cardiac regeneration is that focal modifications of electrical properties suffice for effective treatment. Since the amount of exogenous biological material delivered can be correspondingly reduced, potential problems due to widespread dissemination may be more readily averted. A focal modification, in principle, is reversible; if needed, implanted biological pacemakers can be destroyed by conventional electrophysiological ablation, followed by electronic pacemaker implantation.

Gene delivery vectors could be tailored to different clinical needs. Adenoviral vectors, for example, achieve the peak expression earlier than other vectors albeit in a transient manner. This would be ideal for the temporary pacing needs to treat infected pacemaker devices [10, 35, 38]. On the other hand, the high immunogenicity of these vectors could be circumvented by using helper-dependent adenovirus [51]. Accordingly, adeno-associated virus or lentivirus vectors would be more suitable for long-term applications [52].

Will biological therapies eventually replace drugs and devices? To answer this question, long-term, large animal studies must be performed to compare the efficacy of pacing by biologic versus device. At the very least, current data justify testing of biological pacemakers for transient use in clinical conditions such as infected pacemaker devices [10, 35, 38]. On the other hand, technological advances in devices, e.g., leadless pacemakers [53], will continue to improve the standard of care. Thus, it would be logical to predict a future in which a device and a biological therapy co-exist in a bradycardic patient, with the device functioning as a backup. Further work will be required to optimize the functional efficacies of biological therapies, to evaluate the safety and toxicology, and to establish the long-term stability of therapy. With the recent breakthroughs, biological pacing is now closer to clinical translation than to experimentation of concepts.

Acknowledgments Supported by the Urowsky-Sahr Foundation, NHLBI, and AHA.

References

1. Bernstein, A.D., Parsonnet, V.: Survey of cardiac pacing and implanted defibrillator practice patterns in the United States in 1997. Pacing Clin. Electrophysiol. **24**(5), 842–855 (2001)
2. Ionta, V., Liang, W., Kim, E.H., Rafie, R., Giacomello, A., Marban, E., et al.: SHOX2 overexpression favors differentiation of embryonic stem cells into cardiac pacemaker cells, improving biological pacing ability. Stem Cell Reports **4**(1), 129–142 (2015). doi:10.1016/j.stemcr.2014.11.004
3. Kehat, I., Khimovich, L., Caspi, O., Gepstein, A., Shofti, R., Arbel, G., et al.: Electromechanical integration of cardiomyocytes derived from human embryonic stem cells. Nat. Biotechnol. **22**(10), 1282–1289 (2004). doi:10.1038/nbt1014
4. Xue, T., Cho, H.C., Akar, F.G., Tsang, S.Y., Jones, S.P., Marbán, E., et al.: Functional integration of electrically active cardiac derivatives from genetically engineered human embryonic stem cells with quiescent recipient ventricular cardiomyocytes: insights into the development of cell-based pacemakers. Circulation **111**(1), 11–20 (2005). doi:01.CIR.0000151313.18547.A2. [pii]

5. Miake, J., Marban, E., Nuss, H.B.: Biological pacemaker created by gene transfer. Nature **419**(6903), 132–133 (2002)
6. Cho, H.C., Backx, P.H.: Three-dimensional structure of a K+ channel pore: basis for ion selectivity and permeability. In: Steven, L., Archer, N.J.R. (eds.) Potassium Channels in Cardiovascular Biology, pp. 17–34. Kluwer Academic / Plenum Publishers, New York (2001)
7. Zobel, C., Cho, H.C., Nguyen, T.T., Pekhletski, R., Diaz, R.J., Wilson, G.J., et al.: Molecular dissection of the inward rectifier potassium current (IK1) in rabbit cardiomyocytes: evidence for heteromeric co-assembly of Kir2.1 and Kir2.2. J. Physiol. **550**(Pt 2), 365–372 (2003)
8. Cho, H.C., Tsushima, R.G., Nguyen, T.T., Guy, H.R., Backx, P.H.: Two critical cysteine residues implicated in disulfide bond formation and proper folding of Kir2.1. Biochemistry **39**(16), 4649–4657 (2000)
9. Mangoni, M.E., Nargeot, J.: Genesis and regulation of the heart automaticity. Physiological Reviews **88**(3), 919–982 (2008). doi:10.1152/physrev.00018.2007
10. Cho, H.C., Marbán, E.: Biological therapies for cardiac arrhythmias: can genes and cells replace drugs and devices? Circ. Res. **106**(4), 674–685 (2010). doi:10.1161/CIRCRESAHA.109.212936
11. Lakatta, E.G., Vinogradova, T., Lyashkov, A., Sirenko, S., Zhu, W., Ruknudin, A., et al.: The integration of spontaneous intracellular Ca^{2+} cycling and surface membrane ion channel activation entrains normal automaticity in cells of the heart's pacemaker. Ann. N. Y. Acad. Sci. **1080**, 178–206 (2006)
12. Miake, J., Marban, E., Nuss, H.B.: Functional role of inward rectifier current in heart probed by Kir2.1 overexpression and dominant-negative suppression. The Journal of Clinical Investigation **111**(10), 1529–1536 (2003). doi:10.1172/JCI17959
13. Rohr, S., Kucera, J.P., Fast, V.G., Kleber, A.G.: Paradoxical improvement of impulse conduction in cardiac tissue by partial cellular uncoupling. Science **275**(5301), 841–844 (1997)
14. Unudurthi, S.D., Wolf, R.M., Hund, T.J.: Role of sinoatrial node architecture in maintaining a balanced source-sink relationship and synchronous cardiac pacemaking. Frontiers in Physiology **5**, 446 (2014). doi:10.3389/fphys.2014.00446
15. Lakatta, E.G., DiFrancesco, D.: What keeps us ticking: a funny current, a calcium clock, or both? J. Mol. Cell. Cardiol. **47**(2), 157–170 (2009)
16. Chien, K.R., Domian, I.J., Parker, K.K.: Cardiogenesis and the complex biology of regenerative cardiovascular medicine. Science **322**(5907), 1494–1497 (2008). doi:10.1126/science.1163267
17. Bleeker, W.K., Mackaay, A.J., Masson-Pevet, M., Bouman, L.N., Becker, A.E.: Functional and morphological organization of the rabbit sinus node. Circ. Res. **46**(1), 11–22 (1980)
18. Christoffels, V.M., Smits, G.J., Kispert, A., Moorman, A.F.: Development of the pacemaker tissues of the heart. Circ. Res. **106**(2), 240–254 (2010). doi:106/2/240. [pii]
19. Blaschke, R.J., Hahurij, N.D., Kuijper, S., Just, S., Wisse, L.J., Deissler, K., et al.: Targeted mutation reveals essential functions of the homeodomain transcription factor Shox2 in sinoatrial and pacemaking development. Circulation **115**(14), 1830–1838 (2007). doi:10.1161/CIRCULATIONAHA.106.637819

20. Espinoza-Lewis, R.A., Yu, L., He, F., Liu, H., Tang, R., Shi, J., et al.: Shox2 is essential for the differentiation of cardiac pacemaker cells by repressing Nk2–5. Dev. Biol. **327**(2), 376–385 (2009). doi:S0012-1606(08)01450-4. [pii]
21. Hoogaars, W.M., Engel, A., Brons, J.F., Verkerk, A.O., de Lange, F.J., Wong, L.Y., et al.: Tbx3 controls the sinoatrial node gene program and imposes pacemaker function on the atria. Genes & Development **21**(9), 1098–1112 (2007)
22. Mori, A.D., Zhu, Y., Vahora, I., Nieman, B., Koshiba-Takeuchi, K., Davidson, L., et al.: Tbx5-dependent rheostatic control of cardiac gene expression and morphogenesis. Dev. Biol. **297**(2), 566–586 (2006). doi:10.1016/j.ydbio.2006.05.023
23. Wiese, C., Grieskamp, T., Airik, R., Mommersteeg, M.T., Gardiwal, A., de Gier-de, Vries C., et al.: Formation of the sinus node head and differentiation of sinus node myocardium are independently regulated by Tbx18 and Tbx3. Circ. Res. **104**(3), 388–397 (2009)
24. Kapoor, N., Liang, W., Marban, E., Cho, H.C.: Direct conversion of quiescent cardiomyocytes to pacemaker cells by expression of Tbx18. Nature Biotechnology **31**(1), 54–62 (2013). doi:10.1038/nbt.2465
25. Kapoor, N., Galang, G., Marbán, E., Cho, H.C.: Transcriptional suppression of Connexin43 by Tbx18 undermines cell-cell electrical coupling in postnatal cardiomyocytes. J. Biol. Chem. (2011). doi:M110.185298 [pii]
26. Lakatta, E.G., Maltsev, V.A., Vinogradova, T.M.: A coupled SYSTEM of intracellular Ca^{2+} clocks and surface membrane voltage clocks controls the timekeeping mechanism of the heart's pacemaker. Circ. Res. **106**(4), 659–673 (2010). doi:106/4/659. [pii]
27. Cho, H.C., Kashiwakura, Y., Marbán, E.: Creation of a biological pacemaker by cell fusion. Circ. Res. **100**(8), 1112–1115 (2007). doi:01.RES.0000265845.04439.78. [pii]
28. Kashiwakura, Y., Cho, H.C., Barth, A.S., Azene, E., Marbán, E.: Gene transfer of a synthetic pacemaker channel into the heart: a novel strategy for biological pacing. Circulation **114**(16), 1682–1686 (2006). doi:CIRCULATIONAHA.106.634865. [pii]
29. Takahashi, K., Yamanaka, S.: Induction of pluripotent stem cells from mouse embryonic and adult fibroblast cultures by defined factors. Cell **126**(4), 663–676 (2006)
30. Yu, J., Vodyanik, M.A., Smuga-Otto, K., Antosiewicz-Bourget, J., Frane, J.L., Tian, S., et al.: Induced pluripotent stem cell lines derived from human somatic cells. Science **318**(5858), 1917–1920 (2007). doi:1151526. [pii]
31. Bakker, M.L., Boink, G.J., Boukens, B.J., Verkerk, A.O., van den Boogaard, M., den Haan, A.D., et al.: T-box transcription factor TBX3 reprogrammes mature cardiac myocytes into pacemaker-like cells. Cardiovascular Research **94**(3), 439–449 (2012). doi:10.1093/cvr/cvs120
32. Chardack, W.M., Gage, A.A., Greatbatch, W.: A transistorized, self-contained, implantable pacemaker for the long-term correction of complete heart block. Surgery **48**, 643–654 (1960)
33. Epstein, A.E., DiMarco, J.P., Ellenbogen, K.A., Estes 3rd, N.A., Freedman, R.A., Gettes, L.S., et al.: 2012 ACCF/AHA/HRS focused update incorporated into the ACCF/AHA/HRS 2008 guidelines for device-based therapy of cardiac rhythm abnormalities: a report of the American College of Cardiology Foundation/American Heart Association Task Force on Practice Guidelines and the Heart Rhythm Society. Circulation **127**(3), e283–e352 (2013). doi:10.1161/CIR.0b013e318276ce9b

34. Bucchi, A., Plotnikov, A.N., Shlapakova, I., Danilo Jr, P., Kryukova, Y., Qu, J., et al.: Wild-type and mutant HCN channels in a tandem biological-electronic cardiac pacemaker. Circulation **114**(10), 992–999 (2006)
35. Cingolani, E., Yee, K., Shehata, M., Chugh, S.S., Marban, E., Cho, H.C.: Biological pacemaker created by percutaneous gene delivery via venous catheters in a porcine model of complete heart block. Heart Rhythm J. **9**(8), 1310–1318 (2012). doi:10.1016/j.hrthm.2012.04.020
36. Tse, H.F., Xue, T., Lau, C.P., Siu, C.W., Wang, K., Zhang, Q.Y., et al.: Bioartificial sinus node constructed via *in vivo* gene transfer of an engineered pacemaker HCN Channel reduces the dependence on electronic pacemaker in a sick-sinus syndrome model. Circulation **114**(10), 1000–1011 (2006)
37. Boink, G.J., Nearing, B.D., Shlapakova, I.N., Duan, L., Kryukova, Y., Bobkov, Y., et al.: Ca(2+)-stimulated adenylyl cyclase AC1 generates efficient biological pacing as single gene therapy and in combination with HCN2. Circulation **126**(5), 528–536 (2012). doi:10.1161/CIRCULATIONAHA.111.083584
38. Hu, Y.F., Dawkins, J.F., Cho, H.C., Marban, E., Cingolani, E.: Biological pacemaker created by minimally invasive somatic reprogramming in pigs with complete heart block. Sci. Transl. Med. 6(245), 245ra94 (2014). doi:10.1126/scitranslmed.3008681
39. Cinca, J., Moya, A., Figueras, J., Roma, F., Rius, J.: Circadian variations in the electrical properties of the human heart assessed by sequential bedside electrophysiologic testing. American Heart Journal **112**(2), 315–321 (1986)
40. Narula, O.S., Samet, P., Javier, R.P.: Significance of the sinus-node recovery time. Circulation **45**(1), 140–158 (1972)
41. Montano, N., Ruscone, T.G., Porta, A., Lombardi, F., Pagani, M., Malliani, A.: Power spectrum analysis of heart rate variability to assess the changes in sympathovagal balance during graded orthostatic tilt. Circulation **90**(4), 1826–1831 (1994)
42. Nicolini, P., Ciulla, M.M., De Asmundis, C., Magrini, F., Brugada, P.: The prognostic value of heart rate variability in the elderly, changing the perspective: from sympathovagal balance to chaos theory. Pacing Clin. Electrophysiol. **35**(5), 622–638 (2012). doi:10.1111/j.1540-8159.2012.03335.x
43. McGavigan, A.D., Roberts-Thomson, K.C., Hillock, R.J., Stevenson, I.H., Mond, H.G.: Right ventricular outflow tract pacing: radiographic and electrocardiographic correlates of lead position. Pacing Clin. Electrophysiol. **29**(10), 1063–1068 (2006). doi:10.1111/j.1540-8159.2006.00499.x
44. Rajappan, K.: Permanent pacemaker implantation technique: part I: arrhythmias. Heart **95**(3), 259–264 (2009). doi:10.1136/hrt.2007.132753
45. Rosen, M.R., Robinson, R.B., Brink, P.R., Cohen, I.S.: The road to biological pacing. Nat. Rev. Cardiol. **8**(11), 656–666 (2011). doi:10.1038/nrcardio.2011.120
46. Kozhevnikov, D., Caref, E.B., El-Sherif, N.: Mechanisms of enhanced arrhythmogenicity of regional ischemia in the hypertrophied heart. Heart Rhythm **6**(4), 522–527 (2009). doi:10.1016/j.hrthm.2008.12.021
47. Antzelevitch, C., Oliva, A.: Amplification of spatial dispersion of repolarization underlies sudden cardiac death associated with catecholaminergic polymorphic VT, long QT, short QT and Brugada syndromes. Journal of Internal Medicine **259**(1), 48–58 (2006). doi:10.1111/j.1365-2796.2005.01587.x
48. Losordo, D.W., Henry, T.D., Davidson, C., Sup Lee, J., Costa, M.A., Bass, T., et al.: Intramyocardial, autologous CD34+ cell therapy for refractory angina. Circ. Res. **109**(4), 428–436 (2011). doi:10.1161/CIRCRESAHA.111.245993

49. Perin, E.C., Dohmann, H.F., Borojevic, R., Silva, S.A., Sousa, A.L., Mesquita, C.T., et al.: Transendocardial, autologous bone marrow cell transplantation for severe, chronic ischemic heart failure. Circulation **107**(18), 2294–2302 (2003). doi:10.1161/01.CIR.0000070596.30552.8B
50. Perin, E.C., Dohmann, H.F., Borojevic, R., Silva, S.A., Sousa, A.L., Silva, G.V., et al.: Improved exercise capacity and ischemia 6 and 12 months after transendocardial injection of autologous bone marrow mononuclear cells for ischemic cardiomyopathy. Circulation **110**(11 Suppl. 1), II213–II218 (2004). doi:10.1161/01.CIR.0000138398.77550.62
51. Mitani, K., Graham, F.L., Caskey, C.T., Kochanek, S.: Rescue, propagation, and partial purification of a helper virus-dependent adenovirus vector. Proceedings of the National Academy of Sciences of the United States of America **92**(9), 3854–3858 (1995)
52. Gray, S.J., Samulski, R.J.: Optimizing gene delivery vectors for the treatment of heart disease. Expert Opinion on Biological Therapy **8**(7), 911–922 (2008)
53. Reddy, V.Y., Knops, R.E., Sperzel, J., Miller, M.A., Petru, J., Simon, J., et al.: Permanent leadless cardiac pacing: results of the LEADLESS trial. Circulation **129**(14), 1466–1471 (2014). doi:10.1161/CIRCULATIONAHA.113.006987

Part VI
Robotics and Informatics in Healthcare, Emerging Theranogstics

Robotic Systems for Minimally Invasive Diagnosis and Therapy

Youngjin Moon and Jaesoon Choi

Abstract Minimally invasive surgery and interventional procedures have seen rapid advancement as one of leading trend in medical technology innovation for its distinguished clinical benefit for patient. Robotic systems and technologies have made remarkable contribution to the innovation enabling various innovative devices and procedures including robotic laparoscopic surgery assist system. While robotic systems to assist general surgery seem to become mature technology, robotic systems for interventional procedures and neurological surgery are newly emerging. The minimally invasive procedures have inherent limitations and constraints that make human operation difficult or less optimal. Various medical imaging modalities are utilized as visual sensor for the procedures and each has limitations such as radiation exposure, resolution and sensitivity, real-time imaging capability, electromagnetic interference and etc. Tiny and complex tissue structure through which devices for the minimally invasive procedures perform diagnostic or therapeutic operation is another major limiting condition in terms of dexterity or precision. Robotic systems that converge various electro-mechanical engineering and computer science technologies facilitate human physician overcoming these limitations and achieving better clinical outcome for patient. Computed tomography (CT) or ultrasound guided biopsy is one of long researched applications for robotic system utilization. Several robotic systems for cardiac intervention and neurological surgery are already available for clinical use. The clinical efficacy of the robotic technologies needs further study including large scale randomized clinical study and safety issue with the use of robotic system either in assist or automation manner also need more research. It seems increasingly clear that robotic system technologies will continuously provide answers to many of unmet clinical needs in minimally invasive diagnosis and therapy.

Y. Moon · J. Choi(✉)
Asan Medical Center, Songpa-gu, South Korea
e-mail: jaesoon@gmail.com

© Springer International Publishing Switzerland 2016

H. Jo et al. (eds.), *Biomedical Engineering: Frontier Research and Converging Technologies*, Biosystems & Biorobotics 9,
DOI: 10.1007/978-3-319-21813-7_20

Keywords: Medical robot · Minimally invasive · Surgical robot · Intervention · Biopsy · Ablation · Percutaneous procedure

1 Introduction

Minimally invasive surgery or interventional procedures enable diagnosis and treatment through an assessment of the inside of the human body via resection of the minimum area necessary. Thus, such methods are associated with fast recovery, low risk of pain and infection, and less psychological pressure, and are hence widely used in various fields. However, from the perspective of doctors who perform such procedures, the technique is considerably more difficult as compared to the current standard resection operation, due to the limited working space and the difficulty of using image-guide equipment such as CT or MRI in interventional procedures. Another disadvantage of these methods is that they require much more time and effort to achieve the relevant skill. Robotics or mechatronics technology can provide solutions for these limitations as well as the difficulties of the manual procedure through the integral utilization of various electromechanical devices and technologies mediated by a computer. The surgical robotic system using teleoperation—a concept that was first introduced in 1980s—has rapidly developed within a short period of time and has now been sufficiently enhanced to be used as standard clinical procedure in the fields of urology and obstetrics/gynecology. The Da Vinci surgical system by Intuitive Surgical in the U.S., which is a predominantly used system at present, is a laparoscopic surgical robot with 7 degrees of freedom, and allows the doctors to perform detailed laparoscopic surgery through dexterous movements of surgical instruments with high degree of freedom and stereo endoscope system with high resolution. Although they do not include the laparoscopic surgery robots with teleoperation type control manner, surgical robots facilitate high clinical efficacy through automatic surgery with high degree of precision and accuracy as compared to a manual operation. A major robotic system of such type has been developed in osteoplastic surgery for total joint replacement. The ROBODOC system—an artificial hip joint and knee joint surgical robotic system—was the first automatic surgery robot to be introduced. It was licensed for sale in the Europe in 1994, and it was first predominantly clinically applied in Germany for total hip replacement arthroplasty, total knee replacement arthroplasty, and partial knee replacement arthroplasty. Since its approval by the FDA in the U.S. in 2008, it has been widely available for patients. In 2011, it was estimated that 1,200 robots had been distributed worldwide and have been used in more than 29,000 surgeries. Currently, a product upgraded with the newest technology of image-guided surgical systems is in the process of FDA approval.

In addition to these robots for surgery, research has been conducted on robots for various interventional procedures based on image guidance. The effort to develop novel medical device with better clinical efficacy and patient benefit for diagnosis and treatment of cancer and cardiovascular diseases has been ever increasing due to the high incidence and mortality rate of the diseases. Robotics systems for interventional radiology procedures and cardiovascular intervention procedures that are essential for the diagnosis and treatment of cancer and cardiovascular diseases have seen remarkable technological advancement. Intervention procedures are even less invasive than minimally invasive surgery and preferred for patient when the clinical efficacy is comparable. However, the minimally invasiveness requires more skillful manipulation of the instrument and trained performance of the clinician. Robotics systems can effectively facilitate the procedure augmenting the dexterity and more accurate medical imaging information handling for the clinician. Percutaneous procedure is a common name of diverse interventional diagnosis and treatment procedures using skin-penetrating instrument in minimally invasive manner including needle biopsy, RF-/cryo- ablation, cardiac/gastrointestinal/cerebrovascular stent procedures, angioplasty procedures and etc. Some of the percutaneous procedures that use needle shape instrument can be separately called as needle-insertion type interventional procedure. So called, needle biopsy is representative example, which is to take tissue samples in the chest, abdomen, and various other organs for confirmative diagnosis by insertion of needle shape tiny-diameter sampling device. Treatment instrument also can be constructed in needle shape. Electrical energy is transferred through the needle shape instrument and transformed into heat at the target tissue area to destroy it thermally. In the cryo-ablation, cold liquid is circulated through a tiny hydraulic circuit in the instrument to freeze and destroy the target tissue. Alcohol or other therapeutic agent can be delivered using needle shape instrument. Due to the increasing demand of early cancer diagnosis and less invasive treatment, the number of procedures and the introduction of various medical device technologies are continuously increasing. Consistent with these efforts, various systems using robot technology have been developed to increase the effectiveness and safety of such interventional procedures.

2 Interventional Procedures

Interventional procedures include a variety of medical procedures for diagnosis and treatment. Usually it involves access of internal organ or tissue in patient body using minimally invasive instruments and monitoring of the task using a type of medical imaging device. Some procedures are generally called as percutaneous procedure, since it is by perforation through skin. Through the small tunnel, instruments and treatment materials are delivered and manipulated by clinician at its proximal end to collect tissue samples such as in biopsy for cancer diagnosis, modify the structure of surrounding tissue or blood vessel at the distal end of the instrument such as in stent insertion or chronic tissue obstruction tunneling, deliver energy or chemicals for treatment such as in radio frequency ablation, cryo-ablation, alcohol or cancer drug injection, or to deploy treatment device such as in transcatheter aortic valve implantation, and etc. To minimize tissue damage, the instruments are desirably designed as minimally invasive type, either flexible or hard pipe-like shape with diameter from a few millimeters down to sub millimeter as in the guide wire for cardiovascular procedures. The distal end of the instrument can be directly steered or guided by pre-curved guide wire by clinician's manipulation at the proximal end of the instrument outside of patient body. To guide the motion and task of the instrument inside patient body, various medical imaging or vision devices are used such as ultrasound imaging, CT, MRI, and etc. Although endoscope or various surgical scopes such as laparoscope, arthroscope, and etc. can also be regarded as vision modality, many of the scope-assisted methods are used for surgery and not classified as interventional procedure with exceptions such as gastrointestinal endoscopy based procedures performed either by clinician or surgeon. Although the intervention and surgery can be differently defined and classified, minimally invasive method or concept can be common core characteristic. While CT and MRI have advantage of providing precise and abundant anatomic information, CT requires relatively large dose of X-ray exposure and MRI system can be used with only MR-compatible devices due to electromagnetic interference, which makes intra-procedural or on-line use of the imaging system difficult. In conventional manual procedure, clinician needs to do the registration of pre-operative imaging data and intra-operative actual patient anatomy qualitatively. Robotic system can adopt proper modality of real-time tracking devices such as optical tracker, electromagnetic tracker, and etc. to provide clinician with quantitative registration and precise guide of the instrument. Teleoperation type robot system in which clinician remotely controls the motion of slave robot at patient side through master robot that can be located outside of CT or MRI room, enables real-time parallel use of the imaging system for the clinician to be able to "see" and perform the procedure conveniently with improved precision and accuracy for patient.

Robotic systems to assist intervention can be categorized into largely two types: Systems for procedures with needle type instruments and those for remote manipulation of catheter type instruments. Needle manipulation robots are systems with multi-joint actuation mechanism for the instrument motion controlled either by computer program that automatically analyses real-time image of the interventional scene obtained by ultrasound imaging device, X-ray fluoroscope, MRI and etc., or by human operator that is remotely located and has a master control device of which the motion is conveyed in real-time to the actual procedure robot, slave robot, to follow exact motion of the human operator. Systems for needle manipulation appear to have been long researched, but not yet have reached full-fledged commercialization stage. It seems that more study and innovative development are required to prove clinical efficacy and establish unique modality with clear value for commercialization. A few commercial systems of guide-only type (without instrument insertion by machine) have been introduced very recently.

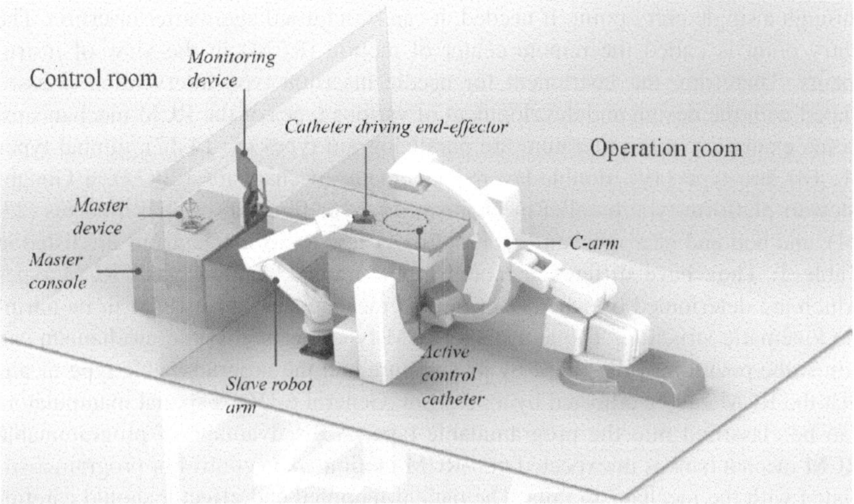

Fig. 1 Typical configuration of image-guided intervention robot system

In contrast, systems for catheter manipulation have seen earlier and active commercialization and been being drawing increasing interest as the robot systems to assist surgery have shown remarkable success in both clinical and industrial aspects. Catheter manipulation robots are mostly master-slave configuration remote motion control system with various degree of robotized motion from a type with simple 3 DOF (insertion/retraction, roll, catheter tip bending) to system with 7 DOF including multi-directional catheter tip bending motion. In cardiology, radio-frequency catheter ablation is one of effective treatment options in cases where arrhythmia is refractory to drug treatment or when the patient requests it (1-3). However, the current procedure performed by human operators has a few disadvantages, which include: 1) prolonged exposure to X-ray in both the clinician and the patient, 2) difficulties in accurately controlling the proximal end tip of catheter, and 3) the absence of proper instruments for novel treatments such as transthoracic epicardial ablation. Identifying and localizing the precise location of the proximal end of a catheter in motion in the anatomical structure in real-time also requires sophisticated methods using computers such as algorithms combining information from pre-procedural imaging, intra-procedural fluoroscopy, electro-anatomical mapping, or 3D atriography.

An issue of the needle insertion type intervention using robot technology is how to adjust the axial direction of the needle before or during insertion. For this problem, various robotic mechanisms have been developed. They basically implement three degrees-of freedom (DOFs): two rotations and one translation. The former is for adjustment of the pose of needles and the latter for insertion of needles. Like laparoscopic instruments, the instrument used in needle insertion type intervention, which is usually called the end-effector, is adjusted so that its axial direction heads a desired path and then is inserted into a patient's body through a single entry point. If needed, it can be adjusted again after insertion. The entry point is called the remote center of motion (RCM) in the view of instruments. Therefore, the instrument for needle insertion type intervention is associated with the design and development of various types of the RCM mechanisms. Some examples in the literature are parallelogram types (8, 14-17), gimbal types (1, 18), arc type (19), double-layered kinematic mechanisms (20-22), a Gough-Stewart platform-type parallel mechanism, four- or five-bars spatial linkages (23, 24), and belt and gear mechanisms (25, 26). A few RCM mechanisms are listed in Table 1. They have difference mount types, mount locations, and RCM types which are determined by whether or not the mechanism has an RCM in its intrinsic kinematic structure. The adjustable RCM type means that the mechanism can adjust the position of the RCM by its structure and the programmable type means that the RCM can be adjusted by a program. General 6-DOFs spatial manipulators can be classified into the programmable type. A disadvantage of programmable RCM mechanisms is unexpected non-RCM motion when control or program associated with the mechanism fails. The patient-mounted end-effector should carefully consider sterilization and needs a measure to resolve unexpected movement and respiration of the patient. Table-mounted end-effector should consider availability to the geometry of CT.

Table 1 Comparison of end-effectors for needle intervention (9)

Institution	Device name	Mount type	RCM type	RCM location
MIT, U.S.A.	Robopsy (4)	patient	intrinsic	same vertical level as the base
Univ. of Strasbourg, France	CT-Bot (50)	patient	adjustable	variable
Johns Hopkins Univ., U.S.A.	Acubot (6)	table	intrinsic	same vertical level as the base
Case Western Reserve Univ., U.S.A.	SABiR (7)	table	programmable	variable
Waseda Univ., Japan	not specified (8)	robot arm	intrinsic	same vertical level as the base

Remote navigation systems for robotic catheterization improved accuracy and precision in treatment, increased success rate, reduced the time for the operator to expose to the radiation, and showed other advantages and feasibility (1-3). Tissue perforation is however a critical risk in catheterization (31-35) even though it does not frequently happen. In robotic catheterization, the remote manipulation of the catheter might make the clinician unaware of insertion force or excessive contact force. Although there are controversial reports on the tissue perforation issue in the robotic catheterization (1, 36), some commercial robotic catheter remote navigation systems are in the market. Among them, the Sensei (Hansen Medical, Inc., MountainView, CA, USA) and the Niobe (Stereotaxis, Inc., St. Louis, MO, USA) are two major systems. The former uses the dedicated catheter with small permanent magnets used to exert magnetic force but the magnets may harm tissue and bring about perforation in the tissue. The latter and most active catheters adopting the push/pull-wire mechanism have stiff elements to form their structures and to orient the proximal end tip toward a desired direction and the elements may perforate tissue (32-34). A clinical case reported tissue perforation during catheterization using the Artisan Sheath (Hansen Medical) (35). For this problem, Hansen Medical improved their system by adding IntelliSense that is a function to give the force signal measured between catheter and tissue to the operator by visual or haptic sense (2, 32-34). Virtual fixture (VF) can be another measure to improve tele-presence in remote manipulation (37). It guides the operator's motion so that the instrument can move to safe directions by limiting direction of the motion and the workspace. Haptic virtual fixture generates position and force signals in software and imposes them on the human operator for safe, precise, and fast manipulation (38). There are two main types in the haptic VF: guidance VF and forbidden-region VF. The former makes the manipulator follow pre-defined trajectories. On the contrary, the latter keeps the manipulator away from the pre-defined forbidden area. The VF technique has been applied to various applications such as surgical robot systems for minimally invasive surgeries and assistive system for rehabilitation exercise (39-41). Force sensing in robotic catheterization can help the operator be aware of contact information, but it is required to embed a force sensor in dedicated catheters. In contrast, fluoroscopy system is regularly used in the clinical cases to visualize the blood vessels and identify the catheters in real-time.

Robots for interventional procedures have unique characteristic and can be compare with other types of surgical robots as summarized in Table 2.

Table 2 Comparison of characteristics of robots for surgery and intervention

Factor	Laparoscopic surgery robot	Arthroplasty assist robot	Percutaneous procedure robot
Intra-op vision	Endoscope	Optical/electromagnetic tracking device	CT, CBCT, MRI, US
Control	Teleoperation	Automatic/shared control	Teleoperation/partial or full autonomy
Imaging information	Pre-op image registration	Planning, registration	Planning, registration
Visual monitoring	Endoscope	Eye observation	Limited (CT fluoroscopy)

3 Robotic Systems for Interventional Radiology

Robots for needle positioning have wide spectrum and can be categorized in many different ways. As a general overview, a few noticeable trends can be summarized as follows: 1) robots to assist needle insertion type intervention guided by CT or fluoroscopy are a recently regaining research interest in accordance with greater interest on technologies for minimally invasiveness-based treatments, although commercialization or clinically applicable system developments are still in its infant stage in spite of long research history in this field, 2) numerous studies are underway for MR-compatible robotic prostate biopsy in response to clinical demand in diagnosis and treatment of the prostate cancer, and 3) one of novel innovations in biopsy robot systems is steerable needle and needle steering implementation technologies.

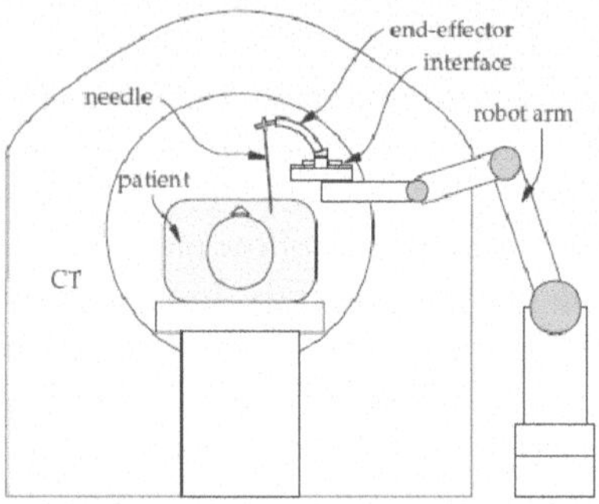

Fig. 2 An example of robotic needle intervention system. (9)

From the aspect of mounting type, the needle manipulation robots can be categorized to bed-mounted robots and patient-mounted robots. (42) The AcuBot (6, 44) from the Johns Hopkins urology robotics group is a representative example of the bed-mounted robots, which is a CT guided needle insertion robot with bridge-like macro-micro architecture. The macro structure is installed like a bridge over patient on the bed of CT machine and a needle manipulation robot, the micro structure, is anchored on the macro structure. The Innomotion system (Innomedic Gmbh, Germany) for MRI guided intervention is another similarly bridge-mounted robot which uses pneumatic actuators coupled with optical sensors to drive the robot (15). A third example is the B-Rob I & II systems based on original research by Austrian Research Center (Seibersdorf, Austria) which use two parallel fingers to align the needle above a target, allowing the clinician to insert the needle manually, with a precision of 0.88mm (46). Patient-mounted systems have features of being smaller and able to adapt to patient's motion such as breathing, which minimizes the effect of motion due to patient discomfort as well as the motion of the insertion point on the patient's skin during breathing. The CT-Bot is one such robot (47) that uses piezoelectric motors to power a parallel-structure with 5 degrees of freedom (DOF) and a single line CT-visible fiducial marker mounted on the base of the robot for registration with the CT images, and can be

placed on the patient's abdomen with total weight of about 3 Kg. The Robopsy is a compact 3-DOF X-ray transparent actuator for needle orientation and insertion

(4), while positioning at the skin insertion point is manual. The Light Puncture Robot (LPR) (49-51) is another example of patient-mounted system, a patient-mounted needle insertion module stretched between four straps that can be pulled individually to control the needle position, and unique in that it can be used for both CT and MRI guided configurations.

In the functional view, the robots for interventional radiology can be categorized as the robots for only guiding the path or pose that needle is inserted, for both guiding and needle insertion, and for full process handling needles in biopsy. The iSYS robot, commercialized version of the aforementioned B-Rob I & II, and Maxio of Perfint Heathcare are in the first category (Fig. 3). These robots linked with image registration software control the pose of a needle so that interventional radiologists just insert the needle along the guided slot in the robot and do the post-process of the interventions. Most needle insertion robots developed for interventional radiology second category belong to the second category. They have two functions of adjustment of the pose and insertion of a needle. These robots are very helpful to improve accuracy in needle insertion and reduce radiologists' exposure to radiation from CT or X-ray fluoroscopy when only one needle is used in intervention procedure. However, for biopsy that needs two different types of needles, a guide needle and biopsy needle (typically biopsy gun), the benefit of reduced radiation exposure may be not guaranteed. The robots in the third category can be an alternative of the issue. Particularly, the biopsy for lung, liver, or kidney is relatively more difficult in the aspect of danger and complexity of needle insertion, compared with the breast biopsy using aspiration needle. Moon and Choi (52) developed a robot for automated biopsy as a part of the slave robot in a teleoperation system. The anther robot developed by them is shown in Fig. 4. For each step in biopsy procedure, available tasks of each type of robot are listed in table 2.

Table 3 Available tasks of each type robot for biopsy procedure

Task / Robot	Type 1	Type 2	Type 3
Anesthesia	No	No	No
Adjusting of pose of guide needle	Yes	Yes	Yes
Insertion of guide needle	No	Yes	Yes
Withdrawal of stylet of guide needle	No	No	Yes
Insertion of sampling needle	No	No	Yes
Tissue sampling	No	No	Yes
Withdrawal of sampling needle	No	No	Yes
Withdrawal of sheath of guide needle	No	No	Yes

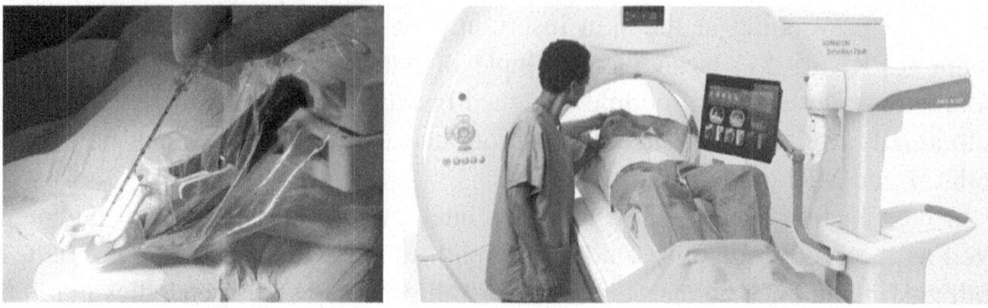

Fig. 3 Commerciallized biopsy assist robots: (Left) iSys (iSys gmbh, Austria), (Right) Maxio (Perfint Healthcare, India)

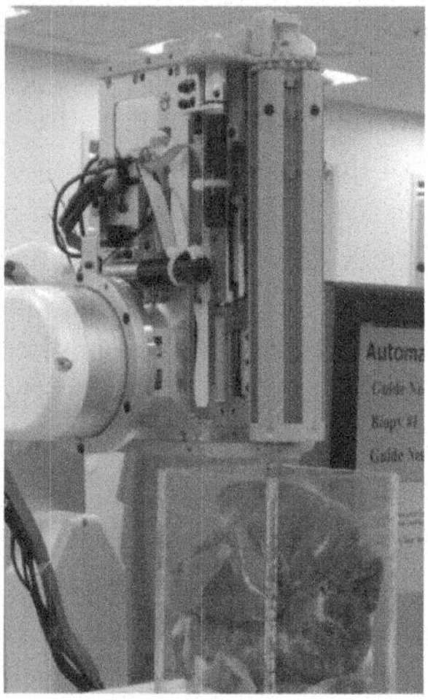

Fig. 4 Automated biopsy robot

The MR-compatible prostate biopsy robots have shown up since MRI-based detection of prostate cancer was used as an alternative method of the conventional transrectal biopsy using ultrasound. These robots are required to have new actuator and body material different from conventional robots consisting of the electric motors and metallic parts because ferromagnetic, conductive, and electronic components interfere MR image. Many studies have attempted to develop various devices for this diagnosis and treatment. APT II from a research group of Johns Hopkins University (53, 54) consists of a manipulator, mounting arm, and targeting software program. In their upgraded version (54), a steerable needle channel in the manipulator is used to adjust the angle of a needle to be inserted into the patient's prostate. Most parts of the manipulator are constructed of plastic material. Another research group in Johns Hopkins University has developed a new pneumatic step motor (60) to actuate their robot (59, 61) for biopsy and brachytherapy in MR environment. The robot has a 5 DOF parallel linkage for three translational and two rotational motions. The prostate biopsy robot developed by a research group of University of Toronto, MRI-P (63) can freely move in space with six

DOF (each three DOF for translation and rotation, respectively) by ultrasonic motors and perform biopsy. A research group of Imperial College has used piezoceramic motors with MR compatibility to control four DOF robotic system for prostate biopsy (64). Navigation and localization of the needle of the prostate biopsy robot require calibration of robots image coordination by registration (65), similarly to other image-based intervention system. Although various robot systems have been developed, there are still important challenges in this field. The robot should be developed to satisfy actual clinical needs and should consider cost issue. Others would be safety measures, sterilization, and so forth.

Table 4 MR-compatible prostate biopsy robot systems

Robot	Researchers	Actuation Method	Accuracy (mm)
APT	Krieger et al (53, 54)	Manual	1.8 (in vivo)
-	Taillant et al (55)	Pneumatic	less than 30
-	Fischer et al (56)	Pneumatic	-
-	Song et al (57)	Pneumatic	less than Max. 6
	Yakar et al (58)	Pneumatic motor	-
MrBot	Muntener et al (59- 61)	Pneumatic motor	less than 0.315
-	Zangos et al (62)	Linkage	-
MRI-P	Goldenberg et al (63)	Ultrasonic motor	less than 2
-	Elhawary et al (64)	Piezoceramic motor	-

A few latest interesting studies are on active steering of the needle tip to have effect of curved insertion that can be utilized to avoid vessels, bones, or dangerous regions during needle insertion. Various methods to achieve needle steering have been proposed, and they can be categorized into a few ways: use of inherent shape such as the bevel tip, use of new controllable needles, needle manipulation by a specific control system, and manipulation of soft tissue including a target. Shoham et al. (11) presented a robotic system for steering of a flexible needle in soft tissues under ultrasound imaging in real-time. The inverse kinematics using a virtual spring model calculates curved trajectories that the needle tip follows. In the experiment, the closed-loop control system updating tissue stiffness parameters demonstrated that the tracking error of the needle tip was less than 1 mm. Tang et al. (12) proposed a new compliant needle model which has a compliant hinge near the magnetized tip and showed the steering simulation. The magnet surrounding the compliant needle can deflect the needle tip and control the needle so that the needle follows a planned path. Okazawa et al. (13) devised a hand-held steerable needle device for percutaneous interventions. They adopted an active steering element instead of the conventional passive stylet in the device. Using this device, they showed 35 mm of lateral deflection for a 100 mm insertion in a phantom. Ayvali et al. (27) developed a multi-DOF discretely actuated steerable cannula. The cannula contains shape memory alloy (SMA) actuators and these make the cannula bent following a desired curve. In the experiment using a tissue sample, the steering cannula resulted in 0.29 mm of the maximum error from a target location. Sears and Dupont (28) proposed a design of steerable needle using curved concentric tubes. The concentric curved tubes were used to construct needles of piecewise constant curvature. It was experimentally found that the four-tube prototype implemented by the concept conformed to curves of almost constant curvature. Torabi et al. (29) developed an automated procedure planning technique to manipulate tissue and then move the target into the path of the needle. Using this system, they could reduce targeting error from more than 20 mm to less than 1 mm. Reed et al. (43) introduced a needle steering system using the bevel-tip needle. The concept of their system is shown in Fig. 5. The control system includes a real-time planner, planar controller, torsion compensator, and pose estimator. The real-time planner gives the information about location of the needle until the needle tip approach the given target. The planar controller controls the needle to stay in a desired plane and the needle tip to follow the desired path using feedback control algorithm. For torsional friction between the needle and tissue, the torsion compensator controls the needle tip using the formulated dynamic model with friction. The control loop is run while the pose estimator confirms that the position error between the needle tip and the target is not within a preset allowable range. The active needle steering has challenging issues. Many studies achieved steering performance but safety such as tissue damage and hemorrhage has not considered. In the path generation of the steerable needle, constant curvature model is somewhat far from practical applications. Tissue structure that consists of multiple heterogeneous layers and membranes also makes the needle steering difficult.

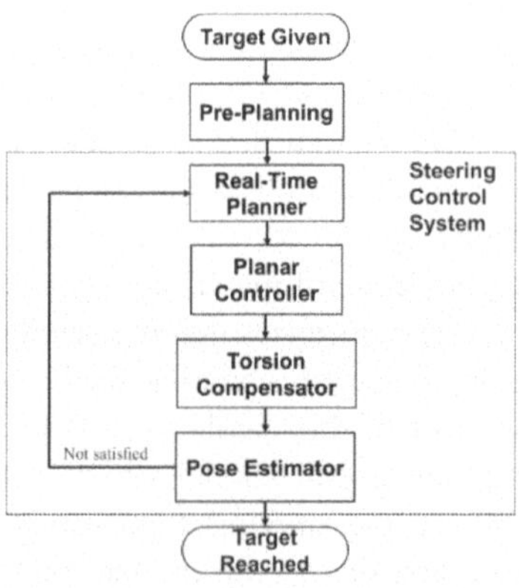

Fig. 5 Concept of needle steering control system

4 Robotic Systems for Cardiac Intervention

There are a few catheter remote control systems in the market. The Niobe applies a wide magnetic field to the patient's body and then controls the metal proximal end tip of the catheter used for ablation. The Sensei implements a high-DOF robotic articulated catheter sheath for improved functionality. The Remote Catheter Manipulation System (Catheter Robotics, Inc., USA) provides a master/slave teleoperation control platform onto which a conventional catheter is simply loaded and controlled. Some limitations of the Niobe are that only one catheter is available due to magnetic field interference and that it needs careful consideration of electromagnetic shielding. The robotic catheter sheath of the Sensei system is stiffer than conventional catheters and thus it may cause to perforate blood vessels or give damage to the tissues. In addition, greater diameter of the sheath-catheter assembly structure than conventional ones may be a problem. The Catheter Robotics system is lack of force reflection and the motion of the proximal end tip of catheters has a single DOF (30-32).

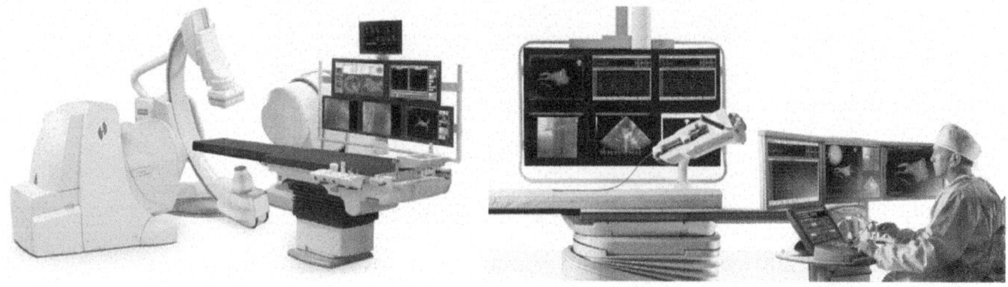

Fig. 6 (Left) Niobe^{TM} (Stereotaxis, Inc.), (right) Sensei^{TM} (Hansen Medical, Inc.)

Some researchers have developed technologies for precise catheter manipulation. Ganji et al. implemented an electromagnetic tracking of the spatial position of the proximal end tip of the catheter. They achieved an automatic control using the electromagnetic signal. In the system, the catheter is automatically guided along a specified path (33). Patel et al. developed a new active controllable catheter embedding a shape memory alloy. The developed catheter can be automatically controlled along the path determined by the visual signal using image processing and a visual-servo technique (34, 35). Plicchi et al. proposed a tele-control technique considerably to reduce the operation time for the catheter to maneuver (36, 45).

Recent advancement in cardiac catheter robot system includes introduction of robot systems for vascular procedures such as percutaneous coronary intervention (PCI). CorPath^{TM} 200 system (Corindus Vascular Robotics, Inc., U.S.A.) is representative example. The robot system is to assist clinician in control of coronary guidewire and stent/balloon catheter in the PCI procedure. The first-in-human clinical study was performed in Colombia. Clinical success rate was 100% with final stenosis diameter of less than 30% after delivering stent to targeted lesion and no occurrences of major adverse coronary events at discharge and 30 day follow-up. Remarkable result is significant (97.1%) reduction in radiation exposure to the operator and reduced contrast utilization (approx. 36% reduction). Aforementioned Hansen Medical has also introduced new system for the vascular procedure, name Magellan ™.

Despite of rapid and remarkable advancement in catheter robot systems, there are still various areas to be improved requiring further research and development.

- Novel steerable catheter with minimal diameter
- Novel mechanism of robotic catheter driving
- Reliable high fidelity haptic feedback
- Fusion of imaging modalities with clinically efficient level of precision
- Systems targeted toward extended applications

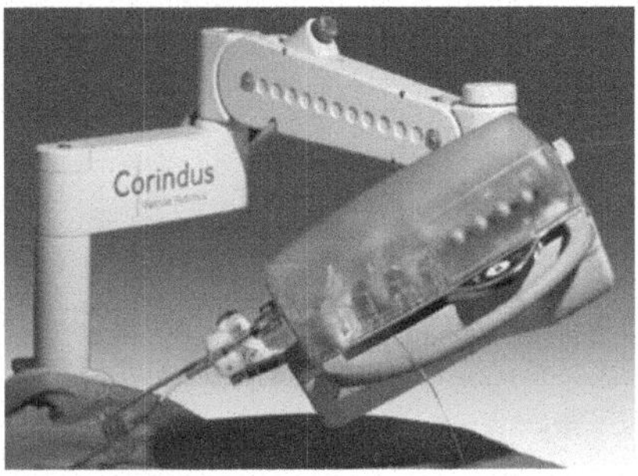

Fig. 7 CorPathTM 200 system (Corindus Vascular Robotics, Inc., U.S.A.)

5 Robotic Systems for Neurological Procedure

Neuromate provides a wide platform of solutions for neurosurgical operations; it is an operation planning system that uses medical imaging information. It can determine the three-dimensional (3D) anatomical structure of the brain and helps effectively visualize the target treatment area. The planning software pairs 3D images obtained from CT and magnetic resonance imaging (MRI) with 2D images obtained from X-ray radiography. The robotic arm of Neuromate is remote controlled and moves along a predetermined path with high accuracy and stability. This device has wide application in neurosurgical operations, including deep brain stimulation (DBS), stereoelectroencephalography (SEEG), neuro-endoscopy, and biopsy. The advantages of this system include the quick setup, easy use, and customized manufacturing. In addition, it has auxiliary functions such as a collision prevention system, precision verification system, and mechanical and electrical components for monitoring.

The robotic device ROSA™ has a consolidated multi-application console, which allows for safe and precise operations and application of previously established surgical procedures in various neurological surgeries. The system is used for processes such as matching patient information, selection of a specific surgical site, and precise manipulation of surgical instruments. ROSA™ is a clinically approved robot used in Europe, the United States, and Canada and possesses the following characteristics:

- The robotic arm of ROSA™ can emulate the movements of a human arm with 6 DOF and can aid in successful completion of complex operations. It also allows selection of various trajectories.
- ROSA™ has a haptic function that allows surgeons to easily guide surgical instruments within set limits at the planning stage. Therefore, the surgeon can interact with the robot using his/her own surgical plan and instruments.
- A non-invasive and touch-free registration system - ROSA™ combines non-invasive laser measurements and precise robotic movements for pairing patient-specific information.

ROSA™ can be used in various areas of surgery such as DBS, SEEG, and biopsies.

With the 6 DOF of its robotic arm, ROSA™ ensures safety via minimally invasive treatment and good accessibility to surgical areas, which in turn increase the accuracy of instrument guidance. In addition, surgeons can identify the surgical site using the pre-surgical image, haptic operation mode of the robotic arm, and real-time navigation system.

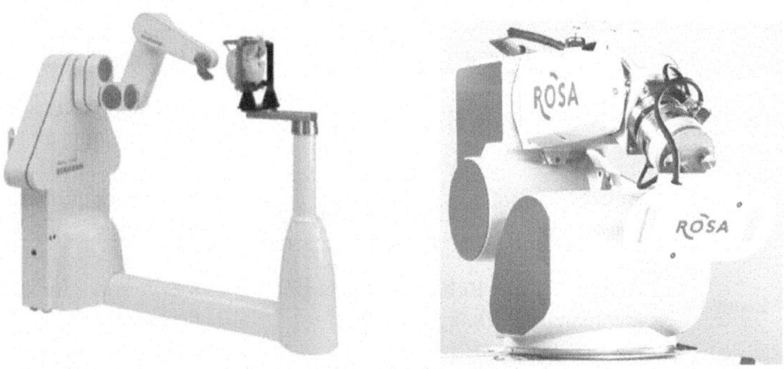

Fig. 8 (Left) Neuromate™ (Renishaw), (right) Rosa™ (MADTECH)

Neuroarm, the first-ever MRI-compatible robot, combines an image-generating device with MRI and ultra-precision surgical instruments. Operations requiring precision to the extent of approximately 1–2 mm are possible because the system and workstation can be controlled independently using two detachable remote-controlled devices. The MR image is transmitted on a real-time basis. The end effector detects tactile signals using a 3D force sensor, and the surgeon is alerted of this through a controller at the workstation during the operation. This system is designed to operate in a 1.5- and 3.0-tesla environment.

Table 5 Neurosurgical robotic project and features

Project	category	Institute, company	Main features
Alpha robot	Active, teleoperated	MicroDexterity Systems Inc.	5 DOF parallel manipulator mounted on the stereotactic frame
Cranio	Active, automated	RWTH-Aachen / Lehrstuhl fur Biomedizinische Technik	Craniectomy with 6 DOF hexapod robot
Evolution	Semi-active, automated	Universal Robot Systems	6 DOF hexapod robot for pedicle screw placement and adenoma dissection
JHU project w/NeuroMATE	Co-operative control	Johns Hopkins University	Skull base drilling with force based co-operative control with virtual Fixtures
KineMedic	Active, teleoperated	DLR / BrainLAB AG	Light-weight, high payload 7 DOF robot for MIS neurosurgery
MARS robot (SmartAssist)	Active, automated	Mazor Surgical Technologies Inc.	FDA approved, light-weight, head mountable robot for needle insertion
Minerva	Active, automated	Lab. of Microengineering, Swiss Federal inst. of Tech	Real time frameless stereotactic instrument guidance in CT scanner
neuroArm	Active, teleoperated	University of Calgary	MRI compatible complete multi-manipulator
NeuRobot	Active, automated	Computer Integrated Medical Intervention Laboratory; Singapore	Instrument guidance, skull-base surgery
NeuroMaster	Active, automated	Robotic Institute Beihang University	6 DOF robot for stereotactic procedures
NeuroMate	Passive, automated	IMMI / ISS / Schaerer Mayfield NeuroMate Sarl	Cannulae positioning for biopsy, neuroendoscopy
PathFinder	Active, automated	Prosugics Ltd.	6 DOF manipulator for instrument guidance
RAMS	Active, teleoperated	NASA JPL	6 DOF manipulator for eye and brain surgery with motion scaling and tremor filtering

6 Automation and Safety

Medical robot is different from conventional medical electrical device or system mainly by its automatic or autonomic nature. Representative standard for medical electrical device safety is IEC 60601, a series of standards on various aspects and types of medical electrical device. The current edition of the standard covers wide area of components in medical electrical device including electromagnetic safety, software safety, biological safety, and etc., as well as fundamental electrical and mechanical safety. Most components and aspects of medical robot as medical device are covered by the standard and representative untreated aspect is the autonomy of medical robot. Currently a technical report is under preparation by a joint working group of the IEC and ISO as a guidance on the safety considerations in medical robot - surgical robot and rehabilitation robot. The level of autonomy in surgical robot has close relation to the safety and risk associated with the device as medical device. Among various methods for definition of the level of autonomy, one example can be a descriptive classification method by Kaber DB and Endsley MR (48) and the level of autonomy definition is listed as below:

- Manual control (MC): The human performs all tasks including monitoring the state of the system, generating performance options, selecting the option to perform (decision making) and physically implementing it.
- Action support (AS): The system assists the operator with performance of the selected action, although some human control actions are required. A teleoperation system is a common example.
- Batch processing (BP): Although the human generates and selects the options to be performed, they then are turned over to the system to be carried out automatically. The automation is, therefore, primarily in terms of physical implementation of tasks.
- Shared control (SHC): Both the human and the computer generate possible decision options. The human still retains full control over the selection of which option to implement, however, carrying out the actions is shared between the human and the system.
- Decision support (DS): The computer generates a list of decision options, which the human can select from, or the operator may generate his or her own options. Once the human has selected an option, it is turned over to the computer to implement.
- Blended decision making (BDM): At this level, the computer generates a list of decision options, which it selects from and carries out if the human consents. The human may approve of the computer's selected option or select one from among those generated by the computer or the operator. The computer will then carry out the selected action.
- Rigid system (RS): This level is representative of a system that presents only a limited set of actions to the operator. The operator's role is to select from among this set.

- Automated decision making (ADM): The system selects the best option to implement and carries out that action, based upon a list of alternatives it generates (augmented by alternatives suggested by the human operator). This system, therefore, automates decision making in addition to the generation of options (as with decision support systems).
- Supervisory control (SC): The system generates options, selects the option to implement and carries out that action. The human mainly monitors the system and intervenes if necessary.
- Full automation (FA): The system carries out all actions. The human is completely out of the control loop and cannot intervene.

An example of application of the autonomy level and associated risk analysis in medical robot can be as follows:

- The medical procedures under consideration are total hip arthroplasty (or replacement) and knee arthroplasty for treatment of diseases in joint such as osteoarthritis. This procedure aims to remodel articular surface of musculoskeletal joint to appropriate shape for artificial joint implantation. Robot system to assist the procedure operates, guides, or adjusts the motion or the speed of the bone milling instrument attached at the end-effector of the robot system. Preoperative imaging such as CT scan is used to plan the proper shape of the remodeled bone. During the treatment, the end-effector is operated autonomously or by human operator holding the end-effector under haptic interactive guide from the robot arm. Optical or certain type of object tracking system can be utilized as vision input for the robot to identify and track the location and movement of the bone under surgery. The haptic guidance gives the operator certain type of force feedback with which the operator can sense virtual wall or boundary of "safe" region or planned region for remodeling. In one implementation, the speed of the bone mill is attenuated or the milling blade is covered as the operator reaches "unsafe" region so that the remodeling can be performed only within the planned boundary.
- Two exemplary systems, automatic bone remodeling system and haptic guidance system, are considered.
- The case of automatic bone remodeling system can be classified as the Batch processing (BP) level. The system performs the bone reshaping as preprogrammed or planned by surgeon. Surgeon can temporarily hold or abort the operation of the robot in the case of unexpected event or emergency such as unexpected movement of patient or the robot system, unexpected interruption of the vision system, unintended damage to surrounding tissue and resultant bleeding, and etc. The haptic guidance system or speed adjustment type system can be classified as the Shared control (SHC) level. The guidance system generates haptic feedback structure or speed adjustment decision and executes the repulsive force generation or speed attenuation. However, the operator holds the bone milling end-effector and can override the guiding reaction from the system as needed. Although it is not that whole execution is shared by the

operator and the robotic system, the robotic system actively participates in the decision making process to determine proper region of operation.

- Tables 5 describes the effect of DoA on risk assessment for the orthopedic surgery assist robot. Autonomy is utilized mainly in the bone milling action of surgeon for improved precision and accuracy of the task.
- Potential risks associated with the DOA in this type of system are mainly related to the surgical object identification and tracking. The Pre-programmed Execution type robot works not much differently as compared to conventional programmable electronic medical device as long as the robot and the patient keep planned position and pose during the operation. As is in the case of conventional drug infusion pump that does programmed drug infusion without human intervention for programmed period with programmed speed value or pattern, the robot performs bone reshaping as planned. Risks associated unexpected change of the patient location and pose relative to the robot, and/or unexpected interruption of the tracking device, and so on should be properly analyzed and dealt with by either software function or hardware device in the system. In the case of the Shared Decision type systems also, mostly the decision made or suggested by the robotic system is related to the geometrical information such as shape, location, and pose of the bone and the planned operating area, and possible risks are similar as in the Pre-programmed Execution type system. One additional consideration may include analysis of risks related to the interpretation process of the geometrical information by the robotic system such as image segmentation and registration process.

Table 6 Effect of degree of autonomy on risk assessment

Function/Task	Hazards	Impact of increasing DoA		Additional considerations
		Reduced risk	Increased risk	
Motion guide (w/ some type of haptic "virtual wall")	Error in object shape/position identification	Improved accuracy in the identification of the shape/position (by quantitative spatial identifier such as optical/electromagnetic tracker)	Taking control from surgeon may introduce critical error and complication	Implementation of proper count measure or exception handling functions in the robot system with certain amount of DoA based on risk assessment
Automatic tool moving	Error in handling the tool in movement	Improved precision/accuracy and safety in the tool movement (robot can suggest "correct" boundary – motion guide)	Unexpected interruption in the machine operation could cause critical damage in patient (ex. vision system error, human interruption, control and etc.)	

7 Convergence

Robotics is one of representative convergent technology and medical robot is based on further convergence with medicine and other engineering technologies. The convergence is driven by eminent clinical unmet needs. Conventional laparoscopic surgery requires long training period for surgeon to reach certain level of skill maturity and some type of the operations are still difficult and meticulous job for even trained surgeon due to the limited dexterity and vision. Robotic laparoscopic surgery system brought breakthrough in these dexterity and vision limitation issue by successfully utilizing conventional robotic mechanism vision technologies. Robotic bone reshaping system could significantly increase the accuracy and precision of the operation, but, the technologies to implement it was not much different from those for conventional industrial robots. Interventional procedures that require smaller invasiveness and workspace and image-guide integration are expected to be benefited by advanced robotic system technologies that enable smaller device with higher dexterity, intelligent imaging information integration, task automation for full teleoperation, and etc. The minimal invasiveness will be ever increasingly pursued for patients' benefit requiring more functions in smaller device more integrated with imaging modalities. Novel material, fabrication method, algorithms and software, and etc. need to be developed and utilized to overcome and resolve the new or extended clinical unmet needs. Some of the problems may not be solved by straightforward assembly or integration of conventional technologies. Fusion or convergence of technologies can be breakthrough instead. Making device smaller with conventional driving mechanism by precision machining technologies must have its hard limit and fusion of novel material technology may enable smaller device with active mechanism. Haptics and virtual reality technologies already have opened a variety of new possibilities and applications such as haptic feedback in teleoperation type surgical robot systems and various virtual simulators for clinical skill training of both medical student and skilled clinician pursuing newly advanced procedure training, and will continue to be a key factor for augmenting clinicians' sense and recognition in the clinical procedure toward better clinical outcome for patient. 3D printing technologies integrated with robotic mechanism and control technique may bring unexpected innovation in surgery or procedure operation room such as in situ patient specific design implant fabrication, real-time direct printing of artificial tissue replacement in custom shape, or real-time fabrication of active electromechanical medical device to be implanted, all implanted or applied in minimally invasive manner through robotic system. Millimeter size devices to be inserted and work in blood vessel are relatively easily devised with current electromagnetic and mechatronic technologies. Some researchers are going further down developing, so called, bacteria robot, which is not only of size of bacteria, but also formed by hybridization of actual biological bacteria and artificial electromechanical structure. Although practically useful functions are yet to be devised and developed, the tiny stand-alone devices based on technology fusion will be one of representative examples of ultimate minimally invasive diagnosis and treatment when it is successfully developed up to clinically applicable stage in hopefully near future.

References

1. Saliba, W., Reddy, V.Y., Wazni, O., Cummings, J.E., Burkhardt, J.D., Haissaguerre, M., Kautzner, J., Peichl, P., Neuzil, P., Schibgilla, V., Noelker, G., Brachmann, J., Di Biase, L., Barrett, C., Jais, P., Natale, A.: Atrial fibrillation ablation using a robotic catheter remote control system: initial human experience and long-term follow-up results. J. Am. Coll. Cardiol. **51**(25), 2407–2411 (2008)
2. Steven, D., Servatius, H., Rostock, T., Hoffmann, B., Drewitz, I., Müllerleile, K., Sultan, A., Aydin, M.A., Meinertz, T., Willems, S.: Reduced fluoroscopy during atrial fibrillation ablation: benefits of robotic guided navigation. J. Cardiovasc. Electrophysiol. **21**(1), 6–12 (2010)
3. Dewire, J., Calkins, H.: State-of-the-art and emerging technologies for atrial fibrillation ablation. Nat. Rev. Cardiol. **7**(3), 129–138 (2010)
4. Walsh, C.J., Hanumara, N.C., Slocum, A.H., Shepard, J.A., Gupta, R.: A patient-mounted, telerobotic tool for ct-guided percutaneous interventions. Journal of Medical Devices **2**(1), 011007–10 (2008)
5. Piccin, O., Bayle, B., Maurin, B., de Mathelin, M.: Kinematic modeling of a 5-dof parallel mechanism for semi-spherical workspace. Mechanism and Machine Theory **44**(8), 1485–1496 (2009)
6. Stoianovici, D., Cleary, K., Patriciu, A., Mazilu, D., Stanimir, A., Craciunoiu, N., Watson, V., Kavoussi, L.: Acubot: a robot for radiological interventions. IEEE Transactions on Robotics and Automation **19**(5), 927–930 (2003)
7. Bebek, O., Hwang, M.J., Cavusoglu, M.C.: Design of a parallel robot for needle-based interventions on small animals. IEEE/ASME Transactions on Mechatronics **18**(1), 62–73 (2013)
8. Kobayashi, Y., Hong, J., Hamano, R., Okada, K., Fujie, M.G., Hashizume, M.: Development of a needle insertion manipulator for central venous catheterization. The International Journal of Medical Robotics and Computer Assisted Surgery **8**(1), 34–44 (2012)
9. Moon, Y., Choi, H.J., Seo, J.B., Choi, J.: Design and kinematic analysis of a new end-effector for robotic needle insertion-type intervention system. International Journal of Advanced Robotic Systems **11**(190), 1–12 (2014)
10. Chung, J, Cha, H.J., Yi, B.J., Kim, W.K.: Implementation of a 4-dof parallel mechanism as a needle insertion device. In: IEEE International Conference on Robotics and Automation, Anchorage, USA, May 3-8, 2010
11. Neubach, Z., Shoham, M.: Ultrasound-Guided Robot for Flexible Needle Steering. IEEE Trans. on Biomed. **57**(4), 799–805 (2010)
12. Tang, L., Chen, Y., He, X.: Compliant needle modeling and steerable insertion simulation. Computer-Aided Design and Applications **5**, 39–46 (2008)
13. Okazawa, S., Ebrahimi, R., Chuang, J., Salcudean, S.E., Rohling, R.: Hand-held steering needle device. IEEE/ASME Transactions on Mechatronics **10**(3), 285–296 (2005)
14. Schueler, B.A.: Operator shielding: how and why. Techniques in Vascular and Interventional Radiology **13**(3), 167–171 (2010)

15. Melzer, A., Gutmann, B., Remmele, T., Wolf, R., Lukoscheck, A., Bock, M., Bardenheuer, H., Fischer, H.: Innomotion for percutaneous image-guided interventions. IEEE Engineering in Medicine and Biology Magazine **27**(3), 66–73 (2008)
16. Salcudean, S.E., Prananta, T.D., Morris, W.J., Spadinger, I.: A robotic needle guide for prostate brachytherapy. In: IEEE International Conference on Robotics and Automation (ICRA), Pasadena, USA, pp. 2975–2981, May 19-23 2008
17. Bassan, H.S., Patel, R.V., Moallem, M.: A novel manipulator for percutaneous needle insertion: Design and experimentation. IEEE/ASME Transactions on Mechatronics **14**(6), 746–761 (2009)
18. Loser, M.H., Navab, N.: A new robotic system for visually controlled percutaneous interventions under ct fluoroscopy. In: Medical Image Computing and Computer-Assisted Intervention (MICCAI), Pittsburgh, USA, pp. 887–896, October 11-14 2000
19. Conradie, J.P.: Fluoroscopy based needle positioning system for percutaneous nephrolithotomy procedures. PhD Dissertation, Stellenbosch: Stellenbosch University (2008)
20. Chen, X.: Instrument guide for MRI-guided percutaneous interventions. PhD Dissertation, Cambridge: Massachusetts Institute of Technology (2010)
21. Lin, M.L., Yang, B.D., Wang, Y.H., Yang, C.L., Wang, J.L.: A miniature patient-mount navigation system for assisting needle placement in ct-guided intervention. The International Journal of Medical Robotics and Computer Assisted Surgery **7**(4), 423–430 (2011)
22. Hungr, N., Fouard, C., Robert, A., Bricault, I., Cinquin, P.: interventional radiology robot for CT and MRI guided percutaneous interventions. In: Fichtinger, G., Martel, A., Peters, T. (eds.) MICCAI 2011, Part I. LNCS, vol. 6891, pp. 137–144. Springer, Heidelberg (2011)
23. Najafi, F., Sepehri, N.: A robotic wrist for remote ultrasound imaging. Mechanism and Machine Theory **46**(8), 1153–1170 (2011)
24. Zoppi, M., Zlatanov, D., Gosselin, C.M.: Analytical kinematics models and special geometries of a class of 4-dof parallel mechanisms. IEEE Transactions on Robotics **21**(6), 1046–1055 (2005)
25. Lum, M., Rosen, J., Sinanan, M.N., Hannaford, B.: Optimization of a spherical mechanism for a minimally invasive surgical robot: theoretical and experimental approaches. IEEE Transactions on Biomedical Engineering **53**(7), 1440–1445 (2006)
26. Zhang, X., Lehman, A., Nelson, C.A., Farritor, S.M., Oleynikov, D.: Cooperative robotic assistant for laparoscopic surgery: Cobrasurge. In: IEEE/RSJ International Conference on Intelligent Robots and Systems (IROS), St. Louis, USA, pp. 5540–5545, October 11-15 2009
27. Ayvali, E., Liang, C.P., Ho, M., Chen, Y., Desia, J.P.: Towards a discretely actuated steerable cannula for diagnostic and therapeutic procedures. International Journal of Robotics Research. **31**(5), 588–603 (2012)
28. Sears, P., Dupont, P.: A steerable needle technology using curved concentric tubes. In: Proceedings of the 2006 IEEE/RSJ International Conference on Intelligent Robots and Systems (2009)
29. Torabi, M., Hauser, K., Alterovitz, R., Duindam, V., Goldberg, K.: Guiding medical needles using single-point tissue manipulation. In: Proc. IEEE Int. Conf. Robot. Autom., pp. 2705–2710 (2009)
30. Burkhardt, J.D., Natale, A.: New technologies in atrial fibrillation ablation. Circulation **120**(15), 1533–1541 (2009)

31. Bunch, T.J., Asirvatham, S.J., Friedman, P.A., Monahan, K.H., Munger, T.M., Rea, R.F., Sinak, L.J., Packer, D.L.: Outcomes after cardiac perforation during radiofrequency ablation of the atrium. J. Cardiovasc. Electrophysiol. **16**(11), 1172–1179 (2005)
32. Fu, Y., Liu, H., Huang, W., Wang, S., Liang, Z.: Steerable catheters in minimally invasive vascular surgery. Int. J. Med. Robotics. Comput. Assist. Surg. **5**, 381–391 (2009)
33. Chun, K.R., Schmidt, B., Köktürk, B., Tilz, R., Fürnkranz, A., Konstantinidou, M., Wissner, E., Metzner, A., Ouyang, F., Kuck, K.H.: Catheter ablation - new developments in robotics. Herz Kardiovaskuläre Erkrankungen **33**(8), 586–589 (2008)
34. Ernst, S.: Robotic approach to catheter ablation. Curr. Opin. Cardiol. **23**, 28–31 (2008)
35. Sensei robotic catheter system [updated 28 March 2011; cited 12 May 2011]. http://en.wikipedia.org/wiki/Sensei_robotic_catheter_system
36. Reddy, V.Y., Neuzil, P., Malchano, Z.J., Vijaykumar, R., Cury, R., Abbara, S., Weichet, J., McPherson, C.D., Ruskin, J.N.: View-synchronized robotic image-guided therapy for atrial fibrillation ablation: experimental validation and clinical feasibility. Circulation **115**(21), 2705–2714 (2007)
37. Rosenberg, L.B.: Virtual fixtures: Perceptual tools for telerobotic manipulation. In: Proceedings of the IEEE Virtual Reality Annual International Symposium, pp. 76–82 (1993)
38. Abbott, J.J., Marayong, P., Okamura, A.M.: Haptic virtual fixtures for robot-assisted manipulation. Springer Tracts in Advanced Robotics **28**, 49–64 (2007)
39. Li, M., Ishii, M., Taylor, R.H.: Spatial motion constraints using virtual fixtures generated by anatomy. IEEE T. Robot. **23**(1), 4–19 (2007)
40. Li, M., Okamura, A.M.: Recognition of operator motions for real-time assistance using virtual fixtures. In: Proceedings of the 11th Symposium on Haptic Interfaces for Virtual Environment and Teleoperator Systems, pp. 125–31(2003)
41. Bettini, A., Marayong, P., Lang, S., Okamura, A.M., Hager, G.D.: Vision-assisted control for manipulation using virtual fixtures. IEEE T. Robot. **20**(6), 953–966 (2004)
42. Hungr, N., Fouard, C., Robert, A., Bricault, I, Cinquin, P.: Interventional Radiology Robot for CT and MRI Guided Percutaneous Interventions. In: Medical Image Computing and Computer-Assisted Intervention, Toronto, Canada (2011)
43. Reed, K.B., Kallem, V., Alterovitz, R., Goldberg, K., Okamura, A.M., Cowan, N.J.: Integrated Planning and Image-Guided Control for Planar Needle Steering. In: Proceedings of the IEEE/RAS-EMBS International Conference on Biomedical Robotics and Biomechatronics, pp. 819–824 (2008). doi:10.1109/BIOROB.2008.4762833
44. Cleary, K., Watson, V., Lindisch, D., Taylor, R.H., Fichtinger, G., Xu, S., White, C.S., Donlon, J., Taylor, M., Patriciu, A., Mazilu, D., Stoianovici, D.: Precision placement of instruments for minimally invasive procedures using a "needle driver" robot. Int. J. Med. Robot. **1**(2), 40–47 (2005)
45. Marcelli, E., Cercenelli, L., Plicchi, G.: A Novel Telerobotic System to Remotely Navigate Standard Electrophysiology Catheters. Computers in Cardiology **35**, 137–140 (2008)
46. Kronreif, G., Fürst, M., Ptacek, W., Kornfeld, M., Kettenbach, J.: Robotic system for image guided Therapie - B-Rob II. In: RAAD Workshop, BFD-022 (2006)
47. Maurin, B., Bayle, B., Piccin, O., Gangloff, J., de Mathelin, M., Doignon, C., Zanne, P., Gangi, A.: A Patient-Mounted Robotic Platform for CT-Scan Guided Procedures. IEEE Transactions on Biomedical Engineering **55**(10), 2417–2425 (2008)

48. Kaber, D.B., Endsley, M.R.: The effects of level of automation and adaptive automation on human performance, situation awareness and workload in a dynamic control task. Theor. Issues in Ergon. Sci. **5**(2), 113–153 (2004)
49. Fischer, G.S., Krieger, A., Iordachita, I.I., Csoma, C., Whitcomb, L.L., Fichtinger, G.: MRI compatibility of robot actuation techniques – A comparative study. In: Metaxas, D., Axel, L., Fichtinger, G., Székely, G. (eds.) MICCAI 2008, Part II. LNCS, vol. 5242, pp. 509–517. Springer, Heidelberg (2008)
50. Zemiti, N., Bricault, I., Fouard, C., Sanchez, B., Cinquin, P.: A CT and MR compatible puncture robot to enhance accuracy and safety of image-guided interventions. IEEE/ASME Transactions on Mechatronics **13**(3), 306–315 (2008)
51. Taillant, E., Avila-Vilchis, J.-C., Allegrini, C., Bricault, I., Cinquin, P.: CT and MR compatible light puncture robot: architectural design and first experiments. In: Barillot, C., Haynor, D.R., Hellier, P. (eds.) MICCAI 2004. LNCS, vol. 3217, pp. 145–152. Springer, Heidelberg (2004)
52. Moon, Y., Choi, J.: Development of a robotic mechanism for teleoperation-based needle interventions. In: 44th International Symposium on Robotics, pp. 1–3, October 2013
53. Krieger, A., Susil, R.C., Menard, C., Coleman, J.A., Fichtinger, G., Atalar, E., Whitcomb, L.L.: Design of a novel MRI compatible manipulator for image guided prostate interventions. IEEE Transactions on Biomedical Engineering **52**(2), 306–313 (2005)
54. Krieger, A., Iordachita, I., Guion, P., Singh, A.K., Kaushal, A., Menard, C., Pinto, P.A., Camphausen, K., Fichtinger, G., Whitcomb, L.L.: An MRI-compatible robotic system with hybrid tracking for MRI-guided prostate intervention. IEEE Transactions on Biomedical Engineering **58**(11), 3049–3060 (2011)
55. Taillant, E., Avila-Vilchis, J.-C., Allegrini, C., Bricault, I., Cinquin, P.: CT and MR Compatible Light Puncture Robot: *Architectural Design and First Experiments*. In: Barillot, C., Haynor, D.R., Hellier, P. (eds.) MICCAI 2004. LNCS, vol. 3217, pp. 145–152. Springer, Heidelberg (2004)
56. Fischer, G., Iordachita, I., DiMaio, S.P., Fichtinger, G.: Design of a robot for transperineal prostate needle placement in MRI scanner. In: IEEE International Conference on Mechatronics, pp. 592–597 (2006)
57. Song, S.E., Cho, N.B., Fischer, G., Hata, N., Tempany, C., Fichtinger, G., Iordachita, I.: Development of a pneumatic robot for MRI-guided transperineal prostate biopsy and brachytherapy: New approaches. In: IEEE International Conference on Robotics and Automation, pp. 2580–2585 (2010)
58. Yakar, D., Schouten, M.G., Bosboom, D.G.H., Barentsz, J.O., Scheenen, T.W.J., Futterer, J.J.: Feasibility of a pneumatically actuated MR-compatible robot for transrectal prostate biopsy guidance. Radiology **260**(1), 241–247 (2011)
59. Muntener, M., Patriciu, A., Petrisor, D., Mazilu, D., Bagga, H., Kavoussi, L., Cleary, K., Stoianovici, D.: Magnetic resonance imaging compatible robotic system for fully automated brachytherapy seed placement. Urology **68**, 1313–1317 (2006)
60. Stoianovici D, Patriciu A, Petrisor D, Mazilu D, Kavoussi L, A new type of motor: Pneumatic step motor. IEEE/ASME Transactions on Mechatronics 12(1): 98–106
61. Stoianovici, D., Song, D., Petrisor, D., Ursu, D., Mazilu, D., Muntener, M., Schar, M., Patriciu, A.: "MRI Stealth" robot for prostate interventions. Minimally Invasive Therapy **16**(4), 241–248 (2007)

62. Zngos, S., Herzog, C., Eichler, K., Hammerstingl, R., Lukoschek, A., Guthmann, S., Gutmann, B., Schoepf, U.J., Costello, P., Vogl, T.J.: MR-compatible assistance system for punction in a high-field system: Device and feasibility of transgluteal biopsies of the prostate gland. Eur. Radiol. **17**, 1118–1124 (2007)
63. Goldenberg, A.A., Trachtenberg, J., Kucharczyk, W., Yi, Y., Haider, M., Ma, L., Weersink, R., Raoufi, C.: Robotic system for closed-bore MRI-guide prostatic interventions. IEEE/ASME Transactions on Mechatronics **13**(3), 374–379 (2008)
64. Elhawary, H., Zivanovic, A., Rea, M., Davies, B., Besant, C., McRobbie, D., de Souza, N., Young, I., Lampérth, M.: The feasibility of mr-image guided prostate biopsy using piezoceramic motors inside or near to the magnet isocentre. In: Larsen, R., Nielsen, M., Sporring, J. (eds.) MICCAI 2006. LNCS, vol. 4190, pp. 519–526. Springer, Heidelberg (2006)
65. Tokuda, J., Fischer, G.S., DiMaio, S.P., Gobbi, D.G., Csoma, C., Mewes, P.W., Fichtinger, G., Tempany, C.M., Hata, N.: Integrated navigation and control software system for MRI-guided robotic prostate interventions. Computerized Medical Imaging and Graphics **34**, 3–8 (2010)

Robotics for Healthcare

Jung Kim, Gwang Min Gu and Pilwon Heo

Abstract Since the first industrial robot was introduced in the 1960s, robotic technologies have contributed to enhance the physical limits of human workers in terms of repeatability, safety, durability, and accuracy in many industrial factories including those of the automobile, consumer electronics and shipbuilding industries. In the 21st century, robots are expected to be further applied in healthcare, which requires procedures that are objective, repetitive, robust and safe for users. Fueled by the rapid improvements of medical imaging and mechatronics technologies, healthcare robots have been rapidly adopted in almost every stage of the medical procedure by surgeons and physical therapists. In this chapter, we describe applications and the state of the art of healthcare robotics developed in the last decade. We focus on research and clinical activities that have followed successful demonstrations of early pioneering robots such as daVinci telesurgical robots and LOKOMAT training robots. First, we categorize major areas of healthcare robotics. Second, we discuss robotics for surgical operating rooms. Third, we review rehabilitation and assistive technologies. Finally, we summarize challenges and limitations of biomedical robotics as assistive tools for medical personnel.

1 Introduction

Ever since the definition of robot as a "reprogrammable machine", people have predicted that the robot will replace the human worker in the near future. This idea has been realized at least in industry, where it is common for robots to assemble automobiles and relocate heavy loads, which might also be dirty or dangerous to human workers. (It is debatable whether robots can replace workers in performing "difficult" jobs.) The definition of the robot has continued to evolve to "professional and personal service robots" and the next area for the application of robots

J. Kim(✉) · G.M. Gu · P. Heo
KAIST, Daejeon, South Korea
e-mail: jungkim@kaist.ac.kr

© Springer International Publishing Switzerland 2016

H. Jo et al. (eds.), *Biomedical Engineering: Frontier Research and Converging Technologies*,
Biosystems & Biorobotics 9,
DOI: 10.1007/978-3-319-21813-7_21

is the medical and healthcare fields, which require safety and reliability for all stock holders from patients to insurance companies [1]. According to the Merriam–Webster dictionary, the definition of healthcare is "the maintaining and restoration of health by the treatment and prevention of disease especially by trained and licensed professionals". Healthcare robots are therefore supposed to provide trained and licensed professionals with advantages (accuracy and sturdiness) and additional information (visual and haptic) and thus improve health conditions without compromising the safety of patients.

Since the first attempts in the 1980s, robots have been used in orthopedic surgery, neurosurgery, laparoscopy, physical therapy treatments, and other several areas in the field of medicine. Several companies including Intuitive Surgical, Hocoma, Hansen Medical and Honda, have launched healthcare robots with notable success. There have been important developments in the application of robotic technologies to new healthcare procedures not limited to laparoscopic and physical training procedures.

1.1 Taxonomy of Healthcare Robotics

Several review papers and edited books have been treated medical robots [2], [3]. The classification of different branches of biomedical robotics is certainly not an easy task. From an engineer's perspective, robots are often classified by their level and type of autonomy into autonomous systems, teleoperated systems and assistive systems. For example, currently used laparoscopic surgical robots are classified as teleoperated robots. In contrast, domain-specific taxonomy can be employed to categorize medical robotics into two main categories: surgical robots and physical therapy/assistant robots. The major difference between the two categories is the invasiveness of the target procedures. Surgical robots are designed to perform or assist in interventional procedures performed in operating rooms. Physical therapy/assistant robots are intended to help recovery by providing therapy and monitoring the progress or assist weakened forces in an activity of daily living. Fig. 1. shows the major categories of and important keywords relating to healthcare robots.

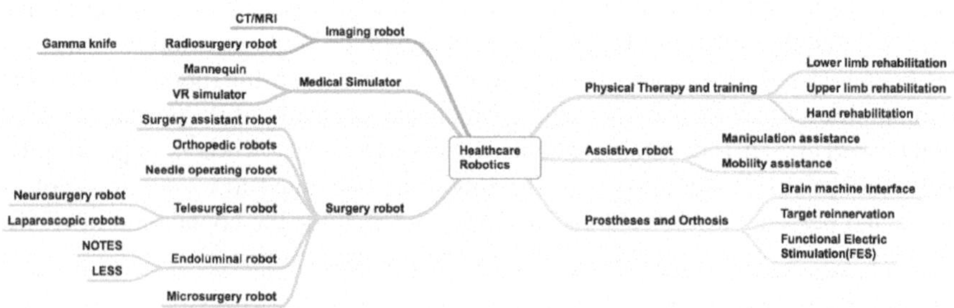

Fig. 1 Major categories of and important keywords relating to healthcare robots

In this chapter, we review the achievements of healthcare robotics in recent decades. We first discuss robotic systems for surgical operations that improve patient safety. We then review physical therapy training/assistive robots for disabled and aging people. Here, we exclude prostheses, orthoses, and robotic transportation assistance devices owing to space limitations but these areas are also considered important in terms of robotics applications. Finally, we conclude the chapter by discussing challenges facing biomedical robotics.

2 Robotics for Patient Safety

Surgical robotics is a relatively young research area, with robot-assisted surgery first being reported in 1985. A Puma 560 industrial robot (Advance Research & Robotics, Oxford, CT, USA) was modified to be used for defining the trajectory of a brain biopsy, taking advantage of the features of industrial robots such as their high repetitive accuracy and absolute positioning accuracy [4].

Minimally invasive surgery, which includes laparoscopic surgery, has been the most important advance in every specialty of surgical medicine in recent decades. Because a minimally invasive procedure uses smaller incisions or openings than conventional surgery, it has several advantages over traditional open surgery including quick recovery, less scarring and shorter hospital stays. Although there are many advantages on the patient side, the surgeon is hampered by, for example, limited vision, a lack of force feedback, a very restricted range of motion and a fulcrum effect that makes it difficult to manipulate tissue. These difficulties are interpreted by engineers as "laboriousness" and a robotic system for minimally invasive surgery was considered to overcome these difficulties. Following the success story of the da Vinci surgical system (Intuitive Surgical, Mountain View, CA, USA), which was founded in 1995, the advantages of using a robot have been well documented and accepted by many clinicians over the last two decades. A camera on an arm allows for improved or three-dimensional visualization via a stereoscopic display. Manipulators having two or three degrees of freedom (DOFs) carry the instruments. Such manipulators include the articulated EndoWrist, which provides greater dexterity, especially in suturing.

Many researchers have pursued other goals. One notable development has been the RAVEN Surgical Robot platform created by a group of researchers led by B. Hannaford, who demonstrated remote surgeries using various network conditions in 2007 [5]. Recently, the group developed the smaller RAVEN 2, which has hands that are more dexterous and can hold surgical tools during operations. They developed their software using open-source codes so that RAVEN can be connected to other devices. The da Vinci Surgical System is expensive and uses proprietary software, whereas the RAVEN robotic system is more accessible to researchers around the world [6].

Although laparoscopic suturing has been made easier by the use of EndoWrist (Intuitive Surgical, Sunnyvale, CA) tools and a single-port stereo vision system, progress in orienting and manipulating curved needles within deformable tissue has been slow. Drake *et al.* investigated an automated suturing method that adopts a single manipulator to implement a suturing task with a standard laparoscopic needle holder and curved suture needle. Three-dimensional images are obtained by a clinical endoscope employing an elliptical/circular pose measurement algorithm, which dynamically tracks the suture needle and surface markers [7].

The invasiveness of laparoscopic surgery has been decreased as the techniques and equipment have been advanced. The concept of minimal invasiveness has resulted in the evolution of new types of surgical robots that require only one opening (single-port surgery) or, in the case of NOTES (Natural Orifice Translumenal Endoscopic Surgery), no artificial opening of the skin. Such procedures mainly use endoscopic instruments, which are slender and more flexible. Single-port surgery is a surgical method performed by inserting several surgical instruments and a laparoscope through an umbilical incision [8]. The research group of D. Kwon proposed a surgical robot system for single-port surgery that uses a special joint mechanism to avoid collisions between surgical tools or arms and approaches the surgical target more easily than a conventional straight surgical tool [8]. Kim *et al.* also developed a surgical robot for single-incision laparoscopic surgery. Their robot system includes a cone-type remote center-of-motion mechanism and two articulated instruments having a flexible linkage-driven elbow. They demonstrated that the robot can carry a payload of more than 10 N and described preliminary experiments on peg transfer and suture motion using the proposed surgical robot [9]. Dario *et al.* developed the Araknes surgical system for single-port laparoscopy through international collaboration. They demonstrated their design frame for surgical robots including software architectures and hardware controllers. They summarized important considerations and steps in the development of surgical robots to be used in operating rooms [10]. The group also developed a robotic platform for laser-assisted transurethral surgery of the prostate for benign prostatic hyperplasia. The system consists of a catheter-like robot equipped with a fiber optic-based sensing system and a cable-driven actuation mechanism [11].

While procedures that are less invasive provide better surgical outcomes than laparoscopic procedures, the difficulties of manipulation and limited workspace become more serious [12]. Whereas single-port surgical robots are considered a special type of laparoscopic robot, the NOTES robot requires new technologies and faces new challenges [13]. Mathelin *et al.* developed STRAS for endoluminal and transluminal surgery. The STRAS robot was designed as a modular robotic system, compatible with the medical environment, allowing an easy setup in the operating room using flexible shafts and continuum bending sections. It provides up to 10 DOFs, allowing the three-dimensional positioning of an endoscopic camera, the positioning of two instruments and grasper opening–closing functionalities [14]. Researchers at the University of Wisconsin–Madison led by M. Zinn proposed a new approach for continuum robotic manipulation that combines flexible, actively actuated continuum segments with small, limited-stroke rigid-link actuators. The small rigid-link joints are interwoven between successive continuum segments and provide a redundant motion and error correction capability. Although they demonstrated their approach only for cases with a limited number of

DOFs, their approach had some potential to compensate for flexible segment motion errors, which the majority of NOTES robots suffer from [15]. Unlike laparoscopic robots, which are nothing but mechatronic versions of traditional instruments, NOTES cannot be easily operated or diversified using traditional endoscopic instruments. The success of NOTES will thus be determined by sufficient improvements of the NOTES robot, which requires sufficient manipulation forces during operation and sufficient flexibility during inletting of the instruments on the same platform, which is difficult to achieve owing to the contradictive nature of the requirements.

2.1 Medical Simulation

The nature of minimally invasive procedures resulted in the development of another new interesting technology, namely virtual reality technologies, in the area of medical training. Because the sense of touch has been reduced by the introduction of minimally invasive techniques when compared to the open surgery, surgeons have to rely more on the sensation of tool–tissue interactions through a long slender instrument and require more training for successful operation on patients. In fact, refinement of the surgeon's sensorimotor system takes an important role in the art of minimally invasive surgery. [16]

Medical simulations based on virtual reality provide an environment identical to that of real medical surgery for the training of surgeons, and therefore the surgeons can obtain numerous medical skills without causing real-world problems. In particular, the emergence of haptic devices with high-fidelity graphics provides a sufficient level of realism in the simulation of laparoscopic surgery, which requires only a limited scope of images and force feedback through long, slender tools. For example, Simbionix (Cleveland, OH, USA) provides an expanding set of modules including those for the training of knot tying, suturing, and gastric bypass along with decision making and teamwork tasks [17]. It also covers various procedures such as endovascular procedures and laparoscopic surgeries. Although there have been extensive discussions on the effectiveness of simulation-based training, a recent systematic study showed that simulation-based training for laparoscopic surgery has great benefits when compared with no intervention and is moderately more effective than instruction without simulation, which used to be the apprenticeship-based model of training [17], [18]. Currently, researchers and medical simulation industries are strongly pushing the development of a more versatile force feedback device, cost-effective tactile feedback and software curriculum design and training metrics.

2.2 Mechatronic Devices for Operating Rooms

Despite the success of da Vinci robots, there are many procedures awaiting advanced technologies. Still, behind the scenes, many robotic devices and technologies have been applied to traditional surgical procedures and minimally invasive surgeries.

First, many mechatronic components were added to traditional surgical instruments to enhance tool–tissue interaction forces or to measure forces critical to the design specifications of surgical device/robots. Recently, Jo *et al.* created two-DOF compliant forceps for the measurement of pulling and grasping forces at the tip of a surgical instrument for improvement of the haptic feedback of a surgical robot [19]. Zhang *et al.* designed and fabricated a multiple-DOF force-sensing instrument to compute operating forces using a cable-driven mechanism [20].

Sutherland *et al.* designed micro force-sensing units that measure the forces exerted by surgeons during neurosurgical procedures. In a human cadaver study, they found that the forces needed to manipulate brain tissue were very low and varied depending on the anatomical structure being manipulated [21]. Yang *et al.* designed a hand-held device capable of amplifying delicate micromanipulation forces during minimal invasive surgical tasks. The device relays measured forces imparted to the user from the surgical instrument. They showed a five-times reduction in the minimum force threshold perceived by the subjects during microsurgery [22].

The group led by I. Iordachita and R. Tayer reported the development of a fiber Bragg grating sensor that has the ability not just to sense forces at the tip of a surgical instrument located inside the eye but also to provide information about the interaction force between the shaft of the instrument and the sclera during vitreoretinal surgery. The sclera section provides vital feedback for cooperative robot control to minimize potentially dangerous forces acting on the eye. The use of the force scaling robot allowed significant reduction of forces on the sclera [23].

Second, instrumented surgical devices with advanced material structures have been actively investigated. Webster *et al.* developed a telerobotic system for transnasal surgery that employs a concentric tube continuum mechanism. The catheter-sized continuum robots are appropriate for minimally invasive surgery within confined body cavities [24]. They showed the potential for surgical procedures such as endonasal skull base surgery, which requires more dexterity in using instruments. A similar mechanism was investigated for a neurosurgical example that involves choroid plexus cauterization for hydrocephalus treatment and closure of a patent foramen ovale inside the beating heart [25]. Recently, Webster's group developed a new hand-held robotic system for new holmium laser enucleation of the prostate (HoLEP), which is known to have better clinical outcomes than TURP (Transurethral Resection of the Prostate) yet is underused because of how challenging it is for the surgeon to perform using conventional endoscopic instruments. The system provides the surgeon with two concentric tube manipulators that can aim the laser and manipulate tissue simultaneously. The manipulators are deployed through a 5-mm working channel in a #26-French (8.66-mm) endoscope clinically used for transurethral procedures [26].

Third, special surgical procedures have specific engineering challenges. Robotics researchers collaborated with clinicians to overcome these challenges.

Howe *et al.* developed a special control algorithm for beating-heart surgery. By controlling a teleoperated robot to continuously follow the heart's vibrating motion, the heart can be made to appear stationary. Using Smith predictor theory, they successfully tracked heart motion during a surgical procedure [27]. A research group at Johns Hopkins University developed an integrated assistive system for membrane peeling in vitreoretinal procedures, combining an active tremor-canceling handheld micromanipulator with force-sensing motorized micro-forceps. Their micro-forceps are a 20-Ga instrument that is mechanically decoupled from its handle and senses the transverse forces at its tips with accuracy of 0.3 mN [28].

Finally, robotic devices were developed to help adopt new imaging modalities or new procedures. Simaan *et al.* designed a robotic manipulator with 11 DOFs for retinal micro-vascular surgery. It uses stents to maintain the structural integrity in artery/vein crossings for ophthalmic microsurgery. The robot also allows for the quick exchange of surgical graspers and the integration of a custom-made B-mode optical coherence tomography probe for image guidance [29]. The SSSA group led by A. Menciassi developed a multi-viewpoint, magnetic actuated laparoscope to implement nine-view autostereoscopic displays in minimally invasive surgery. The system is anchored by a magnetic link to the abdomen and freely moved by magnetic actuation to adjust the points of view and the horizons of the cameras [30].

Gitlin developed the Miniature Anchored Robotic Videoscope for Expedited Laparoscopy (MARVEL) and a wireless laparoscopic camera module that allow wireless robotic laparoscopic imaging. Two MARVEL camera modules were successfully tested *in vivo* in a porcine subject [31].

3 Rehabilitation and Assistive Robots

3.1 Conventional Therapy and Robot-Aided Therapy

Robot-aided therapies and their applications have been proposed to meet the increasing demands of an aging society. The number of therapists is insufficient compared with the number of patients. Physical therapy involves the laborious and repetitive intervention of therapists. In gait rehabilitation, a patient is forced to move his/her legs passively by three therapists. Moreover, physical therapists and occupational therapists work together in conjunction to treat patients. Robotic rehabilitation is expected to support a therapist, allowing more intensive and repetitive training, and to assess patients' motor recovery quantitatively, measuring changes in movement kinematics and forces. Many kinds of robotic rehabilitation systems have been developed but not clinically evaluated thoroughly. In this section, we describe recent studies on clinical evaluation and commercially available robotic systems, from the perspective of conventional rehabilitation. The scope excludes hand rehabilitation systems.

3.1.1 Perspective of Conventional Rehabilitation

Physical therapy focuses on evaluating and diagnosing movement dysfunctions as well as treating a patient injury. As compared with physical therapy, occupational therapy tends to focus more on improving functional abilities. A patient is treated with a combination of the two therapies by the therapist or medical doctor.

An effective therapy protocol is required in consideration of cost and time. Taub *et al.* proposed a concept of constraint-induced movement therapy (CIMT), which involves repetitive exercises of the affected limb under the instruction of therapists [34]. Standard prescription for the hemiplegic upper limb has not involved CIMT due to the cost of resources needed. Participants generally receive up to 6 hours of one-on-one therapy in a day. Electrical stimulation is one of the tools to be used following the therapist's conventional approaches that can be implemented easily. Electrical stimulation is able to improve function of weak muscles of arm and leg [36]. Mirror therapy is also prescribed to the affected side in the case of hemiplegia, especially for improvements of the arms and hands functions. In mirror therapy, a patient uses movements of the unaffected side to trick his/her brain into thinking that the affected arm is moving virtually, utilizing visual feedback [37]. The Bobath concept, also known as neurodevelopmental treatment, is a therapy based on the movement of components that emphasizes the integration of selective movements in the production of coordinated sequences of a movement [38]. Task practice is part of entire motion to assess an individual movement and to identify specific deficits of neurological functions. Finally, proprioceptive neuromuscular facilitation (PNF) stretching is an exercise based on the principles of functional human anatomy and neuromuscular functions to enhance the active and passive range of motion [39]. PNF uses proprioceptive, cutaneous and auditory input to produce functional improvement in motor output, especially for sports related injuries.

3.1.2 Robotic Rehabilitation Approaches

Developing a robotic rehabilitation system is multidisciplinary work involving a robotic engineering group and physical medicine and rehabilitation specialists. A robotic engineering group first suggests a robot-aided therapy as an alternative to a laborious physical therapy. Many robotic devices have been developed for the arm and gait therapy of individuals who have suffered a stroke [40], [41], but only a few have been widely known at clinical field; e.g., MIT-Manus [42]-[44], Mirror Image Movement Enabler (MIME) [45], [46], Bi-Manu-Track [47]-[49], Armin [50], [51], and Lokomat [52]-[55]. Most commercially available robotic systems were first developed at universities and other academic institutes. As mentioned in the previous section, the safety of the robotic system is the most important consideration for the user. The certification of the electrical and mechanical safety is issued by professional organizations.

For rehabilitation of the upper limb, the MIT Manus system provided sufficient workspace for horizontal motion [42] and vertical motion in most workspace in ADL. One of trends in the development of rehabilitation robotics has been the development of an exoskeleton system with a full range of motion employing multiple actuators. ArmeoPower was the first commercially available exoskeleton-type

Table 1 Summary of clinically evaluated studies and commercialized works in respect to perspectives of conventional therapy

Conventional rehabilitation perspectives	Robotics system	Clinical evaluation	Commercialization
Constraint-induced movement therapy [35] (1918)	MIT Manus [42] (Kreps *et al.* 1998)	Kreps *et al.*, Freeman *et al.* [43-44] (2004)	InMotion Arm (Interactive Motion Tech., Inc., US, 2009)
	ARMIN I [50] (Nef *et al.* 2003)	Nef *et al.* [51] (2009)	ArmeoPower (Hocoma AG, Switzer-land, 2013)
	T-WREX [56] (Sanchez *et al.* 2004)	Sanchez *et al.* [57] (2006)	ArmeoSpring [58] (Hocoma AG, Switzer-land, 2011)
Electrical stimulation [36] (1968)	-	Freeman *et al.* [64] (2009)	NeuroMove (Zynex Medical, Inc., US).
Mirror therapy [37] (1998)	Robotic assisted arm trainer [47] (Hesse *et al.* 2003)	Hesse *et al.* [48,49] (2005,2008)	Bi-Manu-Track (Reha-Stim Co., Ger-many)
	Mirror image motion enhancer(MIME) [45] (Lum *et al.* 2002)	Lum *et al.* [46] (2006)	-
Bobath (NDT: Neurodevelop-mental treatment) [38] (1990)	Continuous passive motion (CPM) [63] (Volpe *et al.* 2000)	-	CPM manchines
Proprioceptive neuromascular facilita-tion [39] (1968)	Ankle rehabilitation system [65] (Zhou *et al.* 2014)	-	-
Gait therapy*	Lokomat [53] (Jezernik *et al.* 2003)	Mayr *et al.* [54-56] (2003)	Lokomat (Hocoma AG, Switzer-land, 2009)
	LokoHelp (Freivogel *et al.* 2008)	Freivogel *et al.* (2009)	LokoHelp (LokoHelp Group)

* All the systems are grouped according to the related perspective of conventional therapy. Gait therapy was difficult to classify them into a particular group.

rehabilitation robot for motor abilities and provides ADL-based arm training in a large workspace. The ARMin was designed by Prof R. Riener team at ETH Zurich University for patients suffered traumatic brain injuries, strokes, and neurological disorders to improve hand and arm functions [50], [51]. ArmeoSpring is passive assistive device for upper-limb rehabilitation utilizing spring-based gravity compensation mechanism. The assistive force is adjusted in consideration of muscle strength and arm weight of the user. The efficacy of the ArmeoSpring was reported by Prof. D. Reinkensmeyer at the University of California at the Rehabilitation Institute of Chicago [56], [57]. NeuroMove detects the EMG signals using three electrodes, and the electrical stimulation is provided for the user. The muscle signals are also displayed for the user in real-time for relearning the movement. Bi-Manu-Track provides bimanual arm training on the basis of mirror therapy. Movements of affected side are assisted by the help of the unaffected limb [47]. In this manner, a cortical representation of the hand larger than the representation of the shoulder was observed. ReoGo is a three-dimensional upper-limb rehabilitation system for repetitive arm movements training [59]. The system has advantage of a wide variety of patients, adjusting assistance level. The variety of the assistance level and motion-based games enhance the motivation of the user.

In terms of rehabilitation of the lower limb, Lokomat, combined with a robotic gait orthosis, an advanced body-weight support, and treadmill, was the first commercially available active gait rehabilitation system. Presently, Lokomat is the most well-known lower-limb rehabilitation system, showing clinical evaluation results [53]-[55]. LokoHelp consists of a harness that supports the body weight and a programmable pedal. Although clinical evaluations of the LokoHelp have also performed, the results showed that the efficacy of the system was similar to a manual training. However, the effort and discomfort of therapist is reduced, when using LokoHelp. ReoAmbulator is one of treadmill-based lower-limb rehabilitation systems, including body-weight supporting harness and robotic arms strapped to the thigh and ankle of user [62].

3.2 Assistive Robots

An assistive robot is, by conventional definition, a robot that physically assists disabled people perform activities of daily living, while the broader definition of an assistive robot encompasses social assistance as well as physical assistance [66]. This subsection focuses on physically assistive robots, which can be classified by assistance type, form, and method of intention recognition. The type of assistance that assistive robots provide can be categorized as either manipulation or mobility assistance. Manipulation assistance robots support the functionalities of the arm and hand in diverse tasks such as self-feeding. Manipulation assistance robots can be classified by form as end-effectors, arm supports, and wearable exoskeletons. Mobility assistance robots have the form of a wheeled platform or wearable exoskeleton. The method of intention recognition is an essential technology required for controlling the assistive robot. The user's motion intention can be extracted from signals acquired by either noninvasive or invasive sensing methods. The invasive methods, however, necessitate the implantation of sensors in the user's body, and noninvasive methods are therefore preferred to allow easy application to a wide variety of users [67]. Manual or intuitive control methods are used for noninvasive intention recognition. Manual control is performed using a joystick, trackball, buttons or similar devices that can be handled by the user, while the intuitive control method captures the intention from the user's natural motion or biosignals.

There have been numerous achievements in the area of assistive robots including robotic arm supports and exoskeletons, meal assistance devices, robotic wheelchairs and mobile platforms, and wearable walking assistance robots. Among the achievements, this subsection introduces recently published research and commercially available products. The assistive robots are classified by the prescribed three criteria—assistance type, form, and intention recognition method—as listed in Table 2.

Table 2 Classification of assistive robots based on assistance type, form, and intention recognition method

Assistance type	Form	Intention recognition method	References
Manipulation	End-effector	Manual control (Joystick / button)	Bestic [68] My Spoon [70]
	Arm support	Manual control (Button)	Neater Arm Support [71]
	Wearable exo-skeleton (arm)	Manual control (Joystick / trackball) sEMG	Hasegawa *et al.* [72]
		sEMG	Kiguchi and Hayashi [73]
	Wearable exo-skeleton (hand)	Contact force (Force sensor)	SEM Glove [75] Heo and Kim [77]
Mobility	Wheeled plat-form	Manual control (Joystick)	TEK [78] Nakajima [79]
	Wearable exo-skeleton	Tilt sensor at torso	ReWalk [80]
		Manual control (Button) Hip/leg motion	Ekso [83]
		Manual control (Joystick)	Rex [85]
		Estimated center of pressure	Indego [87]

3.2.1 Manipulation Assistance

Bestic (Bestic AB, Stockholm, Sweden) [68], [69] and My Spoon (SECOM Co., Ltd., Tokyo, Japan) [70] are commercially available meal assistance robots that aid people with disabilities to feed themselves. They have the form of an end-effector that is fixed to a table, and their key functions are picking or scooping food and bringing the food to the user's mouth. My Spoon has three operation modes: manual, semi-automatic, and automatic. The three modes differ in terms of the amount of user intervention during the feeding motion. In the manual mode, the user controls every single movement of the device with a joystick, whereas a button initiates automatic completion of a preconfigured feeding motion in the automatic mode. Bestic can be controlled by buttons or a joystick to set the location for picking food and to initiate a preconfigured motion.

Neater Arm Support (Neater Solutions Ltd., Buxton, UK) is a commercially available motorized arm support device that enables people with muscle weakness to use their arms [71]. It allows free motion within a horizontal plane via a serial link mechanism having three free rotational joints. One additional inclined rotational joint connecting the forearm brace and link mechanism allows automatic lowering of the elbow upon raising the hand to the level of the mouth. Only the vertical motion is motorized and the user manually controls this motion using buttons to raise or lower the arm.

An assistive exoskeleton that can be worn on the user's body is promising because it resembles a person's natural motion while seamlessly combining with the human body. Hasegawa *et al.* developed an exoskeleton having four DOFs for the support of shoulder (flexion/extension, abduction/adduction, and medial/lateral rotation) and elbow (flexion/extension) motion [72]. A direct-current (DC) motor actuates each DOF through a wire-driven mechanism. The exoskeleton can be controlled manually by a joystick or trackball, and can also be controlled by a surface electromyography (sEMG) signal from the muscles of the arm in conjunction with supportive manual control.

Kiguchi and Hayashi developed an exoskeleton that is controlled by an sEMG signal to aid arm motion [73]. It has seven DOFs—three for the shoulder (flexion/extension, abduction/adduction, and internal/external rotation), one for elbow flexion/extension, one for forearm supination/pronation, and two for the wrist (flexion/extension and radial/ulnar deviation)—that are controlled according to the user's joint torques estimated from 16 channels of sEMG signals. All the actuated joints are powered by DC motors mounted to a wheelchair frame or the exoskeleton's arm. A neuro-fuzzy modifier adopting a neural network structure needs to be trained before using the exoskeleton to build a proper model for the estimation of joint torques from the sEMG signals.

Because of the large contribution of the hand function in performing activities of daily living [74], there have been several works on the development of assistive robots for the hand. Among assistive robots for the hand, there is a commercially available assistive glove named the Soft Extra Muscle (SEM) Glove (Bioservo Technologies AB, Kista, Sweden) [75], [76]. The SEM Glove has three actuated fingers—middle and ring fingers and a thumb. The index finger is left unassisted to provide fine motor capability as well as undisturbed tactile sensation for the index finger. As a force sensor for measuring the user's gripping force, a force-sensing resistor is mounted at the fingerpad of each of the actuated fingers. Flexion forces for the actuated fingers are generated by DC motors. A wire-driven mechanism is employed to transmit a force to the fingers.

A hand exoskeleton developed by Heo and Kim has a palmar opening at the fingerpad that allows for direct contact between the user's fingerpad and objects to make use of the user's own tactile sensation [77]. The user's pinch gripping force is estimated by load cells mounted to lateral side walls at the end of the finger module. The assistance force is generated by a pneumatic cylinder and then transmitted to the finger through a link mechanism.

3.2.2 Mobility Assistance

The Tek Robotic Mobilization Device (Matia Robotics, Istanbul, Turkey) is a commercially available, four-wheeled mobile platform that supports disabled people in standing up and sitting down as well as moving around in a standing posture [78]. The standing-up motion is performed by the user him or herself, assisted by the force of a compressed gas spring so that he or she can accomplish the transition with little effort. The user can drive this device using a joystick in an upright position.

Nakajima developed a robotic wheelchair that is capable of negotiating rough terrain using its wheel mechanisms as legs, while traversing smooth terrain with its wheels [79]. The wheelchair is equipped with two axles whose steering and rolling motions can be independently controlled to mimic legged locomotion. At the end of each axle, there is a wheel driven by a DC motor. To negotiate a step or other rough terrain, the axle is controlled to lift one wheel off the ground and then move forward while using the grounded wheel as a pivot point.

There are several walking assistance robots commercially available today. Because they enable walking based on a preprogrammed gait pattern or they aid voluntary movements of leg joints, they can also be used for rehabilitation. ReWalk (ReWalk Robotics, Ltd., Yokneam Ilit, Israel) enables wheelchair users having lower-limb disabilities to sit, stand, walk, and climb/descend stairs [80]. The movements of hip and knee joints in the sagittal plane are assisted by DC motors, while the ankle joint is not powered. The ankle joint is a simple double-action orthotic joint having limited range of motion and spring-assisted dorsiflexion [81]. A tilt sensor located at the torso detects the motion of the upper body to recognize the user's intention and to initiate the preprogrammed gait movement [82]. Owing to the need for crutches to maintain balance and stability, ReWalk is suitable for users with unimpaired upper-limb functionalities.

Ekso (Ekso Bionics, Richmond, CA, USA) can operate in four ways. In the easiest mode, called First Step, a physical therapist usually manipulates buttons of a remote controller to control the steps. The motion of the user's legs is achieved through the actuation of the device's hip and knee joints powered by DC motors. While walking, the user maintains balance using instrumented crutches and by shifting their body weight [83]. At the next level, Active Step, the user controls the steps via buttons installed on crutches or a walker. The next level, Pro Step, uses the forward movement of the user's hip to initiate step motion. The next mode, New Pro Step Plus, recognizes the user's walking intention from the user's weight shift and the initiation of forward leg movement. Ekso has hip and knee joints that can be actuated in the sagittal plane while the other DOFs at these joints are locked out or passively supported by a spring [84]. Like ReWalk, Ekso requires the use of crutches to maintain stability.

REX (Rex Bionics Plc, Auckland, New Zealand) is a robotic mobility device that does not require crutches or a walking frame for stability, and the physical burden on the shoulder, arm, and hand is thus much lower than that for other devices necessitating crutches [85]. The user controls the device with a joystick, and the device thus does not use sensors to estimate the user's motion intention [82].

Indego (Parker Hannifin, Cleveland, OH, USA) is a commercialized version of the Vanderbilt powered lower limb orthosis that assists the motion of hip and knee joints in the sagittal plane [86], [87]. This device does not have a part that is worn under the shoes, and is intentionally designed to be used with a standard ankle–foot orthosis. Each joint is actuated by a DC motor, and the joint angle trajectories are preprogrammed for standing, sitting, and walking movements. The operation mode can be automatically selected according to the location of the estimated center of pressure, defined as the center of mass projection onto the horizontal ground plane [86].

4 Conclusions and Further Research Directions

In this chapter, we presented recent achievements and important works relating to healthcare robotics, the success of which relies on the collaboration between clinicians and engineers in front of real patients in operating/training rooms. In other words, the development of medical robotics is typical translational research relying on collaboration and intercommunication.

From a research perspective, the major types of medical robotics discussed can be categorized as physical human–robot interactions. Ishiguro *et al.* summarized challenges in physical human–robot interaction research, which apply also to biomedical robotics [32]. These challenges are the guarantee of safety at all times including during pre- and post-procedure, designing robot reactions that are appropriate for the intentions of the human interaction partner, and improving human–machine interfaces including those in cognitive and learning stages. One important improvement of current medical robots is the development of systems that provide a realistic level of force feedback (tactile feedback) in terms of the DOFs, magnitude, and safety. Robots that are currently commercially available provide a certain level of force feedback but their performance is far from that expected by the medical community. One challenge facing surgical robots is the manipulation of thin tissues and the suturing of small openings because the lack of force feedback makes these operations more difficult than their open-surgery counterparts.

From a clinical perspective, there remain points to consider before using robots in healthcare applications. Yang *et al.* summarized these barriers, which include the risk of malfunctioning/failure, setup procedures that are yet to be established, such as procedures of patient safety control, and insurance policy [33]. Finally, the further development of new technologies and effective robotic instruments will increase the acceptance of robotic assistance in healthcare to an even higher level. We are certain that healthcare professionals of the future will employ assistive robotics without anxiety or technical barriers. However, the degree of automation between medical practitioners and robotic systems in control of a healthcare procedure should be addressed thoughtfully.

In robot-aided therapy, the actual role of the robotic system remains to be clarified. Clinical studies show that early robotic training of the upper limb can improve ADL significantly more than chronic-phase training. The successful motor rehabilitation of stroke survivors requires early intensive and task-specific training, whereas the efficacy of the rehabilitation protocol for the chronic phase remains controversial. The most accepted result on robotic rehabilitation is that patient attention and effort are critical to the efficacy of the protocol. Neuroimaging techniques will be utilized to rehabilitation engineering field to suggest a guideline for efficient rehabilitation protocol and design method of such rehabilitation devices. Nowadays, a brain activation pattern for a specific therapy can be observed using brain imaging techniques. Among these techniques, functional magnetic resonance imaging is a powerful tool with which to monitor brain activations and to assess the therapy protocol. The monitoring of progress in real time requires actuating and sensing devices that are compatible with magnetic resonance imaging. A rehabilitation robotic system compatible with magnetic resonance imaging is expected to analyze the efficacies of robot-aided therapies.

Assistive robots have to perform appropriate actions according to the user's motion intention, and the intention recognition method is therefore a primary concern in the development of an assistive robot. Conventional manual control methods such as the use of a joystick and buttons are simple to apply and they do not require complicated algorithms for extracting the user's motion intention. However, when close physical coordination with the human body is required, especially in the case of wearable exoskeletons, manual control methods lack intuitiveness because the user's motion for manipulating the controller device does not directly match the resulting action of the assistive robot. Intention recognition methods based on human motion sensing or interaction force measurement at the human–robot interface are widely adopted methods for the control of exoskeleton-type devices. Such methods are intuitive and can thus capture the user's motion intention without the need for intensive learning about how to manipulate the controller devices. Nevertheless, external disturbances acting on the user's body or the robot may result in unwanted behaviors of the robot because the disturbances cannot be easily discriminated from motion intention. Although healthier users may cope with the disturbances through their own sensory feedback, such disturbances are more difficult for people with disabilities to overcome. A biosignal-based approach including the use of sEMG can be applied to estimate the muscle force without measuring the contact force or motion. However, the sEMG signal is very weak and extremely sensitive to skin conditions and electrode locations [88], and it is also susceptible to motion artifacts [89]. Furthermore, estimating forces for multiple DOFs simultaneously from an sEMG signal requires a large number of electrodes and a complex algorithm [90], making it difficult to use sEMG for practical applications.

References

1. Thrun, S.: Toward a framework for human-robot interaction. Human-Computer Interaction **19**(1), 9–24 (2004)
2. Rosen, J., Hannaford, B., Satava, R.M. (eds) Surgical Robotics: Systems Applications and Visions. Springer Science & Business Media (2011)
3. Taylor, R.H., Stoianovici, D.: Medical robotics in computer-integrated surgery. IEEE Transactions on Robotics and Automation **19**(5), 765–781 (2003)
4. Kwoh, Y.S., Hou, J., Jonckheere, E.A., Hayati, S.: A robot with improved absolute positioning accuracy for CT guided stereotactic brain surgery. IEEE Transactions on Biomedical Engineering **35**(2), 153–160 (1988)
5. Lum, M.J.H., Friedman, D.C.W., Sankaranarayanan, G., King, H., Fodero, K., Leuschke, R., Hannaford, B., Rosen, J., Sinanan, M.N.: The RAVEN: Design and validation of a tele-surgery system. International Journal of Robotics Research **28**(9), 1183–1197 (2009)
6. Hannaford, B., Rosen, J., Friedman, D.W., King, H., Roan, P., Cheng, L., Glozman, D., Ma, J., Kosari, S.N., White, L.: RAVEN-II: an open platform for surgical robotics research. IEEE Transactions on Biomedical Engineering **60**(4), 954–959 (2009)
7. Iyer, S., Looi, T., Drake, J.: A single arm, single camera system for automated suturing. In: Proceedings of the IEEE International Conference on Robotics and Automation, pp. 239–244 (2013)
8. Shin, W.-H., Kwon, D.-S.: Surgical robot system for single-port surgery with novel joint mechanism. IEEE Transactions on Biomedical Engineering **60**(4), 937–944 (2013)
9. Choi, H., Kwak, H.S., Lim, Y.A., Kim, H.J.: Surgical robot for single-incision laparoscopic surgery. IEEE Transactions on Biomedical Engineering **61**(9), 2458–2466 (2014)
10. Sánchez, A., Poignet, P., Dombre, E., Menciassi, A., Dario, P.: A design framework for surgical robots: Example of the Araknes robot controller. Robotics and Autonomous Systems **62**(9), 1342–1352 (2014)
11. Russo, S., Dario, P., Menciassi, A.: A Novel Robotic Platform for Laser-Assisted Transurethral Surgery of the Prostate. IEEE Transactions on Biomedical Engineering **62**(2), 489–500 (2014)
12. Koutenaei, B.A., Wilson, E., Monfaredi, R., Peters, C., Kronreif, G., Cleary, K.: Robotic natural orifice transluminal endoscopic surgery (R-NOTES): Literature review and prototype system. Minimally Invasive Therapy & Allied Technologies **24**(1), 18–23 (2015)
13. Zygomalas, A., Kehagias, I., Giokas, K., Koutsouris, D.: Miniature Surgical Robots in the Era of NOTES and LESS Dream or Reality? Surgical Innovation **22**(1), 97–107 (2015)
14. De Donno, A., Zorn, L., Zanne, P., Nageotte, F., de Mathelin, M.: Introducing STRAS: a new flexible robotic system for minimally invasive surgery. In: Proceedings of the IEEE International Conference on Robotics and Automation, pp. 1213–1220 (2013)
15. Conrad, B.L., Jung, J., Penning, R.S., Zinn, M.R.: Interleaved continuum-rigid manipulation: An augmented approach for robotic minimally-invasive flexible catheter-based procedures. In: Proceedings of the IEEE International Conference on Robotics and Automation, pp. 718–724 (2013)

16. Basdogan, C., De, S., Kim, J., Muniyandi, M., Kim, H., Srinivasan, M.A.: Haptics in minimally invasive surgical simulation and training. Computer Graphics and Applications **24**(2), 56–64 (2004)
17. Coles, T.R., Meglan, D., John, N.: The role of haptics in medical training simulators: a survey of the state of the art. IEEE Transactions on Haptics **4**(1), 51–66 (2011)
18. Zendejas, B., Brydges, R., Hamstra, S.J., Cook, D.A.: State of the evidence on simulation-based training for laparoscopic surgery: a systematic review. Annals of Surgery **257**(4), 586–593 (2013)
19. Hong, M.B., Jo, Y.-H.: Design and evaluation of 2-DOF compliant forceps with force-sensing capability for minimally invasive robot surgery. IEEE Transactions on Robotics **28**(4), 932–941 (2012)
20. He, C., Wang, S., Sang, H., Li, J., Zhang, L.: Force sensing of multiple-DOF cable-driven instruments for minimally invasive robotic surgery. International Journal of Medical Robotics and Computer Assisted Surgery **10**(3), 314–324 (2014)
21. Marcus, H.J., Zareinia, K., Gan, L.S., Yang, F.W., Lama, S., Yang, G.-Z., Sutherland, G.R.: Forces exerted during microneurosurgery: a cadaver study. International Journal of Medical Robotics and Computer Assisted Surgery **10**(2), 251–256 (2014)
22. Payne, C.J., Latt, W.T., Yang, G.-Z.: A new hand-held force-amplifying device for micromanipulation. In: Proceedings of the IEEE International Conference on Robotics and Automation, pp 1583–1588 (2012)
23. Gonenc, B., Feldman, E., Gehlbach, P., Handa, J., Taylor R.H., Iordachita, I.: Towards robot-assisted vitreoretinal surgery: Force-sensing micro-forceps integrated with a handheld micromanipulator. In: Proceedings of the IEEE International Conference on Robotics and Automation, pp 1399–1404 (2014)
24. Burgner, J., Rucker, D.C., Gilbert, H.B., Swaney, P.J., Russell 3rd, P.T., Weaver, K.D., Webster III, R.J.: A telerobotic system for transnasal surgery. IEEE/ASME Transactions on Mechatronics **19**(3), 996–1006 (2013)
25. Bedell, C., Lock, J., Gosline, A., Dupont, P.E.: Design optimization of concentric tube robots based on task and anatomical constraints. In: Proceedings of the IEEE International Conference on Robotics and Automation, pp 398–403 (2011)
26. Hendrick, R.J., Herrell, S.D., Webster, III, R.J.: A multi-arm hand-held robotic system for transurethral laser Prostate surgery. In: Proceedings of the IEEE International Conference on Robotics and Automation, pp 2850–2855 (2014)
27. Bowthorpe, M., Tavakoli, M., Becher, H., Howe, R.: Smith predictor-based robot control for ultrasound-guided teleoperated beating-heart surgery. IEEE Journal of Biomedical and Health Informatics **18**(1), 157–166 (2014)
28. He, X., Balicki, M., Gehlbach, P., Handa, J., Taylor, R., Iordachita, I.: A novel dual force sensing instrument with cooperative robotic assistant for vitreoretinal surgery. In: Proceedings of the IEEE International Conference on Robotics and Automation, pp 213–218 (2013)
29. Yu, H., Shen, J.-H., Joos, K.M., Simaan, N.: Design, calibration and preliminary testing of a robotic telemanipulator for OCT guided retinal surgery. In: Proceedings of the IEEE International Conference on Robotics and Automation, pp 225–231 (2013)
30. Ranzani, T., Silvestri, M., Argiolas, A., Vatteroni, M., Menciassi, A.: A novel trocar-less, multi-point of view, magnetic actuated laparoscope. In: Proceedings of the Robotics and Automation, pp 1199–1204 (2013)

31. Castro, C.A., Smith, S., Alqassis, A., Ketterl, T., Sun, Y., Ross, S., Rosemurgy, A., Savage, P.P., Gitlin, R.D.: MARVEL: A wireless miniature anchored robotic videoscope for expedited laparoscopy. In: Proceedings of the IEEE International Conference on Robotics and Automation, pp 2926–2931 (2012)
32. Ikemoto, S., Amor, H.B., Minato, T., Jung, B., Ishiguro, H.: Physical human-robot interaction: Mutual learning and adaptation. IEEE Robotics and Automation Magazine **19**, 24–35 (2012)
33. Marcus, H., Nandi, D., Darzi, A., Yang, G.Z.: Surgical robotics through a keyhole: From today's translational barriers to tomorrow's "disappearing" robots. IEEE Transactions on Biomedical Engineering **60**(3), 674–681 (2013)
34. Oden, R.: Systematic therapeutic exercises in the management of the paralyses in hemiplegia. Journal of the American Medical Association **70**(12), 828–833 (1918)
35. Sterr, A., Saunders, A.: CI therapy distribution: theory, evidence and practice. NeuroRehabilitation **21**(2), 97–105 (2006)
36. Brown, J.A.: Recovery of motor function after stroke. Progress in Brain Research **157**, 223–228 (2006)
37. Ramachandran, V., Hirstein, W.: The perception of phantom limbs. The D. O. Hebb lecture. Brain **121**(Pt9), 1603–1630 (1998)
38. Bobath, B.: Adult hemiplegia evaluation and treatment. Elsevier Health Sciences (1990)
39. Knott, M., Voss, D., Hipshman, H., Buckley, J., Mead, S.: Proprioceptive neuromuscular facilitation: patterns and techniques. Hoeber Medical Division, Harper & Row (1968)
40. Maciejasz, P., Eschweiler, J., Gerlach-Hahn, K., Jansen-Troy, A., Leonhardt, S.: A survey on robotic devices for upper limb rehabilitation. Journal of Neuroengineering and Rehabilitation **11**, 3 (2014)
41. Díaz, I., Gil, J.J., Sánchez, E.: Lower-Limb Robotic Rehabilitation: Literature Review and Challenges. Journal of Robotics **2011**, 1–11 (2011)
42. Hogan, N., Krebs, H.I., Charnnarong, J., Srikrishna, P., Sharon, A.: MIT-MANUS: A workstation for manual therapy and training II. In: Proceedings of the Telemanipulator Technology, vol, 1833. pp, 28–34 (1993)
43. Krebs, H.I., Ferraro, M., Buerger, S.P., Newbery, M.J., Makiyama, A., Sandmann, M., Lynch, D., Volpe, B.T., Hogan, N.: Rehabilitation robotics: pilot trial of a spatial extension for MIT-Manus. Journal of Neuroengineering and Rehabilitation **1**, 5 (2004)
44. Krebs, H.I., Volpe, B.T., Williams, D., Celestino, J., Charles, S.K., Lynch, D., Hogan, N.: Robot-aided neurorehabilitation: A robot for wrist rehabilitation. IEEE Transactions on Neural Systems and Rehabilitation Engineering **15**(3), 327–335 (2007)
45. Lum, P.S., Burgar, C.G., Shor, P.C.: Evidence for improved muscle activation patterns after retraining of reaching movements with the MIME robotic system in subjects with post-stroke hemiparesis. IEEE Transactions on Neural Systems and Rehabilitation Engineering **12**(2), 186–194 (2004)
46. Lum, P.S., Burgar, C.G., Van der Loos, M., Shor, P.C., Majmundar, M., Yap, R.: MIME robotic device for upper-limb neurorehabilitation in subacute stroke subjects: A follow-up study. Journal of Rehabilitation Research and Development **43**(5), 631–642 (2006)
47. Hesse, S., Schulte-Tigges, G., Konrad, M., Bardeleben, A., Werner, C.: Robot-assisted arm trainer for the passive and active practice of bilateral forearm and wrist movements in hemiparetic subjects. Archives of Physical Medicine and Rehabilitation **84**(6), 915–920 (2003)

48. Hesse, S., Werner, C., Pohl, M., Rueckriem, S., Mehrholz, J., Lingnau, M.L.: Computerized arm training improves the motor control of the severely affected arm after stroke: A single-blinded randomized trial in two centers. Stroke **36**(9), 1960–1966 (2005)
49. Hesse, S., Kuhlmann, H., Wilk, J., Tomelleri, C., Kirker, S.G.B.: A new electromechanical trainer for sensorimotor rehabilitation of paralysed fingers: a case series in chronic and acute stroke patients. Journal of Neuroengineering and Rehabilitation **5**, 21 (2008)
50. Riener, R., Nef, T., Colombo, G.: Robot-aided neurorehabilitation of the upper extremities. Medical and Biological Engineering and Computing **43**(1), 2–10 (2005)
51. Nef, T., Quinter, G., Müller, R., Riener, R.: Effects of arm training with the robotic device ARMin I in chronic stroke: Three single cases. Neurodegenerative Diseases **6**(5–6), 240–251 (2009)
52. Jezernik, S., Colombo, G., Keller, T., Frueh, H., Morari, M.: Robotic Orthosis Lokomat: A Rehabilitation and Research Tool. Neuromodulation **6**(2), 108–115 (2003)
53. Krewer, C., Müller, F., Husemann, B., Heller, S., Quintern, J., Koenig, E.: The influence of different Lokomat walking conditions on the energy expenditure of hemiparetic patients and healthy subjects. Gait and Posture **26**(3), 372–377 (2007)
54. Hidler, J., Nichols, D., Pelliccio, M., Brady, K., Campbell, D.D., Kahn, J.H., Hornby, T.G.: Multicenter randomized clinical trial evaluating the effectiveness of the Lokomat in subacute stroke. Neurorehabilitation and Neural Repair **23**(1), 5–13 (2009)
55. Alcobendas-Maestro, M., Esclarin-Ruz, A., Casado-Lopez, R.M., Munoz-Gonzalez, A., Perez-Mateos, G., Gonzalez-Valdizan, E., Martin, J.L.R.: Lokomat robotic-assisted versus overground training within 3 to 6 months of incomplete spinal cord lesion: Randomized controlled trial. Neurorehabilitation and Neural Repair **26**(9), 1058–1063 (2012)
56. Sanchez, R., Reinkensmeyer, D., Shah, P., Liu, J., Rao, S., Smith, R., Cramer, S., Rahman, T., Bobrow, J.: Monitoring functional arm movement for home-based therapy after stroke. In: Proceedings of the Annual International Conference of the IEEE Engineering in Medicine & Biology Society, pp 4787–4790 (2004)
57. Sanchez, R.J., Liu, J., Rao, S., Shah, P., Smith, R., Rahman, T., Cramer, S.C., Bobrow, J.E., Reinkensmeyer, D.J.: Automating arm movement training following severe stroke: functional exercises with quantitative feedback in a gravity-reduced environment. IEEE Transactions on Neural Systems and Rehabilitation Engineering **14**(3), 378–389 (2006)
58. Gijbels, D., Lamers, I., Kerkhofs, L., Alders, G., Knippenberg, E., Feys, P.: The Armeo Spring as training tool to improve upper limb functionality in multiple sclerosis: a pilot study. Journal of Neuroengineering and Rehabilitation **8**, 5 (2011)
59. Bovolenta, F., Sale, P., Dall'Armi, V., Clerici, P., Franceschini, M.: GRobot-aided therapy for upper limbs in patients with stroke-related lesions. Brief report of a clinical experience. Journal of Neuroengineering and Rehabilitation **8**, 18 (2008)
60. Freivogel, S., Mehrholz, J., Husak-Sotomayor, T., Schmalohr, D.: Gait training with the newly developed 'LokoHelp'-system is feasible for non-ambulatory patients after stroke, spinal cord and brain injury. A feasibility study. Brain Injury **22**(7–8), 625–632 (2008)
61. Freivogel, S., Schmalohr, D., Mehrholz, J.: Improved walking ability and reduced therapeutic stress with an electromechanical gait device. Journal of Rehabilitation Medicine **41**(9), 734–739 (2009)

62. West, G.R.: Powered gait orthosis and method of utilizing same. US Patent 6,689 075 (February 10, 2004)
63. Volpe, B.T., Krebs, H., Hogan, N., Otr, L., Diels, C., Aisen, M.: A novel approach to stroke rehabilitation: robot-aided sensorimotor stimulation. Neurology **54**, 1938–1944 (2000)
64. Freeman, C.T., Hughes, A.M., Burridge, J.H., Chappell, P.H., Lewin, P.L., Rogers, E.: A robotic workstation for stroke rehabilitation of the upper extremity using FES. Medical Engineering & Physics **31**(3), 364–373 (2009)
65. Zhihao, Z., Yuan, Z., Ninghua, W., Fan, G., Kunlin, W., Qining., W.: On the design of a robot-assisted rehabilitation system for ankle joint with contracture and/or spasticity based on proprioceptive neuromuscular facilitation. In: Proceedings of the IEEE International Conference on Robotics and Automation, pp 736–741 (2014)
66. Feil-Seifer, D., Mataric, M.J.: Defining socially assistive robotics. In: Proceedings of the IEEE International Conference on Rehabilitation Robotics, pp 465–468 (2005)
67. Lobo-Prat, J., Kooren, P.N., Stienen, A.H.A., Herder, J.L., Koopman, B.F.J.M., Veltink, P.H.: Non-invasive control interfaces for intention detection in active movement-assistive devices. Journal of Neuroengineering and Rehabilitation **11**, 168 (2014)
68. http://www.besticinc.com/home/
69. Dahl, T.S., Boulos, M.N.K.: Robots in Health and Social Care: A Complementary Technology to Home Care and Telehealthcare. Robotics **3**(1), 1–21 (2013)
70. My Spoon. http://www.secom.co.jp/english/myspoon (retrieved February 10, 2015)
71. Neater Arm Support. http://www.neater.co.uk/neater-arm-support (retrieved February 10, 2015)
72. Hasegawa, Y., Oura, S., Takahashi, J.: Exoskeletal meal assistance system (EMAS II) for patients with progressive muscular disease. Advanced Robotics **27**(18), 1385–1398 (2013)
73. Kiguchi, K., Hayashi, Y.: An EMG-based control for an upper-limb power-assist exoskeleton robot. IEEE Transactions on Systems, Man, and Cybernetics, Part B: Cybernetics **42**(4), 1064–1071 (2012)
74. Katz, S., Ford, A.B., Moskowitz, R.W., Jackson, B.A., Jaffe, M.W.: Studies of illness in the aged: The index of ADL: a standardized measure of biological and psychosocial function. Journal of the American Medical Association **185**(12), 914–919 (1963)
75. The Robotic SEM Glove. http://www.bioservo.com/en/the-sem-technology/sem-glove/ (retrieved February 13, 2015)
76. Nilsson, M., Ingvast, J., Wikander, J., von Holst, H.: The soft extra muscle system for improving the grasping capability in neurological rehabilitation. In: Proceedings of the IEEE EMBS Conference on Biomedical Engineering and Sciences, pp 412–417 (2012)
77. Heo, P., Kim, J.: Power-Assistive Finger Exoskeleton With a Palmar Opening at the Fingerpad. IEEE Transactions on Biomedical Engineering **61**(11), 2688–2697 (2014)
78. TEK Robotic Mobilization Device. http://www.matiarobotics.com/index.html (retrieved February 15, 2015)
79. Nakajima, S.: Proposal of a personal mobility vehicle capable of traversing rough terrain. Disability and Rehabilitation: Assistive Technology **9**(3), 248–259 (2013)
80. ReWalk – More than Walking. http://www.rewalk.com/ (retrieved February 15, 2015)
81. Esquenazi, A., Talaty, M., Packel, A., Saulino, M.: The ReWalk powered exoskeleton to restore ambulatory function to individuals with thoracic-level motor-complete spinal cord injury. American Journal of Physical Medicine & Rehabilitation **91**(11), 911–921 (2012)

82. Carpino, G., Accoto, D., Tagliamonte, N.L., Ghilardi, G., Guglielmelli, E.: Lower limb wearable robots for physiological gait restoration: state of the art and motivations. MEDIC Methodology & Education for Clinical Innovation – New Series **21**(2), 72–80 (2013)
83. Ekso Bionics – An exoskeleton bionic suit or a wearable robot that helps people walk again. http://intl.eksobionics.com/ (retrieved February 16, 2015)
84. Strausser, K.A., Kazerooni, H.: The development and testing of a human machine interface for a mobile medical exoskeleton. In: Proceedings of the IEEE International Conference on Intelligent Robots and Systems, pp 4911–4916 (2011)
85. Rex Bionics – Step into the future. http://www.rexbionics.com/. (retrieved February 16, 2015)
86. Quintero, H.A., Farris, R.J., Goldfarb, M.: A method for the autonomous control of lower limb exoskeletons for persons with paraplegia. Journal of Medical Devices **6**(4), 041003 (2012)
87. Indego – Powering People Forward. http://www.indego.com/indego/en/home (retrieved February 17, 2015)
88. Ronsse, R., Vitiello, N., Lenzi, T., van den Kieboom, J., Carrozza, M.C., Ijspeert, A.J.: Humanrobot synchrony: flexible assistance using adaptive oscillators. IEEE Transactions on Biomedical Engineering **58**(4), 1001–1012 (2011)
89. Roy, S., De Luca, G., Cheng, M., Johansson, A., Gilmore, L., De Luca, C.: Electro-mechanical stability of surface EMG sensors. Medical & Biological Engineering & Computing **45**(5), 447–457 (2007)
90. Kamavuako, E.N., Englehart, K.B., Jensen, W., Farina, D.: Simultaneous and proportional force estimation in multiple degrees of freedom from intramuscular EMG. IEEE Transactions on Biomedical Engineering **59**(7), 1804–1807 (2012)

Author Index

The manufacturer's authorised representative in the EU is Springer Nature Customer Service Centre GmbH, Europaplatz 3, 69115 Heidelberg, Germany. If you have any concerns regarding our products, please contact ProductSafety@springernature.com

Printed and bound by CPI Group (UK) Ltd, Croydon, CR0 4YY
25/09/2026
02222851-0011